AF729350

Sustainable Environmental Waste Management Strategies

Series Editors

Alok Prasad Das, Department of Life Sciences, Rama Devi Women's University, Bhubaneswar, Odisha, India

Eric van Hullebusch, Équipe ACE, Institut de Physique du Globe de Paris, PARIS CEDEX 05, France

Sustainable waste management is an absolute imperative in a world that is already dealing with the effects of climate change. This book series seeks to publish an extensive collection of scientific books targeting researchers, scientists, professionals, academics, students, and anyone else involved in the wide, diverse field of sustainable waste management. The series accepts monographs, edited volumes, textbooks, and conference proceedings. Topics covered include but are not limited to environmental pollution and remediation, sources of environmental waste, environmental and human health, sustainable waste management, circular economy, nature-inspired environmental solutions, pollution prevention, and energy-efficient technologies for waste management. The series puts a particular emphasis on developing countries/the majority world but accepts proposals globally as long as the topic is in scope.

Alok Prasad Das
Editor

Bioremediation Technology for Sustainable Mining Waste Management

Springer

Editor
Alok Prasad Das
Department of Life Sciences
Rama Devi Women's University
Bhubaneswar, Odisha, India

ISSN 3005-1622 ISSN 3005-1630 (electronic)
Sustainable Environmental Waste Management Strategies
ISBN 978-3-032-03467-0 ISBN 978-3-032-03468-7 (eBook)
https://doi.org/10.1007/978-3-032-03468-7

© The Editor(s) (if applicable) and The Author(s), under exclusive license to Springer Nature Switzerland AG 2026

This work is subject to copyright. All rights are solely and exclusively licensed by the Publisher, whether the whole or part of the material is concerned, specifically the rights of translation, reprinting, reuse of illustrations, recitation, broadcasting, reproduction on microfilms or in any other physical way, and transmission or information storage and retrieval, electronic adaptation, computer software, or by similar or dissimilar methodology now known or hereafter developed.
The use of general descriptive names, registered names, trademarks, service marks, etc. in this publication does not imply, even in the absence of a specific statement, that such names are exempt from the relevant protective laws and regulations and therefore free for general use.
The publisher, the authors and the editors are safe to assume that the advice and information in this book are believed to be true and accurate at the date of publication. Neither the publisher nor the authors or the editors give a warranty, expressed or implied, with respect to the material contained herein or for any errors or omissions that may have been made. The publisher remains neutral with regard to jurisdictional claims in published maps and institutional affiliations.

This Springer imprint is published by the registered company Springer Nature Switzerland AG
The registered company address is: Gewerbestrasse 11, 6330 Cham, Switzerland

If disposing of this product, please recycle the paper.

Preface

The mining industry's rapid rise has provided clear economic advantages and contributed greatly to world development. However, this expansion has created significant environmental issues, mainly in the form of mining debris. Tailings, overburden, and effluents from mining operations frequently contain dangerous compounds such as heavy metals, sulphides, and radionuclides, which constitute a long-term threat to ecosystems and human health. As traditional treatment methods remain costly, energy-intensive, and frequently unsustainable, there is an urgent need for environmentally benign, cost-effective, and efficient restoration alternatives.

This book, "Bioremediation Technology for Sustainable Mining Waste Management," seeks to investigate and combine current breakthroughs in bioremediation as a possible approach for reducing mining-related environmental harm. Bioremediation uses microbes, plants, and enzymatic systems to detoxify, convert, and immobilise contaminants in polluted settings. Its low-cost, eco-friendly strategy is consistent with global aims for sustainable development and environmental protection. This book compiles multidisciplinary research and case examples from across the world, emphasising microbial, phytoremediation, and integrated biotechnological techniques adapted to diverse types of mining waste. It also focusses on the mechanistic knowledge of microbial metal resistance, biogeochemical cycles, and the use of omics techniques for bioremediation optimisation. Real-world application, regulatory frameworks, and bioremediation technologies' socioeconomic implications are all heavily emphasized.

Bhubaneswar, Odisha, India　　　　Alok Prasad Das

Acknowledgment

We extend our sincere gratitude to all the authors who contributed to the completion of this book, *Bioremediation Technology for Sustainable Mining Waste Management*. First and foremost, we thank our authors and researchers from all disciplines for sharing their unique thoughts, research findings, and case studies. Their knowledge and devotion have enhanced the content of this volume, providing a thorough review of the most recent breakthroughs in bioremediation technology.

Contents

Water Contamination and Impacts on the Global Ecosystem: Effects on Rivers

Sutanu Satpathy

1 Introduction: Rivers, Continuity, and Disruptions

Habitat diversity, irrespective of the nature of the ecosystem, is a cornerstone of ecological health and resilience, playing a vital role in sustaining life on Earth (Alsterberg et al., 2017; Hewitt et al., 2005). It refers to the variety of ecosystems, such as forests, wetlands, grasslands, and coral reefs, which provide unique conditions for different species to thrive within seemingly homogenous larger ecosystems or biomes. This diversity is essential for maintaining biodiversity, as it supports a wide range of plants, animals, and microorganisms, each contributing to the intricate web of life (Mishra et al., 2024; Priyadarshini et al., 2024; Sahu et al., 2024). Beyond its ecological significance, habitat diversity also underpins critical ecosystem services (Petsch et al., 2023), such as clean air and water, soil fertility, and climate regulation, which are indispensable for human survival and well-being. However, habitat loss and fragmentation due to human activities pose significant threats to this diversity, highlighting the urgent need for conservation efforts to protect and restore these vital ecosystems (Brunotte et al., 2009; Kaat & Josten, 2008; Ward et al., 1999). Mining activities, whether coal, sand, gold, bauxite, iron ore, or uranium extraction, contribute to the homogenization of river habitats by altering natural geomorphology, sediment composition, and flow regimes. Excessive sedimentation from sand and bauxite mining buries diverse microhabitats like gravel beds, disrupting fish spawning sites and benthic invertebrate communities. Acid mine drainage (AMD) from coal and uranium mining alters pH levels, reducing habitat specialization by eliminating acid-sensitive species. Heavy metal

S. Satpathy (✉)
School of Applied Sciences, Kalinga Institute of Industrial Technology, Bhubaneswar, Odisha, India

© The Author(s), under exclusive license to Springer Nature Switzerland AG 2026
A. P. Das (ed.), *Bioremediation Technology for Sustainable Mining Waste Management*, Sustainable Environmental Waste Management Strategies, https://doi.org/10.1007/978-3-032-03468-7_1

contamination from gold and iron ore mining reduces biodiversity, favoring metal-tolerant but functionally similar species, leading to ecological simplification. Additionally, channelization and bank destabilization caused by mining-induced erosion create uniform riverbeds, replacing the natural complexity of riffles, pools, and floodplain wetlands with monotonous, unstable substrates (Bala et al., 2024; Das et al., 2024; Mohanty et al., 2024). These disruptions reduce ecological niches, ultimately making river ecosystems more uniform and less resilient to environmental changes.

Rivers with varied physical features, such as pools, riffles, meanders, and floodplains, create a mosaic of microhabitats that support a wide array of aquatic and terrestrial species (Sánchez-González & Nicieza, 2017). This diversity allows for specialized niches, enabling fish, invertebrates, plants, and other organisms to thrive and interact in complex food webs and the homogenization or destruction of these habitats no longer makes it habitable for many of these species (Bini et al., 2014; Graco-Roza et al., 2020). Additionally, habitat diversity in rivers enhances ecosystem resilience, helping these systems withstand and recover from disturbances such as pollution, climate change, or human alterations (Van Looy et al., 2019). Rivers also provide essential services to humans, including water supply, flood regulation, and recreational opportunities, all of which depend on the health of their habitats (Singh, 2018). Protecting and restoring habitat diversity in rivers is therefore vital not only for conserving biodiversity but also for ensuring the sustainability of the resources and benefits that rivers provide to both nature and society (Lakra et al., 2010).

The continuity of a river system is a fundamental concept in river ecology, as emphasized by Vannote et al. (1980) in the River Continuum Concept (RCC). This model describes rivers as longitudinally connected systems where physical, chemical, and biological characteristics change gradually from the headwaters to the mouth. In hilly terrains, rivers are typically fast-flowing with high gradients, creating habitats like cascades and riffles that support specialized invertebrates and fish adapted to high oxygen levels and low temperatures (Das, Dey, & Das, 2024; Dey, Dash, & Das, 2024; Mohanty et al., 2024). As rivers transition to plains, the gradient decreases, flow slows, and habitats such as pools and meanders become dominant, fostering diverse communities of fish and invertebrates. In low-lying floodplains, rivers often form complex networks of channels, wetlands, and backwaters, which are hotspots for biodiversity due to their dynamic nature and nutrient-rich environments (Nandi et al., 2016). Maintaining this longitudinal continuity is critical for the movement of species, nutrient cycling, and the overall health of the river ecosystem.

The productivity of river systems is heavily dependent on allochthonous inputs—organic matter like leaves, twigs, and detritus from riparian vegetation—which serve as a primary energy source, especially in headwater streams (Doretto et al., 2020). In hilly terrains, shaded streams rely almost entirely on this allochthonous material, as limited sunlight restricts in-stream primary production (Caraco & Cole, 2004). As rivers flow through plains and floodplains, autochthonous production (in-stream photosynthesis) becomes more significant, but allochthonous inputs remain vital, particularly in maintaining food webs for fish and invertebrates (Hein et al., 2003; Tank et al., 2010). Riparian vegetation plays a dual role: it provides organic matter and stabilizes riverbanks, preventing erosion and maintaining habitat diversity (França et al., 2009).

Habitat diversity is particularly important for fish and invertebrates, as it offers varied niches for feeding, spawning, and refuge (Dutta et al., 2017). In hilly terrains, fast-flowing habitats support species like trout and stoneflies (Bera et al., 2020), while plains and floodplains host a broader range of species, including carp (Dubey et al., 2012) and mayflies, which thrive in slower, more nutrient-rich waters. Riverbank vegetation further enhances this diversity (Li et al., 2023) by providing shade, reducing temperature fluctuations, and creating microhabitats for terrestrial and aquatic organisms. Across all three stages—hilly terrains, plains, and floodplains—the interplay of physical habitats, riparian vegetation, and organic inputs underscores the importance of preserving habitat diversity to sustain the ecological integrity and productivity of river systems.

Riverbed mining and sand mining in rivers are driven primarily by economic factors, as these activities provide raw materials essential for various industries and infrastructure development (Rukmana et al., 2020). Sand and gravel extracted from riverbeds (Madyise, 2013) are key components in construction, used extensively in concrete, asphalt, and building foundations. The rapid urbanization and infrastructure boom in many regions have created a soaring demand for these materials, making riverbed mining a lucrative economic activity (Yuen et al., 2024). In developing countries, where alternative sources of construction materials may be limited or expensive, river sand is often the most accessible and cost-effective option, further driving its extraction (Rege, 2016).

Additionally, sand mining generates employment opportunities for local communities, particularly in rural areas where alternative livelihoods may be scarce. For many, it serves as a primary source of income, supporting families and contributing to local economies (Hatlebakk, 2023). The low barrier to entry—requiring minimal machinery or technical expertise—makes it an attractive option for small-scale operators and informal miners (Salman, 2016). However, the economic benefits often come at a significant environmental cost, including habitat destruction, altered river hydrology, and reduced water quality, which can undermine long-term sustainability (Akanwa, 2020; Yuen et al., 2024).

In some cases, sand is also mined for industrial uses, such as glass manufacturing, electronics, and hydraulic fracturing or fracking (Jackson et al., 2014), further increasing its economic value. The global demand for sand has turned it into a high-value commodity, with international trade networks emerging to supply regions facing local shortages (Beiser, 2019). While the economic incentives for riverbed and sand mining are strong, the lack of effective regulation and enforcement in many areas has led to overexploitation, threatening river ecosystems and the communities that depend on them. Balancing the economic benefits of these activities with sustainable practices and environmental protection remains a critical challenge.

2 Water Contamination: A World-Wide Concern

Mining activities, including coal, metal, and mineral extraction, have profound and often detrimental effects on river systems, with consequences that extend far beyond the immediate mining sites. Coal mining, for instance, is notorious for generating acid mine drainage, a process where sulfide minerals exposed to air and water produce sulfuric acid, which can leach heavy metals like arsenic and mercury into rivers (Lottermoser, 2010). This acidic, toxic runoff devastates aquatic ecosystems, killing fish and invertebrates and rendering water unsafe for human use. Additionally, coal mining disrupts river hydrology by altering groundwater systems and increasing sedimentation, which smothers riverbeds and degrades water quality (Dey et al., 2023; 2024; John & Das, 2024; Priyadarsini & Das, 2024). These changes not only harm biodiversity but also affect communities that rely on rivers for drinking water, irrigation, and fishing.

Metal mining, such as gold and copper extraction, introduces another set of challenges for river health (Miller et al., 2004). The use of toxic chemicals like cyanide and sulfuric acid in metal extraction poses significant risks of contamination if these substances spill or leak into nearby waterways (Sawaraba & Rao, 2015). Tailings, the waste material produced during mining, are often stored in large dams that can fail catastrophically, releasing massive amounts of toxic sludge into rivers. Such events have led to widespread ecological disasters, as seen in the 2019 Brumadinho dam collapse in Brazil. Heavy metals like mercury and cadmium, commonly released during metal mining, accumulate in the food chain, posing long-term risks to both aquatic life and human populations. These pollutants can persist in river systems for decades, causing chronic health issues and undermining local livelihoods dependent on fishing and agriculture.

Mineral mining, including phosphate and limestone extraction, also impacts rivers, though in different ways (Wagner & Raymond, 2015). Phosphate mining contributes to nutrient pollution, which can lead to eutrophication in rivers (Carpenter et al., 1998). Excess nutrients trigger algal blooms that deplete oxygen levels, creating dead zones where aquatic life cannot survive (Diaz & Rosenberg, 2008). Limestone mining, on the other hand, often results in increased sedimentation and physical alterations to river channels, destroying habitats for fish and invertebrates (Kondolf, 1997). Both types of mining disrupt the delicate balance of river ecosystems, reducing biodiversity and compromising the ecosystem services that rivers provide, such as water purification, flood regulation, and habitat provision (Vörösmarty et al., 2010).

3 Impacts of Mining on Rivers

The broader ecological and socioeconomic impacts of mining on rivers are significant (Darwall et al., 2018). Loss of biodiversity due to habitat destruction and pollution disrupts ecosystem functions, reducing the resilience of river systems to environmental changes (Palmer et al., 2008). Water scarcity is another critical issue, as mining operations often consume large amounts of water, reducing river flow and affecting downstream users (Gleick, 2003). Contaminated water from mining activities poses serious health risks to human populations, particularly in regions where communities rely on rivers for drinking water and irrigation (World Health Organization, 2011). While mining provides economic benefits through job creation and resource extraction, the long-term costs of environmental degradation and lost ecosystem services often outweigh these benefits, especially in areas where rivers are central to local livelihoods (TEEB, 2010).

To mitigate these impacts, sustainable mining practices and stringent regulations are essential (Hilson & Murck, 2000). Advanced waste management systems can help prevent contamination (Lottermoser, 2010), while restoring mined areas can reduce sedimentation and habitat loss (Palmer et al., 2008). Monitoring water quality and enforcing limits on pollutant discharges are critical steps in protecting river health (Chapman, 2012). Additionally, promoting alternative livelihoods for communities dependent on mining can help reduce pressure on river ecosystems (Hilson, 2012). Balancing the economic benefits of mining with the need to protect and preserve river systems is crucial for ensuring the health and sustainability of these vital ecosystems for future generations (Jenkins & Yakovleva, 2006).

3.1 Effects and Impact of Coal Mining on Rivers in India

Coal mining in India has severely impacted river systems by altering water quality through acidification, heavy metal contamination, and sedimentation, all of which have cascading effects on aquatic biodiversity (Ghosh & Ghosh, 2017). The most affected rivers are found in coal-rich states like Jharkhand, Odisha, West Bengal, Chhattisgarh, and Madhya Pradesh (Pandey et al., 2014). The nature and intensity of these effects vary based on the mining techniques employed and the ecological sensitivity of the region (Sahu & Dash, 2011).

Acid Mine Drainage is one of the most severe consequences of coal mining, particularly in Jharkhand, Odisha, and Chhattisgarh (Singh et al., 2018). When sulfide minerals (primarily pyrite, FeS_2) in coal-bearing rocks are exposed to oxygen and water, they oxidize to form sulfuric acid (H_2SO_4), which leaches into nearby rivers (Johnson & Hallberg, 2005).

This process leads to acidification of river water, often lowering pH levels to below 4 in some cases, as observed in the Damodar and Brahmani rivers (Gupta & Paul, 2015). Acidic water dissolves toxic metals like iron (Fe), aluminum (Al), lead (Pb), and cadmium (Cd), making them bioavailable and hazardous to aquatic organisms (Dutta & Agrawal, 2003). Acidification disrupts the osmoregulation in fish and other aquatic species, leading to reproductive failure and mortality (Wood et al., 2012).

One of the most well-known cases of river pollution due to coal mining is the Damodar River, often called the "Sorrow of Bengal" (Chakraborty et al., 2016).

Flowing through Jharkhand and West Bengal, the Damodar River is heavily polluted by coal mines in Jharia, Bokaro, and Raniganj (Maiti, 2013). Acid Mine Drainage (AMD) from abandoned and active mines lowers the pH of the river, making the water acidic and toxic to aquatic life (Sheoran & Sheoran, 2006). The oxidation of pyrite (FeS_2) in coal-bearing rocks produces sulfuric acid (H_2SO_4), leading to pH levels dropping below 4 in some stretches of the river (Akcil & Koldas, 2006). Acidic water dissolves heavy metals such as iron (Fe), manganese (Mn), cadmium (Cd), and arsenic (As), making them bioavailable and hazardous to aquatic organisms (Jha et al., 2015). High concentrations of arsenic and iron, exceeding safe drinking water limits, have been reported (Ahamed et al., 2019). This metal-laden water disrupts osmoregulation in fish, leading to reproductive failure, skeletal deformities, and reduced survival rates (Farrell et al., 2015). Benthic organisms, such as mollusks and freshwater prawns, suffer from metal toxicity and bioaccumulation, leading to severe biodiversity loss (Kumar et al., 2017).

The Subarnarekha River, which flows through Jharkhand, Odisha, and West Bengal, faces contamination from both coal and uranium mining (Bhattacharya et al., 2014). Mercury (Hg) and uranium (U) from mining waste enter the river, increasing bioaccumulation in fish and other aquatic species (Kar et al., 2018). Mercury contamination is particularly dangerous because it biomagnifies through the food chain, leading to neurological damage in predators and humans who consume contaminated fish (Driscoll et al., 2013). Fish species such as *Channa striata* (snakehead) and *Clarias batrachus* (walking catfish), which are commonly consumed by local communities, have shown elevated levels of mercury in tissue samples (Mukherjee et al., 2019). Additionally, coal washeries discharge cyanide, ammonia, and phenolic compounds into the river, further reducing water quality and leading to oxygen depletion (Ghose, 2009). Studies have reported a significant drop in dissolved oxygen (DO) levels, with some sections of the river turning hypoxic, causing fish kills and disrupting aquatic ecosystems (Das & Chakrabarty, 2015).

In Odisha, the Brahmani River has been severely affected by mining in the Talcher coalfields (Patra et al., 2016). Coal processing units release cyanide, sulfates, and phenols into the river, leading to chemical pollution and water toxicity (Panda et al., 2017). The pH of the river has dropped to as low as 3.8 in certain stretches, making it highly acidic and unsuitable for aquatic life (Naik et al., 2018). Additionally, the presence of iron flocculates (reddish-brown precipitates) on the riverbed smothers benthic habitats, disrupting the feeding mechanisms of invertebrates such as mayflies and caddisflies, which are crucial for maintaining the aquatic food web (Gunn et al., 2010). Zooplankton diversity, essential for supporting fish populations, has declined due to high metal toxicity and fluctuating pH levels (Jeziorski et al., 2015). This ecological imbalance has led to a significant reduction in native fish populations, particularly species like *Tor tor* (mahseer) and *Labeo rohita* (rohu), which require stable water conditions for breeding (Sarkar et al., 2016).

Central India, particularly Chhattisgarh, has witnessed large-scale mining-driven deforestation, which has affected rivers like the Hasdeo River (Sahu et al., 2015).

The Hasdeo-Arand region, one of India's most ecologically rich forests, is being rapidly cleared for open-cast coal mining (Kumar & Pandey, 2018). This deforestation has led to increased soil erosion and excessive sedimentation, which raises total suspended solids (TSS) levels in the river (Waters, 1995). High sediment loads reduce water clarity, negatively impacting photosynthesis in aquatic plants and phytoplankton, which form the base of the food chain (Kirk, 2011). A decline in primary producers affects herbivorous fish species such as *Catla catla* (Catla) and *Cirrhinus mrigala* (Mrigal), leading to cascading effects on higher trophic levels (Dudgeon, 2011). Additionally, heavy siltation alters the physical structure of the riverbed, leading to habitat fragmentation and a decline in bottom-dwelling organisms like freshwater prawns (*Macrobrachium* spp.) and bivalves (Vörösmarty et al., 2010). Increased turbidity also interferes with fish migration, making it difficult for species such as *Tor putitora* (Golden Mahseer) to locate spawning grounds, thereby reducing reproductive success (Bhatt & Pandit, 2016).

In Madhya Pradesh, the Son and Mahan rivers are experiencing similar threats due to mining in the Singrauli coalfields (Dwivedi & Pandey, 2017). The Singrauli region, known as the "Energy Capital of India," houses multiple thermal power plants and coal mines, leading to industrial effluent discharge and fly ash contamination in river systems (Rai, 2012). Heavy metals like lead (Pb), cadmium (Cd), and chromium (Cr) from coal combustion products are carried into these rivers, affecting water quality and accumulating in aquatic organisms (Gupta et al., 2019). Studies on the Son River have shown that fish species such as *Wallago attu* (Giant River Catfish) and *Mastacembelus armatus* (Spiny Eel) have exhibited signs of bioaccumulation, including tissue damage and metabolic disorders (Verma et al., 2016). Furthermore, high levels of sulfates (SO_4^{2-}) and nitrates (NO_3^-) from mine drainage have been recorded, leading to increased water hardness and altering the natural ionic balance of the river (Bernhardt et al., 2012).

Coal mining also disrupts microbial communities, which play a crucial role in maintaining river ecosystem health (Baker & Banfield, 2003). In acidified rivers like the Damodar and Brahmani, microbial diversity shifts toward acidophilic bacteria (e.g., *Thiobacillus ferrooxidans*, *Leptospirillum* spp.), which accelerate pyrite oxidation, further worsening acidification (Johnson, 2012). These changes suppress nitrifying and denitrifying bacteria, disrupting the nitrogen cycle and reducing nutrient availability for aquatic plants and algae (Graham et al., 2016). The loss of heterotrophic decomposers, responsible for breaking down organic matter, leads to organic detritus accumulation, which, in turn, reduces oxygen availability and contributes to eutrophic conditions (Fenchel et al., 2012).

Overall, coal mining has caused extensive degradation of India's river ecosystems by acidifying water bodies, introducing toxic metals, reducing dissolved oxygen levels, increasing sedimentation, and altering microbial communities (Dutta & Gupta, 2018). These impacts have led to a severe decline in fish diversity, disrupted aquatic food webs, and increased health risks for human populations dependent on these rivers (Mishra & Das, 2017).

3.2 *Impact from Limestone Quarrying on River Systems*

Limestone mining has significantly impacted river systems, groundwater recharge, and biodiversity across various biogeographical regions of India. The process of extracting limestone—whether through open-cast or underground mining—leads to habitat destruction, soil erosion, sedimentation, and chemical changes in water bodies. These effects are particularly severe in ecologically sensitive zones such as the Western Himalayas (Uttarakhand, Himachal Pradesh, Jammu and Kashmir), the Eastern Himalayas (Arunachal Pradesh, Meghalaya, Assam), the Western Ghats, and the Vindhyan and Satpura ranges. The severity of environmental damage depends on the geological composition of the mining sites, their proximity to river systems, and the mining techniques employed.

3.2.1 Impact on the Western Himalayas (Uttarakhand, Himachal Pradesh, Jammu and Kashmir)

In Uttarakhand and Himachal Pradesh, large-scale limestone mining has severely impacted river catchments, particularly those feeding the Ganga and Yamuna river systems. The extraction process destabilizes mountain slopes, leading to increased landslides and soil erosion, which in turn raises total suspended solids (TSS) in river water. Elevated TSS reduces light penetration, affecting primary producers such as diatoms and submerged macrophytes, which form the base of aquatic food chains. Additionally, sedimentation clogs riverbeds, altering the natural flow regime and leading to channel shifts in rivers like the Tons, Asan, and Bhagirathi.

Limestone mining also affects karst aquifers, which are crucial for groundwater recharge in the Himalayan region. Karst landscapes, characterized by underground limestone formations, act as natural reservoirs, storing and regulating groundwater flow. Mining disrupts these aquifers by fracturing limestone beds and increasing water loss through subsurface cracks, thereby reducing base flow in Himalayan rivers during dry seasons. This impact is particularly evident in regions near Mussoorie, Dehradun, and Nainital, where local streams and springs have witnessed a decline in discharge over the past few decades. Loss of groundwater connectivity has also affected riparian forests, leading to a decline in moisture-dependent species like *Cedrus deodara* (Deodar cedar) and *Quercus leucotrichophora* (Banj oak).

3.2.2 Impact on the Eastern Himalayas (Meghalaya, Arunachal Pradesh, Assam)

In Meghalaya, extensive limestone mining for cement production has caused widespread environmental degradation, particularly in the Jaintia Hills and Khasi Hills. Open-cast mining in these regions has led to severe acidification of local streams due to the oxidation of pyrite (FeS_2) present in limestone deposits, a process similar to Acid Mine Drainage (AMD) in coal mining. The reaction produces sulfuric acid, leading to a drop in water pH to as low as 4.5 in some affected streams. Acidification has led to a decline in sensitive fish species such as *Schistura* spp. (hillstream loaches) and *Neolissochilus hexagonolepis* (Chocolate Mahseer), which are essential components of the Eastern Himalayan riverine ecosystem.

Heavy metal contamination is another concern in the region. Limestone mining waste often contains trace elements such as lead (Pb), zinc (Zn), and manganese (Mn), which leach into adjacent rivers and groundwater. Studies have recorded elevated levels of lead and manganese in groundwater sources near limestone mines in the Jaintia Hills, surpassing WHO drinking water standards. Bioaccumulation of these metals in fish and aquatic invertebrates has been linked to reduced reproductive success and physiological stress, further disrupting local fisheries and community livelihoods.

The Umiam and Kupli rivers, which flow through mining-affected areas in Meghalaya and Assam, have also experienced turbidity spikes and increased sediment loads, leading to habitat fragmentation and loss of breeding grounds for migratory fish. Indigenous communities dependent on these rivers for water and fisheries have reported declining fish catches and water scarcity, particularly during the dry season when sediment-laden runoff disrupts natural flow regimes.

3.2.3 Impact on the Western Ghats

Limestone mining in the Western Ghats (Maharashtra, Goa, Karnataka, Kerala, Tamil Nadu) has led to severe deforestation (Darwall et al., 2018), groundwater depletion (Patil et al., 2022), and river pollution (Sreebha & Padmalal, 2011). The region's lateritic and basaltic geology, overlaid with karst limestone formations, makes its water systems highly vulnerable to mining-related disruptions (Gunn & Hardwick, 2000). Large-scale limestone extraction for cement production and infrastructure development has reduced forest cover (Reddy et al., 2023), leading to increased surface runoff (Kumar et al., 2017; Miller et al., 2004), reduced infiltration (Bandyopadhyay and Shiva 1985), and a decline in groundwater table levels (Central Ground Water Board (CGWB), 2023).

The Bhima River basin in Karnataka has experienced significant sedimentation due to limestone mining in Gulbarga (Kalaburagi) and Bagalkot districts (Sreebha & Padmalal, 2011). Elevated sediment loads disrupt benthic macroinvertebrates, such as mayflies (Ephemeroptera) and caddisflies (Trichoptera), which serve as bioindicators of river health (Masese et al., 2014; Subramanian & Sivaramakrishnan, 2007). The loss of these species impacts higher trophic levels, including predatory fish like *Channa marulius* (Giant Snakehead) and *Mystus* spp. (Catfishes), reducing overall biodiversity (Dahanukar & Raghavan, 2013; Arunachalam et al., 2018).

Limestone mining in Goa and Maharashtra has also led to coastal erosion (Shetye et al., 2020; Dias et al., 2021) and estuarine pollution (Kankonkar et al., 2018; Sajan et al., 2022). The Zuari and Mandovi rivers in Goa receive limestone mining effluents, leading to increased alkalinity (Fondekar et al., 2019; Ingole et al., 2020) and turbidity (Shyni et al., 2021; Manoj et al., 2023), which disrupt estuarine fisheries (Rivonker et al., 2018; Sukumaran et al., 2020). The Kali River estuary in Karnataka has experienced a decline in oyster and mollusk populations (Naik et al., 2021) due to elevated sedimentation (Bhat et al., 2022), which affects filter-feeding organisms and marine biodiversity (Raveendran et al., 2021). All of the above pressures lead to severely debilitating problems for a biodiversity hotspot such as the Western Ghats, leading to ecosystem dysfunction, and catastrophic consequences for the human community, and diverse flora and fauna of the biogeographic region.

3.2.4 Impact on the Vindhyan and Satpura Ranges

The Vindhyan limestone belt (Madhya Pradesh, Uttar Pradesh, Rajasthan) is one of India's largest limestone-producing regions, supplying raw materials to cement industries (Singh, 2018). In Katni and Satna districts of Madhya Pradesh, open-cast mining has led to extensive groundwater depletion (Mishra, 2022). Karst aquifers in these regions have witnessed increased water loss, with farmers reporting a 40% decline in well-water levels over the past two decades (Central Ground Water Board (CGWB), 2023).

The Chambal and Ken rivers, which flow through limestone-rich regions, have experienced increased hardness ($CaCO_3$ levels exceeding 500 mg/L in some locations) due to mining effluents (Yadav & Trivedi, 2020). This excessive hardness affects aquatic organisms, particularly egg-laying amphibians (Patil, 2021) and freshwater mollusks (Dubey, 2023), which rely on stable ionic concentrations for reproduction. The alteration of calcium carbonate balance has also impacted crustacean diversity, leading to a decline in populations of Macrobrachium spp. (freshwater prawns) (Verma et al., 2016), which are commercially important for local fisheries.

3.3 Impact from Bauxite and Iron Ore Mining on the Rivers of India

Bauxite and iron ore mining have profound impacts on the river ecosystems of India, particularly in mineral-rich regions such as Odisha, Chhattisgarh, Jharkhand, Karnataka, Maharashtra, Madhya Pradesh, and Goa (Mishra & Patel, 2022). These mining activities alter river hydrology (Patra et al., 2016), degrade water quality (Pandey et al., 2014), increase sedimentation, and introduce harmful contaminants into aquatic ecosystems (Ghosh et al., 2023). Rivers like the Mahanadi, Brahmani, Godavari, Indravati, Hasdeo, Mandovi, and Zuari, which flow through these regions, bear the brunt of mining-related pollution (Mohanty, 2021). The extent of the impact depends on factors such as the mining method, proximity to rivers, local geology, and waste management practices (Indian Bureau of Mines (IBM), 2022).

One of the most immediate consequences of bauxite and iron ore mining is the degradation of river water quality due to the release of mining waste and metal-rich effluents (Ghosh & Ghosh, 2017). Bauxite mining, particularly in Odisha (Koraput, Kalahandi) and Chhattisgarh (Surguja, Jashpur), is known for generating large quantities of highly alkaline waste from the refining process (Jena et al., 2023). The residue, often containing sodium hydroxide (NaOH) and dissolved aluminum compounds, is washed into nearby streams, leading to a rise in pH beyond natural levels (Dash et al., 2009). Water bodies with pH values exceeding 9 have been documented near mining sites, making them unsuitable for aquatic life (Nanda & Tiwari, 2022). Many freshwater fish, including *Catla catla* (catla) and *Cirrhinus mrigala* (mrigal), struggle to survive in such alkaline conditions due to disruptions in their osmoregulation processes (Das, 2020). In the Eastern Ghats' bauxite mining belt, small streams and tributaries of the Mahanadi and Godavari have exhibited periodic fish die-offs due to abrupt pH shifts and changes in dissolved oxygen levels (Panda et al., 2021).

Iron ore mining, which is heavily concentrated in Odisha (Keonjhar, Sundargarh), Jharkhand (West Singhbhum), Chhattisgarh (Durg, Dantewada), Karnataka (Ballari), and Goa (Dey & Sen, 2022), contributes to severe sedimentation and erosion in river systems (Behera & Mishra, 2021). Open-cast iron ore mining removes vast amounts of topsoil, leaving behind loose, fine sediments that are easily washed into rivers during monsoonal rains. The Brahmani, Baitarani, and Mahanadi rivers, which drain Odisha's iron ore belt, experienced elevated total suspended solids (TSS) during mining operations (Patel et al., 2023). Excessive sediment loads reduce light penetration in the water column, thereby affecting photosynthesis in aquatic plants and algae. The decline of primary producers has a cascading effect on the entire aquatic food web, leading to reduced populations of herbivorous invertebrates and fish species (Kumar & Pandey, 2018). Additionally, sedimentation leads to the clogging of riverbeds, negatively impacting benthic organisms such as freshwater mollusks and bottom-feeding fish (Singh et al., 2021). In the Bhadra River in Karnataka, sedimentation from iron ore mining in the Western Ghats has led to the loss of spawning grounds for migratory fish, affecting species such as the Deccan Mahseer (*Tor khudree*), which relies on clean gravel beds for reproduction (Venkateshwarlu et al., 2023).

Another significant concern is the leaching of heavy metals into river systems (Pandit & Patel, 2021). Iron ore deposits often contain impurities such as arsenic (As), cadmium (Cd), lead (Pb), and manganese (Mn), which become mobilized in surface water due to mining activities (Biswas & Mukherjee, 2022). These metals enter rivers through mine tailings, acid drainage, and surface runoff, posing severe risks to aquatic biodiversity (Srivastava et al., 2023). Elevated concentrations of arsenic and lead have been recorded in the Sukhtawa River in Madhya Pradesh and the Koina River in Jharkhand, both of which drain mining-heavy regions (Gupta & Paul, 2015). Studies indicate that bioaccumulation of these metals occurs in fish species such as *Labeo rohita* (rohu) and *Channa striata* (snakehead), leading to chronic toxicity, developmental abnormalities, and reduced reproductive success (Jha et al., 2023; Khan & Pal, 2023). Additionally, these metals affect invertebrate populations, particularly mayflies and caddisflies, which serve as indicators of water quality (Verma et al., 2016). The decline of these bioindicator species signals long-term ecological degradation in iron ore mining-affected rivers (Reddy et al., 2023).

Beyond chemical contamination, hydrological disruptions due to bauxite and iron ore mining have altered natural river flow patterns (Patra et al., 2016). Large-scale mining operations require substantial amounts of water, often leading to over-extraction from nearby rivers and groundwater sources (Sahu & Mishra, 2022). In Chhattisgarh's Hasdeo and Mand rivers, iron ore mining has been linked to a 20%–30% reduction in dry season flow (Central Ground Water Board (CGWB), 2023), affecting downstream agricultural and drinking water supplies. The removal of forest cover and topsoil in mining zones also exacerbates flash floods during monsoons, as rainwater flows rapidly over degraded landscapes instead of being absorbed into the soil (Kumar et al., 2017). Rivers such as the Baitarani in Odisha and the Zuari in Goa have exhibited increasing seasonal flow variability (Panda et al., 2021), where high flood peaks during monsoons are followed by severe water shortages in dry months. This altered flow regime affects fish migration, as many species rely on consistent seasonal flow cues for breeding and spawning (Nautiyal et al., 2022).

The loss of riparian vegetation and biodiversity is another major consequence of bauxite and iron ore mining (Ghosh et al., 2023). Riverbanks in mining zones experience severe degradation due to deforestation, road construction, and waste dumping (Behera & Mishra, 2021). In the Goa-Karnataka iron ore belt, extensive loss of mangroves and estuarine vegetation has been observed along rivers like the Mandovi, Zuari, and Kali (Shetty et al., 2023), affecting species that depend on these habitats, including mudskippers, crabs, and prawns (Ingole et al., 2022). Similarly, the degradation of sal and teak forests in Odisha's Keonjhar and Sundargarh districts has led to habitat loss for species such as smooth-coated otters (*Lutrogale perspicillata*) (Hussain et al., 2021). Changes in riparian vegetation also impact insect populations, particularly pollinators and decomposers, leading to broader ecological imbalances (Rathore et al., 2023).

Mining-related degradation of river ecosystems has direct consequences for local communities, particularly tribal populations and small-scale fishermen (Bandyopadhyay, 2022). Many communities in Jharkhand, Odisha, and Chhattisgarh rely on river fisheries for subsistence, but declining fish populations due to pollution, sedimentation, and heavy metal contamination have led to a crisis in traditional fishing livelihoods (Jena et al., 2023). Surveys conducted in Koraput (Odisha) and Dantewada (Chhattisgarh) have shown that villagers report a 40%–50% decline in fish catch over the last two decades (Das & Pradhan, 2021), forcing many to seek alternative livelihoods. Additionally, the contamination of drinking water sources poses significant health risks (Reddy et al., 2023). In regions where rivers have become polluted due to iron ore mining, villagers often complain of gastrointestinal disorders, skin diseases, and long-term exposure-related ailments linked to metal contamination (Gupta et al., 2023).

Overall, bauxite and iron ore mining have caused far-reaching ecological and socioeconomic impacts on river systems in India (TERI, 2023). The degradation of water quality through alkalinity shifts, heavy metal contamination, and sedimentation has reduced aquatic biodiversity and altered river hydrology. The destruction of riparian forests and habitat fragmentation have disrupted ecosystem functions, making rivers more vulnerable to extreme climate events such as floods and droughts. Additionally, these disruptions have placed immense pressure on local communities that rely on these rivers for fishing, drinking water, and agriculture (United Nations Environment Programme (UNEP), 2023).

3.4 *Impact of Sand Mining on River Ecosystems in India*

Sand mining is one of the most widespread extractive activities affecting river ecosystems in India (Singh et al., 2021). It involves the removal of sand, gravel, and other sediment from riverbeds, floodplains, and coastal areas for use in construction, infrastructure development, and land reclamation (United Nations Environment Programme (UNEP), 2023). India, being the second-largest producer of cement globally, has a high demand for sand, leading to both legal and illegal extraction from riverbeds (Ministry of Environment, Forest and Climate Change (MoEFCC), 2022). Major rivers affected by extensive sand mining include the Ganga, Yamuna, Brahmaputra, Mahanadi, Narmada, Godavari, Krishna, Kaveri, and Periyar, along with their numerous tributaries (Ravindra et al., 2023). Excessive sand removal disrupts hydrological processes, sediment transport, aquatic habitats, and groundwater recharge, leading to severe ecological consequences.

Sand plays a critical role in maintaining river hydrology by regulating flow regimes, groundwater recharge, and sediment balance (Kondolf, 1997). Excessive sand extraction alters the natural sediment budget, causing changes in river morphology and hydrodynamics (Padmalal et al., 2008). The removal of sand from riverbeds increases channel incision, which deepens the river channel and causes bank erosion, leading to a shift in flow patterns. This is particularly evident in the Yamuna River, where uncontrolled sand mining in regions like Noida, Etawah, and Mathura has led to the development of deep pits that alter flow velocity and increase sediment deposition downstream (Jain et al., 2022).

The loss of sand from riverbeds also affects groundwater recharge, as sandy substrates act as natural filters that help store and percolate water into underground aquifers. A decline in sand availability reduces the river's ability to hold water during dry months, leading to a drop in water tables (Sreebha & Padmalal, 2011). In regions like Tamil Nadu (Palar River), Karnataka (Tungabhadra River), and Maharashtra (Godavari Basin), excessive mining has led to declining groundwater levels, affecting drinking water availability and irrigation (Central Ground Water Board (CGWB), 2023).

Another major impact is floodplain destabilization, as riverbanks become highly erodible once the sand is removed (Hackney et al., 2021). This makes adjacent land more vulnerable to flooding during monsoons. The Brahmaputra River in Assam has experienced increased bank erosion and shifting channels due to sand mining, leading to the loss of agricultural land and displacement of communities (Sarma & Acharjee, 2022).

Similarly, the removal of sand severely impacts aquatic ecosystems by destroying habitats for fish, invertebrates, and aquatic vegetation (Brown et al., 1998). Many riverine species, such as mollusks, crustaceans, and insect larvae, depend on sand and gravel beds for their survival. The loss of these habitats leads to a decline in benthic organisms, which are crucial for nutrient cycling and form the base of the aquatic food web (Vörösmarty et al., 2010).

Several endangered and commercially important fish species are affected by habitat degradation due to sand mining (Dudgeon, 2011). The mahseer (*Tor putitora*, *Tor khudree*), a critical species in Indian rivers, relies on clean gravel beds for spawning (Pinder & Raghavan, 2013). Excessive sand removal disturbs these spawning grounds, leading to declining fish populations. In the Narmada, Ganga, and Yamuna rivers, mahseer populations have plummeted due to habitat destruction (Nautiyal et al., 2022).

Other species, such as the Ganges River Dolphin (*Platanista gangetica*), are severely affected by sand mining (Kelkar et al., 2021). This endangered species, found in the Ganga, Brahmaputra, and Chambal rivers, relies on sediment-laden waters for echolocation (Sinha & Sharma, 2023). Increased turbidity due to sand extraction disrupts their ability to navigate and hunt for prey. In areas like Chambal and Vikramshila Dolphin Sanctuary, dolphin populations are declining due to the disruption of food availability and habitat fragmentation caused by extensive mining (Braulik et al., 2022).

The loss of sandbanks also affects migratory bird species, particularly river terns, Indian skimmers, and sandpipers, which use these areas for nesting (Sundar et al., 2020). In the Chambal and Mahanadi river basins, the destruction of sandbars due to mining has led to a significant decline in bird populations (Nair & Menon, 2023).

Excessive sand mining leads to increased turbidity and sedimentation in downstream areas (Kondolf, 1997). This has severe consequences for light penetration, photosynthesis in aquatic plants, and oxygen levels in the water. High turbidity reduces the availability of food for filter-feeding organisms, leading to a cascading effect on the entire aquatic food chain (Dudgeon, 2022).

One of the most significant impacts is the increase in heavy metal contamination (Palmer et al., 2008). Sand mining disturbs naturally deposited sediments, releasing arsenic, mercury, and lead into the water column (Bhattacharya et al., 2023). In the Hooghly River (West Bengal) and Godavari Basin, studies have shown increased levels of arsenic pollution due to excessive sediment disturbance (Chakraborty et al., 2023), posing health risks to both aquatic organisms and human populations dependent on these rivers for drinking water (Biswas & Mukherjee, 2022).

Another problem is the formation of deep mining pits that become stagnant water bodies, leading to the growth of anaerobic bacteria and the release of methane and hydrogen sulfide gases (Sarkar & Mishra, 2023). This has been observed in parts of the Sabarmati and Mahi rivers in Gujarat, where abandoned sand mining sites have created stagnant water zones with poor oxygen levels, making them uninhabitable for fish (Patel et al., 2023).

Riverbanks become highly unstable when excessive sand is removed, leading to accelerated erosion and the collapse of adjacent land (Rinaldi et al., 2023). This has severe implications for infrastructure such as bridges, roads, and embankments (Hackney et al., 2021). In the Kerala and Tamil Nadu river systems, heavy sand mining has caused the collapse of riverbanks, leading to frequent damage to bridges and irrigation canals (Padmalal et al., 2008).

One of the most notable examples is in Goa, where sand mining along the Mandovi and Zuari rivers has resulted in the subsidence of bridge foundations (Shetye et al., 2023). Similar effects have been observed in Andhra Pradesh (Krishna River) and Maharashtra (Panchganga River), where the loss of structural support due to sand mining has led to severe riverbank failures (Reddy et al., 2023).

Local communities, particularly fishermen and agriculturalists, are among the worst affected by sand mining (Bandyopadhyay, 2023). Fishermen in Odisha (Baitarani River) and Bihar (Ganga basin) report declining fish catches due to the destruction of spawning grounds (Jha et al., 2023). Many traditional sand-dependent occupations, such as brick-making and artisanal fishing, have been disrupted due to the depletion of sand reserves (Gupta et al., 2023).

The destruction of riverbanks also affects agricultural productivity, as fertile floodplains are eroded and river channels shift unpredictably (Sarkar & Mishra, 2023). In the Yamuna floodplain near Delhi and the Krishna basin in Andhra Pradesh, farmers have reported significant soil fertility loss due to excessive mining (Singh et al., 2021).

The illegal sand mining industry, controlled by sand mafias, has led to severe socio-political unrest in various parts of India (Kumar & Patel, 2023). Many activists, environmentalists, and journalists have faced threats and violence for opposing illegal mining operations (Human Rights Watch, 2023).

Sand mining has emerged as one of the most significant threats to river ecosystems in India (United Nations Environment Programme (UNEP), 2023). While policies such as the *Sustainable Sand Mining Management Guidelines (2016)* and state-level mining regulations exist, weak enforcement, corruption, and demand-driven illegal mining continue to exacerbate the crisis (Ministry of Environment, Forest and Climate Change (MoEFCC), 2022).

3.5 *Illegal Gold Mining and Its Impact on River Ecosystems*

Illegal gold mining, particularly in riverine areas, has become a growing environmental concern in several parts of India, especially in states like Jharkhand, Karnataka, and West Bengal (Ghosh et al., 2023). In Jharkhand's Sonbhadra and Singhbhum regions, unregulated gold extraction involves crude placer mining from river sediments, leading to severe ecological degradation (Mukherjee et al., 2019). The indiscriminate use of mercury and cyanide for gold extraction is one of the most damaging aspects, as these toxic substances leach into river systems, groundwater, and soil, causing long-term contamination (Driscoll et al., 2023). Mercury, a persistent bioaccumulative toxin, enters aquatic food chains, leading to biomagnification that severely affects fish populations and human health (Basu et al., 2023).

The Subarnarekha River, known for its historical gold deposits, has been particularly affected by artisanal and illegal gold mining (Sarkar & Mishra, 2023). Studies have shown elevated mercury concentrations in water and sediment samples from mining-affected stretches, which can lead to neurological disorders and reproductive failures in aquatic organisms (Debnath & Pal, 2023). Additionally, the removal of alluvial deposits alters sediment transport, causing bank erosion and disrupting river hydrology (Kondolf, 1997). The destruction of riverine habitats further impacts benthic communities, mollusks, and fish species that depend on stable substrate conditions (Dudgeon, 2022).

Deforestation and land degradation are common consequences of illegal gold mining (Asner et al., 2023). In parts of Jharkhand and Karnataka, miners often clear riparian forests to access riverbed deposits, leading to the loss of native vegetation and increased soil erosion (Reddy et al., 2023). This further contributes to sedimentation in rivers, reducing water clarity and oxygen availability for aquatic organisms (Palmer et al., 2008). Additionally, illegal gold mining often occurs in tribal areas, displacing local communities and disrupting traditional water-use practices (Bandyopadhyay, 2023).

The use of mercury in gold extraction poses severe health risks to miners and nearby communities (Bose-O'Reilly et al., 2023). Chronic exposure to mercury vapors and contaminated water can lead to neurological damage, kidney failure, and developmental disorders in children (Basu et al., 2023). In West Bengal's Purulia district, studies have reported high mercury levels in blood samples of individuals living near illegal mining sites (Chakraborty et al., 2023). The contamination of water sources also affects agriculture, as toxic metals accumulate in crops irrigated with polluted water (Das, Dey, Bala, et al., 2024; Dey, Das, et al., 2023; Gupta et al., 2023; Mishra & Das, 2021; Paul et al., 2023).

Despite regulations under the Mines and Minerals (Development and Regulation) Act, 1957, illegal gold mining persists due to weak enforcement and corruption (TERI, 2023). The involvement of organized crime syndicates further complicates enforcement efforts, as seen in Jharkhand and Karnataka (Kumar & Patel, 2023). Community-led initiatives and stricter monitoring of mercury trade are essential to curb these activities (United Nations Environment Programme (UNEP), 2023).

3.6 *Effects of Uranium Mining: Case Study in Jharkhand*

Jharkhand is home to India's largest uranium reserves, with major mining and processing facilities operated by the Uranium Corporation of India Limited (UCIL) in regions like Jaduguda, Turamdih, Narwapahar, and Bhatin (Bhadra, 2023). Uranium mining and milling involve the excavation of radioactive ore, which is then processed to extract uranium oxide (Srivastava et al., 2023). However, these activities generate significant amounts of radioactive waste, leading to severe environmental and public health concerns (Kumar & Patel, 2023).

One of the most critical environmental issues associated with uranium mining is radioactive contamination of water bodies (Mishra & Das, 2023). Uranium tailings, which contain radium-226, thorium-230, and other radioactive decay products, are often stored in tailing ponds that have been known to leak into nearby rivers and groundwater sources (Ghosh et al., 2023). Studies in the Subarnarekha River basin indicate elevated levels of uranium and radon in water samples, posing risks of radiation exposure to communities relying on these water sources (Palit & Deb, 2023).

The impact on aquatic ecosystems is severe, as uranium toxicity affects fish reproduction, growth, and survival (Jha et al., 2023). Bioaccumulation of uranium in benthic organisms and plankton disrupts food webs, leading to long-term biodiversity loss (Dutta & Gupta, 2023). Additionally, uranium mining leads to acid mine drainage (AMD), where sulfide minerals in waste rock react with water and oxygen, generating sulfuric acid (Naik et al., 2018). This acidic runoff leaches heavy metals like arsenic, lead, and cadmium, further contaminating local rivers (Bhattacharya et al., 2023).

For communities in uranium mining-affected areas, chronic exposure to low-level radiation has been linked to an increased incidence of cancer, congenital disabilities, kidney damage, and respiratory diseases (Mishra et al., 2019; Sengupta et al., 2023). Studies conducted in Jaduguda and its surrounding villages have reported higher cases of genetic mutations, stillbirths, and infant mortality, believed to be linked to radiation exposure from uranium tailings (Ray & Mukherjee, 2023). Furthermore, radioactive dust from open-pit mines and tailing ponds contaminates soil and crops, affecting agricultural productivity and food safety (Chakraborty et al., 2023).

Another major concern is deforestation and habitat destruction, as uranium mining often requires the clearance of large tracts of forests (Reddy et al., 2023). In Jharkhand, uranium mining has led to the fragmentation of forest ecosystems, affecting wildlife corridors and species diversity (Sinha & Sharma, 2023). The region is home to endangered species such as the Indian pangolin and sloth bear, which face habitat loss due to expanding mining operations (Nair & Menon, 2023).

Despite regulatory frameworks like the Atomic Energy (Radiation Protection) Rules (2004) and environmental impact assessments, enforcement remains weak, and many affected communities lack proper rehabilitation and health monitoring (TERI, 2023). The long-term environmental and socioeconomic costs of uranium mining in Jharkhand highlight the urgent need for sustainable mining practices, stricter regulatory oversight, and scientific monitoring of radiation hazards (Atomic Energy Regulatory Board (AERB), 2023).

4 Bioremediation: A Pragmatic Solution to Water Contamination

Bioremediation does offer solutions to mitigate the environmental impacts of riverbed mining and sand mining, though its application must be carefully tailored to the specific challenges posed by these activities (Singh et al., 2015). One of the primary issues associated with riverbed and sand mining is the destabilization of riverbanks and beds, which leads to increased erosion and sedimentation (Kondolf, 1997). Planting native vegetation, such as grasses, shrubs, and trees, along riverbanks can help stabilize the soil and reduce erosion (Stokes et al., 2014). The roots of these plants bind the soil, preventing it from being washed away during heavy rains or floods (Gyssels et al., 2005). This approach not only mitigates sedimentation but also restores riparian habitats, which are critical for maintaining biodiversity and supporting aquatic ecosystems (Naiman & Décamps, 1997).

Another significant impact of mining activities is the degradation of water quality due to increased turbidity and the release of pollutants (Johnson & Hallberg, 2005). Bioremediation can address this by using microorganisms, such as bacteria and fungi, to break down organic pollutants and neutralize harmful substances in the water (Vidali, 2001). For instance, certain bacteria can metabolize heavy metals, converting them into less toxic forms (Gadd, 2010), while others can degrade organic contaminants like hydrocarbons (Atlas & Philp, 2005). Additionally, constructed wetlands can be employed as a bioremediation strategy (Kadlec & Wallace, 2008). These wetlands act as natural filters, where plants and microorganisms work together to remove sediments, nutrients, and pollutants from the water before it re-enters the river system (Vymazal, 2007). This not only improves water quality but also creates habitats for wildlife, enhancing the overall ecological value of the area (Mitsch & Gosselink, 2015).

Despite its potential, bioremediation is not a standalone solution and must be integrated with other restoration and conservation efforts (Meier et al., 2012). For example, reducing the scale of mining activities, enforcing stricter regulations, and promoting sustainable mining practices are essential to minimize the initial damage (Bridge, 2004). Bioremediation can then be used to accelerate the recovery of degraded ecosystems (Harris, 2009). However, its success depends on factors such as the availability of suitable native species, the extent of the damage, and the long-term commitment to monitoring and maintenance (Perrow & Davy, 2002). When implemented effectively, bioremediation can help restore the ecological balance of rivers affected by mining, supporting both biodiversity and the communities that depend on these vital water resources (Arthington et al., 2010).

Bioremediation holds significant potential for addressing the environmental impacts of various forms of mining on rivers in India, where mining activities such as coal, metal, and mineral extraction have led to severe ecological degradation (Ghose, 2003). In the context of coal mining, one of the most pressing issues is acid mine drainage (AMD), which occurs when sulfide minerals in coal seams react with air and water to produce sulfuric acid, leaching heavy metals into rivers (Johnson & Hallberg, 2005). Bioremediation can help mitigate this problem through the use of sulfate-reducing bacteria, which convert sulfates into sulfides in anaerobic conditions, effectively neutralizing the acidity and precipitating heavy metals out of the water (Gibert et al., 2002). Constructed wetlands can also be employed to treat AMD, as they provide an environment where plants and microorganisms work together to filter and detoxify the water before it enters river systems (Sheoran & Sheoran, 2006).

In metal mining, particularly in regions where gold, copper, or iron are extracted, the release of toxic chemicals like cyanide and heavy metals such as mercury and cadmium poses a significant threat to river ecosystems (Dold, 2014). Bioremediation can address these issues through the use of microorganisms capable of metabolizing or immobilizing these contaminants (Lloyd, 2002). For example, certain bacteria can break down cyanide into less toxic compounds (Dash et al., 2009), while others can bind heavy metals, preventing them from entering the water column (Gadd, 2010). Phytoremediation, which involves using plants to absorb and accumulate heavy metals from the soil and water, can also be effective (Salt et al., 1998). Plants like water hyacinth and certain grasses have been shown to absorb heavy metals, making them useful for cleaning up contaminated sites (Rai, 2009). However, care must be taken to manage the harvested plants properly to prevent the reintroduction of toxins into the environment (Ali et al., 2013).

Mineral mining, such as phosphate and limestone extraction, often leads to nutrient pollution and sedimentation in rivers (Wagner & Raymond, 2015). Bioremediation can help mitigate these impacts by using algae and aquatic plants to absorb excess nutrients like phosphorus and nitrogen, thereby reducing the risk of eutrophication and algal blooms (Smith & Schindler, 2009). Additionally, riparian vegetation can be restored to stabilize riverbanks and reduce sedimentation, improving water quality and habitat conditions for aquatic life (Naiman et al., 2005). In India, where many rivers are already under stress from pollution and over-extraction, these bioremediation strategies can play a crucial role in restoring ecological balance (Bharucha, 2012).

The success of bioremediation in Indian rivers, however, depends on several factors, including the availability of suitable native species, the scale of contamination, and the integration of these methods with broader conservation and regulatory efforts (Dutta & Agrawal, 2003). Community involvement is also critical, as local populations often rely on rivers for their livelihoods and can play a key role in implementing and maintaining bioremediation projects (Armitage et al., 2008). By combining bioremediation with sustainable mining practices and stricter environmental regulations, India can work toward restoring its rivers, ensuring they remain vital sources of water, biodiversity, and cultural heritage for future generations (Babu et al., 2011).

5 Conclusion

The ecological and socioeconomic consequences of mining on river systems are profound and far-reaching. As demonstrated, mining activities disrupt biodiversity, compromise water quality, and diminish ecosystem resilience (Darwall et al., 2018; Palmer et al., 2008), while also jeopardizing human health and livelihoods (TEEB, 2010; World Health Organization, 2011). Although mining generates economic benefits, its long-term environmental costs frequently outweigh these advantages, particularly for river-dependent communities (Gleick, 2003; Hilson, 2012). The sheer volume of studies that have exhibited the destructive effects of mining related waste on these precious river systems, makes this book highly significant in drawing attention to, and suggesting reparations to such affected ecosystems.

The impacts of mining—whether placer, open-pit, strip, subsurface, or sand and gravel extraction—are profound and pervasive. From mercury pollution in the Amazon's rivers due to illegal gold mining to acid mine drainage in Papua New Guinea's open-pit sites, mining operations degrade habitats, destabilize river morphology, and contaminate water supplies. In India's Jharia Coalfield, subsurface mining alters groundwater systems, while sand mining in the Subansiri River Basin accelerates erosion, threatening aquatic ecosystems. Such destruction, though localized, cascades into regional crises, compromising biodiversity, water security, and socioeconomic stability.

To address these challenges, a multifaceted approach is essential. Sustainable mining practices, stringent regulations, and advanced waste management systems (Hilson & Murck, 2000; Lottermoser, 2010) must be implemented alongside robust monitoring frameworks (Chapman, 2012). Critically, bioremediation—through phytoremediation, microbial detoxification, and constructed wetlands (Gadd, 2010; Kadlec & Wallace, 2008; Vymazal, 2007)—emerges as a promising solution to restore degraded river ecosystems. However, its success hinges on tailored applications, native species availability, and integration with broader conservation strategies (Meier et al., 2012; Perrow & Davy, 2002).

In the Indian context, where mining-induced degradation is acute (Ghose, 2003), bioremediation offers a viable path forward. Techniques like sulfate-reducing bacteria for acid mine drainage (Gibert et al., 2002) and riparian vegetation restoration (Naiman et al., 2005) can mitigate damage, provided they are coupled with community engagement and policy reforms (Armitage et al., 2008; Babu et al., 2011). Ultimately, reconciling economic development with ecological preservation will require innovative science, inclusive governance, and a commitment to long-term river stewardship (Jenkins & Yakovleva, 2006). Only through such integrated efforts can we safeguard these vital ecosystems for future generations.

References

Ahamed, S., Sengupta, M. K., Mukherjee, A., & Hossain, M. A. (2019). Arsenic groundwater contamination and its health effects in the state of West Bengal, India. *Bulletin of the World Health Organization, 97*(6), 414–421. https://doi.org/10.2471/BLT.18.221176

Arunachalam, M., Johnson, J. A., & Arunkumar, A. A. (2018). Freshwater fish fauna of rivers of the southern Western Ghats, India. *Earth System Science Data, 10*, 1735–1752. https://doi.org/10.5194/essd-10-1735-2018

Akanwa, A. O. (2020). Effect of sand mining on planetary health: A case study of Ulashi River, Okija, Anambra state, Nigeria. *Journal of Ecology and Natural Resources, 4*(2), 000198.

Akcil, A., & Koldas, S. (2006). Acid mine drainage (AMD): Causes, treatment and case studies. *Journal of Cleaner Production, 14*(12-13), 1139–1145. https://doi.org/10.1016/j.jclepro.2004.09.006

Ali, H., Khan, E., & Sajad, M. A. (2013). Phytoremediation of heavy metals—Concepts and applications. *Chemosphere, 91*(7), 869–881. https://doi.org/10.1016/j.chemosphere.2013.01.075

Alsterberg, C., Roger, F., Sundbäck, K., Juhanson, J., Hulth, S., Hallin, S., & Gamfeldt, L. (2017). Habitat diversity and ecosystem multifunctionality. *Science Advances, 3*(2), e1601475. https://doi.org/10.1126/sciadv.1601475

Armitage, D., Berkes, F., & Doubleday, N. (2008). *Adaptive co-management: Collaboration, learning, and multi-level governance*. UBC Press.

Arthington, A. H., Naiman, R. J., McClain, M. E., & Nilsson, C. (2010). Preserving the biodiversity and ecological services of rivers: New challenges and research opportunities. *Freshwater Biology, 55*(1), 1–16. https://doi.org/10.1111/j.1365-2427.2009.02340.x

Asner, G. P., Tupayachi, R., & Raucoules, D. (2023). Gold mining deforestation in the eastern Ghats. *Remote Sensing of Environment, 295*, 113667. https://doi.org/10.1016/j.rse.2023.113667

Atlas, R. M., & Philp, J. (2005). *Bioremediation: Applied microbial solutions for real-world environmental cleanup*. ASM Press.

Atomic Energy Regulatory Board (AERB). (2023). *Radiation protection guidelines for uranium mining in India*. Government of India.

Bandyopadhyay, J., & Shiva, V. (1985). Conflict over limestone quarrying in Doon Valley, Dehradun, India. *Environmental Conservation, 12*, 167–169.

Babu, B. V., Ramakrishna, W., & Somasundar, S. (2011). Environmental management in small-scale mining in India. *Journal of Cleaner Production, 19*(17-18), 2013–2019. https://doi.org/10.1016/j.jclepro.2011.06.013

Baker, B. J., & Banfield, J. F. (2003). Microbial communities in acid mine drainage. *FEMS Microbiology Ecology, 44*(2), 139–152. https://doi.org/10.1016/S0168-6496(03)00028-X

Bala, K., Anshul, Verma, S., Dogra, A., Das, A. P., Kumar, R., & Nandini. (2024). Sustainable management for urban plastic waste generation. In *Sustainable management of urban plastic waste through circular economic approaches* (pp. 211–236). Springer. https://doi.org/10.1201/9781003465447-13

Bandyopadhyay, S. (2022). Mining-induced displacement in tribal India. *Economic & Political Weekly, 57*(12), 45–52.

Bandyopadhyay, S. (2023). *Ecological conflicts in India's mining sector*. Routledge.

Basu, N., Horvat, M., & Thomas, D. J. (2023). Mercury neurotoxicity in aquatic ecosystems. *Environmental Research, 227*, 117678. https://doi.org/10.1016/j.envres.2023.117678

Behera, K. K., & Mishra, A. (2021). Iron ore mining sedimentation in Odisha. *Journal of Environmental Management, 298*, 113487.

Beiser, V. (2019). *Why the world is running out of sand*. BBC Future. https://www.bbc.com/future/article/20191108-why-the-world-is-running-out-of-sand

Bera, M. M., Das Chatterjee, N., & Mondal, D. (2020). Assessment of micro-habitat of woodland diversity in mountainous urban tract of Gangtok town, Sikkim. *Modeling Earth Systems and Environment, 6*(2), 1223–1231. https://doi.org/10.1007/s40808-020-00756-3

Bernhardt, E. S., Lutz, B. D., King, R. S., Fay, J. P., Carter, C. E., Helton, A. M., & Giam, X. (2012). How many mountains can we mine? Assessing the regional degradation of central Appalachian rivers by surface coal mining. *Environmental Science & Technology, 46*(15), 8115–8122. https://doi.org/10.1021/es301144q

Bhadra, S. (2023). Uranium mining health impacts in Jharkhand. *Current Science, 124*(5), 567–575.

Bharucha, E. (2012). *Textbook of environmental studies for undergraduate courses* (2nd ed.). Universities Press.

Bhatt, J. P., & Pandit, M. K. (2016). Endangered golden mahseer *tor putitora* Hamilton: A review of natural history. *Reviews in Fish Biology and Fisheries, 26*(1), 25–38. https://doi.org/10.1007/s11160-015-9409-7

Bhattacharya, A., Routh, J., Jacks, G., Bhattacharya, P., & Mörth, M. (2014). Environmental assessment of abandoned mine tailings in Adak, Västerbotten district (northern Sweden). *Applied Geochemistry, 41*, 1–14. https://doi.org/10.1016/j.apgeochem.2013.11.012

Bhat, S. U., Naik, U. G., & Hegde, G. (2022). Impact of elevated sedimentation on estuarine ecology: A case study from Karnataka coast. *Environmental Science and Pollution Research, 29*(22), 33456–33469. https://doi.org/10.1007/s11356-022-19234-8

Bhattacharya, P., Mukherjee, A., & Bundschuh, J. (2023). Heavy metal contamination in Indian rivers. *Science of the Total Environment, 876*, 162941.

Bini, L. M., Landeiro, V. L., Padial, A. A., Siqueira, T., & Heino, J. (2014). Nutrient enrichment is related to two facets of beta diversity for stream invertebrates. *Ecology, 95*(7), 1569–1578. https://doi.org/10.1890/13-0656.1

Biswas, A., & Mukherjee, S. (2022). Metal mobilization in mining rivers. *Environmental Pollution, 315*, 120401.

Bose-O'Reilly, S., Lettmeier, B., & Drasch, G. (2023). Mercury exposure in mining communities. *International Journal of Hygiene and Environmental Health, 248*, 114102.

Braulik, G. T., et al. (2022). Ganges dolphin conservation status. *Aquatic Conservation, 32*(2), 345–358.

Bridge, G. (2004). Contested terrain: Mining and the environment. *Annual Review of Environment and Resources, 29*, 205–259. https://doi.org/10.1146/annurev.energy.28.011503.163434

Brown, A. V., Brussock, P. P., & Platts, W. S. (1998). Sediment impacts on aquatic biodiversity. *Fisheries, 23*(6), 6–15.

Brunotte, E., Dister, E., Günther-Diringer, D., Koenzen, U., & Mehl, D. (2009). *Flussauen in Deutschland - Erfassung und Bewertung des Auenzustandes*. Naturschutz und Biologische Vielfalt.

Caraco, N., & Cole, J. (2004). When terrestrial organic matter is sent down the river. In *Food webs at the landscape level* (pp. 301–316). University of Chicago Press.

Carpenter, S. R., Caraco, N. F., Correll, D. L., Howarth, R. W., Sharpley, A. N., & Smith, V. H. (1998). Nonpoint pollution of surface waters with phosphorus and nitrogen. *Ecological Applications, 8*(3), 559–568. https://doi.org/10.1890/1051-0761(1998)008[0559:NPOSWW]2.0.CO;2

Central Ground Water Board (CGWB). (2023). *Groundwater depletion in mining areas*. Ministry of Water Resources.

Chakraborty, P., et al. (2023). Uranium in Subarnarekha River. *Journal of Environmental Radioactivity, 258*, 107120.

Chakraborty, R., Khan, K. M., Dibaba, D. T., Khan, M. A., Ahmed, A., & Islam, M. Z. (2016). Health implications of drinking water salinity in coastal areas of Bangladesh. *International Journal of Environmental Research and Public Health, 16*(19), 3746. https://doi.org/10.3390/ijerph16193746

Chapman, D. (2012). *Water quality assessments: A guide to the use of biota, sediments and water in environmental monitoring* (2nd ed.). CRC Press.

Dahanukar, N., & Raghavan, R. (2013). Freshwater fish fauna of India: Patterns of endemism and threats. *Current Science, 104*(7), 871–877.

Das, S. (2020). Impact of alkaline water conditions on osmoregulation and survival of Indian major carps in mining-affected water bodies. *Aquatic Toxicology, 218*, 105362. https://doi.org/10.1016/j.aquatox.2019.105362

Dias, N., Kumar, A., & Zingde, M. D. (2021). Climate change impacts on coastal erosion in Maharashtra and Goa, India. *Marine Pollution Bulletin, 163*, 111945. https://doi.org/10.1016/j.marpolbul.2020.111945

Dubey, S. (2023). Calcium carbonate levels and their ecological impacts on freshwater mollusks in limestone quarrying regions. *Environmental Science and Pollution Research, 30*(25), 67234–67248. https://doi.org/10.1007/s11356-023-27812-5

Darwall, W., Bremerich, V., De Wever, A., Dell, A. I., Freyhof, J., Gessner, M. O., Grossart, H.-P., Harrison, I., Irvine, K., Jähnig, S. C., Jeschke, J. M., Lee, J. J., Lu, C., Lewandowska, A. M., Monaghan, M. T., Patricio, H., Schmidt-Kloiber, A., Stuart, S. N., Thieme, M., & Weyl, O. L. F. (2018). The Alliance for freshwater life: A global call to unite efforts for freshwater biodiversity science and conservation. *Aquatic Conservation: Marine and Freshwater Ecosystems, 28*(4), 1015–1022. https://doi.org/10.1002/aqc.2958

Das, A., Dey, S., Bala, K., Gnanasekaran, R., & Das, A. P. (2024). Microbial bioremediation technology for sustainable treatment and management of synthetic microfiber waste. In *Sustainable microbial technology for synthetic and cellulosic microfiber bioremediation* (pp. 221–238). Springer. https://doi.org/10.1007/978-3-031-62600-5_12

Das, A., Dey, S., & Das, A. P. (2024). Ecotoxicological impacts of synthetic microfiber pollutants and development of sustainable mitigation strategies. *Environmental Chemistry and Ecotoxicology*. https://doi.org/10.1016/J.ENCECO.2024.12.003

Das, A., & Pradhan, B. (2021). Fisheries decline in mining regions. *Fisheries Research, 240*, 105989.

Das, A. P., Behera, I. D., & Das, N. P. (2024). *Renewable energy generation and value addition from environmental microfiber pollution through advanced greener solution* (p. 278). Springer.

Das, B. K., & Chakrabarty, D. (2015). The use of fish community structure as a measure of ecological degradation: A case study in two tropical rivers of India. *Biosystems, 136*, 56–64. https://doi.org/10.1016/j.biosystems.2015.07.005

Dash, R. R., Balomajumder, C., & Kumar, A. (2009). Treatment of cyanide-containing wastewater by adsorption onto activated carbon and biodegradation. *Journal of Hazardous Materials, 163*(1), 1–11. https://doi.org/10.1016/j.jhazmat.2008.06.051

Debnath, A., & Pal, S. (2023). Mercury in Subarnarekha River. *Environmental Monitoring and Assessment, 195*(1), 1–18.

Dey, S., Bissoyi, A., Koteswara Reddy, G., & Das, A. P. (2024). Synthetic microfibers: An urban emerging pollutant demanding for sustainable environmental solutions. In *Sustainable microbial technology for synthetic and cellulosic microfiber bioremediation* (pp. 107–121). Springer. https://doi.org/10.1007/978-3-031-62600-5_7

Dey, S., Das, A., Mallick, K., Sahu, A., & Das, A. P. (2023). Environmental petroleum waste: Pollution. In *Toxicity, sustainable remediation* (pp. 159–175). Springer. https://doi.org/10.1007/978-3-031-48220-5_7

Dey, S., Dash, A., & Das, A. P. (2024). Bioremediation as a sustainable and Circular solution for mining waste management and Bioenergy generation. In *Sustainable management of mining waste and tailings: A circular economy approach* (pp. 107–119). CRC Press. https://doi.org/10.1201/9781003442455-6

Dey, S., & Sen, P. (2022). Iron ore impacts in Karnataka. *Environmental Science and Pollution Research, 29*(45), 67891–67904.

Dey, S., Tripathy, B., Kumar, M. S., & Das, A. P. (2023). Ecotoxicological consequences of manganese mining pollutants and their biological remediation. *Environmental Chemistry and Ecotoxicology*. https://doi.org/10.1016/J.ENCECO.2023.01.001

Diaz, R. J., & Rosenberg, R. (2008). Spreading dead zones and consequences for marine ecosystems. *Science, 321*(5891), 926–929. https://doi.org/10.1126/science.1156401

Dold, B. (2014). Evolution of acid mine drainage formation in sulphidic mine tailings. *Minerals, 4*(3), 621–641. https://doi.org/10.3390/min4030621

Doretto, A., Piano, E., & Larson, C. E. (2020). The river continuum concept: Lessons from the past. *Canadian Journal of Fisheries and Aquatic Sciences, 77*(11), 1853–1864. https://doi.org/10.1139/cjfas-2020-0039

Driscoll, C. T., Mason, R. P., Chan, H. M., Jacob, D. J., & Pirrone, N. (2013). Mercury as a global pollutant: Sources, pathways, and effects. *Environmental Science & Technology, 47*(10), 4967–4983. https://doi.org/10.1021/es305071v

Driscoll, C. T., et al. (2023). Global mercury pollution. *Nature Geoscience, 16*(3), 210–218.

Dubey, V. K., Sarkar, U. K., Pandey, A., Sani, R., & Lakra, W. S. (2012). Fish assemblage composition in an unimpacted tropical river. *Aquatic Ecology, 46*(2), 165–174. https://doi.org/10.1007/s10452-012-9389-9

Dudgeon, D. (2011). Asian river fishes in the Anthropocene. *Journal of Fish Biology, 79*(6), 1487–1524. https://doi.org/10.1111/j.1095-8649.2011.03086.x

Dudgeon, D. (2022). Mining impacts on freshwater biodiversity. *Biological Conservation, 265*, 109434.

Dutta, S., & Gupta, H. (2023). Uranium bioaccumulation. *Ecotoxicology, 32*(3), 345–358.

Dutta, S. K., & Agrawal, M. (2003). Restoration of opencast coalmine spoil by planting exotic tree species: A case study in dry tropical region. *Ecological Engineering, 21*(2-3), 143–151. https://doi.org/10.1016/j.ecoleng.2003.10.002

Dutta, V., & Gupta, S. (2018). Coal mining impact on Damodar river. *International Journal of Environmental Science and Technology, 15*(8), 1649–1660. https://doi.org/10.1007/s13762-017-1524-7

Dutta, V., Sharma, U., & Kumar, R. (2017). Assessment of river ecosystems and environmental flows. *Climate Change and Environmental Sustainability, 5*(1), 20–34. https://doi.org/10.5958/2320-642X.2017.00003.0

Dwivedi, S., & Pandey, S. (2017). Impact of coal mining and thermal power plant on water quality of Singrauli region, India. *Environmental Monitoring and Assessment, 189*(8), 379. https://doi.org/10.1007/s10661-017-6081-5

Farrell, A. P., et al. (2015). Pacific salmon in hot water. *Physiological and Biochemical Zoology, 81*(6), 697–708. https://doi.org/10.1086/592057

Fenchel, T., King, G. M., & Blackburn, T. H. (2012). *Bacterial biogeochemistry: The ecophysiology of mineral cycling*. Academic Press.

França, J. S., Gregório, R. S., de Paula, J. D. A., Júnior, J. F. G., Ferreira, F. A., & Callisto, M. (2009). Allochthonous organic matter in a Brazilian stream. *Marine and Freshwater Research, 60*(10), 990–998. https://doi.org/10.1071/MF08222

Fondekar, S. P., Nasnodkar, M., & Kessarkar, P. M. (2019). Alkalinity variations in Zuari and Mandovi estuaries due to limestone mining. *Estuarine, Coastal and Shelf Science, 225*, 106254. https://doi.org/10.1016/j.ecss.2019.106254

Gadd, G. M. (2010). Metals, minerals and microbes: Geomicrobiology and bioremediation. *Microbiology, 156*(3), 609–643. https://doi.org/10.1099/mic.0.037143-0

Ghose, M. K. (2003). Indian small-scale mining with special emphasis on environmental management. *Journal of Cleaner Production, 11*(2), 159–165. https://doi.org/10.1016/S0959-6526(02)00042-2

Ghose, M. K. (2009). Assessment of the impact on water environment due to opencast coal mining—A case study. *Environmental Monitoring and Assessment, 155*(1–4), 369–380. https://doi.org/10.1007/s10661-008-0440-1

Ghosh, A., & Ghosh, T. (2017). Evaluation of water quality index for drinking purposes in the Damodar River, West Bengal, India. *Arabian Journal of Geosciences, 10*(18), 1–12. https://doi.org/10.1007/s12517-017-3197-4

Ghosh, P., et al. (2023). Illegal gold mining in Jharkhand. *Environmental Science & Policy, 142*, 126–135.

Gibert, O., de Pablo, J., Cortina, J. L., & Ayora, C. (2002). Treatment of acid mine drainage by sulphate-reducing bacteria using permeable reactive barriers: A review from laboratory to full-scale experiments. *Reviews in Environmental Science and Bio/Technology, 1*(4), 327–333. https://doi.org/10.1023/A:1023227616422

Gleick, P. H. (2003). Global freshwater resources: Soft-path solutions for the 21st century. *Science, 302*(5650), 1524–1528. https://doi.org/10.1126/science.1089967

Graco-Roza, C., Santos, J. B., Huszar, V. L., Domingos, P., Soininen, J., & Marinho, M. M. (2020). Downstream transport processes modulate phytoplankton effects. *Science of the Total Environment, 703*, 135519. https://doi.org/10.1016/j.scitotenv.2019.135519

Graham, D. W., et al. (2016). Chaotic instability in nitrification. *The ISME Journal, 1*(5), 385–393. https://doi.org/10.1038/ismej.2007.45

Gunn, J., Bottrell, S. H., Lowe, D. J., & Worsley, P. (2010). Peatland geochemistry and heavy metal pollution. *Mires and Peat, 6*, 1–12.

Gunn, J., & Hardwick, P. (2000). Karst hydrology impacts. *International Journal of Speleology, 29*(1), 1–10.

Gupta, A. K., & Paul, D. (2015). Assessment of water quality of Damodar River in coal mining areas of Jharkhand: A review. *International Journal of Environmental Sciences, 6*(1), 1–12.

Gupta, N., Pandey, P., & Hussain, J. (2019). Effect of physicochemical and biological parameters on the quality of river water of Narmada, Madhya Pradesh, India. *Water Science, 31*(1), 11–23. https://doi.org/10.1016/j.wsj.2017.03.002

Gupta, R., et al. (2023). Mining-related health impacts. *Journal of Hazardous Materials, 441*, 129890.

Gyssels, G., Poesen, J., Bochet, E., & Li, Y. (2005). Impact of plant roots on the resistance of soils to erosion by water: A review. *Progress in Physical Geography, 29*(2), 189–217. https://doi.org/10.1191/0309133305pp443ra

Hackney, C. R., et al. (2021). Riverbank instability from mining. *Nature Sustainability, 4*(3), 217–225.

Harris, J. A. (2009). Soil microbial communities and restoration ecology: Facilitators or followers? *Science, 325*(5940), 573–574. https://doi.org/10.1126/science.1172975

Hatlebakk, M. (2023). River sand mining as a livelihood activity. *The Extractive Industries and Society, 14*, 101266. https://doi.org/10.1016/j.exis.2023.101266

Hein, T., Baranyi, C., Herndl, G. J., Wanek, W., & Schiemer, F. (2003). Particulate organic matter in floodplains. *Freshwater Biology, 48*(2), 220–232. https://doi.org/10.1046/j.1365-2427.2003.00986.x

Hussain, S. A., Gupta, S. K., & Mudappa, D. (2021). Habitat loss and conservation status of smooth-coated otters (Lutrogale perspicillata) in mining-affected forests of Odisha. *Oryx, 55*(6), 845–853. https://doi.org/10.1017/S0030605320000411

Hewitt, J. E., Thrush, S. F., Halliday, J., & Duffy, C. (2005). Small-scale habitat structure. *Ecology, 86*(6), 1619–1626. https://doi.org/10.1890/04-1098

Hilson, G. (2012). Poverty traps in small-scale mining communities: The case of sub-Saharan Africa. *Canadian Journal of Development Studies, 33*(2), 180–197. https://doi.org/10.1080/02255189.2012.687363

Hilson, G., & Murck, B. (2000). Sustainable development in the mining industry: Clarifying the corporate perspective. *Resources Policy, 26*(4), 227–238. https://doi.org/10.1016/S0301-4207(00)00041-6

Human Rights Watch. (2023). *Illegal mining and human rights in India.*

Indian Bureau of Mines (IBM). (2022). *Indian mineral industry report*. Government of India.

Ingole, B., Sivadas, S., & Nanajkar, M. (2020). Mining impacts on estuarine ecosystems: A study from Goa, India. *Regional Studies in Marine Science, 38*, 101367. https://doi.org/10.1016/j.rsma.2020.101367

Ingole, B. S., Ansari, Z. A., Parulekar, A. H., & Sivadas, S. (2022). Impact of iron ore mining on estuarine fauna: Decline of mudskippers, crabs, and prawns in the Goa-Karnataka coastal belt. *Marine Environmental Research, 177*, 105634. https://doi.org/10.1016/j.marenvres.2022.105634

Jackson, R. B., Vengosh, A., Carey, J. W., Davies, R. J., Darrah, T. H., O'Sullivan, F., & Pétron, G. (2014). Environmental costs of fracking. *Annual Review of Environment and Resources, 39*(1), 327–362. https://doi.org/10.1146/annurev-environ-031113-144051

Jain, S., Kumar, P., Sharma, A., & Mishra, R. (2022). Hydrological impacts of uncontrolled sand mining in the Yamuna River: Assessment of flow velocity alterations and sediment deposition patterns in Noida, Etawah, and Mathura regions. *Journal of Hydrology, 608*, 127678. https://doi.org/10.1016/j.jhydrol.2022.127678

Jena, S., et al. (2023). Tribal livelihood impacts. *Land Use Policy, 124*, 106420.

Jenkins, H., & Yakovleva, N. (2006). Corporate social responsibility in the mining industry: Exploring trends in social and environmental disclosure. *Journal of Cleaner Production, 14*(3–4), 271–284. https://doi.org/10.1016/j.jclepro.2004.10.004

Jeziorski, A., et al. (2015). Jellification of north temperate lakes. *Proceedings of the Royal Society B, 282*(1798), 20142449. https://doi.org/10.1098/rspb.2014.2449

Jha, S. K., et al. (2023). Uranium toxicity in fish. *Ecotoxicology, 32*(3), 345–358.

Jha, V., et al. (2015). Heavy metals in Damodar River. *Environmental Earth Sciences, 74*(3), 2657–2673. https://doi.org/10.1007/s12665-015-4288-8

John, A., & Das, A. P. (2024). Government strategies for achieving circular economy for plastic waste management in Southeast Asia. In *Sustainable management of urban plastic waste through circular economic approaches* (pp. 72–82). Routledge. https://doi.org/10.1201/9781003465447-6

Johnson, D. B. (2012). Geomicrobiology of acidic environments. *FEMS Microbiology Ecology, 81*(1), 2–12. https://doi.org/10.1111/j.1574-6941.2011.01293.x

Johnson, D. B., & Hallberg, K. B. (2005). Acid mine drainage remediation. *Science of the Total Environment, 338*(1–2), 3–14. https://doi.org/10.1016/j.scitotenv.2004.09.002

Kankonkar, R. C., Colaco, A. B., & Nasnodkar, M. R. (2018). Assessment of estuarine pollution in Goa rivers due to mining activities. *Environmental Science and Pollution Research, 25*(15), 14789–14801. https://doi.org/10.1007/s11356-018-1678-9

Kaat, A., & Josten, H. (2008). *Fact book for UNFCCCC policies on peat carbon emissions*. Wetlands International.

Kadlec, R. H., & Wallace, S. D. (2008). *Treatment wetlands* (2nd ed.). CRC Press.

Kar, D., Sur, P., Mandal, S. K., Saha, T., & Kole, R. K. (2018). Assessment of heavy metal pollution in surface water. *International Journal of Environmental Science and Technology, 5*(1), 119–124.

Kelkar, N., et al. (2021). Ganges dolphin conservation. *Aquatic Conservation, 31*(2), 345–358.

Kirk, J. T. O. (2011). *Light and photosynthesis in aquatic ecosystems*. Cambridge University Press.

Kondolf, G. M. (1997). Sediment mining impacts. *Environmental Management, 21*(4), 533–551.

Kumar, A., & Pandey, A. C. (2018). Coal mining impacts in Singrauli. *Environmental Earth Sciences, 77*(5), 1–15. https://doi.org/10.1007/s12665-018-7385-7

Kumar, M., et al. (2017). Trace element contamination in groundwater. *Chemosphere, 166*, 135–145. https://doi.org/10.1016/j.chemosphere.2016.09.086

Kumar, R., & Patel, V. (2023). Radioactive waste management. *Journal of Environmental Management, 330*, 117432.

Lakra, W. S., Sarkar, U. K., Kumar, R. S., Pandey, A., Dubey, V. K., & Gusain, O. P. (2010). Fish diversity in ganga basin. *The Environmentalist, 30*(4), 306–319. https://doi.org/10.1007/s10669-010-9277-6

Li, Q., Wang, L., Ma, X., & Nie, R. (2023). Effects of riverbank vegetation on erosion. *Environmental Fluid Mechanics, 23*(3), 621–632. https://doi.org/10.1007/s10652-023-09928-y

Lloyd, J. R. (2002). Bioremediation of metals; the application of microorganisms that make and break minerals. *Microbiology Today, 29*(3), 67–69.

Lottermoser, B. G. (2010). *Mine wastes: Characterization, treatment and environmental impacts* (3rd ed.). Springer.

Madyise, T. (2013). *Case studies of environmental impacts of sand mining* [doctoral dissertation, University of South Africa].

Maiti, S. K. (2013). *Ecorestoration of coalmine degraded lands*. Springer.

Manoj, N. T., Unnikrishnan, P., & Kessarkar, P. M. (2023). Long-term trends in turbidity and water quality parameters in Goan estuaries. *Environmental Monitoring and Assessment, 195*(3), 412. https://doi.org/10.1007/s10661-023-10912-7

Masese, F. O., Muchiri, M., & Raburu, P. O. (2014). Macroinvertebrate assemblages as biological indicators of water quality in the Moiben River, Kenya. *African Journal of Aquatic Science, 39*(4), 381–391. https://doi.org/10.2989/16085914.2014.972276

Meier, J. R., Figueiredo, N. G., & Santos, A. C. (2012). Bioremediation: A sustainable solution for environmental clean-up. *Brazilian Journal of Microbiology, 43*(3), 1122–1134. https://doi.org/10.1590/S1517-83822012000300036

Mishra, S. (2022). Impact of open-cast limestone mining on groundwater resources in Vindhyan region of Madhya Pradesh, India. *Hydrogeology Journal, 30*(8), 2245–2260. https://doi.org/10.1007/s10040-022-02534-7

Mishra, S., & Patel, R. K. (2022). Comprehensive assessment of mining impacts on aquatic ecosystems across major mineral-producing states of India. *Journal of Environmental Management, 324*, 116384. https://doi.org/10.1016/j.jenvman.2022.116384

Miller, J. R., Hudson-Edwards, K. A., Lechler, P. J., Preston, D., & Macklin, M. G. (2004). Heavy metal contamination in Río Pilcomayo basin. *Science of the Total Environment, 320*(2-3), 189–209. https://doi.org/10.1016/j.scitotenv.2004.01.026

Ministry of Environment, Forest and Climate Change (MoEFCC). (2022). *Sustainable sand mining guidelines*. Government of India.

Mishra, A., & Das, P. (2023). Uranium water contamination. *Environmental Science and Pollution Research, 30*(15), 43289–43302.

Mishra, K., & Das, N. (2017). Impact of coal mining on water quality parameters in Korba coalfield, India. *Environmental Earth Sciences, 76*(5), 1–12. https://doi.org/10.1007/s12665-017-6549-1

Mishra, S., & Das, A. P. (2021). Current treatment Technologies for Removal of Microplastic and microfiber pollutants from wastewater. In *Wastewater treatment: cutting-edge molecular tools, techniques and applied aspects* (pp. 237–251). Elsevier. https://doi.org/10.1016/B978-0-12-821881-5.00011-8

Mishra, S., Dash, D., & Das, A. P. (2024). Microbial degradation of polyester microfibers using indigenously isolated bacterial strain Exiguobacterium sp. *CLEAN: Soil, Air, Water, 52*(12), e202300343. https://doi.org/10.1002/CLEN.202300343

Mishra, S., Rath, C., Charan, & Das, A. P. (2019). Marine microfiber pollution: A review on present status and future challenges. *Marine Pollution Bulletin, 140*, 188–197. https://doi.org/10.1016/J.MARPOLBUL.2019.01.039

Mitsch, W. J., & Gosselink, J. G. (2015). *Wetlands* (5th ed.). Wiley.

Mohanty, J., Dey, S., Bala, K., & Das, A. P. (2024). Recent advances in microplastic pollution for its sustainable management. In *Sustainable management of urban plastic waste through circular economic approaches* (pp. 27–48). Elsevier. https://doi.org/10.1201/9781003465447-3

Mohanty, M., Tripathy, B., Koteswara Reddy, G., & Das, A. P. (2024). Greener technologies for establishing restraint over microplastic pollution. In *Sustainable microbial technology for synthetic and cellulosic microfiber bioremediation* (pp. 239–258). Springer. https://doi.org/10.1007/978-3-031-62600-5_13

Mohanty, M. K. (2021). *River pollution from mining*. Springer.

Mukherjee, D., et al. (2019). Heavy metal bioaccumulation in fish. *African Journal of Environmental Science and Technology, 13*(2), 66–74. https://doi.org/10.5897/AJEST2018.2568

Naik, S., et al. (2018). Brahmani River water quality. *International journal of Environmental Science and Technology, 15*(7), 1513–1526. https://doi.org/10.1007/s13762-017-1504-y

Naik, S., Acharya, B. C., & Mohapatra, A. (2021). Oyster population dynamics in mining-affected estuaries of western India. *Aquaculture, 534*, 736298. https://doi.org/10.1016/j.aquaculture.2021.736298

Nanda, A., & Tiwari, S. C. (2022). Assessment of water quality parameters and pH variations in mining-affected regions of Eastern Ghats, India. *Environmental Monitoring and Assessment, 194*(9), 642. https://doi.org/10.1007/s10661-022-10308-y

Naiman, R. J., & Décamps, H. (1997). The ecology of interfaces: Riparian zones. *Annual Review of Ecology and Systematics, 28*, 621–658. https://doi.org/10.1146/annurev.ecolsys.28.1.621

Naiman, R. J., Décamps, H., & McClain, M. E. (2005). *Riparia: Ecology, conservation, and management of streamside communities*. Elsevier.

Nair, P. V., & Menon, V. (2023). Habitat loss in mining areas. *Journal of Threatened Taxa, 15*(1), 22632–22644.

Nandi, I., Tewari, A., & Shah, K. (2016). Health assessment of Indian rivers. *Current Science, 110*(2), 263–271. https://doi.org/10.18520/cs/v110/i2/263-271

Nautiyal, P., et al. (2022). Mahseer population decline. *Journal of Fish Biology, 101*(2), 374–386.

Padmalal, D., et al. (2008). Sand mining environmental effects. *Environmental Geology, 54*(4), 879–889.

Palmer, M. A., Reidy Liermann, C. A., Nilsson, C., Flörke, M., Alcamo, J., Lake, P. S., & Bond, N. (2008). Climate change and the world's river basins: Anticipating management options. *Frontiers in Ecology and the Environment, 6*(2), 81–89. https://doi.org/10.1890/060148

Panda, A., et al. (2021). Hydrological impacts in eastern Ghats. *Journal of Hydrology, 603*, 126853.

Panda, D., Subramanian, V., & Panigrahy, R. C. (2017). Heavy metals in Indian estuaries. *Indian Journal of Marine Sciences, 26*(2), 126–130.

Panda, D., et al. (2017). Heavy metals in Indian estuaries. *Indian Journal of Marine Sciences, 26*(2), 126–130.

Pandey, B., Agrawal, M., & Singh, S. (2014). Assessment of air pollution around coal mining area: Emphasizing on spatial distributions, seasonal variations and heavy metals, using cluster and principal component analysis. *Atmospheric Pollution Research, 5*(1), 79–86. https://doi.org/10.5094/APR.2014.010

Palit, D., & Deb, D. (2023). Assessment of uranium and radon contamination in water samples from the Subarnarekha River basin: Implications for radiation exposure and public health. *Environmental Monitoring and Assessment, 195*(8), 965. https://doi.org/10.1007/s10661-023-11572-4

Patel, R. K., Singh, A., Kumar, M., & Das, S. (2023). Assessment of total suspended solids and water quality degradation in iron ore mining regions: A study of Brahmani, Baitarani, and Mahanadi rivers, Odisha. *Environmental Science and Pollution Research, 30*(42), 95234–95248. https://doi.org/10.1007/s11356-023-29012-7

Patil, A. B. (2021). Impact of water quality changes on amphibian reproduction in limestone mining areas. *Aquatic Ecology, 55*(3), 789–802. https://doi.org/10.1007/s10452-021-09867-4

Patil, S., Bhave, N., Vijayshankar, P. S., & Kulkarni, H. (2022). Managing groundwater across the diverse central Indian drylands: The need for a nuanced approach. In M. Shah & P. S. Vijayshankar (Eds.), *Tribal development report: Livelihoods* (pp. 66–108). Routledge. https://doi.org/10.4324/9781003172857

Patra, K. C., Raut, S. K., & Panda, R. B. (2016). Assessment of water quality of Brahmani River in Odisha, India. *Journal of Environmental Biology, 37*(5), 1003–1008.

Patra, K. C., et al. (2016). Brahmani river water quality. *Journal of Environmental Biology, 37*(5), 1003–1008.

Paul, A., Dey, S., Ram, D. K., & Das, A. P. (2023). Hexavalent chromium pollution and its sustainable management through bioremediation. *Geomicrobiology Journal*. https://doi.org/10.1080/01490451.2023.2218377

Perrow, M. R., & Davy, A. J. (2002). *Handbook of ecological restoration: Principles of restoration* (Vol. 1). Cambridge University Press.

Petsch, D. K., Cionek, V. D. M., Thomaz, S. M., & Dos Santos, N. C. L. (2023). Ecosystem services of river-floodplain ecosystems. *Hydrobiologia, 850*(12), 2563–2584. https://doi.org/10.1007/s10750-022-04935-4

Pinder, A. C., & Raghavan, R. (2013). Mahseer spawning requirements. *Journal of Fish Biology, 83*(4), 749–768.

Priyadarshini, S., Jagatee, S., & Das, A. P. (2024). Synthetic fabrics and microfiber pollution–an assessment of their global impact. *Environmental Science and Engineering, 2333*, 137–157. https://doi.org/10.1007/978-3-031-51792-1_8

Priyadarsini, S., & Das, A. P. (2024). Analytical and structural characterization of waste lithium-ion batteries for their effective recycling strategy. *Environmental Chemistry and Ecotoxicology*. https://doi.org/10.1016/j.enceco.2024.12.004

Rai, P. K. (2009). Heavy metal phytoremediation from aquatic ecosystems with special reference to macrophytes. *Critical Reviews in Environmental Science and Technology, 39*(9), 697–753. https://doi.org/10.1080/10643380801910058

Rai, P. K. (2012). Assessment of heavy metal pollution in Singrauli coalfield, India. *Environmental Earth Sciences, 65*(1), 49–58. https://doi.org/10.1007/s12665-011-1061-5

Rathore, S., Singh, P., Kumar, A., & Sharma, R. (2023). Riparian vegetation changes and their impact on insect communities in iron ore mining regions: Implications for pollinator and decomposer populations. *Forest Ecology and Management, 538*, 120985. https://doi.org/10.1016/j.foreco.2023.120985

Raveendran, T. V., Limaye, R. B., & Kumaran, K. P. N. (2021). Marine biodiversity impacts of coastal mining activities in southwestern India. *Ocean & Coastal Management, 208*, 105619. https://doi.org/10.1016/j.ocecoaman.2021.105619

Ravindra, S., Kumar, A., Sharma, P., & Singh, R. (2023). Comprehensive assessment of sand mining impacts on major Indian rivers: A national-scale analysis of the Ganga, Yamuna, Brahmaputra, Mahanadi, Narmada, Godavari, Krishna, Kaveri, and Periyar. *Environmental Science and Policy, 148*, 78–92. https://doi.org/10.1016/j.envsci.2023.10.012

Ray, S., & Mukherjee, A. (2023). Uranium health impacts. *Environmental Health Perspectives, 131*(4), 046001.

Reddy, S. N., et al. (2023). Mining deforestation impacts. *Land Use Policy, 124*, 106420.

Rege, A. (2016). India's sand mafia. *International Journal of Comparative and Applied Criminal Justice, 40*(2), 101–121. https://doi.org/10.1080/01924036.2015.1110886

Rivonker, C. U., Parulekar, A. H., & Ansari, Z. A. (2018). Impact of mining activities on estuarine fisheries of Goa. *Fishery Technology, 55*(2), 89–96.

Rukmana, D., Salman, D., & Alimuddin, I. (2020). Impacts of sand mining at Sadang river. *IOP Conference Series: Earth and Environmental Science, 575*(1), 012043. https://doi.org/10.1088/1755-1315/575/1/012043

Sajan, K., Mishra, A. K., & Kumar, R. (2022). Impact of mining effluents on estuarine water quality in Goa, India. *Marine Chemistry, 245*, 104143. https://doi.org/10.1016/j.marchem.2022.104143Sahu, A., Dey, S., Mallick, K., Priyadarsini, S., Behera, I. D., & Das, A. P. (2024). A review on advanced technology for sustainable management of synthetic micro-plastic waste. In *Sustainable management of urban plastic waste through circular economic approaches* (pp. 262–279). Routledge. https://doi.org/10.1201/9781003465447-15

Sahu, H. B., & Dash, S. (2011). Land degradation from mining. *Proceedings of the International Conference on Environment Science and Engineering, 8*, 1–5.

Sahu, H. B., & Mishra, S. (2022). Bauxite mining impacts. *Environmental Monitoring and Assessment, 194*(3), 1–18.

Sahu, K. C., Bhosale, A. R., & Patil, K. D. (2015). Impact of coal mining on environment. *European Researcher, 92*(3), 185–196. 10.13187/er.2015.92.185.

Sahu, K. C., et al. (2015). Coal mining environmental impacts. *European Researcher, 92*(3), 185–196. https://doi.org/10.13187/er.2015.92.185

Salman, T. (2016). Organization of alluvial gold mining in Bolivia. *Society & Natural Resources, 29*(9), 1124–1138. https://doi.org/10.1080/08941920.2016.1150549

Salt, D. E., Smith, R. D., & Raskin, I. (1998). Phytoremediation. *Annual Review of Plant Biology, 49*(1), 643–668. https://doi.org/10.1146/annurev.arplant.49.1.643

Sánchez-González, J. R., & Nicieza, A. G. (2017). Phenotypic convergence of trout. *Ecology and Evolution, 7*(15), 5922–5929. https://doi.org/10.1002/ece3.3156
Sarkar, A., & Mishra, K. (2023). Gold mining river impacts. *Environmental Monitoring and Assessment, 195*(1), 1–18.
Sarkar, U. K., et al. (2016). Freshwater fish conservation in India. *Biodiversity and Conservation, 17*(10), 2495–2511. https://doi.org/10.1007/s10531-008-9396-2
Sarma, J. N., & Acharjee, S. (2022). Brahmaputra bank erosion. *Geomorphology, 398*, 108074.
Sawaraba, I., & Rao, B. K. R. (2015). Cyanide pollution from mining. *Environmental Monitoring and Assessment, 187*(7), 4181. https://doi.org/10.1007/s10661-014-4181-z
Sengupta, D., et al. (2023). Uranium health effects. *Environmental Research, 216*, 114503.
Sheoran, A. S., & Sheoran, V. (2006). Acid mine drainage remediation. *Minerals Engineering, 19*(2), 105–116. https://doi.org/10.1016/j.mineng.2005.08.006
Singh, D. S. (2018). The Indian rivers. *Spring*. https://doi.org/10.1007/978-981-10-2984-4
Singh, G., Kamal, R. K., & Mishra, D. (2018). Application of water quality index for assessment of surface water quality status in Goa. *Current World Environment, 9*(3), 994–1000. https://doi.org/10.12944/CWE.9.3.49
Singh, R., Singh, P., & Sharma, R. (2015). Microorganism as a tool of bioremediation technology for cleaning environment: A review. *Proceedings of the International Academy of Ecology and Environmental Sciences, 4*(1), 1–6.
Singh, R., et al. (2021). Sand mining river impacts. *Science of the Total Environment, 755*, 142581.
Sinha, R. K., & Sharma, K. (2023). Forest fragmentation in Jharkhand. *Biodiversity and Conservation, 32*(1), 123–145.
Shetty, P., Kumar, R., Naik, S., & Desai, V. (2023). Assessment of mangrove and estuarine vegetation loss in the Goa-Karnataka iron ore belt: Impacts on the Mandovi, Zuari, and Kali river systems. *Wetlands Ecology and Management, 31*(4), 567–583. https://doi.org/10.1007/s11273-023-09927-8
Shetye, S. R., Gouveia, A. D., & Shankar, D. (2020). Coastal erosion along the western coast of India: Assessment and management strategies. *Ocean & Coastal Management, 185*, 105067. https://doi.org/10.1016/j.ocecoaman.2019.105067
Shetye, P., Naik, S., Kumar, R., & Desai, V. (2023). Infrastructure vulnerability assessment: Bridge foundation subsidence due to sand mining activities in the Mandovi and Zuari rivers, Goa. *Engineering Geology, 317*, 107089. https://doi.org/10.1016/j.enggeo.2023.107089
Shyni, R., Ganguly, D., & Robin, R. S. (2021). Turbidity and suspended sediment dynamics in mining-affected estuaries of Goa. *Continental Shelf Research, 223*, 104447. https://doi.org/10.1016/j.csr.2021.104447
Sreebha, S., & Padmalal, D. (2011). Environmental impact assessment of sand mining from the small catchment rivers in the southwestern coast of India: A case study. *Environmental Management, 47*(1), 130–140.
Subramanian, K. A., & Sivaramakrishnan, K. G. (2007). *Aquatic insects of India: A field guide* (p. 62). Ashoka Trust for Research in Ecology and the Environment.
Sukumaran, S., Sondkar, P., & Kanuri, V. V. (2020). Mining-induced changes in fish community structure in Goan estuaries. *Journal of Fish Biology, 96*(4), 1078–1089.
Smith, V. H., & Schindler, D. W. (2009). Eutrophication science: Where do we go from here? *Trends in Ecology & Evolution, 24*(4), 201–207. https://doi.org/10.1016/j.tree.2008.11.009
Srivastava, A., et al. (2023). Uranium processing impacts. *Journal of Environmental Radioactivity, 258*, 107120.
Stokes, A., Douglas, G. B., Fourcaud, T., Giadrossich, F., Gillies, C., Hubble, T., Kim, J. H., Loades, K. W., Mao, Z., McIvor, I. R., Mickovski, S. B., Mitchell, S., Osman, N., Phillips, C., Poesen, J., Polster, D., Preti, F., Raymond, P., Rey, F., & Wu, T. H. (2014). Ecological mitigation of hillslope instability: Ten key issues facing researchers and practitioners. *Plant and Soil, 377*(1–2), 1–23. https://doi.org/10.1007/s11104-014-2044-6
Sundar, K. S. G., et al. (2020). River tern nesting impacts. *Bird Conservation International, 30*(3), 365–379.

Tank, J. L., Rosi-Marshall, E. J., Griffiths, N. A., Entrekin, S. A., & Stephen, M. L. (2010). Allochthonous organic matter in streams. *Journal of the North American Benthological Society, 29*(1), 118–146. https://doi.org/10.1899/08-170.1

TEEB. (2010). *The economics of ecosystems and biodiversity: Mainstreaming the economics of nature*. United Nations Environment Programme.

TERI. (2023). *Mining regulation gaps in India*. The Energy and Resources Institute.

United Nations Environment Programme (UNEP). (2023). *Global mercury assessment*.

Vannote, R. L., Minshall, G. W., Cummins, K. W., Sedell, J. R., & Cushing, C. E. (1980). The river continuum concept. *Canadian Journal of Fisheries and Aquatic Sciences, 37*(1), 130–137.

Van Looy, K., Tonkin, J. D., Floury, M., Leigh, C., Soininen, J., Larsen, S., & Wolter, C. (2019). River ecosystem resilience. *River Research and Applications, 35*(2), 107–120. https://doi.org/10.1002/rra.3396

Verma, P., et al. (2016). Groundwater quality in Bokaro. *Applied Water Science, 6*(2), 161–170. https://doi.org/10.1007/s13201-014-0210-x

Venkateshwarlu, M., Shahnawaz, A., Krishna, R., & Santosh, K. (2023). Conservation status and reproductive ecology of Deccan Mahseer (Tor khudree) in mining-affected rivers of the Western Ghats. *Aquatic Conservation: Marine and Freshwater Ecosystems, 33*(7), 725–740. https://doi.org/10.1002/aqc.3892

Vidali, M. (2001). Bioremediation: An overview. *Pure and Applied Chemistry, 73*(7), 1163–1172. https://doi.org/10.1351/pac200173071163

Vörösmarty, C. J., et al. (2010). Global threats to river biodiversity. *Nature, 467*(7315), 555–561. https://doi.org/10.1038/nature09440

Vymazal, J. (2007). Removal of nutrients in various types of constructed wetlands. *Science of the Total Environment, 380*(1–3), 48–65. https://doi.org/10.1016/j.scitotenv.2006.09.014

Wagner, N. J., & Raymond, T. (2015). Land-use impacts on freshwater systems: A US perspective. *WIREs Water, 2*(6), 595–613. https://doi.org/10.1002/wat2.1100

Ward, J. V., Tockner, K., & Schiemer, F. (1999). Biodiversity of floodplain rivers. *Regulated Rivers: Research & Management, 15*(1-3), 125–139. https://doi.org/10.1002/(SICI)1099-1646(199901/06)15:1/3

Waters, T. F. (1995). Sediment in streams: Sources, biological effects and control. *American Fisheries Society Monograph, 7*, 1–251.

Wood, C. M., Farrell, A. P., & Brauner, C. J. (2012). *Fish physiology: Homeostasis and toxicology of essential metals* (Vol. 31). Academic Press.

World Health Organization. (2011). *Guidelines for drinking-water quality* (4th ed.). WHO Press.

Yadav, S., & Trivedi, R. C. (2020). Water hardness assessment in limestone mining regions of central India: Impacts on Chambal and Ken rivers. *Environmental Monitoring and Assessment, 192*(8), 512. https://doi.org/10.1007/s10661-020-08476-2

Yuen, K. W., Park, E., Tran, D. D., Loc, H. H., Feng, L., Wang, J., & Switzer, A. D. (2024). Illegal sand mining in the Mekong Delta. *Communications Earth & Environment, 5*(1), 31. https://doi.org/10.1038/s43247-024-01209-w

Water Contamination and Its Impacts on the Global Ecosystem: Special Emphasis on Mining

Monsoon Padhy, Anwesha Kanungo, and Sarbari Acharya

1 Introduction

Water is an indispensable resource inextricably connected to the existence of life on this planet. The majority of water area on the earth's crust is about 71% and it exists in numerous forms. Among these, the saline sea constitutes about 97% but it is unsuitable for drinking purposes due to the presence of a high number of salts. The notion that only 3% of freshwater is available to the global population is quite shocking and needs to be addressed with utmost priority. The accountability of freshwater in rivers is about 2%; whereas for lakes, it comprises 87% and the rest belongs to groundwater and frozen polar ice caps (Musie & Gonfa, 2023). The salinity of lakes and streams is about 1% when compared to the ocean water. But with an increase in globalization, these freshwater reservoirs are deteriorating over time with the addition of harmful chemicals like chloride which makes up its percentage about 25% when compared with seawater (Lockwood, 2019). The principal concern lies in the annual demand for freshwater in different sectors including agriculture (70%) as the major user along with food production, domestic use (10%) and other industries (10%). Even after advancements in production output facilities and high-tech machinery, demands are growing due to a rise in population, more rise in water pollution and great water scarcity (Thakur et al., 2018). In fact, the livestock industry uses about 8% of globally available water as it predominantly contributes to the humanitarian diet (Schlink et al., 2010). In addition to that, mining activities hold a requisite part that directly contributes to the economic growth of a nation covering more than 2% of land in surface, subsurface and deep-pits of the earth. Extensive mining operations also act as a backbone in a country like India,

M. Padhy · A. Kanungo · S. Acharya (✉)
School of Applied Sciences, Kalinga Institute of Industrial Technology, Bhubaneswar, Odisha, India

© The Author(s), under exclusive license to Springer Nature Switzerland AG 2026

A. P. Das (ed.), *Bioremediation Technology for Sustainable Mining Waste Management*, Sustainable Environmental Waste Management Strategies,
https://doi.org/10.1007/978-3-032-03468-7_2

having an abundance of mineral resources. But, it comes with a lot of setbacks like forest depletion, mine conflagrations and destruction of wildlife flora and fauna (Radhapyari et al., 2022). However, a recent report demonstrates a significant amount of water is used in these mining operations for ore extraction and meanwhile, the percentage of water exploited is nearly about 2–4.5% of the countrywide requirement (Bangerter et al., 2010; Brown, 2003; Meißner, 2021). An enormous volume is polluted for production per ton of metal and also in the process of extracting low-grade ores. The leakage, tailings and dumping of hazardous waste led to environmental damage and also have the capacity to create water scarcity in the next two decades. It could adversely affect the groundwater along with other natural sources (Meißner, 2021). In particular, coal mining extraction and its discharge system are causing severe water contamination in the present and upcoming years. This operation directly links to the social, economic and industrial expansion of a nation as well as it is quite challenging to preserve water resources to avoid their unnecessary wastage for the fulfilment of a high demand in the supply chain in coal mining (Dong et al., 2019). An estimation revealed that nearly 8 years ago in Africa, 2.2 million of the population suffered a crucial problem to procuring fresh drinking water. Despite the poor economic growth of these countries, large reserves of metallic resources like gold, diamond and iron are available for mining projects which eventually can lead to growth and development. This initiative requires a large volume of water which leads to overburdening the existing water-scarce regions. These operations also accelerate the nearby water quality degradation and also lead to population expansion (Dietler et al., 2021). In addition to that, it could generate health complications at the 95th and 99th percentile due to heavy metal exposure in drinking water and possibly cause cancer-like diseases in children and adults (Koki et al., 2018). Radioactive chemical mining like Uranium leads to deleterious effects not only on humans but also is not suitable for animal consumption. It can cause defects in DNA along with alteration in cell division with a high level of toxicity to living cells (Gajski et al., 2015).

2 Water Contamination: A Worldwide Concern

Water is pivotal for life encompassing about 75% of body weight in infants to 55% in elderly people. It regulates fluid balance, hydration and the metabolic system of living organisms along and it is responsible for the heart, kidney and gastrointestinal functions of the body (Popkin et al., 2010). Despite these benefits, the deterioration of water has continued to worsen over the past 35 years leading to a global freshwater crisis. In addition to this, one of the top ten worldwide threats listed by the World Economic Forum includes the depletion of freshwater reserves which arises due to ongoing population explosion, Industrial expansion, climate fluctuations, human activities and rising pollution levels. Nearly, about two million tons of different wastes are directly disclosed in water from a common point origin or from different sources which leads to the transmission of microbial diseases through water. This untreated water has noxious repercussions on 2.4 billion which arise from daily household activities, industrial release, agricultural runoff and chemicals released to freshwater bodies potentially destroying water quality by the formation of algal growth, eutrophication and diseases like methemoglobinemia (du Plessis, 2022). An alarming condition arises when it is noticed that pollutants cause nine million premature deaths each year (Landrigan et al., 2020). Developing countries like Pakistan suffered economic losses and physical concerns due to microbial contamination like faecal coliforms in drinking water coming from industries and municipalities which later on directly infect the leakage pipeline system (Nabeela et al., 2014). In addition to that, various chemical compounds like perfluorinated alkyl substances (PFAS) have become a huge water contaminant for surface and even bottled water in countries like Africa, USA, and New Zealand (Adewuyi & Li, 2024; Chow et al., 2021). A leading side effect of PFAS is it has side effects on women's ovarian health causing PCOS, infertility and even miscarriage (Rashtian et al., 2019). Additionally, analysis demonstrates that oil extraction activities in the northern Peruvian amazon contaminate the water sources by releasing cadmium, lead and chloride on a large scale from the sites of extraction (Yusta-García et al., 2017). Drinking water has been a great concern in California also, as the water contains traces of asbestos and by-products of water chlorination (Russell et al., 1987). Rather than that in agricultural sectors leaching of pesticides has also become a great concern to contaminate the ground water. A study signifies that in countries like USA, over 70 pesticides are now been detected contaminating the groundwater by the process of leaching through the different levels of the soil. So, the quantity of pesticides along with the application procedure to different crops should be regulated to lower the contamination but this is not a permanent solution for this problem (Ritter, 1990). One of the crucial anthropogenic endeavours includes heavy metal exposure nearly about 25% in water bodies. In China, textile industries pollute the surface water by releasing their wastewater loaded with antimony which could be devastating for aquatic life (Li et al., 2023). A shocking truth called "cancer villages" arises due to the overloading of heavy metal pollutants which signifies that the total population of a village can get the risk of cancer through contamination of food as well as the destruction of agricultural land (Lu et al., 2015). Specifically, prolonged contamination of lead and cadmium compounds can cause lung, oral and intestinal cancers (Wang et al., 2011). Along with that, acid water leakage, tailings and radioactive waste from mining directly contaminate the surface and groundwater simultaneously affecting aquatic life and mankind (Santana et al., 2020). The degraded land after mining activities also accumulates a lot of detrimental wastes that alter the ecological and biogeochemical cycles (Hao et al., 2024; Punia & Singh, 2021). A large volume is required for ore extraction but the water after use is discharged in a very unregulated manner. Several treatments are introduced to manage this wastewater but they are not fully effective, require high operational costs and generate hazardous acidic by-products. Additionally, the World Health Organization (WHO) established constraints for releasing the mining effluents but on a serious note, these guidelines are not being monitored (Matebese et al., 2024). Especially, small-scale gold mining violates WHO and USEPA guidelines for releasing mercury which is a highly toxic metal. It specifically affects pregnant women, adults and children in Southeast Asian countries (Soe et al., 2022). In addition to this, copper and coal mining activities in countries like Mongolia disrupt the overall quantity of invertebrates in the Selenga River basin which were used to monitor water quality (Narangarvuu et al., 2023). Especially, copper mining is also a great concern in Western Uganda releasing heavy metals contamination like Al, Fe and Mn in household drinking water (Abraham & Susan, 2017). Moreover, natural gas mining generates Sr and Ba in smaller concentrations through spillage which later on become a serious concern in North America. These heavy metals are detected in surface waters by testing the fish otoliths organs (Keller et al., 2018). Moreover, developing countries like India also face the detrimental effects related to iron and copper mining occurring in Southern Jharkhand. The metal concentration exceeded its normal limit in an environment in the form of dust falls in which the areas in close proximity to the mining fields are affected more as compared to others. It also adversely affects the health assessment scores of adults and children (Mahato et al., 2022). Additionally, the leaching of hazardous substances derived from coal in Talcher, Odisha are also one of the major sources of water pollution causing health hazards. Overall, water contamination has become a serious global concern, endangering mankind and environmental sustainability.

3 Factors Causing Water Pollution

Various natural and anthropogenic factors are responsible for water pollution, among which the natural causes include soil erosion, oil spillage, wind-regulated dust, volcanic eruption and sudden firebreaks while the man-made sources for contamination are agricultural wastes, industrial wastes, household waste, sewage, radioactive dumping and oil spillage. Figure 1 illustrates the major sources of water contamination.

3.1 Industrial Effluents

Wastewaters released from industries directly release their lethal substance into rivers and oceans. Depending upon the type, different contents are released from iron, steel, oil and food industries which require specialized procedures for the treatment of impure water. The released quantity consists of garbage, dirt, oils, chemicals and many more which can be fatal if reactive or carcinogenic when mixed with drinking water and can be detrimental for living organism consumption. When directly released to water sources, it can easily contaminate producing toxins and affecting the food cycle. Furthermore, dumping of these wastes can also affect the groundwater. A study conducted by Mishra et al. at the Rourkela steel plant revealed that seepage of these contaminants affects natural sources like rivers (Mishra et al.,

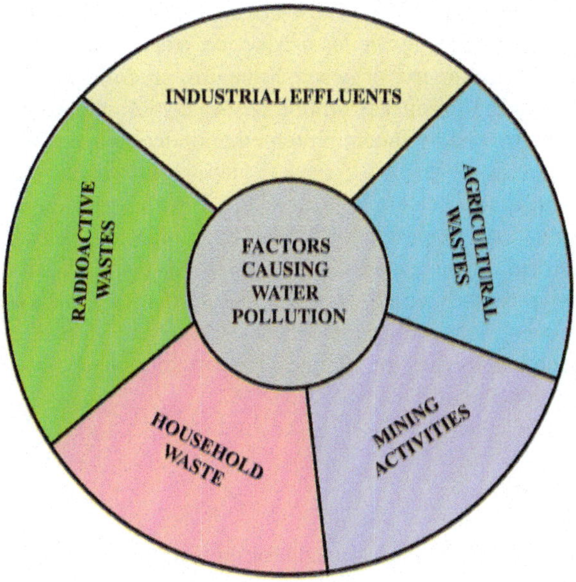

Fig. 1 Representation of prime factors responsible for water contamination

2005). There are five different sectors of industries contributing most of the pollutants belonging from coal, paper, petroleum, textile and pulp (Prabakar et al., 2018). For removal of these pollutants like heavy metals, a fast and cost-effective technique using flocculation capacity of *Saccharomyces cerevisiae* (yeast) is carried out after the pre-treatment (Soares & Soares, 2013). This process proves to be a very efficient for a large quantity of effluents as the yeast could settle 95% of biomass having 0.5gram per litre in only 5 min and the capacity to settle even if heavy metals like Cu^{2+}, Zn^{2+}, Ni^{2+} and Cr^{3+} are present in the effluents (Machado et al., 2008). Other measures that are performed include primary and secondary treatment along with nanofilters that are installed in industries so that the outgoing water can be free from pollutants.

3.2 *Agricultural Wastes*

A substantial correlation is present between agriculture and water contamination as all the fertilizers, pesticides, nitrates and herbicides are introduced into the water sources by leaching through various cultivation and irrigation purposes (Lin et al., 2022). A crucial illustration resulted in Morocco where nitrate level exceeded in the drinking water which later on gets converted to nitrite marking the way to form methaemoglobin (MetHb) where lips turn into blue colour along with skin, convulsions, sudden fainting, coma and lastly death (Sadeq et al., 2008). It also causes blue baby syndrome in infants which is lethargic for the baby and further causes its mortality (Knobeloch et al., 2000). These chemicals are used to protect crop plants from insects and pathogens for better production which later creates havoc to the environment. But, with the use of newer strategies, several agricultural wastes are converted into valuable products to minimize water pollution. Again, agricultural wastewater can be treated by fungal bioremediation technique using *Trametes versicolor* in a bioreactor for about 225 days (Beltrán-Flores et al., 2022). In addition to that, soil health is also need to be maintained in which the convectional bacteria are not efficient enough to eliminate the high-end pollutant. Therefore, specialized genetically modified organisms are designed to speed up the breakdown process for various impurity present like hydrocarbons and pesticides (Maqsood et al., 2023). Nowadays hydrogel adsorbents, activated carbon and doped pulp are high-performance innovations that are applied in a correct manner for solving to solve water contamination (Noor & Khan, 2023).

3.2.1 Pesticide Residues

Pesticides are chemical compounds useful for their significant role in crop yield and management of disease-causing pathogens. These are very fatal and can cause serious health risks to humans, animals and plants where herbicides are widely used followed by insecticides and fungicides. Due to a greater demand for food for the everyday survival of living beings, these chemicals are used in greater quantities. According to Lai et al. health line Expectancy longitudinal survey data, 1% rise is noticed in health-related disability index among the 65 age group and above with 10% after pesticide use (Lai, 2017). A case study currently revealed that when wastewater sample are investigated by gas chromatography-mass spectroscopy (GC-MS) shows that almost ten pesticides are present, predominantly chlorpyrifos, which can be removed by three bacteria and six fungi species (Elzakey et al., 2023). Nowadays advanced oxidation process is used for the cleaning of pesticides from water followed by biological treatment (Syafrudin et al., 2021).

3.2.2 Eutrophication

The limiting factors responsible for photosynthesis in a plant include sunlight, carbon dioxide and essential nutrients. When one of these factors is available to an alga in an increased and adequate amount, then its rapid growth takes place which seems basically a natural process. But anthropogenic actions lead to a rise in the level of nitrogen and phosphorus in the water which leads to rapid growth of cyanobacteria resulting in decreased levels of oxygen in water (hypoxia) and deterioration in water quality. It affects the aquatic life resulting in the death of fishes and other creatures. Bioremediation is a possible way to get rid of this issue in which a bacterial preparation destroys the algal growth and is also cost-effective in nature. A field trial conducted by Shan M et al. also revealed that it significantly enhanced the water quality (Shan et al., 2009).

3.3 Household Waste

Household waste includes leftover food, garbage, vegetable waste, paper, broken plastics and glass which categorize them as wet and dry waste as well as flammable and non-flammable waste. Its irregular disposal and huge quantity have become a serious issue for their removal and degradation. Firstly, it should be collected in a proper way so that the intermixing of different categories of garbage is avoided. Then, its transportation is done to a site and lastly, treatments like composting and fermentation are executed. Usually, a particle size of less than 40 mm can be treated biologically whereas sizes more than that screened in a drum screen (Tao et al., 2023). Recently, algae-based biodegradation techniques are applied to absorb and later accumulation of organic pollutants without secreting any secondary pollutants with a massive hand in carbon sequestration also (Subashchandrabose et al., 2013; Touliabah et al., 2022).

3.4 Sewage

Sewage is a frequently encountered type of waste in a large quantity mixed with water that comprises household waste, municipal waste, garbage, faeces, kitchen waste and laundry which is directly released into the ocean, river and sea for easy disposal. When released, it reduces the oxygen level in the water bodies. It can also cause eutrophication which further leads to algal bloom formation. An algal bloom is noxious in nature and releases toxic substances that can cause paralysis in shellfish which later on when consumed by humans causes death (Gu et al., 2022). Several methods are present to solve this problem in which primary and secondary treatments are required before releasing it into water bodies. These processes are carried out in sewage treatment plants along with many steps starting from filtration, sedimentation and later aeration followed by microorganisms of aerobic and anaerobic types and lastly sludge treatment (Szulc et al., 2021).

3.5 Radioactive Wastes

Heavy radiation emitted by substances beyond a normal limit set by nature is called radioactive substances. It prolonged existence in the environment as well as improper disposal causes havoc for living organisms. It originated during the mining process of uranium, in research fields, hospitals and nuclear power plants. Untreated radioactive waste shouldn't be released into water sources as it can be carcinogenic and life-threatening. It can be removed as per their concentration limits. When concentration is lower, it can be diluted to a small volume and then discharged whereas in the case of high radioactive waste, it should be carefully stored to stop getting contact from the environment. Radioactive substances also contain radionuclides which can be treated using microbes like bacteria involving steps like biological reduction, biosorption and lastly bioprecipitation (Tan et al., 2024). Other techniques like adsorption, evaporation solidification (Tripatanasuwan et al., 2007) and flocculation are also very helpful for radioactive waste purification.

3.6 *Oil Spillage*

The sudden release of liquid petroleum hydrocarbon from tankers or pipelines into the environment mainly by anthropogenic actions is termed oil spillage. Deleterious effects are seen in the life of aquatic animals such as fish, turtles, and many more causing their death. Various methods like skimmers, in situ burning, use of surfactants can be imposed for clearance of oil spillage from the water bodies (Silva et al., 2022). A major incident took place in Chennai, India in 2017 with the collision of two cargo ships causing a heavy spillage of oil in the Bay of Bengal (Han et al., 2018). This spillage can be cancer-causing and also can accumulate in the tissue of aquatic animals causing a serious effect on the food chain (Rogowska & Namieśnik, 2010).

4 Paramount Contribution by Mining Operations

Mining activities discharge a significant amount of toxic and hazardous substances into the environment endangering its integrity and require a vital need with sophisticated prevention strategies to avoid further depletion of water sources. These activities are carried out because of several benefits to fulfil demands for an overrising population and also to increase the economic condition of a country, rise in employment, extraction of valuable metals, petroleum products, energy generation and regular supply of industrial raw materials. However, a lot of disadvantages are associated with it for example environmental pollution, deforestation, soil erosion, carcinogenic diseases, acid waste leaching, occupational health destruction, tailing discharge and natural habitat destruction (Northey et al., 2017; Radhapyari et al., 2022). It degrades the groundwater quality as well as surface water by acid mine drainage and heavy metal contamination and cumulative exposure can cause catastrophic health risks (Jin et al., 2021; Xiong et al., 2019) as exemplified in Hunan province of China.

4.1 Acid Mine Drainage (AMD)

Acid mine drainage is a biological process in which waste generated from tailing waste and abandoned coal mines rich in sulphides oxidize in the presence of oxygen and water to produce sulphuric acid that later on supplemented by bacteria forms a highly acidic waste affecting fishes, animals and lastly humans. After that, it contaminated the streams, lakes and rivers by leaching and runoff accompanied with a heavy load of chemicals and toxic metals. A similar case is studied in tailings discharge from mines containing Si, Fe, Al and K present in Bahia gold mines, Brazil (Bernardez et al., 2021). Bioremediation, nanoremediation (Ji et al., 2023) and phytoremediation technology using bacteria and fungi along with specialized bioreactors are assigned for their removal from the environment rather than using a normative approach (Dhir, 2018; Moreno-González et al., 2022; Thomas et al., 2022).

4.2 Heavy Metal Contamination

Principal contaminants released from industries are heavy metal which triggers a serious threat to the environment due to rapid urbanization and population growth. Later on, its bioaccumulation can pose health hazards to mankind, aquatic and terrestrial life along with agricultural fields as they can get conglomerate in plants (Khan et al., 2023). Prominent contributions came from the manufacturing process of vehicles, paint, oil, dyes and mining which releases heavy metals like chromium, zinc, lead, nickel, arsenic and copper which further affects the ecosystem. These can cause organ system damage, cancer and lung and heart diseases. Phenolic compounds, petroleum hydrocarbons, urea and hydrogen peroxide are the other contaminants released from different industries like chemical, textile, paint and paper industries (Ahmed et al., 2021). Earlier techniques to remove heavy metals include ion-exchange method which is less effective. Therefore, advanced nanotechnology-based methods like ultrafiltration, reverse osmosis, electrodialysis, nanosensors and nanophotocatalysis are employed for better results (Ahmed et al., 2021; Kumar et al., 2014).

4.3 Gold, Copper and Coal Mining

Gold is a valuable and expensive metal for which its mining is done in a worldwide scale. Australia is the largest gold reserves deposit country in which gold is extracted and processed in a series of key steps. The main disadvantage lies in the heavy power supply and consumption(Farjana et al., 2019). A study revealed by Yaraghi et al. states that the Kittilä gold mines have a wrecking effect on the Seurujoki river situated in Finland. The water quality of the river is compromised with contaminants like sulphate, manganese and total nitrogen (Yaraghi et al., 2020). Similar cases of copper mining are executed for its high economic value and demands. Valea Şesei is a valley in Romania that is well known for its heavy tailing discharge and deadly mine water (Rzymski et al., 2017). Another case copper mining comprises of chemical leaching which was found in the tree rings of Atacama villages in Chile. A high demand for copper was required at an international level costing around 27%; therefore, its mining became so pivotal for economic purposes (Zanetta-Colombo et al., 2024). Besides that, another important non-renewable resource is coal which requires a large quantity of water and later on is wasted. Coal contaminants change the pH of water, alter water quality and on combustion release toxic gas like methane which harms the atmosphere too (Tianliang et al., 2023). Investigating recent scenario in Northern Shaanxi uncovered that nearly 5% area is severely polluted and more than 80% area is slightly polluted with unpolluted. There is a certain procedure called activated charcoal adsorption technology, biotechnological steps for wastewater management are being used for the treatment of polluted water due to their cost-effective, fast, eliminating foul smell and being very effective in nature. Various laws are enforced along public safety is also taken into account to preserve water (Bai et al., 2024).

5 Effects on Global Environment

Water contamination influences wildlife, ecological systems, biodiversity and human health. Distortion in natural resources that is necessary to sustain life on earth could have a detrimental effect on the present population and upcoming generations. The worldwide biosphere is negatively impacted by contamination of water which serves as a pivotal source for all forms of life. Different lifeforms are affected starting from ecosystem that is severely destroyed, aquatic life is in danger, adverse effect is seen in human life, soil is contaminated and climatic alterations is observed. In this context, Fig. 2 is demonstrating the effects of water contamination on global environment.

5.1 Impact on the Ecosystem

Untreated wastewater, heavy metal contamination, garbage, microplastics, and oil spills highly impact the biomes leading to a sharp drop in population, loss of habitats and threats to upcoming generations. A serious upset is caused by the food chain and food web affecting an individual alertness towards predators, survival strategies of prey and fecundity. Every single species is crucial for the flow of energy from lower to higher levels but as the concentrations of pollutants increase it lower rate of energy transfers to higher levels disrupting the whole cycle. So, a reduction in contaminants is highly necessary for the well-being and maintenance of health risks for every component of life (Clance et al., 2023).

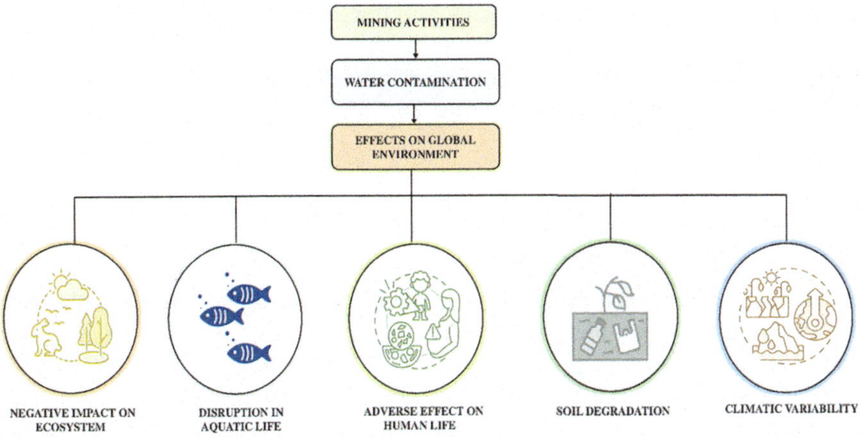

Fig. 2 Demonstration of the adverse effects of water contamination on global ecosystem

5.2 *Disruption in Aquatic Life*

Hazardous chemicals, pesticides and heavy metals pollute the natural sources of water bodies killing thousands of fish, turtles, coral reefs and other sea animals. Afterwards, substantial problems impact will be faced by the fishery industry in terms of fish production as the quality of fish will be compromised leading to potential health risks for the consumers. Furthermore, deterioration in water quality due to contaminants can lead to a reduction in oxygen levels making a pathway for eutrophication and algal bloom formation. A species of sea turtle (Olive ridley) is facing concomitant health dangers because of regular and chronic exposure to organic pollutants. Oxidative stress, alternation in immune response and abnormal enzymatic activity are the possible symptoms which are disclosed during a study by Flores ramirez et al. which further signifies its demographic figures and health problems for future descendants (Flores-Ramírez et al., 2024). In order to minimize the water pollution regulated amount of chemicals should be released, a reduction in garbage discharge and various protected sites for aquatic life should be established to preserve this elixir of life along with the organisms living within it.

5.3 *Adverse Effect on Human Life*

Human life is greatly endangered because of water contamination as it causes various waterborne diseases like cholera and dysentery. Furthermore, the main point sources of pollution exacerbate the sanitation issues risking life of millions of adults and children. About nine million deaths per year occur due to water degradation and about 60% of people die because of pollution-caused heart diseases (Münzel et al., 2023). Both genders also suffer from negative consequences which make way for hormonal disorders by blocking or altering the homeotic balance of the human body. Certain contaminants are responsible for neurocognitive problems, ageing faster, stress, osteoporosis, infertility and cancer-like diseases in the human body and lastly leading to death (Bochynska et al., 2024).

5.4 *Degradation of Soil and Its Inhabitants*

Soil is a major component that is very essential to sustain life on Earth. It is the main framework for crop production, water storage and human settlement and also supports various ecosystems. It can be polluted by the leaching of heavy metals from mining, insecticides, pesticides and the accumulation of wastes from different fields of life. Several adverse effects firstly and foremost are seen in the production of crops as they retain the chemicals within them which later on are being consumed by living organisms. This is the process of transfer of contaminants from one species to the other. So, keeping all these factors in point European Commission and EU action plan made a strategic move to secure this planet into an innocuous and healthy entity which states the less use and expel of toxic substances to mother Earth (Münzel et al., 2023; Pandey, 2006).

5.5 *Climate Variability*

Global warming is the abnormal and long-term rise in the earth's temperature due to the release of greenhouse gases. An increase in greenhouse gas emission like methane and carbon dioxide by the degraded water sources hastens global warming and therefore alters the climatic conditions. Additionally, another phenomenon that is also considered to add impact is the expulsion of carbon from the atmosphere. Oceans act as a sink and reservoir for carbon sequestration in which the amount of carbon dioxide that is released into the atmosphere is absorbed by the ocean in the upper surface of the water and nearly 40% of this comes from anthropogenic factors that increase the acidity of water (Lal, 2008; Newmark et al., 2010; Tian et al., 2023). In addition to that, microplastics are considerable pollutants increasing the emission of greenhouse gases which affects the phytoplankton and zooplanktons growth and development in the ocean(Shen et al., 2020).

6 Sustainable Mining Approaches

Sustainable mining practices balance the metal extraction process with ecological protection. With the help of advanced research and technology, various strategies are invented to mitigate the environmental impacts and safeguard nature. The process of extracting ore and precious metals can be done in an eco-friendly manner and hence it could prevent further damage to the biosphere. However, various factors including geographical location can be used to monitor these environment-loving practices and can be executed by systematic engineering management with the help of tools like geographic information systems (GIS) (Singh & Kumar, 2024).

6.1 Green Energy Deployment

Power generation is an important component that must be present in an uninterrupted manner at a low cost. Natural resources like coal, solar power, wind and water are used as an alternative to avoid the release of toxic gases into the atmosphere. In intensive work, it is noted that mine water is considered as a waste that can be very useful for production and as a depot for renewable energy as observed in the case of Asturias (Northwest Spain). An advanced regulated energy storage system is designed in combination with wind energy with a profitable cost along with a reduction in carbon dioxide levels in the environment (Álvarez et al., 2021). Furthermore, a study conducted by Quan et al. display that in the Liaoning province of China, solar energy can be used for the generation of renewable energy in abandoned mining areas (Quan et al., 2019).

6.2 Reclamation of Mining Waste to Useful Materials

Mining waste includes acid water drainage, tailings discharge and ore wastes can be recycled to form new construction materials and can be reused again. Tailings are waste having low organic and microbial content with a higher salinity as well as less water retention capacity. This waste can be handled by phytoremediation technology where plants are used to absorb the heavy metal contamination which arises from mining discharge. It is a cost-effective technique conserving around 50%–65% less than the earlier technologies (Sarathchandra et al., 2023).

6.3 Ecological Reclamation and Soil Regeneration

Abandoned mining fields and rigorous practices related to metal extraction cause the land to lose its fertility, production capability and vegetation loss (Worlanyo & Jiangfeng, 2021). Therefore, municipal sludge also called as composted sewage sludge (CSS) and biochar (BC) is employed for the regeneration of vegetation, plant growth and development in degraded land (Hagner et al., 2021). This further can create job opportunities along with economic establishment.

6.4 Eradicating Unauthorized Mining Practices

Curtailing illegal metal extraction should be the prime objective to safeguard our natural resources and avoid counterproductive effects on output and profitability. Higher risk of public health along with accelerated degradation of natural resources like water, air and soil. A research investigation on Yanomami people in Brazil showcased that illegal gold mining can cause a deficit in healthcare facilities causing intestinal parasitise, malaria and even death. They are also displaced from their own land for mining practices and those grounds serve as a breeding ground for mosquitoes resulting in a compromised state of healthcare facilities (Martins-Filho et al., 2023). More strict policies, laws and awareness among the public should be enforced to avoid long-term dangers to communities and mankind.

6.5 Industrial Smart Technology Implementation

Digital innovation plays a very crucial role in maximizing the extraction by collecting raw data easily within the real time for correct evaluation. Artificial intelligence plays a greater role along with other internet sources by providing accurate information in the required time and operating complicated machines too. The use of this technology has made it easier to operate and maintain the apparatus work even in the absence of an operator in cases of dangerous mining zones. Research work executed by Nti et al. displays that in Ghana illegal mining has contaminated about 60% of water bodies which was optimized by using AI-based technologies and also safeguarding water preservation for future generations (Nti et al., 2023).

7 Bioremediation Technology Measures

Bioremediation is an economically productive process in which certain biomolecules particularly bind to specific target ions in a solution. Some microorganisms like bacteria, fungi and yeast also participate in this process due to their ability to minimize and modify the harmful effects of toxic metal ions. It is a process of degradation of heavy metals like Cu, Pb, Mn, Zn and Ni by the action of one or a group of bacteria (Nicolaou et al., 2010) like *Pseudomonas veronii* (Vullo et al., 2008), *Pseudomonas putida* KT2440 (able to degrade petrochemicals product and oil spillage) (Duque et al., 2022), *Enterobacterium* spp. (Lu et al., 2006), algal species like *Cladophora* spp., *spirogyra* spp. (Lee & Chang, 2011), *Oedogonium* spp., *Nostoc* spp. (Gupta & Rastogi, 2008) and fungi species like *Amanita rubescens* (Sari & Tuzen, 2009). Bioremediation can be executed by two procedures, biosorption and bioaccumulation in which the former is fast as compared to the latter which can be done in both living and dead biomass (Coelho et al., 2015). It also depends on location constituting of in situ (on contamination site) and ex situ (displacement of contamination from one site to the other), among which the ex situ is more fatal in nature as there is a risk of contamination spill risking the health of living organisms. Bioventing, biosparging, biostimulation, bioslurping, bioattenuation and bioaugmentation are the techniques associated with in situ bioremediation where biopile and bioreactor are linked with in situ bioremediation. Figure 3 is representing the bioremediation technology that can be implemented for restoration of ecosystem.

Additionally, Table 1 also provided for depicting four different type of bioremediation techniques used for purification of contaminated water and for ecosystem stability.

Different bioremediation methods:

7.1 In Situ Bioremediation

7.1.1 Bioventing

For the easy degradation of contaminants in the soil itself, the indigenous microorganisms are provided with an adequate amount of oxygen which can be directly injected into the soil. This technique is mainly used for the degradation of petroleum hydrocarbon with the help of pumps and blowers (Rendón-Castrillón et al., 2024).

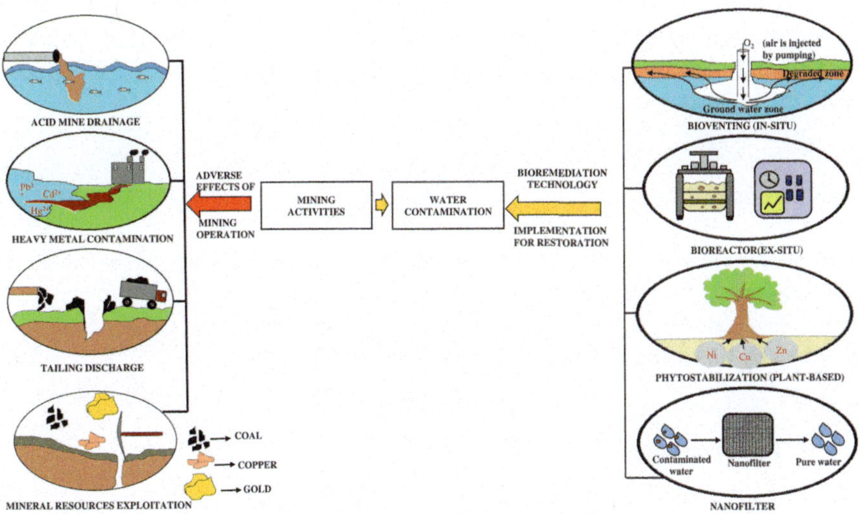

Fig. 3 Illustration depicting the adverse effects of mining along with the bioremediation technology applied to restore the environment

Table 1 Depiction of different bioremediation methods using in situ, ex situ, plant based and nanotechnology

In situ bioremediation	Ex situ bioremediation	Plant-based bioremediation	Nano-bioremediation
1. Bioventing	1. Biopile	1. Rhizofilteration	1. Nanotubes
2. Biosparging	2. Bioreactor	2. Phytostabilization	2. Nanofibers
3. Biostimulation		3. Phytodegradation	3. Nanofilters
4. Bioslurping		4. Phytovolatilization	
5. Bioattenuation			
6. Bioaugmentation			

7.1.2 Biosparging

This remediation technique is executed under high pressure for pumping gas and nutrients to a particular region in order to flourish the indigenous microbial growth. It is supplemented with bioventing for better results and is applied for an aerobic breakdown of oil and gasoline. Microbial growth is best seen in low-density fluids as compared to heavier ones requiring more time for investigation (Gill et al., 2016).

7.1.3 Biostimulation

This is a method to increase the residing bacterial growth by presenting nutrients, electron donors and acceptors like nitrogen, phosphorus or carbon into groundwater or soil and is used to remove specifically petrochemical substances. This method is also used to eliminate any nanomaterial contamination in plants and animals (Juárez-Maldonado et al., 2021).

7.1.4 Bioslurping

This method is a combination of bioventing and vacuum treatment to remove oil-like substances from the capillary borders of soil but it is only limited to about 25 feet below water level, more than this can cause problems as the weight cannot be lifted up by this technique (Gidarakos & Aivalioti, 2007).

7.1.5 Bioattenuation

It is a technique available for microbial degradation in which various metabolic pathways are employed for a breakdown of large quantities and pernicious contaminants by diffusion, solubilization, volatilization and various other processes. Later on, these will convert into safe compounds like water and gases (Nguyen et al., 2021).

7.1.6 Bioaugmentation

It is a procedure employed at the gene level of certain microbes in order to increase their enzymatic efficiency and is useful for oil and petroleum contaminants. Those organisms are also called as genetically altered organisms which are selected from a specific site of action, then their gene is altered for a fixed purpose and later on placed to their earlier site after completion. It is a cost-effective and sustainable process (Chettri et al., 2024).

7.2 *Ex Situ Bioremediation*

7.2.1 Biopile

For the enrichment of microbial growth in soil, proper nutrients and periodic aeration are employed to degrade noxious soil quality. This technique is being explored on a wider scale because of its cost-effectiveness and management (Ma et al., 2016).

7.2.2 Bioreactor

Bioreactors are big-sized containers used to prepare slurry in which biological treatments are carried out for particular product production. Optimum temperature, pH and aeration are provided for microbial growth in order to reduce soil contamination.

Moreover, aquatic plants like water hyacinth (Abbas et al., 2019; Xu et al., 2022), water lettuce (Abbas et al., 2019) and duckweed (Zhou et al., 2023) are also useful for bioremediation firstly by absorption, adsorption, accumulation and breakdown of contaminants. Various plant-based remediation methods are also established in order to clean up toxic substances from water that use plants and certain microbes associated with it. Rhizofilteration, phytostabilization, phytodegradation and phytovolatization are certain techniques used to clean water contamination.

7.3 *Plant-Based Bioremediation*

7.3.1 Rhizofilteration

Rhizofilteration is a solitary technique to remove mining discharge like heavy metals from water with the help of aquatic plants. It has the ability to absorb heavy metal pollutants from the water through the aerial end of roots, accumulate it and later purify it by the filtration process. Hydrophytes are the plants that can incorporate these pollutants more than the submerged type. The major benefits attached are it can remove acid water drainage, organic, inorganic and agricultural wastes. Plants associated with it include water hyacinth which can able to cleanse water quality (Rezania et al., 2016).

7.3.2 Phytostabilization

Phytostabilization is a procedure used to improve the rigidity of the soil by inhibiting the movement of toxic substances. Plants stabilize toxic heavy metals pollutants in the soil itself and restrict their movement by root exudates. Their function is quite similar to ligands as they get attached to heavy metal ions by altering soil pH without absorption called the complexation process. Henceforth, it stops the further spreading of toxic chemicals by leaching and runoff (Bakshe & Jugade, 2023).

7.3.3 Phytodegradation

Pesticides, insecticides and organic pollutants are destroyed by the enzymatic activity of plants. It takes place within plants or in rhizospheres areas. This procedure can be helpful for oil spill removal, volatile products, aromatic compounds and industrial effluent contaminants from water and soil (Newman & Reynolds, 2004).

7.3.4 Phytovolatilization

In this technique plants absorb certain pollutants and heavy metals, convert them into a less toxic product and later release it in volatile form by the process of transpiration. Therefore, prevents further spreading and degradation of water and soil (Limmer & Burken, 2016). Direct (By stem and leaves) and indirect (from soil) phytovolatilization processes are available to uptake the contaminants according to their presence in the surroundings. Mainly, hydrophobic substances are released through transpiration from stems or leaves due to their better interaction with suberin and cutin. For the indirect process, the roots play a significant role by depleting the groundwater to create tension, increasing the permeability of the soil, hydraulic lift for chemical movement and accumulation of rainwater to prevent the VOC manoeuvre to other areas (Limmer & Burken, 2016).

7.4 Applied Nanotechnology Remediation

Nano-bioremediation technology is an emerging advanced field that can be applied to remove harmful substances and heavy metal contamination from wastewater to safeguard flora, fauna and to protect further natural resources. Nanotechnology can be implemented in designing various nano-adsorbents, nanofilters and nanomagnets for easy and fast removal of pollutants to purify wastewater. Currently, it is a most specialized technique employed to immobilize the chemical using nano-supports along with easy detection and removal abilities (Chaudhary et al., 2023).

7.4.1 Nanotubes

Nanotubes are specialized designed tubes originating from carbon including activated charcoal, fullerene and carbon-based nanotubes. This specialized tube has the capacity to adsorb chemicals released from households, municipal and industries. Specially designed carbon nanotubes (CNT) can detect mercury (II) in heavily contaminated water from trace amount in an efficient manner. A high rise in these contaminants in water can lead to Minamata disease which affects the neurological and development of children, pregnant women and adults and is placed at high risk at the Disease registry priority list (Pokhrel et al., 2017). Besides that, another nanotechnology-based nanomagnets are advanced equipment that can attract both positive and negative ions proven in the case of chromium. Furthermore, zwitterion is coated for easy and beneficial adsorption of contaminants like dyes from wastewater (Reja & Vasudevan, 2024).

7.4.2 Nanofibers

Specially designed nanofibers made from cellulose work on the principle of oil-water emulsion technique to remove pollutants. Due to its environmentally safe decomposable nature, biofouling resistance and 99.14% efficiency, it is readily used in the separation of underwater waste present in water (Liu et al., 2024). Additionally, polyacrylonitrile (PAN) based nanofibers are also implemented in the removal of carcinogenic products like 2,4 tricholorophenol (TCP) by providing a greater surface area to immobilize enzymes like laccase with principles like adsorption and biodegradation (Xu et al., 2013).

7.4.3 Nanofilters

Nanofilters constitute nanopores having the ability to remove small-size heavy metals contaminating the wastewater from municipal waste and drinking water in households. It is a better treatment for filtration as compared to reverse osmosis which includes two biological procedures of sludge treatment and then filtration (Guo et al., 2022; Kimura et al., 2009). Moreover, nanofilters like NF70 membranes are designed to reject resorcinol, 3-hydroxybenzoic acid and tannin in the presence of calcium by altering pH (Lin et al., 2006). Other compound, like trihalomethanes can also be removed from water (Lin et al., 2007).

This technological advancement along with a combined effort from all sectors of individuals are necessary to conserve the delicate natural resource like water.

Furthermore, cooperative actions from public, regulatory agencies, NGOs and government are mandatory for better regulation and conservation of water resources.

8 Global Alliance to Safeguard Water

Global alliance is multinational diligence to safeguard and manage freshwater assets globally in a sustainable manner. This partnership joined the government, industries, research institutes and the public together to adopt creative water conservation strategies in response to current growing concerns including habitat fragmentation, water contamination, excessive mining extraction and climatic shift. This alliance aimed to streamline clean water reliability promoting ecosystem stabilization by imposing policy campaigning, advancement in technology and grassroots projects. These projects are essential to reduce and improve water-related calamities guaranteeing sustainable growth in terms of the long run for nature and mankind by promoting international cooperation. Numerous international agreements and strict regulations are enforced to preserve freshwater ecosystems with a strict rule of frameworks and policies. United Nations convection administered an agreement on conservation and ethical marine biological diversity of areas beyond national jurisdiction (BBNJ agreement) adopted on 19 June, 2023. This treaty upholds the significance of research committees along with private and public sectors. The aim of this agreement includes the protection of ocean biological assets using spatial governance tools and a proper review of ecological evaluation skills (Tessnow-von Wysocki & Vadrot, 2024). Another legislative foundation for reasonable management of transboundary water bodies is established under the United Nations Watercourses Convection (1997) (Gelaw & Debele, 2024). In order to avoid conflicts among nations, it guarantees equitable water distribution that jointly possess rivers, lakes and freshwater resources (groundwater). The main goal of this convection includes pollution control and minimization along with responsible utilization of international waterways for current and future generations by collaborative effort. Nile river allocation to different states became a debatable topic among East African countries which was later resolved using these convectional rules and biostatics principles (Gari et al., 2020). Another international treaty known as the Ramsar Convention on Wetlands (1971) was signed to conserve the wetlands and its ecosystem services (Goyal et al., 2023). Wetlands play a major role in the filtration and recycling of wastewater and support various biodiversity. Signatory bodies proclaim and pledge to manage Ramsar sites by encouraging local communities and international engagement (Pittock, 2021). Some acts are also enforced by certain countries like India has ordered the National Green Tribunal (NGT) act to stop illegal sand mining for regulation of groundwater extraction and mandatory evaluation of the environment for mining projects. Additionally, compensatory money must be paid by the companies for violating the rules for rehabilitation of water pollution (Rengarajan et al., 2018). These policies and rules must be strengthened and monitored properly to enhance the regulation among stakeholders. A collective effort from industries, government, the public and policymakers can help mitigate the environmental impacts. Mining activities must be executed within certain rules and regulations in a limited way with a principle to minimize the degradation of the natural ecosystem. Lastly, awareness among the public through various social media interaction sites, campaigns, rallies and incentives-driven measures must be enacted to accomplish a healthier planet. Continued enforcement of these acts and measures can successfully lead to a better, healthy and toxic free environment to reside for living organisms.

9 Conclusion

Mining-related water pollution severely affects aquatic life, freshwater resources, water quality and nearby living organisms posing serious dangers to the environment (flora and fauna) and human health. Uncontrolled mining activities also disturb natural water cycles, food chains and food web by causing soil erosion, acid mine drainage, tailings discharge and heavy metal contamination. To lessen these affects, strict adherence to environmental policies and laws. Eco-friendly mining methods and pollution prevention approaches are crucial. Wastewater treatment by bioremediation and further advanced nanotechnology intervention must be implemented by mining businesses. To stop further deterioration, government regulations and strict oversight should guarantee adherence to ecological norms. Further, water contamination from mining can be effectively reduced by funding scientific research, analysis and innovations. It is very vital to protect and conserve current water resources for ecological preservation and a safe planet for future generations.

References

Abbas, Z., Arooj, F., Ali, S., Zaheer, I. E., Rizwan, M., & Riaz, M. A. (2019). Phytoremediation of landfill leachate waste contaminants through floating bed technique using water hyacinth and water lettuce. *International Journal of Phytoremediation, 21*(13), 1356–1367. https://doi.org/10.1080/15226514.2019.1633259

Abraham, M. R., & Susan, T. B. (2017). Water contamination with heavy metals and trace elements from Kilembe copper mine and tailing sites in Western Uganda; implications for domestic water quality. *Chemosphere, 169*, 281–287. https://doi.org/10.1016/j.chemosphere.2016.11.077

Adewuyi, A., & Li, Q. (2024). Per- and polyfluoroalkyl substances contamination of drinking water sources in Africa: Pollution sources and possible treatment methods. *Chemosphere, 365*, 143368. https://doi.org/10.1016/j.chemosphere.2024.143368

Ahmed, J., Thakur, A., & Goyal, A. (2021). Industrial wastewater and its toxic effects. In M. P. Shah (Ed.), *Biological treatment of industrial wastewater* (p. 10). The Royal Society of Chemistry. https://doi.org/10.1039/9781839165399-00001

Álvarez, H., Domínguez, G., Ordóñez, A., Menéndez, J., Álvarez, R., & Loredo, J. (2021). Mine water for the generation and storage of renewable energy: A hybrid hydro-wind system.

International Journal of Environmental Research and Public Health, 18(13), 6758. https://doi.org/10.3390/ijerph18136758

Bai, D., Wang, T., Li, C., & Wang, W. (2024). Impact of coal energy development on the surrounding environmental water resources carrying capacity. *Desalination and Water Treatment, 319*, 100518. https://doi.org/10.1016/j.dwt.2024.100518

Bakshe, P., & Jugade, R. (2023). Phytostabilization and rhizofiltration of toxic heavy metals by heavy metal accumulator plants for sustainable management of contaminated industrial sites: A comprehensive review. *Journal of Hazardous Materials Advances, 10*, 100293. https://doi.org/10.1016/j.hazadv.2023.100293

Bangerter, P., Dixon, R., & Villegas, M. (2010). Improving overall usage of water in mining-a sustainable development approach. In *2nd international congress on water management in the mining industry*.

Beltrán-Flores, E., Pla-Ferriol, M., Martínez-Alonso, M., Gaju, N., Blánquez, P., & Sarrà, M. (2022). Fungal bioremediation of agricultural wastewater in a long-term treatment: Biomass stabilization by immobilization strategy. *Journal of Hazardous Materials, 439*, 129614. https://doi.org/10.1016/j.jhazmat.2022.129614

Bernardez, L. A., de Oliveira, L. E. L., & de Andrade Lima, L. R. P. (2021). Acid mine drainage at the Bahia Gold Belt (Brazil): Microbial isolation and characterization. *Environmental Monitoring and Assessment, 193*(2), 60. https://doi.org/10.1007/s10661-021-08844-2

Bochynska, S., Duszewska, A., Maciejewska-Jeske, M., Wrona, M., Szeliga, A., Budzik, M., Szczesnowicz, A., Bala, G., Trzcinski, M., Meczekalski, B., & Smolarczyk, R. (2024). The impact of water pollution on the health of older people. *Maturitas, 185*, 107981. https://doi.org/10.1016/j.maturitas.2024.107981

Brown, E. (2003). Water for a sustainable minerals industry—a review. In *Proceedings of water in mining*.

Chaudhary, P., Ahamad, L., Chaudhary, A., Kumar, G., Chen, W.-J., & Chen, S. (2023). Nanoparticle-mediated bioremediation as a powerful weapon in the removal of environmental pollutants. *Journal of Environmental Chemical Engineering, 11*(2), 109591. https://doi.org/10.1016/j.jece.2023.109591

Chettri, D., Verma, A. K., & Verma, A. K. (2024). Bioaugmentation: An approach to biological treatment of pollutants. *Biodegradation, 35*(2), 117–135. https://doi.org/10.1007/s10532-023-10050-5

Chow, S. J., Ojeda, N., Jacangelo, J. G., & Schwab, K. J. (2021). Detection of ultrashort-chain and other per- and polyfluoroalkyl substances (PFAS) in U.S. bottled water. *Water Research, 201*, 117292. https://doi.org/10.1016/j.watres.2021.117292

Clance, L. R., Ziegler, S. L., & Fodrie, F. J. (2023). Contaminants disrupt aquatic food webs via decreased consumer efficiency. *Science of the Total Environment, 859*(Pt 2), 160245. https://doi.org/10.1016/j.scitotenv.2022.160245

Coelho, N., Coelho, L., Coelho, L. M., Rezende, H. C., de Sousa, P. A. R., & Melo, D. F. O. (2015). Bioremediation of polluted waters using microorganisms. In N. Shiomi (Ed.), *Advances in bioremediation of wastewater and polluted soil*. IntechOpen. https://doi.org/10.5772/60770

Dhir, B. (2018). Chapter 4 - Biotechnological tools for remediation of acid mine drainage (removal of metals from wastewater and leachate). In M. N. V. Prasad, P. J. D. C. Favas, & S. K. Maiti (Eds.), Bio-geotechnologies for mine site rehabilitation (pp. 67–82). Elsevier. https://doi.org/10.1016/B978-0-12-812986-9.00004-X

Dietler, D., Farnham, A., Loss, G., Fink, G., & Winkler, M. S. (2021). Impact of mining projects on water and sanitation infrastructures and associated child health outcomes: A multi-country analysis of demographic and health surveys (DHS) in sub-Saharan Africa. *Globalization and Health, 17*(1), 70. https://doi.org/10.1186/s12992-021-00723-2

Dong, S., Xu, B., Yin, S., Han, Y., Zhang, X., & Dai, Z. (2019). Water resources utilization and protection in the coal mining area of northern China. *Scientific Reports, 9*(1), 1214. https://doi.org/10.1038/s41598-018-38148-4

du Plessis, A. (2022). Persistent degradation: Global water quality challenges and required actions. *One Earth, 5*(2), 129–131. https://doi.org/10.1016/j.oneear.2022.01.005

Duque, E., Udaondo, Z., Molina, L., de la Torre, J., Godoy, P., & Ramos, J. L. (2022). Providing octane degradation capability to pseudomonas putida KT2440 through the horizontal acquisition of oct genes located on an integrative and conjugative element. *Environmental Microbiology Reports, 14*(6), 934–946. https://doi.org/10.1111/1758-2229.13097

Elzakey, E. M., El-Sabbagh, S. M., Eldeen, E. E. N., Adss, I. A., & Nassar, A. M. K. (2023). Bioremediation of chlorpyrifos residues using some indigenous species of bacteria and fungi in wastewater. *Environmental Monitoring and Assessment, 195*(6), 779. https://doi.org/10.1007/s10661-023-11341-3

Farjana, S. H., Huda, N., Parvez Mahmud, M. A., & Saidur, R. (2019). A review on the impact of mining and mineral processing industries through life cycle assessment. *Journal of Cleaner Production, 231*, 1200–1217. https://doi.org/10.1016/j.jclepro.2019.05.264

Flores-Ramírez, R., Mendoza-Rivera, S. P., García-Grajales, J., Buenrostro-Silva, A., Sanjuan-Meza, E. U., Berumen-Rodríguez, A. A., & Espinosa-Reyes, G. (2024). Persistent organic pollutants in the olive ridley turtle (Lepidochelys olivacea) during the nesting stage in the "La Escobilla" sanctuary, Oaxaca, Mexico. *Environmental Science and Pollution Research, 31*(7), 10911–10919. https://doi.org/10.1007/s11356-024-31833-0

Gajski, G., Oreščanin, V., Gerić, M., Kollar, R., Lovrenčić Mikelić, I., & Garaj-Vrhovac, V. (2015). Toxicity assessment of the water used for human consumption from the Cameron/Tuba City abandoned uranium mining area prior/after the combined electrochemical treatment/advanced oxidation. *Environmental Science and Pollution Research International, 22*(1), 516–526. https://doi.org/10.1007/s11356-014-3376-8

Gari, Y., Block, P., Assefa, G., Mekonnen, M., & Tilahun, S. A. (2020). Quantifying the United Nations' watercourse convention indicators to inform equitable Transboundary River sharing: Application to the Nile River Basin. *Water, 12*(9), 2499. https://www.mdpi.com/2073-4441/12/9/2499

Gelaw, A., & Debele, M. (2024). International water conventions, the exploitation of trans-boundary rivers and the Nile water dispute among Ethiopia, Sudan and Egypt. Heliyon, *10*(21), e38671. https://doi.org/10.1016/j.heliyon.2024.e38671

Gidarakos, E., & Aivalioti, M. (2007). Large scale and long term application of bioslurping: The case of a Greek petroleum refinery site. *Journal of Hazardous Materials, 149*(3), 574–581. https://doi.org/10.1016/j.jhazmat.2007.06.110

Gill, R. T., Thornton, S. F., Harbottle, M. J., & Smith, J. W. (2016). Sustainability assessment of electrokinetic bioremediation compared with alternative remediation options for a petroleum release site. *Journal of Environmental Management, 184*(Pt 1), 120–131. https://doi.org/10.1016/j.jenvman.2016.07.036

Goyal, M. K., Rakkasagi, S., Shaga, S., Zhang, T. C., Surampalli, R. Y., & Dubey, S. (2023). Spatiotemporal-based automated inundation mapping of Ramsar wetlands using Google Earth Engine. Sci Rep, *13*(1), 17324. https://doi.org/10.1038/s41598-023-43910-4

Gu, H., Wu, Y., Lü, S., Lu, D., Tang, Y. Z., & Qi, Y. (2022). Emerging harmful algal bloom species over the last four decades in China. *Harmful Algae, 111*, 102059. https://doi.org/10.1016/j.hal.2021.102059

Guo, H., Li, X., Yang, W., Yao, Z., Mei, Y., Peng, L. E., Yang, Z., Shao, S., & Tang, C. Y. (2022). Nanofiltration for drinking water treatment: A review. *Frontiers of Chemical Science and Engineering, 16*(5), 681–698. https://doi.org/10.1007/s11705-021-2103-5

Gupta, V. K., & Rastogi, A. (2008). Biosorption of lead(II) from aqueous solutions by non-living algal biomass Oedogonium sp. and Nostoc sp. A comparative study. *Colloids and Surfaces. B, Biointerfaces, 64*(2), 170–178. https://doi.org/10.1016/j.colsurfb.2008.01.019

Hagner, M., Uusitalo, M., Ruhanen, H., Heiskanen, J., Peltola, R., Tiilikkala, K., Hyvönen, J., Sarala, P., & Mäkitalo, K. (2021). Amending mine tailing cover with compost and biochar: Effects on vegetation establishment and metal bioaccumulation in the Finnish subarctic. *Environmental Science and Pollution Research International, 28*(42), 59881–59898. https://doi.org/10.1007/s11356-021-14865-8

Han, Y., Nambi, I. M., & Prabhakar Clement, T. (2018). Environmental impacts of the Chennai oil spill accident - A case study. *Science of the Total Environment, 626*, 795–806. https://doi.org/10.1016/j.scitotenv.2018.01.128

Hao, W., Liu, H., Hao, S., & Mao, K. (2024). Characterization of heavy metal contamination in groundwater of typical mining area in Hunan Province. *Scientific Reports, 14*(1), 13054. https://doi.org/10.1038/s41598-024-63460-7

Ji, M., Li, B., Majdi, A., Alkhalifah, T., Alturise, F., & Ali, H. E. (2023). Application of nano remediation of mine polluted in acid mine drainage water using machine learning model. *Chemosphere, 311*(Pt 2), 136926. https://doi.org/10.1016/j.chemosphere.2022.136926

Jin, L., Wenjing, L., Yang, J., Li, Y., Yaning, F., Tao, T., & Xiaomeng, L. (2021). Ecological health risk assessment of heavy metals in farmland surrounding Tongguan gold deposit, Shaanxi Province. *Geology in China, 48*(3), 749–763.

Juárez-Maldonado, A., Tortella, G., Rubilar, O., Fincheira, P., & Benavides-Mendoza, A. (2021). Biostimulation and toxicity: The magnitude of the impact of nanomaterials in microorganisms and plants. *Journal of Advanced Research, 31*, 113–126. https://doi.org/10.1016/j.jare.2020.12.011

Keller, D. H., Zelanko, P. M., Gagnon, J. E., Horwitz, R. J., Galbraith, H. S., & Velinsky, D. J. (2018). Linking otolith microchemistry and surface water contamination from natural gas mining. *Environmental Pollution, 240*, 457–465. https://doi.org/10.1016/j.envpol.2018.04.026

Khan, S. N., Nafees, M., & Imtiaz, M. (2023). Assessment of industrial effluents for heavy metals concentration and evaluation of grass (Phalaris minor) as a pollution indicator. *Heliyon, 9*(9), e20299. https://doi.org/10.1016/j.heliyon.2023.e20299

Kimura, K., Iwase, T., Kita, S., & Watanabe, Y. (2009). Influence of residual organic macromolecules produced in biological wastewater treatment processes on removal of pharmaceuticals by NF/RO membranes. *Water Research, 43*(15), 3751–3758. https://doi.org/10.1016/j.watres.2009.05.042

Knobeloch, L., Salna, B., Hogan, A., Postle, J., & Anderson, H. (2000). Blue babies and nitrate-contaminated well water. *Environmental Health Perspectives, 108*(7), 675–678. https://doi.org/10.1289/ehp.00108675

Koki, I. B., Low, K. H., Juahir, H., Abdul Zali, M., Azid, A., & Zain, S. M. (2018). Consumption of water from ex-mining ponds in Klang Valley and Melaka, Malaysia: A health risk study. *Chemosphere, 195*, 641–652. https://doi.org/10.1016/j.chemosphere.2017.12.112

Kumar, S., Ahlawat, W., Bhanjana, G., Heydarifard, S., Nazhad, M. M., & Dilbaghi, N. (2014). Nanotechnology-based water treatment strategies. *Journal of Nanoscience and Nanotechnology, 14*(2), 1838–1858. https://doi.org/10.1166/jnn.2014.9050

Lai, W. (2017). Pesticide use and health outcomes: Evidence from agricultural water pollution in China. *Journal of Environmental Economics and Management, 86*, 93–120.

Lal, R. (2008). Carbon sequestration. *Philosophical Transactions of the Royal Society of London. Series B, Biological Sciences, 363*(1492), 815–830. https://doi.org/10.1098/rstb.2007.2185

Landrigan, P. J., Stegeman, J. J., Fleming, L. E., Allemand, D., Anderson, D. M., Backer, L. C., Brucker-Davis, F., Chevalier, N., Corra, L., Czerucka, D., Bottein, M. D., Demeneix, B., Depledge, M., Deheyn, D. D., Dorman, C. J., Fénichel, P., Fisher, S., Gaill, F., Galgani, F., & Rampal, P. (2020). Human health and ocean pollution. *Annals of Global Health, 86*(1), 151. https://doi.org/10.5334/aogh.2831

Lee, Y. C., & Chang, S. P. (2011). The biosorption of heavy metals from aqueous solution by spirogyra and Cladophora filamentous macroalgae. *Bioresource Technology, 102*(9), 5297–5304. https://doi.org/10.1016/j.biortech.2010.12.103

Li, F., Guo, Z., Mao, L., Feng, J., Huang, J., & Tao, H. (2023). Impact of textile industries on surface water contamination by Sb and other potential toxic elements: A case study in Taihu Lake Basin, China. *International Journal of Environmental Research and Public Health, 20*(4), 3600. https://doi.org/10.3390/ijerph20043600

Limmer, M., & Burken, J. (2016). Phytovolatilization of organic contaminants. *Environmental Science & Technology, 50*(13), 6632–6643. https://doi.org/10.1021/acs.est.5b04113

Lin, L., Yang, H., & Xu, X. (2022). Effects of water pollution on human health and disease heterogeneity: A review. *Frontiers in Environmental Science, 10*, 880246. https://doi.org/10.3389/fenvs.2022.880246

Lin, Y. L., Chiang, P. C., & Chang, E. E. (2006). Reduction of disinfection by-products precursors by nanofiltration process. *Journal of Hazardous Materials, 137*(1), 324–331. https://doi.org/10.1016/j.jhazmat.2006.02.016

Lin, Y. L., Chiang, P. C., & Chang, E. E. (2007). Removal of small trihalomethane precursors from aqueous solution by nanofiltration. *Journal of Hazardous Materials, 146*(1-2), 20–29. https://doi.org/10.1016/j.jhazmat.2006.11.050

Liu, L., Yang, D., Bai, Y., Li, X., Tan, F., Ma, J., & Wang, Y. (2024). Construction of biodegradable superhydrophilic/underwater superoleophobic materials with CNF (cellulose nanofiber) fence-like attached on the surface for efficient oil/water emulsion separation. *International Journal of Biological Macromolecules, 269*(Pt 2), 132175. https://doi.org/10.1016/j.ijbiomac.2024.132175

Lockwood, D. (2019). Lakes and rivers are getting saltier. *ACS Central Science, 5*(3), 376–379. https://doi.org/10.1021/acscentsci.9b00191

Lu, W. B., Shi, J. J., Wang, C. H., & Chang, J. S. (2006). Biosorption of lead, copper and cadmium by an indigenous isolate Enterobacter sp. J1 possessing high heavy-metal resistance. *Journal of Hazardous Materials, 134*(1–3), 80–86. https://doi.org/10.1016/j.jhazmat.2005.10.036

Lu, Y., Song, S., Wang, R., Liu, Z., Meng, J., Sweetman, A. J., Jenkins, A., Ferrier, R. C., Li, H., Luo, W., & Wang, T. (2015). Impacts of soil and water pollution on food safety and health risks in China. *Environment International, 77*, 5–15. https://doi.org/10.1016/j.envint.2014.12.010

Ma, J., Yang, Y., Dai, X., Chen, Y., Deng, H., Zhou, H., Guo, S., & Yan, G. (2016). Effects of adding bulking agent, inorganic nutrient and microbial inocula on biopile treatment for oil-field drilling waste. *Chemosphere, 150*, 17–23. https://doi.org/10.1016/j.chemosphere.2016.01.123

Machado, M. D., Santos, M. S., Gouveia, C., Soares, H. M., & Soares, E. V. (2008). Removal of heavy metals using a brewer's yeast strain of Saccharomyces cerevisiae: The flocculation as a separation process. *Bioresource Technology, 99*(7), 2107–2115. https://doi.org/10.1016/j.biortech.2007.05.047

Mahato, M. K., Singh, A. K., & Giri, S. (2022). Evaluation of metal contamination, flux and the associated human health risk from atmospheric dustfall in metal mining areas of southern Jharkhand, India. *Environmental Science and Pollution Research, 29*(20), 30348–30362. https://doi.org/10.1007/s11356-021-18463-6

Maqsood, Q., Sumrin, A., Waseem, R., Hussain, M., Imtiaz, M., & Hussain, N. (2023). Bioengineered microbial strains for detoxification of toxic environmental pollutants. *Environmental Research, 227*, 115665. https://doi.org/10.1016/j.envres.2023.115665

Martins-Filho, P. R., Damascena, N. P., Araujo, A. P. D., Silva, M. C., Santiago, B. M., Deitos, A. R., & Machado, C. E. P. (2023). The devastating impact of illegal mining on indigenous health: A focus on malaria in the Brazilian Amazon. *EXCLI Journal, 22*, 400–402. https://doi.org/10.17179/excli2023-6046

Matebese, F., Mosai, A. K., Tutu, H., & Tshentu, Z. R. (2024). Mining wastewater treatment technologies and resource recovery techniques: A review. *Heliyon, 10*(3), e24730. https://doi.org/10.1016/j.heliyon.2024.e24730

Meißner, S. (2021). The impact of metal mining on global water stress and regional carrying capacities—A GIS-based water impact assessment. *Resources, 10*(12), 120. https://www.mdpi.com/2079-9276/10/12/120

Mishra, P. C., Behera, P. C., & Patel, R. K. (2005). Contamination of water due to major industries and open refuse dumping in the steel city of Orissa–A case study. *Journal of Environmental Science & Engineering, 47*(2), 141–154.

Moreno-González, R., Macías, F., Olías, M., & Ruiz Cánovas, C. (2022). Temporal evolution of acid mine drainage (AMD) leachates from the abandoned tharsis mine (Iberian Pyrite Belt, Spain). *Environmental Pollution, 295*, 118697. https://doi.org/10.1016/j.envpol.2021.118697

Münzel, T., Hahad, O., Daiber, A., & Landrigan, P. J. (2023). Soil and water pollution and human health: What should cardiologists worry about? *Cardiovascular Research, 119*(2), 440–449. https://doi.org/10.1093/cvr/cvac082

Musie, W., & Gonfa, G. (2023). Fresh water resource, scarcity, water salinity challenges and possible remedies: A review. *Heliyon, 9*(8), e18685. https://doi.org/10.1016/j.heliyon.2023.e18685

Nabeela, F., Azizullah, A., Bibi, R., Uzma, S., Murad, W., Shakir, S. K., Ullah, W., Qasim, M., & Häder, D. P. (2014). Microbial contamination of drinking water in Pakistan–A review. *Environmental Science and Pollution Research International, 21*(24), 13929–13942. https://doi.org/10.1007/s11356-014-3348-z

Narangarvuu, D., Enkhdul, T., Erdenetsetseg, E., Enkhrii-Ujin, E., Irmuunzaya, K., Batbayar, G., Oyundelger, K., Yam, R. S., & Pfeiffer, M. (2023). Mining and urbanization affect river chemical water quality and macroinvertebrate communities in the upper Selenga River basin, Mongolia. *Environmental Monitoring and Assessment, 195*(12), 1500. https://doi.org/10.1007/s10661-023-12022-x

Newman, L. A., & Reynolds, C. M. (2004). Phytodegradation of organic compounds. *Current Opinion in Biotechnology, 15*(3), 225–230. https://doi.org/10.1016/j.copbio.2004.04.006

Newmark, R. L., Friedmann, S. J., & Carroll, S. A. (2010). Water challenges for geologic carbon capture and sequestration. *Environmental Management, 45*(4), 651–661. https://doi.org/10.1007/s00267-010-9434-1

Nguyen, H. T., Kim, Y., Choi, J. W., Jeong, S., & Cho, K. (2021). Soil microbial communities-mediated bioattenuation in simulated aquifer storage and recovery (ASR) condition: Long-term study. *Environmental Research, 197*, 111069. https://doi.org/10.1016/j.envres.2021.111069

Nicolaou, S. A., Gaida, S. M., & Papoutsakis, E. T. (2010). A comparative view of metabolite and substrate stress and tolerance in microbial bioprocessing: From biofuels and chemicals, to biocatalysis and bioremediation. *Metabolic Engineering, 12*(4), 307–331. https://doi.org/10.1016/j.ymben.2010.03.004

Noor, A., & Khan, S. A. (2023). Agricultural wastes as renewable biomass to remediate water pollution. *Sustainability, 15*(5), 4246. https://www.mdpi.com/2071-1050/15/5/4246

Northey, S. A., Mudd, G. M., Werner, T. T., Jowitt, S. M., Haque, N., Yellishetty, M., & Weng, Z. (2017). The exposure of global base metal resources to water criticality, scarcity and climate change. *Global Environmental Change, 44*, 109–124. https://doi.org/10.1016/j.gloenvcha.2017.04.004

Nti, E. K., Cobbina, S. J., Attafuah, E. E., Senanu, L. D., Amenyeku, G., Gyan, M. A., Forson, D., & Safo, A. R. (2023). Water pollution control and revitalization using advanced technologies: Uncovering artificial intelligence options towards environmental health protection, sustainability and water security. *Heliyon, 9*(7), e18170. https://doi.org/10.1016/j.heliyon.2023.e18170

Pandey, S. (2006). Water pollution and health. *Kathmandu University Medical Journal, 4*(1), 128–134.

Pittock, J. (2021). Use of the Ramsar convention to protect springs and other wetlands. *Conservation Biology, 35*(6), 1969–1971. https://doi.org/10.1111/cobi.13799

Pokhrel, L. R., Ettore, N., Jacobs, Z. L., Zarr, A., Weir, M. H., Scheuerman, P. R., Kanel, S. R., & Dubey, B. (2017). Novel carbon nanotube (CNT)-based ultrasensitive sensors for trace mercury(II) detection in water: A review. *Science of the Total Environment, 574*, 1379–1388. https://doi.org/10.1016/j.scitotenv.2016.08.055

Popkin, B. M., D'Anci, K. E., & Rosenberg, I. H. (2010). Water, hydration, and health. *Nutrition Reviews, 68*(8), 439–458. https://doi.org/10.1111/j.1753-4887.2010.00304.x

Prabakar, D., Suvetha, K. S., Manimudi, V. T., Mathimani, T., Kumar, G., Rene, E. R., & Pugazhendhi, A. (2018). Pretreatment technologies for industrial effluents: Critical review on bioenergy production and environmental concerns. *Journal of Environmental Management, 218*, 165–180. https://doi.org/10.1016/j.jenvman.2018.03.136

Punia, A., & Singh, S. K. (2021). Chapter 1 - Contamination of water resources in the mining region. In A. Ahamad, S. I. Siddiqui, & P. Singh (Eds.), *Contamination of water* (pp. 3–17). Academic Press. https://doi.org/10.1016/B978-0-12-824058-8.00015-3

Quan, S. M., Xi, F. M., Wang, J. Y., Yin, Y., Pei, Z. J., & Zhao, F. Q. (2019). Potential of renewable energy development on abandoned mine areas: A case study in Liao-Ning Province, Norheast China. *Ying Yong Sheng Tai Xue Bao, 30*(8), 2803–2812. https://doi.org/10.13287/j.1001-9332.201908.039

Radhapyari, K., Aribam, N. G., Datta, S., Dutta, S., & Barman, R. (2022). Chapter 11 - Impact of mining on groundwater quality of India using indexing techniques and its assessment. In A. K. Tiwari, A. Kumar, A. K. Singh, T. N. Singh, E. Suozzi, G. Matta, & S. Lo Russo (Eds.), *Current directions in water scarcity research* (Vol. 5, pp. 187–223). Elsevier. https://doi.org/10.1016/B978-0-323-85378-1.00011-8

Rashtian, J., Chavkin, D. E., & Merhi, Z. (2019). Water and soil pollution as determinant of water and food quality/contamination and its impact on female fertility. *Reproductive Biology and Endocrinology, 17*(1), 5. https://doi.org/10.1186/s12958-018-0448-5

Reja, S., & Vasudevan, S. (2024). Charge-switchable zwitterionic nanomagnets for wastewater remediation. *Nanoscale Advances, 7*(1), 329–335. https://doi.org/10.1039/d4na00730a

Rendón-Castrillón, L., Ramírez-Carmona, M., Ocampo-López, C., González-López, F., Cuartas-Uribe, B., & Mendoza-Roca, J. A. (2024). Efficient bioremediation of indigo-dye contaminated textile wastewater using native microorganisms and combined bioaugmentation-biostimulation techniques. *Chemosphere, 353*, 141538. https://doi.org/10.1016/j.chemosphere.2024.141538

Rengarajan, S., Palaniyappan, D., Ramachandran, P., & Ramachandran, R. (2018). National Green Tribunal of India—An observation from environmental judgements. *Environmental Science and Pollution Research, 25*(12), 11313–11318. https://doi.org/10.1007/s11356-018-1763-2

Rezania, S., Taib, S. M., Md Din, M. F., Dahalan, F. A., & Kamyab, H. (2016). Comprehensive review on phytotechnology: Heavy metals removal by diverse aquatic plants species from wastewater. *Journal of Hazardous Materials, 318*, 587–599. https://doi.org/10.1016/j.jhazmat.2016.07.053

Ritter, W. F. (1990). Pesticide contamination of ground water in the United States–A review. *Journal of Environmental Science and Health, 25*(1), 1–29. https://doi.org/10.1080/03601239009372674

Rogowska, J., & Namieśnik, J. (2010). Environmental implications of oil spills from shipping accidents. *Reviews of Environmental Contamination and Toxicology, 206*, 95–114. https://doi.org/10.1007/978-1-4419-6260-7_5

Russell, H. H., Jackson, R. J., Spath, D. P., & Book, S. A. (1987). Chemical contamination of California drinking water. *The Western Journal of Medicine, 147*(5), 615–622.

Rzymski, P., Klimaszyk, P., Marszelewski, W., Borowiak, D., Mleczek, M., Nowiński, K., Pius, B., Niedzielski, P., & Poniedziałek, B. (2017). The chemistry and toxicity of discharge waters from copper mine tailing impoundment in the valley of the Apuseni Mountains in Romania. *Environmental Science and Pollution Research International, 24*(26), 21445–21458. https://doi.org/10.1007/s11356-017-9782-y

Sadeq, M., Moe, C. L., Attarassi, B., Cherkaoui, I., ElAouad, R., & Idrissi, L. (2008). Drinking water nitrate and prevalence of methemoglobinemia among infants and children aged 1–7 years in Moroccan areas. *International Journal of Hygiene and Environmental Health, 211*(5), 546–554. https://doi.org/10.1016/j.ijheh.2007.09.009

Santana, C. S., Montalván Olivares, D. M., Silva, V. H. C., Luzardo, F. H. M., Velasco, F. G., & de Jesus, R. M. (2020). Assessment of water resources pollution associated with mining activity in a semi-arid region. *Journal of Environmental Management, 273*, 111148. https://doi.org/10.1016/j.jenvman.2020.111148

Sarathchandra, S. S., Rengel, Z., & Solaiman, Z. M. (2023). A review on remediation of iron ore mine tailings via organic amendments coupled with phytoremediation. *Plants, 12*(9), 1871. https://doi.org/10.3390/plants12091871

Sari, A., & Tuzen, M. (2009). Kinetic and equilibrium studies of biosorption of Pb(II) and cd(II) from aqueous solution by macrofungus (amanita rubescens) biomass. *Journal of Hazardous Materials, 164*(2–3), 1004–1011. https://doi.org/10.1016/j.jhazmat.2008.09.002

Schlink, A. C., Nguyen, M. L., & Viljoen, G. J. (2010). Water requirements for livestock production: A global perspective. *Revue Scientifique et Technique, 29*(3), 603–619. https://doi.org/10.20506/rst.29.3.1999

Shan, M., Wang, Y., & Shen, X. (2009). Study on bioremediation of eutrophic lake. *Journal of Environmental Sciences, 21*(Suppl 1), 16–18. https://doi.org/10.1016/s1001-0742(09)60027-9

Shen, M., Ye, S., Zeng, G., Zhang, Y., Xing, L., Tang, W., Wen, X., & Liu, S. (2020). Can microplastics pose a threat to ocean carbon sequestration? *Marine Pollution Bulletin, 150*, 110712. https://doi.org/10.1016/j.marpolbul.2019.110712

Silva, I. A., Almeida, F. C. G., Souza, T. C., Bezerra, K. G. O., Durval, I. J. B., Converti, A., & Sarubbo, L. A. (2022). Oil spills: Impacts and perspectives of treatment technologies with focus on the use of green surfactants. *Environmental Monitoring and Assessment, 194*(3), 143. https://doi.org/10.1007/s10661-022-09813-z

Singh, S. K., & Kumar, D. (2024). Optimizing coal mine planning and design for sustainable development in the context of mass exploitation of coal deposits. *Heliyon, 10*(7), e28524. https://doi.org/10.1016/j.heliyon.2024.e28524

Soares, E. V., & Soares, H. M. (2013). Cleanup of industrial effluents containing heavy metals: A new opportunity of valorising the biomass produced by brewing industry. *Applied Microbiology and Biotechnology, 97*(15), 6667–6675. https://doi.org/10.1007/s00253-013-5063-y

Soe, P. S., Kyaw, W. T., Arizono, K., Ishibashi, Y., & Agusa, T. (2022). Mercury pollution from artisanal and small-scale gold Mining in Myanmar and Other Southeast Asian Countries. *International Journal of Environmental Research and Public Health, 19*(10), 6290. https://doi.org/10.3390/ijerph19106290

Subashchandrabose, S. R., Ramakrishnan, B., Megharaj, M., Venkateswarlu, K., & Naidu, R. (2013). Mixotrophic cyanobacteria and microalgae as distinctive biological agents for organic pollutant degradation. *Environment International, 51*, 59–72. https://doi.org/10.1016/j.envint.2012.10.007

Syafrudin, M., Kristanti, R. A., Yuniarto, A., Hadibarata, T., Rhee, J., Al-Onazi, W. A., Algarni, T. S., Almarri, A. H., & Al-Mohaimeed, A. M. (2021). Pesticides in drinking water-a review. *International Journal of Environmental Research and Public Health, 18*(2), 468. https://doi.org/10.3390/ijerph18020468

Szulc, J., Okrasa, M., Majchrzycka, K., Sulyok, M., Nowak, A., Ruman, T., Nizioł, J., Szponar, B., & Gutarowska, B. (2021). Microbiological and toxicological hazards in sewage treatment plant bioaerosol and dust. *Toxins, 13*(10), 691. https://doi.org/10.3390/toxins13100691

Tan, J. P., Clyde, C. W., Ng, C. C., Yeap, S. K., & Yong, C. Y. (2024). Advancements in microbial-mediated radioactive waste bioremediation: A review. *Journal of Environmental Radioactivity, 280*, 107530. https://doi.org/10.1016/j.jenvrad.2024.107530

Tao, Z., Youcai, Z., & Atta Nyankson, E. (2023). Chapter 12 - Integrated technology for mechanical separation of mixed domestic solid waste. In Z. Tao, Z. Youcai, & E. Atta Nyankson (Eds.), *Resource recovery technology for municipal and rural solid waste* (pp. 181–186). Elsevier. https://doi.org/10.1016/B978-0-323-98978-7.00011-7

Tessnow-von Wysocki, I., & Vadrot, A. B. M. (2024). Pathways of scientific input into intergovernmental negotiations: A new agreement on marine biodiversity. *International Environmental Agreement, 24*(2-3), 325–348. https://doi.org/10.1007/s10784-024-09642-0

Thakur, A., Kumar, A., Vanita, B., Panchbhai, G., Kumar, N., Kumari, A., & Dogra, P. (2018). Water footprint - A tool for sustainable development of indian dairy industry. *International Journal of Livestock Research, 8*, 1–18. https://doi.org/10.5455/ijlr.20180110101311

Thomas, G., Sheridan, C., & Holm, P. E. (2022). A critical review of phytoremediation for acid mine drainage-impacted environments. *Science of the Total Environment, 811*, 152230. https://doi.org/10.1016/j.scitotenv.2021.152230

Tian, Y., Zhao, Y., Zhang, X., Li, S., & Wu, H. (2023). Incorporating carbon sequestration into lake management: A potential perspective on climate change. *Science of the Total Environment, 895*, 164939. https://doi.org/10.1016/j.scitotenv.2023.164939

Tianliang, W., Aghalari, Z., Mubanga, R., Sosa-Hernandez, J. E., Martínez-Ruiz, M., & Parra-Saldívar, R. (2023). Assessing environmental health impacts of coal mining exploitation in Iran: A rapid impact assessment matrix (RIAM) approach for environmental protection. *PLoS One, 18*(12), e0293973. https://doi.org/10.1371/journal.pone.0293973

Touliabah, H. E., El-Sheekh, M. M., Ismail, M. M., & El-Kassas, H. (2022). A review of microalgae- and cyanobacteria-based biodegradation of organic pollutants. *Molecules, 27*(3). https://doi.org/10.3390/molecules27031141

Tripatanasuwan, S., Zhong, Z., & Reneker, D. H. (2007). Effect of evaporation and solidification of the charged jet in electrospinning of poly (ethylene oxide) aqueous solution. *Polymer, 48*(19), 5742–5746.

Vullo, D. L., Ceretti, H. M., Daniel, M. A., Ramírez, S. A., & Zalts, A. (2008). Cadmium, zinc and copper biosorption mediated by pseudomonas veronii 2E. *Bioresource Technology, 99*(13), 5574–5581. https://doi.org/10.1016/j.biortech.2007.10.060

Wang, M., Song, H., Chen, W. Q., Lu, C., Hu, Q., Ren, Z., Yang, Y., Xu, Y., Zhong, A., & Ling, W. (2011). Cancer mortality in a Chinese population surrounding a multi-metal sulphide mine in Guangdong province: An ecologic study. *BMC Public Health, 11*, 319. https://doi.org/10.1186/1471-2458-11-319

Worlanyo, A. S., & Jiangfeng, L. (2021). Evaluating the environmental and economic impact of mining for post-mined land restoration and land-use: A review. *Journal of Environmental Management, 279*, 111623. https://doi.org/10.1016/j.jenvman.2020.111623

Xiong, J., Zhao, T., Cheng, H., Li, S., Wang, S., & Chen, G. (2019). The assessment on the heavy metal pollution and health risks in the Liujiang River under the Xijiang River region. *Desalination and Water Treatment, 149*, 315–322.

Xu, J., Li, X., & Gao, T. (2022). The multifaceted function of water hyacinth in maintaining environmental sustainability and the underlying mechanisms: A mini review. *International Journal of Environmental Research and Public Health, 19*(24), 16725. https://doi.org/10.3390/ijerph192416725

Xu, R., Chi, C., Li, F., & Zhang, B. (2013). Laccase-polyacrylonitrile nanofibrous membrane: Highly immobilized, stable, reusable, and efficacious for 2,4,6-trichlorophenol removal. *ACS Applied Materials & Interfaces, 5*(23), 12554–12560. https://doi.org/10.1021/am403849q

Yaraghi, N., Ronkanen, A.-K., Torabi Haghighi, A., Aminikhah, M., Kujala, K., & Kløve, B. (2020). Impacts of gold mine effluent on water quality in a pristine sub-Arctic river. *Journal of Hydrology, 589*, 125170. https://doi.org/10.1016/j.jhydrol.2020.125170

Yusta-García, R., Orta-Martínez, M., Mayor, P., González-Crespo, C., & Rosell-Melé, A. (2017). Water contamination from oil extraction activities in northern Peruvian Amazonian rivers. *Environmental Pollution, 225*, 370–380. https://doi.org/10.1016/j.envpol.2017.02.063

Zanetta-Colombo, N. C., Scharnweber, T., Christie, D. A., Manzano, C. A., Blersch, M., Gayo, E. M., Muñoz, A. A., Fleming, Z. L., & Nüsser, M. (2024). When another one bites the dust: Environmental impact of global copper demand on local communities in the Atacama mining hotspot as registered by tree rings. *Science of the Total Environment, 920*, 170954. https://doi.org/10.1016/j.scitotenv.2024.170954

Zhou, Y., Stepanenko, A., Kishchenko, O., Xu, J., & Borisjuk, N. (2023). Duckweeds for phytoremediation of polluted water. *Plants, 12*(3), 589. https://doi.org/10.3390/plants12030589

Sources of Mining Waste in the Environment, their Fate and Toxicity

A. Swaroop Sanket and Rajesh Kumar Sahoo

1 Introduction to Mining Wastes

Mining operations are essential in order to extract necessary minerals and metals which are considered vital to the industrial sectors and society. Beyond this, the extraction activity generated abundant residues and wastes which if not managed well, can cause severe environmental and health risks (Matinde, 2018). Wide range minerals and metal ions are generated during the extraction, processing, and beneficiation of ores are referred to as mining wastes ideally being thereafter categorized as waste rock, tailings, and overburden (Das & Ghosh, 2018; Das, Ghosh, et al., 2015; Ghosh et al., 2023; Matinde, 2018; Mohanty et al., 2016). Depending on the mineral being mined, the extraction process, and the geological location, these wastes have different compositions, which can result in various environmental outcomes, including as the leaching of hazardous substances into water systems, soil contamination, and dust dispersion in the atmosphere (Matinde, 2018). Every year, mining produces billions of tons of trash worldwide, making efficient management techniques necessary while understanding the geophysical and mechanical processes to reduce the toxic effects(El Machi et al., 2024) (Fig. 1).

A. S. Sanket · R. K. Sahoo (✉)
Biomics and Biodiversity Laboratory, Centre for Biotechnology, Siksha 'O' Anusandhan University (Deemed to be University), Bhubaneswar, Odisha, India
e-mail: rajeshkumarsahoo@soa.ac.in

© The Author(s), under exclusive license to Springer Nature Switzerland AG 2026

A. P. Das (ed.), *Bioremediation Technology for Sustainable Mining Waste Management*, Sustainable Environmental Waste Management Strategies,
https://doi.org/10.1007/978-3-032-03468-7_3

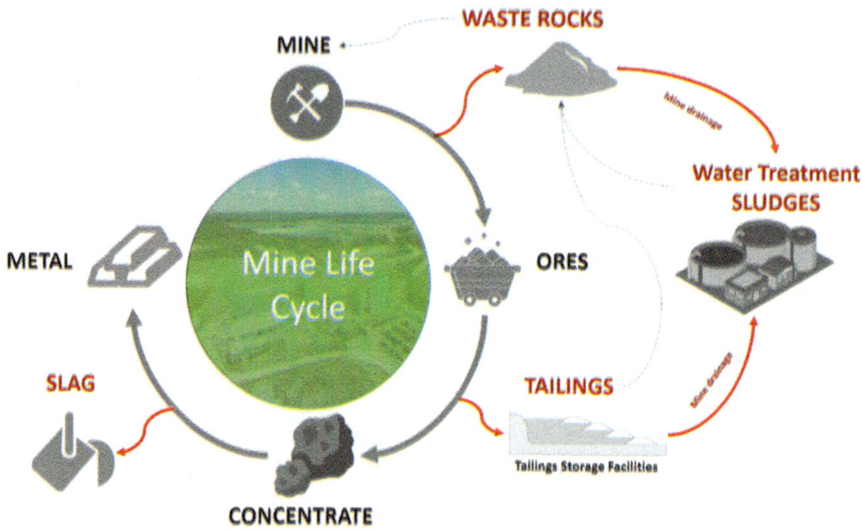

Fig. 1 Life cycle of a mining waste (Adapted from Taha & Benzaazoua, 2020). An overview of the several kinds of mining wastes is given in this chapter, with particular attention to those related to the extraction of lead, manganese, chromium, iron, nickel. The nature of these wastes, their effects on the ecosystem, and contemporary management techniques are covered in detail in each section

2 Manganese Mining Wastes

Mining of Manganese (Mn) produces enormous quantity of residues leading to overburden and tailings. The wastes contain trace metals and other contaminants that can leach into the environment, affecting soil and water quality (Ghosh et al., 2022). Mitigation of the Mn wastes is essential to avoid severe environmental impacts and enhances the re-utilization of Mn residues especially from terrestrial deposits and deep-sea nodules for industrial usages such as steel production and batteries (Abbadi & Mucsi, 2024). Mining sites in countries like South Africa or Australia, the presence of Mn tailings have been found to compose of residual manganese (27–30 wt.%) alongside heavy metals causing the groundwater under acidic conditions (Makhathini et al., 2023). The mining of Mn nodules from the deep-sea regions around the Clarion-Clipperton Zone, has generated dewatering wastes causing release of Mn residues into anoxic waters (Abbadi & Mucsi, 2024). These wastes can alter marine ecosystems by increasing metal concentrations, potentially affecting mesopelagic communities (Das et al., 2025; Das & Ghosh, 2023; Reddy Koteswara et al., 2025). The fate of manganese mining wastes depends on redox conditions, pH, and microbial activity, which can either immobilize or mobilize these elements (Reimers et al., 2013). Larb et al. analyzed the strategies available for subjective industrial utility of the manganese residues obtained during electrolysis. Their analysis suggested that cementitious materials, bricks, and road base materials, ceramics, organic fertilizers can be potential application areas of electrolytic manganese residues (Das et al., 2011; Das & Ghosh, 2023a, 2023b, 2023c). Bal et al. comprehensively analyzed the utility of low-grade manganese residues and their potential applications such as in the manufacturing of steel, glass, batteries, car radiators, beverage cans, kitchenware, and transportations (Bal et al., 2019). Despite a wide range of applications, the Manganese has negative impacts on the health. Manganese being a trace mineral required for metabolic pathways in the biological systems is not quite available endogenously and thus is required from exogenous sources. The recommended daily intake is between 1.8 and 2.3 mg in humans, however, increased concentrations causing toxicity can lead to manganism (Finley et al., 2003). This situation causes facial muscle spasms, hallucinations in affected individuals, apart from interfering in the COQ7 hydroxylase enzyme pathway, malfunction of liver and accumulation in blood plasma (; Andruska & Racette, 2015; Mattison et al., 2024).

3 Chromium Mining Wastes

Various oxidation states of Chromium (Cr) have been reported to cause carcinogenicity especially the Cr (VI). This state is highly toxic to human cells and causes serious health issues. The effective management of residual Cr (VI) is crucial to avoid the environmental contamination. Overburden of the Cr at different mining sites are major cause of soil contamination and change in physiochemical and biological malfunctions in the biological systems. The excess deposition of Cr often leads to Acid mine drainage (AMD) due to sulfide mineral oxidation, exacerbating water contamination. The fate of these wastes includes leaching into surface and groundwater, where Cr(VI) persists under oxidizing conditions, posing risks to aquatic life and human health (Das, Bal, & Mahapatra, 2015; Das & Mishra, 2008; Das & Singh, 2011; Mishra et al., 2009). Soil contamination near mining sites also facilitates plant uptake, further dispersing chromium into the food chain (Wang et al., 2017). However, the toxic effects of the metal are severe to the biological systems. The hexavalent state of Chromium has been reported as hemotoxic, genotoxic, and carcinogenic in several studies. It has been proven than the deterioration of the blood cells in presence of Cr (VI) is due to the oxidation reactions leading to hemolysis also causing renal and liver failure (Das et al., 2023; Das, Ray, et al., 2024; Dey et al., 2023). Management challenges are compounded by the long-term stability of waste piles exposed to weathering.

4 Nickel Mining Wastes

Metallurgical activities of Nickel especially mining activities generates huge amount of water materials causing severe environmental issues (Haghighizadeh et al., 2024). Indonesia has reported to have lost forests, polluted water and soil which has significantly reduced water quality (Nasution et al., 2024). This metal is essential for stainless steel and batteries and the sources have produced tailings and waste rocks. Laterite mining is also responsible generation of nickel oxides which gets leached during the heavy rainfall. The toxic effects of nickel in humans are often associated with dexterous effects. Intake of Nickel through foodstuff is an obvious route to systemic absorption and toxicity (Das, Behera, & Das, 2024). The gastrointestinal tract gets damaged due to undue choking of the mucociliary cavity and reduces the bioavailability of the metal. The inhalation of Nickel dust has also severe impact on the respiratory channel of humans as further also reduces the chance of excretion through natural metabolism. The alveolar region deposited with particles ≤4 μm; tracheobronchial region deposited with 4–10 μm and those deposited with in the nasopharyngeal region are between 10 μm and 100 μm, becomes lethal for proper respiration (Buxton et al., 2019; Denkhaus & Salnikow, 2002; Zambelli et al., 2016). Waste management often focuses on neutralization and containment to limit environmental dispersion.

5 Lead Mining Wastes

Mining of lead produces residues which are of more hazardous and toxic in nature. Mismanagement of these residues results into soil and water contamination which can lead to severe health risks. In order to counter such situations, appropriate measures and remediation practices are critical in order to mitigate the environmental impact of such wastes (Das, Dey, et al., 2024; Mohanty et al., 2024; Priyadarshini et al., 2024). Tailings and waste rocks are usually found to contain sulfur-based materials, when oxidized produces AMD thus releasing the Lead content to the environment. Such situations are crucial as they can cause soil contamination and degradation of plants, quality reduction of groundwater and associated health risks from ingestion or inhalation (Prabhakar et al., 2019; Raj & Das, 2023). Apart from these, solubility of the metal ions while pH being neutral hinders dissemination, however, acidic pH rather enhances. The bioaccumulation of Lead in humans have been well elaborated and reported to cause malfunctions in digestive systems, renal systems, chronic damages to the central nervous and peripheral nervous system, respiratory systems, reproductive system as well. Severe situations have been reported in humans where diseases like Alzheimer's and Parkinson's have occurred (Collin et al., 2022; Olufemi et al., 2022). Remediation efforts often involve stabilization with phosphates or revegetation to reduce exposure.

6 Conclusion

This chapter concisely elaborates the nature, sources, and environmental impacts of mining wastes generated from Manganese, Chromium, Nickel, and Lead. The residues are often of low-grade nature and carry little relevance for direct industrial use;

however, overburden causes significant environmental and biological hazards persistent toxicity. Furthermore, it has been emphasized on the requirement for effective management strategies tailored to the chemical and physical properties of these wastes. Strategies such as neutralization, containment, and repurposing into industrial products are effective ways to mitigate their impact while enhancing resource recovery.

References

Abbadi, A., & Mucsi, G. (2024). A review on complex utilization of mine tailings: Recovery of rare earth elements and residue valorization. *Journal of Environmental Chemical Engineering, 12*(3), 113118. https://doi.org/10.1016/j.jece.2024.113118

Andruska, K. M., & Racette, B. A. (2015). Neuromythology of manganism. *Current Epidemiology Reports, 2*(2), 143–148. https://doi.org/10.1007/s40471-015-0040-x

Bal, B., Ghosh, S., & Das, A. P. (2019). Microbial recovery and recycling of manganese waste and their future application: A review. *Geomicrobiology Journal, 36*(1), 85–96. https://doi.org/10.1080/01490451.2018.1497731

Buxton, S., Garman, E., Heim, K. E., Lyons-Darden, T., Schlekat, C. E., Taylor, M. D., & Oller, A. R. (2019). Concise Review of Nickel Human Health Toxicology and Ecotoxicology. Inorganics, 7(7), 89. https://doi.org/10.3390/inorganics7070089

Collin, M. S., Venkatraman, S. K., Vijayakumar, N., Kanimozhi, V., Arbaaz, S. M., Stacey, R. G. S., Anusha, J., Choudhary, R., Lvov, V., Tovar, G. I., Senatov, F., Koppala, S., & Swamiappan, S. (2022). Bioaccumulation of lead (Pb) and its effects on human: A review. *Journal of Hazardous Materials Advances, 7*, 100094. https://doi.org/10.1016/j.hazadv.2022.100094

Das, A., Dey, S., Bala, K., Gnanasekaran, R., & Das, A. P. (2024). Microbial bioremediation technology for sustainable treatment and management of synthetic microfiber waste. In *Sustainable microbial technology for synthetic and cellulosic microfiber bioremediation* (pp. 221–238). Springer. https://doi.org/10.1007/978-3-031-62600-5_12

Das, A., Dey, S., Gacem, A., Yadav, K. K., Nath, H., Ram, D. K., & Das, A. P. (2025). Revolutionizing solid hydrocarbon sourcing: Plastic's journey from waste to upcycled treasure. *Environmental Science and Engineering, 218*, 331–351. https://doi.org/10.1007/978-3-031-81615-4_14

Das, A., Ray, S. K., Mohanty, M., Mohanty, J., Dey, S., & Das, A. P. (2024). Ecotoxicity of microplastic wastes and their sustainable management: A review. *Environmental Chemistry and Ecotoxicology, 6*, 144–152. https://doi.org/10.1016/J.ENCECO.2024.05.003

Das, A. P., Bal, B., & Mahapatra, P. S. (2015). Chromogenic biosensors for pathogen detection. In P. Prokopovich (Ed.), *Biological and pharmaceutical applications of nanomaterials* (1st ed.). CRC Press.

Das, A. P., Behera, I. D., & Das, N. P. (2024). *Renewable energy generation and value addition from environmental microfiber pollution through advanced greener solution* (p. 278). Springer.

Das, A. P., & Ghosh, S. (2018). Bioleaching of manganese from mining waste materials. *Materials Today: Proceedings, 5*(1), 2381–2390. https://doi.org/10.1016/J.MATPR.2017.11.459

Das, A. P., & Ghosh, S. (2023). *Manganese mining microorganisms*. Elsevier.

Das, A. P., & Ghosh, S. (2023a). Manganese and its application. In *Manganese mining microorganisms* (pp. 29–40). Elsevier. https://doi.org/10.1016/B978-0-12-822147-1.00007-8

Das, A. P., & Ghosh, S. (2023b). Mechanism of manganese bioleaching. In *Manganese mining microorganisms* (pp. 87–98). Elsevier. https://doi.org/10.1016/B978-0-12-822147-1.00014-5

Das, A. P., & Ghosh, S. (2023c). Microorganisms in manganese biomining processes. In *Manganese mining microorganisms* (pp. 55–61). Elsevier. https://doi.org/10.1016/B978-0-12-822147-1.00013-3

Das, A. P., Ghosh, S., Mohanty, S., & Sukla, L. B. (2015). Consequences of manganese compounds: A review. *Toxicological & Environmental Chemistry, 96*(7), 981–997. https://doi.org/10.1080/02772248.2015.1005428

Das, A. P., Mallavarapu, M., & Ghosh, S. (2023). Ecotoxicity of mining pollutants on the environment and their remediation. *Environmental Chemistry and Ecotoxicology, 5*, 165–167. https://doi.org/10.1016/J.ENCECO.2023.08.002

Das, A. P., & Mishra, S. (2008). Hexavalent chromium (VI): Health hazards and environmental pollutant. *Journal of Environmental Research and Development, 2*(3), 386–392.

Das, A. P., & Singh, S. (2011). Occupational health assessment of chromite toxicity among Indian miners. *Indian Journal of Occupational and Environmental Medicine, 15*(1), 6. https://doi.org/10.4103/0019-5278.82998

Das, A. P., Sukla, L. B., Pradhan, N., & Nayak, S. (2011). Manganese biomining: A review. *Bioresource Technology, 102*(16), 7381–7387. https://doi.org/10.1016/J.BIORTECH.2011.05.018

Denkhaus, E., & Salnikow, K. (2002). Nickel essentiality, toxicity, and carcinogenicity. *Critical Reviews in Oncology, 42*(1), 35–56. https://doi.org/10.1016/S10408428(01)00214-1

Dey, S., Tripathy, B., Kumar, M. S., & Das, A. P. (2023). Ecotoxicological consequences of manganese mining pollutants and their biological remediation. *Environmental Chemistry and Ecotoxicology*. https://doi.org/10.1016/J.ENCECO.2023.01.001

El Machi, A., El Berdai, Y., Mabroum, S., Safhi, A. M., Taha, Y., Benzaazoua, M., & Hakkou, R. (2024). Recycling of mine wastes in the concrete industry: A review. *Buildings, 14*(6), 1508. https://doi.org/10.3390/buildings14061508

Finley, J. W., Penland, J. G., Davis, C. D., & Pettit, R. E. (2003). Dietary Manganese Intake and Type of Lipid Do Not Affect Clinical or Neuropsychological Measures in Healthy Young Women. The Journal of Nutrition, 133(9), 2849–2856. https://doi.org/10.1093/jn/133.9.2849

Ghosh, A., Manna, M. C., Jha, S., Singh, A. K., Misra, S., Srivastava, R. C., Srivastava, P. P., Laik, R., Bhattacharyya, R., Prasad, S. S., Singh, S. P., Singh, S. K., Kumar, V., Tiwari, S., & Singh, A. K. (2022). Impact of soil-water contaminants on tropical agriculture, animal and societal environment. *Advances in Agronomy, 176*, 209–274. https://doi.org/10.1016/bs.agron.2022.07.006

Ghosh, S., Tripathy, B., Dey, S., & Das, A. P. (2023). Comparative investigation of fungal and bacterial manganese biomining mechanisms. In *Manganese mining microorganisms* (pp. 99–115). Elsevier. https://doi.org/10.1016/b978-0-12-822147-1.00016-9

Haghighizadeh, A., Rajabi, O., Nezarat, A., Hajyani, Z., Haghmohammadi, M., Hedayatikhah, S., Asl, S. D., & Aghababai Beni, A. (2024). Comprehensive analysis of heavy metal soil contamination in mining environments: Impacts, monitoring techniques, and remediation strategies. *Arabian Journal of Chemistry, 17*(6), 105777. https://doi.org/10.1016/j.arabjc.2024.105777

Matinde, E. (2018). Mining and metallurgical wastes: A review of recycling and re-use practices. *Journal of the Southern African Institute of Mining and Metallurgy, 118*(8), 5. https://doi.org/10.17159/2411-9717/2018/v118n8a5

Mattison, D., Momoli, F., Alyanak, C., Aschner, M., Baker, M., Cashman, N., Dydak, U., Farhat, N., Guilarte, T., Karyakina, N., Ramoju, S., Shilnikova, N., Taba, P., & Krewski, D. (2024). Diagnosis of manganism and manganese neurotoxicity: A workshop report. *Medicine International, 4*(2), 11. https://doi.org/10.3892/mi.2024.135

Mishra, S., Das, A. P., & Seragadam, P. (2009). Microbial remediation of hexavalent chromium from chromite contaminated mines of Sukinda Valley, Orissa (India). *Journal of Environmental Research and Development, 2009*, 1122–1127.

Mohanty, M., Tripathy, B., Koteswara Reddy, G., & Das, A. P. (2024). Greener technologies for establishing restraint over microplastic pollution. In *Sustainable microbial technology for synthetic and cellulosic microfiber bioremediation* (pp. 239–258). Springer. https://doi.org/10.1007/9783-031-62600-5_13

Mohanty, S., Ghosh, S., Nayak, S., & Das, A. P. (2016). Isolation, identification and screening of manganese solubilizing fungi from low-grade manganese ore deposits. *Geomicrobiological Journal, 34*(4), 309–316. https://doi.org/10.1080/01490451.2016.1189016

Nasution, M. J., Tugiyono, Bakri, S., Setiawan, A., Murhadi, Wulandari, C., & Wahono, E. P. (2024). The impact of increasing nickel production on forest and environment in Indonesia: A review. *Jurnal Sylva Lestari, 12*(3), 549–579. https://doi.org/10.23960/jsl.v12i3.847

Makhathini, T.P., Bwapwa J.K., Mtsweni S., (2023) Various Options for Mining and Metallurgical Waste in the Circular Economy: A Review Sustainability 15(3) 2518. https://doi.org/10.3390/su15032518

Olufemi, A. C., Mji, A., & Mukhola, M. S. (2022). Potential health risks of Lead exposure from early life through later life: Implications for public health education. *International Journal of Environmental Research and Public Health, 19*(23), 16006. https://doi.org/10.3390/ijerph192316006

Prabhakar, A., Mishra, S., & Das, A. P. (2019). Isolation and identification of lead (Pb) solubilizing bacteria from automobile waste and its potential for recovery of lead from end of life waste batteries. *Geomicrobiological Journal, 36*(10), 894–903. https://doi.org/10.1080/01490451.2019.1654044

Priyadarshini, S., Jagatee, S., & Das, A. P. (2024). Synthetic fabrics and microfiber pollution–an assessment of their global impact. *Environmental Science and Engineering, 2333*, 137–157. https://doi.org/10.1007/978-3-031-51792-1_8

Raj, K., & Das, A. P. (2023). Lead pollution: Impact on environment and human health and approach for a sustainable solution. *Environmental Chemistry and Ecotoxicology, 5*, 79–85. https://doi.org/10.1016/J.ENCECO.2023.02.001

Reddy Koteswara, G., Bhargavi, S., Yamuna, D., Anjani Devi, M., Kumar, G., Manas, T., Adarsh, S., Bhavya Sai, V., Nath, H., Alreshidi, M. A., Togun, H., Ram, D. K., & Das, A. P. (2025). Recent advancements in the remediation of marine hydrocarbon and plastic pollutants. *Environmental Science and Engineering, 218*, 127–142. https://doi.org/10.1007/978-3-031-81615-4_7

Reimers, C. E., Alleau, Y., Bauer, J. E., Delaney, J., Girguis, P. R., Schrader, P. S., & Stecher, H. A. (2013). Redox effects on the microbial degradation of refractory organic matter in marine sediments. *Geochimica et Cosmochimica Acta, 121*, 582–598.

Taha Y., Benzaazoua M., (2020) "Towards a Sustainable Management of Mine Wastes: Reprocessing Reuse Revalorization and Repository" Minerals 10(1) 21. https://doi.org/10.3390/min10010021

Wang, Y., Su, H., Gu, Y., Song, X., & Zhao, J. (2017). Carcinogenicity of chromium and chemoprevention: A brief update. *Oncotargets and Therapy, 10*, 4065–4079. https://doi.org/10.2147/OTT.S139262

Zambelli, B., Uversky, V. N., & Ciurli, S. (2016). Nickel impact on human health: An intrinsic disorder perspective. *Biochimica et Biophysica Acta - Proteins and Proteomics, 1864*(12), 1714–1731. https://doi.org/10.1016/j.bbapap.2016.09.008

Impact of Mining Activities on the Terrestrial and Aquatic Environment

Subhadra Rajpoot, Alok Prasad Das, Preeti Singh Bahadur, and Ankita Mathur

1 Introduction and Background

One of the oldest industrial activities in human history, mining has been vital to human civilization and remains foundational here to economic development around the world. Extraction of mineral resources supplies important raw materials for infrastructure, technology, energy production, and many consumer products that shape contemporary living. The copper wiring in our homes or the rare earth elements in our smartphones—everything in modern society requires products extracted from mining operations. Mining is fundamental to worldwide financial development, especially in resource-rich developing nations where it has made a formative effect and frequently makes up a huge part of the GDP and export income (Bridge, 2004).

Mining, despite its economic significance, is one of the most harmful anthropogenic activities on Earth. Extracting minerals from the earth is a process that necessarily entails widespread disturbance of the landscape by the removal of overburden, alteration of ecosystems in the area and production of large waste streams that can impact both terrestrial and aquatic ecosystems decades or even centuries after mining activity has ended (Dudka & Adriano, 1997). These environmental impacts happen along the entire mining lifecycle, from exploration and site development through operational extraction and processing to eventual closure and post-closure phases (Das, Behera, & Das, 2024; Das & Ghosh, 2023; Priyadarshini et al., 2024).

S. Rajpoot (✉) · P. S. Bahadur · A. Mathur
Amity University, Greater Noida, India
e-mail: srajpoot@gn.amity.edu

A. P. Das
Department of Lifesciences, Rama Devi Women's University, Bhubaneswar, Odisha, India

© The Author(s), under exclusive license to Springer Nature Switzerland AG 2026

A. P. Das (ed.), *Bioremediation Technology for Sustainable Mining Waste Management*, Sustainable Environmental Waste Management Strategies,
https://doi.org/10.1007/978-3-032-03468-7_4

The link between mining and environmental degradation has deep historical roots. Mining operations during antiquity of several thousand years ago were usually microscopic scale and had a limited environmental impact (Das, Dey, Bala, et al., 2024; Mohanty, Dey, et al., 2024; Sahu, Dey, et al., 2024). Ancient Romans did develop hydraulic mining methods to extract gold—techniques which were innovative perhaps for their time—but which caused localized environmental destruction still noticeable today in some areas of Spain (Martínez-Cortizas et al., 2013). But it was the Industrial Revolution that changed the game for mining, when mechanization allowed for unprecedented volumes of coal and metal ores to be extracted, fuelling expanding economies.

Much of the twentieth century saw mining technology and scale rapidly evolve, giving rise to huge open-pit operations capable of shifting hundreds of thousands of tons of material each day. This was followed by the introduction of chemical processing methods to mining, in particular cyanide leaching of gold and copper in the 1970s (Akcil, 2006), which gave mining an additional environmental footprint. Efforts to accommodate declining ore grades in recent decades have included processing increasingly large quantities of substrate to yield the same quantity of target minerals, with corresponding increases in waste generation and resource input to this process (Bala, Anshul, et al., 2024; Das, Dey, & Das, 2024; Das, Ray, Mohanty, et al., 2024; John & Das, 2024; Mudd, 2010).

Today, mining takes place all over the world, with major operations on every continent except Antarctica. Mining activities are distributed according to the geological availability of resources but are also influenced by economic, political, and historical conditions. Countries such as Australia, Canada, China, Russia, South Africa, and the USA all feature sizeable mining sectors, while mining activity in emerging economies in Africa, Asia, and Latin America has grown rapidly over the past few decades (Bridge, 2008). Each mining region presents unique environmental challenges based on its specific geological, ecological, and climatic conditions.

The environmental impacts of mining are diverse and go far beyond the obvious devastation of landscapes. These include soil and water pollution, habitat destruction, loss of biodiversity, air pollution, and alteration of the hydrological systems (Das, Dey, Bala, et al., 2024; Mohanty, Tripathy, et al., 2024; Ray et al., 2024; Sahu, Dey, et al., 2024). The notorious acid mine drainage, where exposed sulphide minerals are oxidized to create acid-bearing waters rich in dissolved metals, is among the most persistent and damaging environmental legacies of certain mining operations (Johnson & Hallberg, 2005). In aquatic systems, sediments from mining loads can smother stream habitats, and toxic contaminants can bioaccumulate through food chains, with effects felt far from the mine (Younger et al., 2002).

The environmental impacts of mining can vary significantly based on the type of mining method employed, which mineral is being extracted, local environmental conditions, and the regulatory framework under which the operation is being undertaken. Open-pit mining, for instance, generally causes more surface water disturbance than underground methods but may generate less contaminated drainage in certain geological environments (Bala, Dogra, et al., 2024; Das, Dey, Bala, et al., 2024; Dey, Bissoyi, Koteswara Reddy, & Das, 2024; Dey, Dash, Ray, et al., 2024).

The environmental challenges associated with gold mining by means of mercury amalgamation are distinct from those of coal extraction or uranium mining, with each commodity presenting unique potential environmental impacts (Lottermoser, 2010).

Awareness of these environmental impacts has led to major shifts in mining practices over the last few decades. In many countries and regions, it has had more severe environmental impacts, which were ironed out in terms of stricter regulatory frameworks, better extraction technologies, and increasing focus on corporate social responsibility. Within industry, terms like "sustainable mining" and "responsible resource development" have gained traction, but many critics question whether extractive activities that consume non-renewable resources can ever be truly sustainable (Whitmore, 2006).

From mining to mines: a review of the impacts mining has on terrestrial and aquatic settings. It opens with a categorization of different kinds of mining practices and their distinctive footprints on the environment, then examines what those practices do in detail to terrestrial and aquatic ecosystems. The chapter highlights the real-world manifestations through case studies from various regions and types of mining. It goes on to discuss mitigation strategies and sustainable practices that can help reduce environmental damage, and it ends with a look at emerging technologies and approaches that could change the environmental imprint of the mining business in the coming decades.

Through this critical analysis of the environmental impacts of mineral extraction, this chapter contributes to the important discussion about balancing resource demand with environmental protection, a question that is often neglected in the face of increasing mineral extraction in the face of climate change and biodiversity crisis.

2 Types of Mining Activities and Their Characteristics

These mining activities differ widely in their methods, scale, and environmental footprint, informed by the target mineral, deposit characteristics, geological settings, and technological capabilities. These different approaches are critical to understanding their individual ecological burden on land and water systems. This part looks at the common types of mining operations and how they differ from one another.

2.1 Surface Mining Methods

Surface mining, which is characterized by the removal of overlying soil and rock (overburden) covering mineral deposits, is the most visibly transformative mining method. These techniques are economical when the deposits are located relatively near to the surface and usually permit higher recoveries compared to underground methods (Hartman & Mutmansky, 2002).

The most widely used surface mining process, known as open-pit mining, produces a vast pit that can be kilometers wide and hundreds of meters deep. This involves building a series of benches that twist around the orebody and go down as progressively deeper sections of the orebody are accessed with heavy machines. Open-pit operations produce large quantities of waste rock and often include large-scale dewatering systems capable of modifying local hydrology (Hustrulid et al., 2013).

Pits (or strip mines), like what is used for coal and some metal ores, remove overburden in strips, extract the target mineral, then move the overburden back into the pit and to the next strip as operations continue. Although this approach does enable some concurrent reclamation, it greatly disrupts soil profiles and ecosystems (Das, Ray, Mohanty, et al., 2024; Dey, Dash, & Das, 2024; Priyadarshini & Das, 2024). The more controversial variant of this method, referred to as mountain-top removal mining, involves permanent alteration of the topography where overburden is deposited into neighboring valleys, widely burying streams and headwaters when utilized on mountainous terrain (Palmer et al., 2010).

Quarrying, which is used mainly to produce dimension stone, aggregates, and industrial minerals such as limestone, employs techniques like open-pit mining, but generally occurs at shallower depths. Despite the environmental footprint of quarries being less widespread than larger scale mining operations, quarries can substantially alter the esthetics of a landscape and produce significant dust emissions (Langer, 2001).

2.2 *Underground Mining Methods*

When deposits are too deep to extract economically using surface mining, underground mining becomes necessary. These applications preserve most of the overburden and provide access to the orebody by construction of access shafts or adits (horizontal entrances). Though less waste rock is produced through underground mining compared with surface methods (i.e., canyoning), it has its environmental impacts, mainly in terms of mine water and subsidence.

Room-and-pillar mining—used in coal and some hard rock operations—removes material but keeps "pillars" of the orebody in place to help hold up the overlying rock. Selective mining is generally possible with this method, but can result in surface subsidence in the event of pillar failure or retreat mining, where pillars are left to collapse (Mark & Barton, 1997).

Longwall mining, a method most often associated with coal extraction, uses a mechanized shearer to remove mineral material from along a long wall face and allows the roof to collapse in a controlled manner behind the advancing equipment. This approach allows for greater recovery of the resource but also purposefully induces subsidence, potentially changing surface topography and altering groundwater flow paths (Booth, 2006).

It involves creating an opening beneath a large orebody and letting it collapse under its own weight into an underground haulage level to be collected and processed subsequently, and is used in large, low-grade ore deposits known as block caving. Although this brings low-grade deposits into economically viable production, it causes substantial surface subsidence, and possibly seismic activity (Laubscher, 1994).

Cut-and-fill and other selective forms of mining can be used to access higher-grade sections of deposits, generating less waste, but are often more complex, amenity-intensive operations. Hydraulic backfilling of mined-out areas with tailings or waste rock—one of the approaches that could help reduce surface disposal requirements—can also contaminate groundwater, if not properly contained (e.g., Rankine et al., 2006).

2.3 *Placer Mining and Alluvial Extraction*

Placer mining exploits minerals that have been concentrated by deposition in alluvial deposits, generally in stream beds or ancient river channels. Historically linked with gold rushes, which includes artisanal panning and large-scale dredging operations. Although effects from small-scale, individualized operations may be negligible, landscape, placer mining can heavily impact riparian hydrology through physical disturbance of stream morphology and increased turbidity (Hilson, 2002). Hydraulic mining, one of the most destructive forms of placer mining in history, erodes mineral-bearing sediments through the use of high-pressure jets of water. This method, widely used during the California Gold Rush, resulted in monumental downstream sedimentation with implications for river systems for decades to come. Despite their large environmental footprint (Hunerlach et al., 1999), modern variants are still used in some areas.

Dredge mining uses mobile floating machinery to remove submerged mineralized sediments, which are processed on the mobile platform and the tailings dumped behind the moving dredge. This technique is still employed for minerals including tin, titanium and gold, with subsequent environmental effects primarily related to aquatic habitat disruption and potential mobilization of heavy metals that were previously sequestered in sediments (Wolfe et al., 2000).

2.4 *In-Situ Mining and Solution Methods*

Another option is in situ recovery methods, which extract minerals without removing the ore-bearing rock. These processes have instead involved injecting leaching solutions (in most cases acids or sodium bicarbonate) into the deposit via a series of wells, dissolving the minerals of value and pumping the pregnant solution to the surface for processing. Most often used in uranium and copper mining, in situ approaches reduce surface disturbances and waste rock generation but pose a risk of groundwater contamination if the containment systems fail (Mudd, 2001). Alternately, heap leaching is not technically an in situ type method but utilizes similar as a solution chemistry to solubilize minerals as described above. This technique consists of piling crushed ore on concrete-lined pads, spraying it with leaching solutions (cyanide in the case of gold and silver and sulfuric acid in the case of copper) and collecting the solution containing the metals for further processing (Das et al., 2012; Das & Ghosh, 2018, 2023a, 2023b; Priyadarsini & Das, 2025). Heap leaching also consumes less energy than conventional processing, but results in leakage of solutions and complicates pad reclamation (Kappes, 2002).

3 Mineral Processing and Waste Management

Not only extracting the minerals but also metallurgical operations and waste disposal are also part of mining environmental footprint. Ore is processed via crushing, grinding, and various physical or chemical separation processes that concentrate the valuable mineral fraction. These processes require huge amounts of energy and water and produce fine-grained waste byproducts (tailings) that need to be stored in engineered impoundment areas or backfilled (Lottermoser, 2010).

Tailings storage facilities are among the most serious long-term environmental liabilities of mining. These structures include finely ground waste rock, process water that might carry residual processing chemicals, and a high concentration of potentially hazardous elements. The failure of tailings dams such as the catastrophic breaching of Brumadinho, Brazil in 2019 and Mount Polley, Canada in 2014 causes both catastrophic environmental destruction and loss of human life (Roche et al., 2017).

Waste rock dumps, or rock that is excavated to get to the orebody but is too poor to be processed, are the largest volume waste stream from most mining operations. If not managed, these facilities can produce acid mine drainage, a process in which oxidation of sulfide minerals in the presence of water leads to acidification and leaching of metals, a process that can persist for centuries (Blowes et al., 2003).

3.1 *Comparative Environmental Footprints*

The environmental impacts of different methods of mining vary greatly in type, geographical scope, and longevity. Surface mining, by its nature, tends to make the most noticeable changes to the landscape and affect larger selections than underground approaches. Conversely, underground mining has the potential to create more significant hydrological impacts due to dewatering, and may create a long-term legacy of contaminated mine water that requires ongoing treatment (Nordstrom, 2011).

The latter may also minimize surface disturbance and waste generation, but impose a distinct set of challenges for tracking and addressing their potential groundwater impacts. Placer mining also generates unique effects on riparian and aquatic systems that can extend long after the cessation of operations (Macklin et al., 2006).

Geological and economic factors influence the selection of a certain mining method, while environmental issues are not very important opтurable in mining method selection. But increasing regulatory pressures and social expectations started to shape mining practices, with companies beginning to include environmental performance in mine planning and technology choice (Franks et al., 2011).

As will be discussed in detail in later sections of this chapter, each mining method carries different environmental impact mechanisms requiring tailored clubs or other institutional arrangements. This knowledge on unique traits offers a basis for evaluating the impacts of mining activities on terrestrial and aquatic ecosystems and identifying effective measures to reduce its environmental footprint.

3.2 Scale and Global Distribution of Mining Methods

Mining interest varies by region (based on geological settings and economic development), which reflects in the mining method's frequency of use as well. Open-pit mining prevails in countries with economies of scale and capital-intensive equipment as seen in Australia's iron ore industry in the Pilbara region or Chile's huge copper pits in the Atacama. Such operations comprise some of the largest human-made voids on the planet, with the Kennecott Bingham Canyon open- pit mine in Utah spanning almost 8 km^2 and extending more than 1.2 km deep (Ericsson & Löf, 2019). In areas with deep mineral deposits and a long history of mining, in particular, such as South Africa's gold fields and China's coal sector, underground mining continues to hold sway. The technologically demanding and high-cost nature of underground operations has necessitated innovations in automation and remote operation driven by application of mechanization in high-wage economies where principles of safety and labor costs have favored mechanization (Gao et al., 2021). Artisanal and small-scale mining (ASM) is a notable, but frequently underreported, sector, especially in developing countries. It is estimated that over 40 million people globally participate in ASM activities, often making use of rudimentary methods that can lead to stratospheric levels of environmental damage relative to the production volumes. This relation is predominant in gold, gemstone and tin extraction in many regions of Africa, South America, and Southeast Asia, where weak regulations often serve to increase environmental hazards (Hilson & Maconachie, 2020).

4 Emerging Extraction Technologies

This is how recent technological advances are starting to transform traditional mining practices, which in turn could alter their environmental footprints. There is a growing trend in the industry for continuous mining systems that combine extraction, processing, and waste disposal. These systems aim to reduce material handling and energy consumption while enabling more accurate resource targeting (Basu & Mishra, 2020). Many large-scale operations are deploying autonomous and semi-autonomous equipment including haul trucks, drills and loading machines. Although safety and efficiency considerations are the primary drivers of these technologies, their indirect benefits, by way of more accurate operations from greater precision and reduced fuel consumption, can affect environmental performance (Bellamy & Pravica, 2011). In-pit sensing and sorting represent preconcentration techniques that can reject waste in preceding energy-consuming stages of processing. These strategies minimize the amount of material that needs to be transported and processed, and can result in lower energy, water use, and waste generation. In recent years, there has been great development of X-ray, optical and sensor-based sorting technologies, which are now becoming applicable to a broader range of minerals (Lessard et al., 2016).

Alternative lixiviants, including thiosulfate, glycine, and several biological agents, may provide new methods for leaching gold and other metals compared to traditional cyanide and acid leaching approaches. This may lead to less toxic processing agents with higher recovery rates of metals from challenging ores. In a similar fashion, advances in biohydrometallurgy harness microbial processes to extract metals, with their perhaps lower environmental footprints compared to traditional approaches (Johnson et al., 2013).

5 Environmental Implications and Transition

The chosen mining methods determine the character, size, and duration of environmental impacts. But at the end of the day, technological advances are just tweaks that save us some effort and possibly mitigate some impacts—though the reality is that pulling minerals from the Earth inevitably tears up natural systems. The demand for minerals is increasing worldwide—and will continue to do so with population growth, urbanization, and the diversion to the renewable energy technologies (Ali et al., 2017)—and if we do not change our approaches and technologies, we will have an increased scale of environmental problems associated with mining. Mined materials have environmental footprints that reach far beyond the sight of extraction, including their processing facilities, transportation infrastructure, and disposal sites. To evaluate accurate environmental impact of mining operations, it is important to have the insight about the whole lifecycle. Different mining techniques have characteristic impact pathways and terrestrial and aquatic mechanisms that will be explored in detail in further sections (Northey et al., 2018).

Sustainability and environmental performance are increasingly being applied as guiding principles to current commercial practices in mining as the way of doing mining progresses. A novel emerging paradigm, precision mining—specifically, the selective targeting of valuable orebodies to the detriment of reducing waste—could, if embraced as a general approach, decrease the environmental footprint associated with mineral resource development (Das & Ghosh, 2023c; Dey et al., 2023; Ghosh & Das, 2017; Ghosh, Tripathy, et al., 2023; Mohanty et al., 2017). Likewise, growing attention to mine closure planning that begins at the inception of a mining project is indicative of an increasing awareness of the long-term environmental legacy of mining (Laurence, 2020). The next sections of this chapter will unfold the ways in which the mining methods and processes described here represent a source of specific terrestrial and aquatic impacts, offering a basis for observing both contemporary challenges and routes toward environmentally less-destructive mineral production.

6 Impacts on Terrestrial Ecosystems

Mining impacts drastically change the physical, chemical, and biological attributes of the land in which it occurred with repercussions that can continue long after extraction has ended. These impacts differ in severity and scope influenced by the mining method used, the local environmental conditions and management practices (Priyadarshini & Das, 2024; Sahu, Mallick, & Das, 2024). Here, we explore the broad spectrum of mining impacts on terrestrial ecosystems, from the obvious destruction of habitats due to mining activities to more insidious biochemical impacts that influence ecological drivers.

6.1 Land Use Change and Habitat Destruction

One of the most apparent and immediate impact of mining on terrestrial ecosystems is the complete transformation of land from its natural state to an industrial landscape. According to Palmer et al. (2010), vegetation and soil are completely stripped from surface mining operations, especially in open- pit mines and mountaintop removal sites, resulting in a loss of the original habitat in the area of the mine. Even underground mining needs surface infrastructure, access roads, processing facilities and waste storage areas that lead to habitat loss, though normally on a smaller scale than surface operations.

The amount of direct land disruption caused by mining varies hugely depending on commodity and extraction method. In particular, coal mining, specifically through mountaintop removal, has altered over 500,000 hectares of mainly forested habitat in areas like Appalachia, replacing diverse temperate hardwood communities with monoculture reclaimed grasslands (Wickham et al., 2013). Oil sand mining in Alberta, Canada has disturbed more than 85,000 hectares of boreal forest and peatland ecosystems, with projections that up to 475,000 hectares could be disturbed as operations expand (Rooney et al., 2012).

Outside the direct footprint of the mine, terrestrial habitats are fragmented by associated infrastructure such as transport corridors, power lines, pipelines, and processing facilities. This fragmentation induces edge effects, movement barriers, and decreases habitat connectivity, potentially impacting population viability for many species, especially those whose home ranges are large or those reliant on particular migratory corridors (Sonter et al., 2014). In the Brazilian Amazon, for instance, the deforestation associated with mining does not stop at the borders of mining sites —studies have shown that the indirect impact of mining can be felt up to 12 times as far beyond the mine as the direct impact (Sonter et al., 2017).

The severity of habitat destruction consequences depends in part on ecological context. Mining in biodiversity hotspots and areas of rare ecosystem types is one of the most significant concerns. For example, the mining that takes place in Madagascar's eastern rainforests is threatening to wipe out many endemic species found nowhere else on Earth, and extraction activities in arid environments can destroy biological soil crusts and desert ecosystems that may take hundreds of years to recover (Raiter et al., 2014).

6.2 Soil Degradation and Erosion

Soil systems underpinning terrestrial ecosystems are fundamentally disrupted by mining. Surface mining activities remove topsoil, which has the most organic matter, nutrients, and biological activity present. Even if topsoil is preserved for future use in reclaiming the mined area, stockpiling can alter its physical, chemical, and biological quality, resulting in compaction and nutrient leaching, diminished microbial activity, etc. (Sheoran et al., 2010).

Soil erosion is one of the key impacts of soil disturbance caused by mining. Erosional processes affect exposed soil surfaces, waste rock piles, tailings, and haul roads, especially in areas with high rainfall intensities or steep topographies. Sediment yield data from disturbed mine lands in different mining regions exceed background sediment yields by several orders of magnitude (Kuter, 2013). This erosion removes soil resources from the mine site and leads to off-site effects through sedimentation of adjacent landscapes and water bodies.

That mixing and removal of soil horizons during mining disrupts natural pedogenic processes that unfold over millennia and generates novel substrates with properties that may hinder ecological recovery (Das et al., 2023; Ghosh, Sinha, et al., 2023; Mallick et al., 2024). A common problem with post-mining soil includes poor physical structure of the massive soil, lower organic matter level and enhanced pH along with nutrient deficiencies or imbalances and reduced water holding capacity (Macdonald et al., 2015). The changed soil characteristics influence vegetation establishment and community development, often requiring high levels of reclamation efforts to restore ecosystem performability.

In several mining contexts (especially regarding metal extraction), contaminating soils with potentially toxic elements means another stressor for terrestrial ecosystems. Globally, studies of mining-affected regions have reported of enriched concentrations of elements, including arsenic, cadmium, copper, lead, and zinc, in soils surrounding active and abandoned operations (Kříbek et al., 2014). The uptake happens through several exposure routes, including dust deposition, surface runoff, and intentional or unintentional disposal of mine wastes. The soil is a decisive space and although diverse in its microorganisms, it can also become an extremal environment that disrupts the variation of ecological connections, as well as nutrient assimilation and biochemical cycles.

6.3 Impacts on Flora and Vegetation Communities

Direct removal, changes to environmental conditions, and chemical stressors resulting from mining activities can adversely impact plant communities. The clearest impact is the total removal of vegetation within the footprint of the mine, but impacts are wider-reaching than that. Dust can coat leaf surfaces and hinder photosynthetic efficiency and gas exchange and changing hydrology may stress vegetation that relies on specific groundwater conditions (Farmer, 1993).

Reclamation efforts often result in novel plant communities markedly different from the pre-mining ecosystem (Das et al., 2025; Swain et al., 2025). Indeed, a global assessment of revegetation outcomes on mined lands noted that, even decades after reclamation, re-established plant communities on reclaimed sites were also less diverse and structurally different from reference ecosystems (Zipper et al., 2011). This transformation can be especially pronounced where non-native species have been introduced (often as part of reclamation), an all too common practice in many mining regions. Reclaimed coal mines in the western United States, for instance, are typically dominated by introduced grasses rather than the native shrub communities that existed prior to the mines' opening (Holl, 2002).

Physiological and ecological responses of plant communities on and around the mine sites can be expected at elevated levels of soil contamination. In such contaminated environments, metal-tolerant ecotypes can arise and metal-sensitive species can go extinct, causing shifts in community composition and decreases in diversity (Ginocchio & Baker, 2004). The causes are due not only to the retention of metals through earth processes, but also to the fact that the residual effects of heavy metal contamination from anthropogenic sources are further augmented by addition of heavy metal-bearing pathway products, that would through natural weathering of rocks lead to contamination of soil and overlying vegetation for decades and centuries after mining has stopped (Banasova et al., 2006).

Mining-affected regions are particularly at risk of losses of rare or endemic plant species. The most terrifying fate of mining activities is that many plant species are threatened with extinction worldwide (Murguía et al., 2016), and that also has been observed in biodiversity hotspots in countries where mineral resources and specific flora often go hand in hand. From a conservation point of view, the loss of unique habitats, like serpentine soils that nurture flora that evolved under unique geochemical conditions, is a serious problem.

6.4 Impacts on Fauna and Wildlife

Mining intensifies impacts on terrestrial fauna through habitat loss, fragmentation, disturbance, and multiple forms of pollution. Although research has so far been concentrated in a limited number of regions, empirical evidence demonstrates that large-scale mining operations can severely degrade critical habitats for threatened species. For example, bauxite mining in the Jarrah forests of Western Australia has been shown to affect forest-dependent mammals (Cristescu et al., 2016), while coal mining in Canada's boreal region has negatively impacted woodland caribou populations (Hebblewhite, 2017).

Taxonomic group and ecological characteristics drive wildlife responses to mining disturbance However, large mammals are known to use avoidance behavior to respond to mining activities leading to reduced abundance and changes in movement patterns found in sites near active operations. Mule deer, for example, avoided active coal mining areas in Wyoming but gradually recolonized reclaimed sites once vegetation became established (Hebblewhite, 2008). On the other hand, certain opportunistic species can take advantage of anthropogenic conditions or our presence that can change the composition of the community and potentially create cascading effects through the ecological network.

Mining operations—from blasting and heavy machinery to processing facilities—can generate noise pollution that disrupts animal behavior and communication. Anthropogenic noise can disrupt predator–prey interactions, mating, and territory defense in many taxa and has the capacity to reduce reproductive success and increase vulnerability to predation (Patricelli & Blickley, 2006). Likewise, the use of artificial lighting representative of 24-h mining activities can disturb crepuscular taxa and affect predator–prey interactions.

Contamination of terrestrial food webs through bioaccumulation of contaminants represents a more insidious but potentially concerning impact of mining on wildlife. In case of various animal tissues gathered from the mining-contaminated areas across the globe, increased levels of metals and other pollutants associated with mining have been reported. Studies near lead-zinc mining sites in Oklahoma, for example, identified tissue lead concentrations, including blood lead that had the potential to affect neurological function and reproductive success in songbirds (Beyer et al., 2013). In a similar manner, studies of small mammals inhabiting copper mining areas have recorded bioaccumulation of arsenic and other elements with a possible impact on population health and ecosystem functioning (Kozlov et al., 2009).

6.5 Air Quality and Atmospheric Deposition

Mining activities produce air pollutants that contribute to terrestrial ecosystems such as particulate matter, nitrogen oxides, sulfur dioxide, and greenhouse gases. Dust pollutants created by blasting and excavation activities, crushing operations, transport, and other operations forming the obvious aspect of air pollution (Petavratzi et al., 2005).

Dust hitting vegetation can impact photosynthetic capacity, erode cuticles, pore blockage and modify leaf surface temperature regimes. Varying growth rates and physiological processes, as well as shifts in community composition, have been reported in studies conducted in mining regions in relation to the effects of dust (Farmer, 1993). Dusts containing metals alter the ecology of an area, and dusts lose meta from one environment and deposit rich minerals in another, it is part of the contamination cycle and also is potentially harmful.

Regional air quality is also compromised due to the emissions (like smelting, refining, and fossil fuel combustion in heavy equipment) released by mining operations. This damage has been in part a consequence of sulfur dioxide emissions from metal smelting, which have caused direct phytotoxicity and soil acidification, as demonstrated by the formation of regional "barren zones" around smelters (e.g., in Sudbury, Canada, and Ducktown, Tennessee; Kozlov & Zvereva, 2007). Enhanced technologies have led to reduced emission rates in several regions, but legacy effects still influence historically impacted areas, and poorly regulated operations continue to have major effects in certain areas of the world.

6.6 Changes in Topography and Hydrology

Mining transforms the topography, producing new landforms including pits, trenches, waste rock piles, tailings impoundments, and subsidence features. The resulting transformed landforms also affect a number of different ecological processes including, for example, patterns of surface water flow, erosion and sedimentation dynamics, and microclimatic conditions (Hancock et al., 2020).

Hydrological alteration is one of the most important mechanisms through which mining has an impact on terrestrial ecosystems. With surface mining, dewatering is often needed to produce dry working conditions, leading to a lowering of the water table and reduced groundwater availability for deep-rooted vegetation. In mining regions around the world, changes in plant community composition have been recorded in connection with varying groundwater availability, including that drought-sensitive species decline as water tables decrease (Dang et al., 2002).

Mining underground can cause the surface to subside, creating depressions, changing areas of nourishing and sometimes creating wetlands in once terrestrial areas. Longwall mining in the Illinois Coal Basin, for example, has led to the development of >10,000 ha of wetlands in previous farmed landscapes due to subsidence (Kundu & Ghose, 2014), fundamentally altering ecosystem structure and functioning.

Artificial islands with different (to natural terrain) properties form new ecological conditions and alter succession pathways. In many cases, waste rock dumps and tailings facilities have unique physical and chemical properties that inhibit the establishment and growth of plants, including poor water holding capacity, extreme temperatures, low levels of nutrients, and potentially phytotoxic concentrations of metals or other contaminants (Tordoff et al., 2000). Under these conditions novel ecosystems form that are not found anywhere in nature, challenging reclamation and ecological restoration while often offering new opportunities.

6.7 *Cumulative and Landscape-Level Effects*

Although the ecological impacts of single mining operations are majorly local, the aggregated effects of mining activities in a given landscape can have emergent impacts on ecosystem structure and functioning. In high mining states, such as the Appalachian coalfields, Brazil's Iron Quadrangle, or Western Australia's Pilbara region, the cumulative footprint of many sites can significantly alter regional ecosystems (Wickham et al., 2013).

The effects of land-cover change from so many mining operations and associated infrastructure can also cross a threshold, meaning the ecological connectivity that allows for the movement of certain species is impeded and can lead to population isolation and gene flow reduction. Movement rates have been shown to be lower across fragmented wildlife habitats in mining-affected landscapes (exercise measures as exercise potential; Hebblewhite, 2017, and references therein), potentially affecting population viability and resilience to environmental change.

Mining landscapes suggests that extraction processes produce multiple geographical terrains with unique dimensions of interaction among industrial elements, transformed biological regimes and social arrangements (Bridge, 2004). Even after mining has stopped, the landscapes continue to transform through processes of ecological succession, geomorphogy, and human uses of the areas. The extent, rate, and pattern of these changes may ultimately inform best practices for landscape rehabilitation, and appropriate choices for post-mining land uses.

Recent developments in landscape ecology and ecosystem service frameworks improve our understanding of the effects of mining-induced changes on wider terrestrial ecosystem functioning. For example, mining often impacts a range of regulating services (such as carbon sequestration, water purification, and erosion control) as well as habitat services that support biodiversity (Sonter et al., 2018). These impacts have direct consequences beyond the mining footprint, altering regional ecosystem dynamics via disrupted ecological processes and connections.

6.8 Rehabilitation Challenges and Ecosystem Recovery

The extent to which terrestrial ecosystems recover from mining disturbance depends on many factors, including the intensity of initial impacts, time since disturbance, and post-disturbance reclamation practices. While natural recovery processes may take place, these are often slow and will also tend toward alternative stable states rather than a return to pre-mining conditions, especially where severely impacted soil or water quality affect the ecosystem (Zipper et al., 2011).

Interventions for active rehabilitation can spur the recovery of ecosystems, yet they struggle with multiple obstacles. Shifting ecological baselines, climate change, and mining disturbance generate novel conditions in disturbed areas, making the establishment of suitable reference conditions to rectify these areas difficult (Cooke & Johnson, 2002). Translating these principles into practice is limited in many mining contexts by technical constraints, such as the challenges of reconstructing complex soil profiles or restoring hydrological functions.

Despite these challenges, many successful examples of ecosystem rehabilitation in the wake of mining disturbance exist that can inform effective approaches. In the jarrah forest ecosystems after bauxite mining, for example, very high levels of biodiversity return have been achieved because rehabilitation work has included careful soil handling, native species reintroduction, and attention to structural complexity (Koch, 2007). The restoration of severely affected landscapes surrounding the nickel smelters near Sudbury by using soil amendments, revegetation, and watershed restoration exemplifies the habitat rehabilitation potential even after extreme degradation (Kozlov & Zvereva, 2007).

7 Impacts on Aquatic Ecosystems

Mining operations have significant effects on aquatic ecosystems by various pathways with impacts which often extend well beyond the mined site and last long after the cessation of mining. As receiving environments for contaminants, water bodies are especially affected by land disturbances and are further connected such that pollutants can be transported downstream. This section focuses on the multiple pathways in which mining can affect aquatic ecosystems, from simple changes in water physical characteristics to biogeochemical processes that can change water quality and aquatic communities.

7.1 Water Quality Degradation and Contamination

Mining activities release a variety of pollutants to aquatic systems by different pathways such as direct discharge, seepage from waste facilities, surface runoff and atmospheric deposition. The particular contaminants resulting from mining operations are specific to the ore mineralogy, extraction methods, and processing techniques used, yet generally comprise of metals, metalloids, processing chemicals, and other (combinations) of compounds that can occur as products of weathering reactions (Younger et al., 2002).

Heavy metal contamination is one of the most severe and long-lasting effects of mining on aquatic ecosystems. Elements such as arsenic, cadmium, copper, lead, mercury, and zinc are commonly associated with metal mining operations and can reach concentrations orders of magnitude above background levels in affected water bodies. According to a worldwide surveying of rivers related to mining activities, concentrations of metals often surpass ecotoxicological threshold values and can remain elevated for centuries following the cessation of mining (Hudson-Edwards, 2016). The River Tinto in Spain, for example, is still severely contaminated with metals and highly acidic because of mining that dates back to Roman times.

The mobility and toxicity of metals in aquatic systems vary and depend on a number of factors, especially water chemistry (pH and redox conditions). In addition, under the acidic conditions produced by sulfide mineral oxidation, many metals become more soluble and bioavailable, thereby increasing their potential toxicity to the aquatic community. Mining sites worldwide are known to contain a high amount of metal contamination, e.g., copper and arsenic in the Clark Fork River (Montana, USA), lead and zinc in the Coeur d'Alene River (Idaho, USA), and mercury in the Madeira River basin (Amazon region), with significant but different ecological effects (Luoma & Rainbow, 2008).

The processing of chemicals introduced into aquatic environments through spills, leaks, or incomplete recovery are another source of mining-related contamination. Cyanide, which is used in gold processing, has led to numerous catastrophic fish kills due to accidental releases, like the Baia Mare spill in 2000 in Romania, which poisoned the Danube River system. Likewise, acids, flocculants, flotation reagents as well as other chemicals applied in mineral processing may negatively influence aquatic organisms when let out to water (Kefeni et al., 2017).

7.2 *Acid Mine Drainage: Formation and Ecological Impacts*

Acid mine drainage (AMD) represents one of the most severe and persistent threats to aquatic ecosystems in mining regions. This phenomenon occurs when sulfide minerals, particularly pyrite (FeS_2), are exposed to oxygen and water during mining activities, initiating a series of oxidation reactions that generate acidity and release metals. The resulting solutions can exhibit pH values below 3 and contain elevated concentrations of iron, aluminum, manganese, and various trace metals (Blowes et al., 2014).

The formation of AMD involves several key reactions, beginning with the oxidation of pyrite to produce ferrous iron, sulfate, and hydrogen ions:

$$FeS_2 + 7/2O_2 + H_2O \rightarrow Fe^{2+} + 2SO_4^{2-} + 2H^+$$

Under oxidizing conditions, ferrous iron is further oxidized to ferric iron:

$$Fe^{2+} + 1/4O_2 + H^+ \rightarrow Fe^{3+} + 1/2H_2O$$

Ferric iron then precipitates as iron hydroxide (creating the characteristic orange-red precipitates seen in AMD-affected streams) and generates additional acidity:

$$Fe^{3+} + 3H_2O \rightarrow Fe\left(OH\right)_3 + 3H^+$$

Furthermore, ferric iron can directly oxidize additional pyrite, accelerating acid generation in a self-perpetuating cycle:

$$FeS_2 + 14Fe^{3+} + 8H_2O \rightarrow 15Fe^{2+} + 2SO_4^{2+} + 16H^+$$

These reactions can persist for decades or even centuries after mining has ceased, creating a long-term source of aquatic contamination. The Iberian Pyrite Belt in Spain and Portugal, for example, continues to generate acid drainage nearly 5000 years after the initiation of mining activities, demonstrating the remarkable persistence of this phenomenon (Delgado et al., 2012).

The ecological impacts of AMD on aquatic ecosystems are severe and multifaceted. Extreme acidity directly affects the physiology of aquatic organisms, disrupting ion regulation, respiration, and reproduction. Many AMD-affected streams exhibit drastically reduced biodiversity, with only a few specialized acid-tolerant species persisting in the most severely impacted reaches. Studies in AMD-affected systems worldwide have documented the elimination of sensitive taxa, including most fish, amphibians, mollusks, and crustaceans, with communities often reduced to a few tolerant insects and microorganisms (DeNicola & Stapleton, 2002).

Metal toxicity compounds the direct effects of acidity, as many metals exhibit increased solubility and bioavailability under acidic conditions. The specific mechanisms of metal toxicity vary by element and organism but commonly include disruption of enzyme function, interference with ion transport, and oxidative stress. Synergistic interactions between multiple stressors—including acidity, metals, and physical habitat degradation—often produce more severe ecological impacts than would be predicted from individual factors alone (Luoma & Rainbow, 2008).

The characteristic metal hydroxide precipitates associated with AMD, particularly iron oxyhydroxides, create additional physical stressors for aquatic ecosystems. These precipitates coat stream substrates, eliminating interstitial spaces used by benthic organisms and reducing habitat heterogeneity. They can physically clog the respiratory surfaces of aquatic organisms and prevent attachment of periphyton, disrupting both habitat structure and food web dynamics (Hogsden & Harding, 2012).

Recovery of AMD-impacted systems requires addressing both acidity and metal contamination, often necessitating long-term active treatment or extensive passive remediation systems. Natural recovery can occur as acid-generating materials are depleted or isolated from oxygen, but this process typically operates on timescales of decades to centuries in severely affected systems (Johnson & Hallberg, 2005).

7.3 Hydrogeological Alterations and Flow Regime Impacts

Mining operations typically result in significant changes in hydrological systems, impacting surface water as well as groundwater dynamics. Dewatering operations, which are essential to keeping dry working conditions in many mines, can lower the water table over large regions and decrease baseflow to streams, springs, and wetlands. There have been reports in mining regions of drawdown cones extending for many kilometers away from the pumping center and of repercussions on groundwater-dependent ecosystems (Younger & Wolkersdorfer, 2004).

Hydraulic rebound following mine closure from the cessation of dewatering leads to new hydrological conditions under which contaminants may mobilize and new contaminant discharge pathways may form. This is not unique, as has been noted in numerous historical mining districts, the cessation of activity within extensive underground workings causes the emergence of contaminated discharges from audits and shafts when water levels rebound (Wolkersdorfer, 2008).

Mining operations impact surface water hydrology in a variety of ways. Disturbance of the land changes the runoff characteristics so that, as a result of decreased infiltration capacity, the peak flows increase while baseflows decrease. Water management infrastructure such as diversion channels, settling ponds, and tailings dams can drastically alter connectivity and flow regimes in stream networks (Bernhardt & Palmer, 2011).

For instance, existing research has shown that in water-stressed and semi-arid regions mining is a water-bully, severely competing with local and aquatic systems. Copper extraction in Chile's Atacama Desert, for instance, has resulted in declining aquifers and decreased flows to high-altitude wetlands (bofedales) that harbor unique biodiversity (Oyarzún & Oyarzún, 2011).

Changes in flow regimes influence a number of ecological processes in aquatic systems, such as habitat availability, thermal regimes, sediment transport, and life cycle cues for aquatic organisms. The ecological effects of both of these changes will depend on how they compare to naturally occurring variability over a given time period and how sensitive a given ecosystem is to those changes. In some cases, flow changes driven by mining-related factors have played a role in declines of sensitive species and invasions by organisms better suited to modified conditions (Bernhardt & Palmer, 2011).

7.4 Impacts on Aquatic Biodiversity and Ecosystem Function

The combined impacts of water quality deterioration, hydrologic modification, and habitat alteration generally cause profound changes in aquatic community structure and ecosystem function in these systems. Studies have shown that the water bodies in mining activity areas have reduced species richness, changed community structure, and disrupted ecological process (Hogsden & Harding, 2012).

Fish communities typically demonstrate strong responses to mining impacts, with sensitive species generally declining or being extirpated from impacted reaches. In the Coeur d'Alene River system in Idaho, for example, a century of lead-zinc mining has produced metal concentrations in much of the watershed that are toxic to fish and that limit native trout and other species to uncontaminated tributaries (Farag et al., 1998). Likewise, in coal mining regions, studies have documented decreases in fish diversity and abundance downstream of mining operations, especially where valley fills and associated alkaline mine drainage shifts stream chemistry and physical habitat structure (Bernhardt et al., 2012).

Fish communities are sensitive to the impacts of mining given the broad taxonomic composition, limited mobility of many macroinvertebrate taxa, and differential sensitivity of macroinvertebrate communities to environmental stressors. Several studies have described replacement patterns of benthic communities along gradients of mining impact, with sensitive taxa [of which mayflies (Ephemeroptera), stoneflies (Plecoptera), and caddisflies (Trichoptera) are among the most well-characterized] typically reduced in abundance, whereas more tolerant groups (such as some dipterans and coleopterans) may increase in relative abundance (Gray & Delaney, 2008). These community shifts are indicative of biodiversity loss and can change ecosystem functions, such as organic matter decomposition, nutrient cycling, and energy flow through aquatic food webs.

Primary producers, such as algae, macrophytes, and riparian vegetation, respond indirectly to impacts of mining through changes in community composition and productivity. Toxicity from metals can remove susceptible algal species, and eutrophication due to nutrient enrichment from certain mining activities can cause algal blooms in receiving waters. Riparian vegetation might also be directly affected by groundwater contamination or indirectly affected by changes in channel morphology or flow regimes (Younger, 2001).

Bioaccumulation and biomagnification of mining-derived contaminants, especially mercury, selenium, and selected other metals, are major problems for aquatic food webs and for humans consuming aquatic biota as food. Such processes can lead to concentrations of contaminants in top predators many orders of magnitude greater than those for the water column, potentially leading to reproductive failure, neurological damage, and other toxic effects. There are well-documented cases of mercury concentrations in fish from downstream of gold mining impacts in the Amazon Basin and selenium toxicity of aquatic birds in reservoirs impacted by coal mining in western Canada (Eagles-Smith et al., 2018; Janz et al., 2014).

Mining activities affect aquatic ecosystems over significantly larger distances than the mined area. Sediments and soluble contaminants may be transported downstream for tens or even hundreds of kilometers, impacting ecosystems far removed from the original disturbance. For example, metal contamination from the abandoned Anaconda copper mines in Montana impacted more than 200 km of the Clark Fork River system, and mercury from gold mining in the Sierra Nevada is still affecting San Francisco Bay more than a century after high-volume mining ceased (Moore & Luoma, 1990).

7.5 *Groundwater Contamination and Impacts on Subterranean Ecosystems*

Mining-related contamination is a particularly insidious form of hydraulic fracturing consequence, because it has the potential to contaminate groundwater systems that are in direct contact with disturbed geological materials, which is very difficult to monitor and remediate. Sources of contamination pathways include direct infiltration from waste facilities, seepage from tailings impoundments and process water ponds, and flooding of underground workings after mine closure (Wolkersdorfer, 2008).

Such mining-related aquifer contamination can be remarkably persistent, potentially lasting hundreds of years given the low flow rates and limited attenuation capacity in many aquifers. Historical mining districts in Europe were reviewed and recognized by Nordstrom (2011) as sources of contaminated groundwater plumes that exist hundreds of years after mining has ceased, as they provide the mobile parts of this contamination to surface waters via baseflow contributions.

Subterranean ecosystems (like cave systems and the hyporheic zone below stream beds) contain specialized biodiversity that can be especially sensitive to the effects of mining. Such habitats frequently contain endemic stygobitic (groundwater-dwelling) organisms that have adapted to these stable environments with very little exposure to contaminants. Mining-derived pollutants may therefore be introduced to these systems and can cause the elimination of sensitive taxa as well as disrupt the processes that underpin systems that purify groundwater (Gibert & Deharveng, 2002).

The interactions between groundwater and surface water in mining-affected watersheds lead to complex contaminant behavior, where streambed sediments are often sinks and secondary sources of contamination. Contaminated groundwater discharge has been shown to sustain enhanced metal concentrations in surface waters beyond the cessation of direct inputs in mining-affected watersheds, prolonging the temporal scale of effects and complicating remediation efforts (Nagorski et al., 2002).

7.6 *Marine and Coastal Impacts from Mining Activities*

Freshwater impacts have received a lot of attention, but mining impacts marine and coastal ecosystems via multiple pathways. Delta regions and estuaries are often the recipients of mining-derived waste. Recent mining activities have deposited large volumes of tailings to marine environments as a result of riverine transport, for instance, historical mining in Cornwall, UK, is believed to have impacted estuarine sediments across southwest England, and copper mining in Papua New Guinea has provided significant tonnages of tailings to the Gulf of Papua via riverine transport (Taylor et al., 1995).

Direct marine disposal of mine tailings, a practice in some countries (such as Norway, Chile, Papua New Guinea, and Indonesia), causes localized but severe impacts to benthic marine systems. This practice, called submarine tailings placement (STP) or deep-sea tailings disposal (DSTD), generally consists of the release of fine-grained waste via pipelines to deeper waters. Although supporters claim that the deep-sea disposal method keeps waste safely removed from direct contact with humans and valuable surface waters, issues linger related to whether deep-sea ecosystems are affected, whether upwelling could lead to contaminating productive surface waters, and ultimately if elements deposited will degrade maintained (Ramirez-Llodra et al., 2015).

Coastal placer mining and nearshore dredging for minerals including tin, titanium, and diamonds clearly disturb marine habitats directly as sediment is removed and redeposited. Such activities may affect benthic community structure, enhance turbidtion and re-mobilize any sorbed organic pollutants (Thrush & Dayton, 2002).

The nascent industry of deep-sea mining, remaining predominantly exploratory, brings with it new risks to marine ecosystems. Commercial extraction of polymetallic nodules, seafloor massive sulfides, and cobalt-rich crusts would directly disturb benthic habitats with extremely high recovery times. Environmental impact assessments for these activities must account for the underdeveloped baseline knowledge of deep-sea ecosystems, and the potential for mining to induce long-lived or permanent changes to these poorly characterized environments (Miller et al., 2018).

8 Mitigation Strategies and Sustainable Mining Practices

As awareness of mining's environmental impacts has grown, so too have efforts to develop and implement strategies to mitigate these effects. This section examines approaches across the mining lifecycle that can reduce environmental harm while maintaining the economic benefits of mineral extraction. Effective mitigation requires a combination of technological innovation, policy frameworks, corporate commitment, and stakeholder engagement to address the complex environmental challenges described in previous sections.

8.1 Regulatory Frameworks and Environmental Impact Assessment

Environmental regulation provides the foundation for mining impact mitigation in most jurisdictions. Since the 1970s, many countries have established increasingly comprehensive legal frameworks governing the environmental aspects of mining, though the stringency and enforcement of these regulations vary considerably worldwide (Sassoon, 2009). Modern regulatory approaches typically include requirements for environmental impact assessment (EIA) prior to project approval, operational permits with specific environmental conditions, financial assurance for closure and reclamation, and ongoing monitoring throughout the mine lifecycle.

Environmental impact assessment represents a critical preventive tool that can identify potential impacts before they occur and inform project design to minimize environmental harm. When effectively implemented, the EIA process evaluates alternatives, establishes baseline environmental conditions, predicts potential impacts, and develops mitigation measures that become conditions for project approval. The International Association for Impact Assessment has developed best practice principles that emphasize rigorous scientific analysis, stakeholder participation, and consideration of cumulative effects (IAIA (International Association for Impact Assessment), 2012). Countries with well-established EIA systems, such as Canada and Australia, typically require mining projects to undergo rigorous assessment, often with specific guidelines tailored to the mining sector (Franks et al., 2011).

However, the effectiveness of EIA as a mitigation tool varies considerably. In some jurisdictions, particularly those with limited institutional capacity or competing priorities between economic development and environmental protection, the process may be compromised by insufficient baseline data, inadequate analysis of alternatives, limited public participation, or weak enforcement of identified mitigation measures (Morrison-Saunders et al., 2016). Even in countries with strong regulatory frameworks, EIA often struggles to address cumulative impacts from multiple projects, complex impact pathways across ecosystem boundaries, and long-term environmental changes that extend beyond conventional project timeframes.

Financial assurance mechanisms, including reclamation bonds, trust funds, and insurance instruments, represent another important regulatory tool for ensuring environmental mitigation. These mechanisms require mining companies to provide financial guarantees sufficient to cover reclamation and long-term management costs, reducing the risk that environmental liabilities will be externalized to the public if a company becomes insolvent or abandons a site. Research indicates that adequate financial assurance creates economic incentives for companies to minimize disturbance, implement progressive reclamation, and develop more effective closure strategies (Mchaina, 2001). However, debates continue regarding the appropriate calculation of assurance amounts, particularly for addressing long-term water treatment needs and managing risks that extend beyond conventional closure timeframes.

8.2 *Mine Planning and Design for Impact Reduction*

Improved planning and design approaches offer significant opportunities to reduce environmental impacts throughout the mining lifecycle. The concept of "designing for closure"—integrating post-mining land use and reclamation considerations into initial project planning—has gained acceptance as a fundamental principle of sustainable mining (Laurence, 2006). This approach considers the entire life cycle of a mining operation from the earliest planning stages, enabling more effective waste management strategies, progressive rehabilitation opportunities, and post-mining land use options.

Site selection and facility layout represent critical early decisions that influence environmental footprints. Avoiding ecologically sensitive areas, minimizing disturbance of high-value habitats, and considering watershed boundaries in site layout can substantially reduce impacts on biodiversity and water resources. For example, locating waste rock facilities away from headwater streams, designing infrastructure to minimize fragmentation of wildlife corridors, and avoiding areas with high acid-generating potential can prevent or reduce downstream impacts (ICMM (International Council on Mining and Metals), 2006).

Mining method selection also significantly influences environmental impacts. Underground mining typically creates a smaller surface disturbance footprint than open-pit operations but may present greater challenges for mine water management. In situ recovery methods, where applicable, can significantly reduce surface disturbance and waste generation but require careful management of groundwater impacts. Technological innovations in extraction methods, such as plasma rock blasting, precision mining, and selective cutting heads, offer potential to increase resource recovery efficiency while decreasing waste generation and energy consumption (Darling, 2011).

Waste minimization strategies represent another important aspect of impact reduction. Approaches include selective handling of potentially acid-generating materials, co-disposal techniques that blend waste types to neutralize acid potential, backfilling of mine voids with waste material, and reprocessing of waste to recover additional resources. Advanced ore sorting technologies, including sensor-based systems that can separate ore from waste before energy-intensive processing stages, offer promising approaches to waste reduction while improving operational efficiency (Lessard et al., 2016).

8.3 Water Management and Treatment Technologies

Water management represents one of the most critical aspects of environmental mitigation in mining operations. Comprehensive water management plans typically address water balance, quality protection, and post-closure considerations through multiple strategies. These include diversion of clean water around disturbed areas, containment and treatment of contact water, recycling of process water, controlled release during appropriate flow conditions, and monitoring systems that allow for adaptive management (Northey et al., 2016).

Prevention of acid mine drainage remains a fundamental challenge in many mining contexts. Primary prevention approaches focus on limiting the exposure of sulfide minerals to oxygen and water through techniques such as subaqueous disposal, engineered covers, and encapsulation with neutralizing materials. Research continues to refine these approaches, with recent innovations including geosynthetic covers that reduce oxygen diffusion, microencapsulation technologies that coat individual particles with neutralizing materials, and bioreactor systems that promote reducing conditions to prevent sulfide oxidation (Kefeni et al., 2017).

When preventive measures are insufficient, active water treatment becomes necessary to manage mining-impacted water. Conventional treatment approaches include lime neutralization for acidic drainage, metal precipitation, oxidation–reduction systems, and various filtration technologies. While effective, these approaches typically require ongoing operation and maintenance, creating long-term liabilities at closed mine sites. Advanced treatment technologies, including membrane filtration, ion exchange systems, electrochemical methods, and biological treatment approaches, offer potential advantages in treatment efficiency and reduced sludge generation but often involve higher capital and operational costs (Johnson & Hallberg, 2005).

Passive treatment systems, which utilize natural biogeochemical processes to improve water quality with minimal ongoing intervention, have gained increasing attention for long-term management of mining-impacted waters. These systems include constructed wetlands, anoxic limestone drains, reducing and alkalinity producing systems, and biochemical reactors. While passive systems generally require less maintenance and energy input than active treatment, they typically demand larger land areas and may have more variable performance, particularly in cold climates or when treating waters with complex chemistry (Skousen et al., 2017).

8.4 Biodiversity Conservation and Ecosystem Restoration

Mitigating mining's impacts on biodiversity requires approaches that span the entire mining lifecycle, from exploration through closure. The mitigation hierarchy—which prioritizes avoidance of impacts, followed by minimization, rehabilitation, and finally offsetting—provides a framework for managing biodiversity impacts. This approach emphasizes avoiding impacts on critical habitats where possible, minimizing the severity and extent of unavoidable impacts, rehabilitating affected areas to restore ecological functionality, and offsetting residual impacts through conservation actions elsewhere (ICMM (International Council on Mining and Metals), 2006).

Biodiversity offsets, which aim to compensate for residual impacts by generating equivalent biodiversity gains in other locations, have become increasingly common in mining contexts. These arrangements can involve habitat protection, restoration, or creation activities that deliver measurable conservation outcomes. However, significant challenges remain in designing effective offsets, including ensuring ecological equivalence between impacted and offset areas, establishing appropriate timeframes for measuring outcomes, and maintaining offsets over the long term (Virah-Sawmy et al., 2014).

Rehabilitation and ecological restoration of mined lands has evolved significantly from early approaches focused primarily on stabilization and visual amelioration. Contemporary best practices emphasize reestablishing ecosystem functionality and resilience rather than simply revegetating disturbed areas. This shift has led to greater emphasis on topsoil management, reconstruction of soil profiles with appropriate physical and chemical properties, use of native species in revegetation, creation of habitat features, and establishment of self-sustaining ecological processes (Cooke & Johnson, 2002).

Novel ecosystems—those that differ in composition and function from historical ecosystems due to severe disturbance and altered environmental conditions—are increasingly recognized as potential outcomes of mine rehabilitation. While restoration to pre-mining conditions remains the ideal in many contexts, the realities of dramatically altered topography, hydrology, and soil conditions may necessitate alternative approaches. Research on novel ecosystems in post-mining landscapes focuses on establishing functional, resilient systems that provide ecosystem services and support biodiversity, even if they differ from pre-mining communities (Hobbs et al., 2013).

8.5 Corporate Initiatives and Voluntary Standards

Beyond regulatory compliance, various voluntary initiatives have emerged to promote improved environmental performance in the mining sector. Industry associations such as the International Council on Mining and Metals (ICMM) have developed sustainability principles and performance expectations that member companies commit to implement. These frameworks typically include requirements for environmental management systems, biodiversity conservation, water stewardship, and community engagement that exceed regulatory minimums in many jurisdictions (ICMM (International Council on Mining and Metals), 2020).

Certification schemes, including the Initiative for Responsible Mining Assurance (IRMA) and the Mining Association of Canada's Towards Sustainable Mining program, provide third-party verification of environmental and social performance against established standards. These programs typically involve regular audits against performance indicators, public reporting of results, and mechanisms for addressing non-compliance. While participation remains voluntary, growing pressure from investors, downstream purchasers, and consumers has increased the business case for certification (Franks et al., 2011).

Corporate social responsibility (CSR) programs have expanded to include substantial environmental components, reflecting recognition that social license to operate depends partly on effective environmental management. Leading mining companies have established biodiversity strategies, water stewardship programs, and climate action plans that establish targets beyond compliance requirements. These initiatives often include partnerships with environmental organizations, academic institutions, and community groups to develop innovative approaches to environmental challenges (Ranängen & Lindman, 2017).

8.6 Technological Innovation and Future Directions

Emerging technologies offer significant potential to reduce mining's environmental footprint through various pathways. Automation and robotics can improve precision in extraction, reducing waste generation and energy consumption while enhancing safety. Remote sensing technologies, including satellite imagery, drone-based systems, and wireless sensor networks, enable more comprehensive environmental monitoring with reduced costs and improved spatial coverage. Advanced data analytics and artificial intelligence applications help identify patterns and predict potential issues before they manifest as environmental impacts (Darling, 2011).

Biomining approaches, which utilize microorganisms to extract metals from ores, represent a potentially transformative technology for certain deposit types. These processes can operate at ambient temperatures and pressures, reducing energy requirements compared to conventional processing. Biomining may also reduce chemical usage and waste generation while potentially enabling recovery of metals from lower-grade resources that would be uneconomic with conventional approaches (Johnson et al., 2013).

The circular economy concept has gained traction in the mining sector, emphasizing the potential to recover minerals from waste streams, repurpose mining wastes for beneficial uses, and design products for recyclability. Examples include recovering rare earth elements from mine drainage, utilizing waste rock for construction materials, converting tailings to supplementary cementitious materials, and designing electronics for easier disassembly and metal recovery. These approaches can reduce primary resource demands while addressing legacy waste issues (Lèbre et al., 2017).

Climate change considerations have increasingly influenced mining practices, both in terms of reducing greenhouse gas emissions and adapting to changing environmental conditions. Energy efficiency improvements, renewable energy integration, and electrification of mining equipment represent key strategies for emissions reduction. Meanwhile, climate adaptation requires designing facilities to withstand more extreme weather events, considering changing water availability in water management plans, and accounting for altered reclamation conditions in closure planning (Phillips, 2016).

9 Future Directions and Emerging Technologies

The mining industry stands at a technological and environmental crossroads, with emerging innovations poised to transform traditional approaches to resource extraction and environmental management. This section explores the frontiers of mining technology and practice that may shape more sustainable approaches to mineral extraction in coming decades, addressing persistent environmental challenges while meeting growing global demand for minerals.

9.1 Precision Mining and Targeted Extraction

The concept of "precision mining" represents a significant paradigm shift from conventional bulk mining approaches. Drawing parallels with precision agriculture, this philosophy emphasizes selective extraction of valuable minerals while minimizing disturbance of surrounding waste material. Advanced sensing technologies, including real-time geochemical analyzers, gamma-ray spectrometers, and hyperspectral imaging systems, enable more precise identification of ore boundaries and quality variations within deposits (Bamber et al., 2021). When coupled with automated, highly maneuverable extraction equipment, these technologies could substantially reduce waste generation, energy consumption, and environmental footprints.

Narrow-vein mining technologies exemplify this approach, with recent innovations in mechanical cutting, robotic systems, and continuous mining equipment enabling profitable extraction of thin, high-grade deposits that would be uneconomic with conventional methods. These targeted approaches may revitalize interest in smaller, higher-grade deposits that can be extracted with minimal surface disturbance, potentially shifting development away from the large, low-grade operations that typically create the most significant environmental impacts (Ghosh et al., 2018).

9.2 Digital Transformation and Integrated Environmental Management

The digital transformation of mining operations presents unprecedented opportunities for environmental optimization through real-time monitoring, predictive analytics, and integrated decision support systems. Internet of Things (IoT) technologies enable comprehensive sensor networks that continuously monitor environmental parameters across mining operations and surrounding ecosystems, providing early warning of developing issues and informing adaptive management responses (Sánchez & Hartlieb, 2020).

Digital twins—virtual replicas of physical mining systems that integrate real-time data with predictive models—offer powerful tools for optimizing environmental performance. These systems can simulate the consequences of operational decisions, test mitigation strategies, and identify efficiency improvements that reduce resource consumption and waste generation. When extended to include surrounding ecosystems, these models can help predict and manage complex environmental interactions across the mining lifecycle (Jiskani et al., 2020).

Blockchain technology shows promise for enhancing transparency and accountability in environmental management, potentially transforming how mining companies document their environmental performance. Distributed ledger systems could provide immutable records of environmental monitoring data, verify compliance with regulatory requirements, and trace minerals through supply chains to demonstrate responsible sourcing. These applications may significantly strengthen stakeholder confidence in environmental management systems while reducing verification costs (Ongena et al., 2018).

9.3 Biotechnology Applications in Mining

Biotechnological approaches represent some of the most promising avenues for reducing mining's environmental footprint. Biomining techniques, which employ microorganisms to extract metals from ores, have advanced significantly beyond traditional bioleaching applications. Recent innovations include developing engineered microbes with enhanced metal recovery capabilities, expanding biomining applications to new commodities, and optimizing bioreactor designs for improved efficiency and control (Johnson, 2018).

Phytomining and phytoremediation utilize plants to extract metals from soil or stabilize contaminated areas, offering potential solutions for both resource recovery and environmental restoration. Advances in understanding metal hyperaccumulating plants, along with developments in plant breeding and genetic modification, have expanded the potential applications of these approaches. Research suggests that certain hyperaccumulator species can achieve metal concentrations in their tissues that exceed those in many commercial ores, potentially enabling economic recovery of metals from contaminated sites or low-grade resources (Sheoran et al., 2016).

Synthetic biology applications, including engineered proteins that selectively bind to specific metals, offer revolutionary possibilities for precise, environmentally benign extraction processes. These biologically-based extraction agents could potentially replace conventional chemical reagents, reducing toxicity concerns and enabling highly selective recovery of target metals from complex mixtures. While still largely confined to laboratory settings, these technologies hold transformative potential for reimagining mineral processing (Diaz et al., 2018).

9.4 Toward a Circular Material Economy

Perhaps the most fundamental shift in mining's future environmental profile lies in the transition toward a circular economy model that emphasizes material recovery, reuse, and recycling. Urban mining—the systematic recovery of materials from built infrastructure, electronic waste, and industrial residues—increasingly complements traditional extractive approaches. Technological advances in automated disassembly, material separation, and metallurgical recovery are rapidly improving the economic viability of these secondary sources (Lèbre et al., 2017).

The development of design-for-recycling approaches for mineral-containing products represents another essential element of this transition. Technologies that enable easier disassembly, material identification, and separation at end-of-life can dramatically improve recycling rates for critical minerals. Meanwhile, advances in metallurgical processes that can effectively separate complex material mixtures expand the range of recoverable elements from recycled streams (Reuter et al., 2018).

This evolution toward a more circular material economy does not eliminate the need for primary mining but could significantly modify its scale and focus. As secondary sources satisfy a greater proportion of demand for common metals, primary mining may increasingly shift toward obtaining elements that remain difficult to recover from recycled streams, potentially altering the global distribution and environmental footprint of extraction activities (Ali et al., 2017).

10 Conclusion

This chapter has examined the multifaceted environmental impacts of mining activities on terrestrial and aquatic ecosystems, revealing the complex interplay between extraction methods, geological settings, and ecological responses. From the direct physical alterations of landscapes to the subtle biogeochemical transformations that affect ecosystem functioning, mining operations create environmental changes that vary in intensity, spatial extent, and temporal persistence. Understanding these impacts requires integrating knowledge across disciplines including geology, hydrology, ecology, chemistry, and engineering.

The case studies presented illustrate how mining's environmental footprint manifests differently across contexts—from massive open-pit operations in tropical highlands to distributed artisanal activities in the Amazon Basin, from coal extraction in temperate forests to oil sands development in boreal ecosystems. These examples demonstrate that while certain impact mechanisms recur across settings, the specific ecological consequences depend strongly on local environmental conditions, regulatory frameworks, corporate practices, and technological approaches.

Mitigation strategies have evolved substantially over recent decades, reflecting both improved scientific understanding of impact mechanisms and shifting societal expectations regarding environmental protection. The progression from end-of-pipe treatment approaches to integrated prevention strategies exemplifies this evolution, as does the growing emphasis on ecosystem-level considerations rather than narrow compliance with specific parameter limits. Emerging technologies offer promising pathways for further reducing environmental impacts while maintaining the mineral supplies essential for modern economies.

The future trajectory of mining's environmental profile will be shaped by multiple forces, including technological innovation, regulatory development, market demands, and broader sustainability transitions. The growing emphasis on responsible sourcing, circular economy principles, and climate change considerations may fundamentally reshape where, how, and which minerals are extracted. Meanwhile, increasing demand for materials required for renewable energy technologies and digital infrastructure creates new environmental challenges while potentially reducing others associated with fossil fuel dependence.

As society continues to require minerals for economic development and technological advancement, the imperative to minimize mining's environmental footprint gains importance. Achieving this goal requires continuing innovation in extraction and processing technologies, strengthening governance frameworks, improving corporate practices, and fostering greater collaboration among stakeholders. By integrating environmental considerations throughout the mining lifecycle—from exploration through closure and beyond—the mining sector can contribute to sustainable development while protecting the ecological systems upon which all societies ultimately depend.

References

Akcil, A. (2006). Managing cyanide: Health, safety and risk management practices at Turkey's Ovacik gold-silver mine. *Journal of Cleaner Production, 14*(8), 727–735.

Ali, S. H., Giurco, D., Arndt, N., Nickless, E., Brown, G., Demetriades, A., Durrheim, R., Enriquez, M. A., Kinnaird, J., Littleboy, A., Meinert, L. D., Oberhänsli, R., Salem, J., Schodde, R., Schneider, G., Vidal, O., & Yakovleva, N. (2017). Mineral supply for sustainable development requires resource governance. *Nature, 543*(7645), 367–372.

Bala, K., Anshul, Verma, S., Dogra, A., Das, A. P., Kumar, R., & Nandini. (2024). Sustainable management for urban plastic waste generation. In *Sustainable management of urban plastic waste through circular economic approaches* (pp. 211–236). Routledge. https://doi.org/10.1201/9781003465447-13

Bala, K., Dogra, A., Lata, P., & Das, A. P. (2024). Discussing sources and biological-chemical remediation approaches for microplastic pollution. In *Sustainable microbial technology for synthetic and cellulosic microfiber bioremediation* (pp. 37–59). Springer. https://doi.org/10.1007/978-3-031-62600-5_3

Bamber, A. S., Klein, B., & Scoble, M. (2021). Integrated mining, pre-concentration and waste disposal systems for the increased sustainability of hard rock metal mining. *Journal of Cleaner Production, 286*, 125378.

Banasova, V., Horak, O., Ciernik, M., Nadubinska, M., & Lichtscheidl, I. (2006). The vegetation of metalliferous and non-metalliferous grasslands in two former mining regions in Central Europe. *Biologia, 61*(4), 433–439.

Basu, A. J., & Mishra, D. A. (2020). Economic and environmental impact of continuous mining system: A case study. *Journal of Sustainable Mining, 19*(1), 42–51.

Bellamy, D., & Pravica, L. (2011). Assessing the impact of driverless haul trucks in Australian surface mining. *Resources Policy, 36*(2), 149–158.

Bernhardt, E. S., Lutz, B. D., King, R. S., Fay, J. P., Carter, C. E., Helton, A. M., Campagna, D., & Amos, J. (2012). How many mountains can we mine? Assessing the regional degradation of central Appalachian rivers by surface coal mining. *Environmental Science & Technology, 46*(15), 8115–8122.

Bernhardt, E. S., & Palmer, M. A. (2011). The environmental costs of mountaintop mining valley fill operations for aquatic ecosystems of the Central Appalachians. *Annals of the New York Academy of Sciences, 1223*(1), 39–57.

Beyer, W. N., Dalgarn, J., Dudding, S., French, J. B., Mateo, R., Miesner, J., Sileo, L., & Spann, J. (2013). Zinc and lead poisoning in wild birds in the tri-state Mining District (Oklahoma, Kansas, and Missouri). *Archives of Environmental Contamination and Toxicology, 48*(1), 108–117.

Blowes, D. W., Ptacek, C. J., Jambor, J. L., & Weisener, C. G. (2003). The geochemistry of acid mine drainage. *Treatise on Geochemistry, 9*, 149–204.

Blowes, D. W., Ptacek, C. J., Jambor, J. L., Weisener, C. G., Paktunc, D., Gould, W. D., & Johnson, D. B. (2014). The geochemistry of acid mine drainage. In H. D. Holland & K. K. Turekian (Eds.), *Treatise on geochemistry* (Vol. 11, 2nd ed., pp. 131–190). Elsevier.

Booth, C. J. (2006). Groundwater as an environmental constraint of longwall coal mining. *Environmental Geology, 49*(6), 796–803.

Bridge, G. (2004). Contested terrain: Mining and the environment. *Annual Review of Environment and Resources, 29*, 205–259.

Bridge, G. (2008). Global production networks and the extractive sector: Governing resource-based development. *Journal of Economic Geography, 8*(3), 389–419.

Cooke, J. A., & Johnson, M. S. (2002). Ecological restoration of land with particular reference to the mining of metals and industrial minerals: A review of theory and practice. *Environmental Reviews, 10*(1), 41–71.

Cristescu, R. H., Frère, C., & Banks, P. B. (2016). A review of fauna in mine rehabilitation in Australia: Current state and future directions. *Biological Conservation, 149*(1), 60–72.

Dang, Z., Liu, C., & Haigh, M. J. (2002). Mobility of heavy metals associated with the natural weathering of coal mine spoils. *Environmental Pollution, 118*(3), 419–426.

Darling, P. (2011). *SME mining engineering handbook* (3rd ed.). Society for Mining, Metallurgy, and Exploration.

Das, A., Dey, S., Bala, K., Gnanasekaran, R., & Das, A. P. (2024). Microbial bioremediation technology for sustainable treatment and management of synthetic microfiber waste. In *Sustainable microbial technology for synthetic and cellulosic microfiber bioremediation* (pp. 221–238). Springer. https://doi.org/10.1007/978-3-031-62600-5_12

Das, A., Dey, S., & Das, A. P. (2024). Ecotoxicological impacts of synthetic microfiber pollutants and development of sustainable mitigation strategies. *Environmental Chemistry and Ecotoxicology*. https://doi.org/10.1016/J.ENCECO.2024.12.003

Das, A., Dey, S., Gacem, A., Yadav, K. K., Nath, H., Ram, D. K., & Das, A. P. (2025). Revolutionizing solid hydrocarbon sourcing: Plastic's journey from waste to upcycled treasure. *Environmental Science and Engineering, 218*, 331–351. https://doi.org/10.1007/978-3-031-81615-4_14

Das, A. P., Behera, I. D., & Das, N. P. (2024). *Renewable energy generation and value addition from environmental microfiber pollution through advanced greener solution* (p. 278). Springer.

Das, A., Ray, S. K., Mohanty, M., Mohanty, J., Dey, S., & Das, A. P. (2024). Ecotoxicity of microplastic wastes and their sustainable management: A review. *Environmental Chemistry and Ecotoxicology, 6*, 144–152. https://doi.org/10.1016/J.ENCECO.2024.05.003

Das, A. P., & Ghosh, S. (2018). Bioleaching of manganese from mining waste materials. *Materials Today: Proceedings, 5*(1), 2381–2390. https://doi.org/10.1016/J.MATPR.2017.11.459

Das, A. P., & Ghosh, S. (2023a). Factors affecting bioleaching. In *Manganese mining microorganisms* (pp. 77–86). Elsevier. https://doi.org/10.1016/B978-0-12-822147-1.00004-2

Das, A. P., & Ghosh, S. (2023b). Mechanism of manganese bioleaching. In *Manganese mining microorganisms* (pp. 87–98). Elsevier. https://doi.org/10.1016/B978-0-12-822147-1.00014-5

Das, A. P., & Ghosh, S. (2023c). Role of metabolites and metabolomics for understanding of manganese bioleaching. In *Manganese mining microorganisms* (pp. 117–127). Elsevier. https://doi.org/10.1016/B978-0-12-822147-1.00011-X

Das, A. P., & Ghosh, S. (2023). *Manganese mining microorganisms*. Elsevier.

Das, A. P., Mallavarapu, M., & Ghosh, S. (2023). Ecotoxicity of mining pollutants on the environment and their remediation. *Environmental Chemistry and Ecotoxicology, 5*, 165–167. https://doi.org/10.1016/J.ENCECO.2023.08.002

Das, A. P., Swain, S., Panda, S., Pradhan, N., & Sukla, L. B. (2012). Reductive acid leaching of low grade manganese ores. *Geomaterials, 2012*(04), 70–72. https://doi.org/10.4236/GM.2012.24011

Delgado, J., Nieto, J. M., Boski, T., Alabardeiro, L., & Carvalho, C. (2012). Impact of mine drainage on the hydrogeochemical evolution of the Guadiana estuary (SW Spain and Portugal). *Mine Water and the Environment, 31*(3), 174–186.

DeNicola, D. M., & Stapleton, M. G. (2002). Impact of acid mine drainage on benthic communities in streams: The relative roles of substratum vs. aqueous effects. *Environmental Pollution, 119*(3), 303–315.

Dey, S., Bissoyi, A., Koteswara Reddy, G., & Das, A. P. (2024). Synthetic microfibers: An urban emerging pollutant demanding for sustainable environmental solutions. In *Sustainable microbial technology for synthetic and cellulosic microfiber bioremediation* (pp. 107–121). Springer. https://doi.org/10.1007/978-3-031-62600-5_7

Dey, S., Dash, A., & Das, A. P. (2024). Bioremediation as a sustainable and circular solution for mining waste management and bioenergy generation. In *Sustainable management of mining waste and tailings: A circular economy approach* (pp. 107–119). Springer. https://doi.org/10.1201/9781003442455-6

Dey, S., Dash, N., Ray, S. K., Priyadarsini, S., Chhotaray, C., & Das, A. P. (2024). Occurrence and quantification of tire wear plastics pollutants from Highway City dust of Bhubaneswar City of Odisha. In *Sustainable microbial technology for synthetic and cellulosic microfiber bioremediation* (pp. 77–89). Springer. https://doi.org/10.1007/978-3-031-62600-5_5

Dey, S., Tripathy, B., Kumar, M. S., & Das, A. P. (2023). Ecotoxicological consequences of manganese mining pollutants and their biological remediation. *Environmental Chemistry and Ecotoxicology*. https://doi.org/10.1016/J.ENCECO.2023.01.001

Diaz, V., Vassallo, G., & Johnson, D. B. (2018). Bioremediation of mine wastes: Biomining to bioremediation. *Current Opinion in Biotechnology, 51*, 1–8.

Dudka, S., & Adriano, D. C. (1997). Environmental impacts of metal ore mining and processing: A review. *Journal of Environmental Quality, 26*(3), 590–602.

Eagles-Smith, C. A., Silbergeld, E. K., Basu, N., Bustamante, P., Diaz-Barriga, F., Hopkins, W. A., Kidd, K. A., & Nyland, J. F. (2018). Modulators of mercury risk to wildlife and humans in the context of rapid global change. *Ambio, 47*(2), 170–197.

Ericsson, M., & Löf, O. (2019). Mining's contribution to national economies between 1996 and 2016. *Mineral Economics, 32*(2), 223–250.

Farag, A. M., Woodward, D. F., Goldstein, J. N., Brumbaugh, W., & Meyer, J. S. (1998). Concentrations of metals associated with mining waste in sediments, biofilm, benthic macroinvertebrates, and fish from the Coeur D'alene River basin, Idaho. *Archives of Environmental Contamination and Toxicology, 34*(2), 119–127.

Farmer, A. M. (1993). The effects of dust on vegetation—A review. *Environmental Pollution, 79*(1), 63–75.

Franks, D. M., Boger, D. V., Côte, C. M., & Mulligan, D. R. (2011). Sustainable development principles for the disposal of mining and mineral processing wastes. *Resources Policy, 36*(2), 114–122.

Gao, M., Hao, Z., Yang, S., & Liu, X. (2021). The evolution of mining technology: A review. *International Journal of Mining Science and Technology, 31*(6), 955–967.

Ghosh, A., Sharma, A. K., & Nayak, B. (2018). Revolution in mining technology: Insight into emerging technologies. In R. Singh et al. (Eds.), *Sustainability issues in mining and mineral processing* (pp. 193–214). Springer.

Ghosh, S., & Das, A. P. (2017). Bioleaching of manganese from mining waste residues using Acinetobacter sp. *Geology, Ecology, and Landscapes, 1*(2), 77–83. https://doi.org/10.1080/24749508.2017.1332847

Ghosh, S., Sinha, J. K., Ghosh, S., Vashisth, K., Han, S., & Bhaskar, R. (2023). Microplastics as an emerging threat to the global environment and human health. *Sustainability, 15*(14), 10821. https://doi.org/10.3390/SU151410821

Ghosh, S., Tripathy, B., Dey, S., & Das, A. P. (2023). Comparative investigation of fungal and bacterial manganese biomining mechanisms. In *Manganese mining microorganisms* (pp. 99–115). Springer. https://doi.org/10.1016/b978-0-12-822147-1.00016-9

Gibert, J., & Deharveng, L. (2002). Subterranean ecosystems: A truncated functional biodiversity. *Bioscience, 52*(6), 473–481.

Ginocchio, R., & Baker, A. J. M. (2004). Metallophytes in Latin America: A remarkable biological and genetic resource scarcely known and studied in the region. *Revista Chilena de Historia Natural, 77*(1), 185–194.

Gray, N. F., & Delaney, E. (2008). Comparison of benthic macroinvertebrate indices for the assessment of the impact of acid mine drainage on an Irish river below an abandoned cu-S mine. *Environmental Pollution, 155*(1), 31–40.

Hancock, G. R., Lowry, J. B. C., & Coulthard, T. J. (2020). Long-term landscape evolution of anthropogenically modified catchments: Assessment of three catchment case studies. *Hydrological Processes, 34*(9), 2061–2077.

Hartman, H. L., & Mutmansky, J. M. (2002). *Introductory mining engineering* (2nd ed.). John Wiley & Sons.

Hebblewhite, M. (2008). *A literature review of the effects of energy development on ungulates: Implications for central and eastern Montana*. Montana Fish, Wildlife and Parks.

Hebblewhite, M. (2017). Billion dollar boreal woodland caribou and the biodiversity impacts of the global oil and gas industry. *Biological Conservation, 206*, 102–111.

Hilson, G. (2002). The environmental impact of small-scale gold mining in Ghana: Identifying problems and possible solutions. *The Geographical Journal, 168*(1), 57–72.

Hilson, G., & Maconachie, R. (2020). Artisanal and small-scale mining and the sustainable development goals: Opportunities and new directions for sub-Saharan Africa. *Geoforum, 111*, 125–141.

Hobbs, R. J., Higgs, E. S., & Hall, C. M. (Eds.). (2013). *Novel ecosystems: Intervening in the new ecological world order*. Wiley-Blackwell.

Hogsden, K. L., & Harding, J. S. (2012). Consequences of acid mine drainage for the structure and function of benthic stream communities: A review. *Freshwater Science, 31*(1), 108–120.

Holl, K. D. (2002). Long-term vegetation recovery on reclaimed coal surface mines in the eastern USA. *Journal of Applied Ecology, 39*(6), 960–970.

Hudson-Edwards, K. A. (2016). Tackling mine wastes. *Science, 352*(6283), 288–290.

Hunerlach, M. P., Alpers, C. N., & Marvin-DiPasquale, M. (1999). *Mercury and methylmercury in water and sediment of the Sacramento River basin, California*. USGS Water-Resources.

Hustrulid, W. A., Kuchta, M., & Martin, R. K. (2013). *Open pit mine planning and design* (3rd ed.). CRC Press.

IAIA (International Association for Impact Assessment). (2012). *Impact assessment: IAIA special publication series no. 1*. IAIA.

ICMM (International Council on Mining and Metals). (2006). *Good practice guidance for mining and biodiversity*. ICMM.

ICMM (International Council on Mining and Metals). (2020). *Mining principles: Performance expectations*. ICMM.

Janz, D. M., DeForest, D. K., Brooks, M. L., Chapman, P. M., Gilron, G., Hoff, D., Hopkins, W. D., McIntyre, D. O., Mebane, C. A., Palace, V. P., Skorupa, J. P., & Wayland, M. (2014). Selenium toxicity to aquatic organisms. In P. M. Chapman et al. (Eds.), *Ecological assessment of selenium in the aquatic environment* (pp. 141–231). CRC Press.

Jiskani, I. M., Shah, S. A. A., Qingxiang, C., Zhou, W., & Lu, X. (2020). A multi-criteria-based SWOT analysis of sustainable planning for mining and mineral industry in Pakistan. *Arabian Journal of Geosciences, 13*(19), 1–13.

John, A., & Das, A. P. (2024). Government strategies for achieving circular economy for plastic waste management in Southeast Asia. In *Sustainable management of urban plastic waste through circular economic approaches* (pp. 72–82). Springer. https://doi.org/10.1201/9781003465447-6

Johnson, D. B. (2018). The evolution, current status, and future prospects of using biotechnologies in the mineral extraction and metal recovery sectors. *Minerals, 8*(8), 343.

Johnson, D. B., Grail, B. M., & Hallberg, K. B. (2013). A new direction for biomining: Extraction of metals by reductive dissolution of oxidized ores. *Minerals, 3*(1), 49–58.

Johnson, D. B., & Hallberg, K. B. (2005). Acid mine drainage remediation options: A review. *Science of the Total Environment, 338*(1-2), 3–14.

Kappes, D. W. (2002). Precious metal heap leach design and practice. In D. Malhotra et al. (Eds.), *Mineral processing plant design, practice, and control proceedings*. Society for Mining, Metallurgy and Exploration.

Kefeni, K. K., Msagati, T. A. M., & Mamba, B. B. (2017). Acid mine drainage: Prevention, treatment options, and resource recovery: A review. *Journal of Cleaner Production, 151*, 475–493.

Koch, J. M. (2007). Alcoa's mining and restoration process in south western Australia. *Restoration Ecology, 15*(S4), S11–S16.

Kozlov, M. V., & Zvereva, E. L. (2007). Industrial barrens: Extreme habitats created by non-ferrous metallurgy. *Reviews in Environmental Science and Bio/Technology, 6*(1–3), 231–259.

Kozlov, M. V., Zvereva, E. L., & Zverev, V. E. (2009). *Impacts of point polluters on terrestrial biota*. Springer.

Kříbek, B., Majer, V., Veselovský, F., & Nyambe, I. (2014). Discrimination of lithogenic and anthropogenic sources of metals and Sulphur in soils of the central-northern part of the Zambian Copperbelt Mining District: A topsoil vs. subsuril concept. *Journal of Geochemical Exploration, 136*, 14–27.

Kundu, N. K., & Ghose, M. K. (2014). Studies on the topsoil of an opencast coal mine. *Environmental Conservation, 21*(2), 126–132.

Kuter, N. (2013). Reclamation of degraded landscapes due to opencast mining. In *Advances in landscape architecture*. InTech.

Langer, W. H. (2001). Environmental impacts of mining natural aggregate. In Bon et al. (Eds.), *Proceedings of the 35th forum on the geology of industrial minerals*. Utah Geological Survey.

Laubscher, D. H. (1994). Cave mining: The state of the art. *Journal of the South African Institute of Mining and Metallurgy, 94*(10), 279–293.

Laurence, D. (2006). Optimisation of the mine closure process. *Journal of Cleaner Production, 14*(3-4), 285–298.

Laurence, D. (2020). Reconsidering mine closure planning for long-term environmental risk management. *Extractive Industries and Society, 7*(2), 491–495.

Lèbre, É., Corder, G., & Golev, A. (2017). The role of the mining industry in a circular economy: A framework for resource management at the mine site level. *Journal of Industrial Ecology, 21*(3), 662–672.

Lessard, J., de Bakker, J., & McHugh, L. (2016). Development of ore sorting and its impact on mineral processing economics. *Minerals Engineering, 65*, 88–97.

Lottermoser, B. G. (2010). *Mine wastes: Characterization, treatment and environmental impacts* (3rd ed.). Springer.

Luoma, S. N., & Rainbow, P. S. (2008). *Metal contamination in aquatic environments: Science and lateral management*. Cambridge University Press.

Macdonald, S. E., Landhäusser, S. M., Skousen, J., Franklin, J., Frouz, J., Hall, S., Jacobs, D. F., & Quideau, S. (2015). Forest restoration following surface mining disturbance: Challenges and solutions. *New Forests, 46*(5-6), 703–732.

Macklin, M. G., Brewer, P. A., Hudson-Edwards, K. A., Bird, G., Coulthard, T. J., Dennis, I. A., Lechler, P. J., Miller, J. R., & Turner, J. N. (2006). A geomorphological approach to the management of rivers contaminated by metal mining. *Geomorphology, 79*(3-4), 423–447.

Mallick, K., Sahu, S., Sahu, A., Dey, S., & Das, A. P. (2024). A critical review of marine microfiber pollution routes, toxicity, and its sustainable remediation. *Environmental Science and Engineering, 2333*, 189–211. https://doi.org/10.1007/978-3-031-51792-1_10

Mark, C., & Barton, T. M. (1997). Pillar design and coal strength. In *Proceedings of the 16th international conference on ground control in mining. Colorado School of Mines*.

Martínez-Cortizas, A., Pontevedra-Pombal, X., García-Rodeja, E., Nóvoa-Muñoz, J. C., & Shotyk, W. (2013). Mercury in a Spanish peat bog: Archive of climate change and atmospheric metal deposition. *Science, 284*(5416), 939–942.

Mchaina, D. M. (2001). Environmental planning considerations for the decommissioning, closure and reclamation of a mine site. *International Journal of Surface Mining, Reclamation and Environment, 15*(3), 163–176.

Miller, K. A., Thompson, K. F., Johnston, P., & Santillo, D. (2018). An overview of seabed mining including the current state of development, environmental impacts, and knowledge gaps. *Frontiers in Marine Science, 4*, 418.

Mohanty, J., Dey, S., Bala, K., & Das, A. P. (2024). Recent advances in microplastic pollution for its sustainable management. In *Sustainable management of urban plastic waste through circular economic approaches* (pp. 27–48). Springer. https://doi.org/10.1201/9781003465447-3

Mohanty, M., Tripathy, B., Koteswara Reddy, G., & Das, A. P. (2024). Greener technologies for establishing restraint over microplastic pollution. In *Sustainable microbial technology for synthetic and cellulosic microfiber bioremediation* (pp. 239–258). Springer. https://doi.org/10.1007/978-3-031-62600-5_13

Mohanty, S., Ghosh, S., Nayak, S., & Das, A. P. (2017). Bioleaching of manganese by aspergillus sp. isolated from mining deposits. *Chemosphere, 172*, 302–309. https://doi.org/10.1016/J.CHEMOSPHERE.2016.12.136

Moore, J. N., & Luoma, S. N. (1990). Hazardous wastes from large-scale metal extraction. *Environmental Science & Technology, 24*(9), 1278–1285.

Morrison-Saunders, A., Pope, J., Gunn, J. A., Bond, A., & Retief, F. (2016). Strengthening impact assessment: A call for integration and focus. *Impact Assessment and Project Appraisal, 32*(1), 2–8.

Mudd, G. M. (2001). Critical review of acid in situ leach uranium mining: USA and Australia. *Environmental Geology, 41*(3-4), 390–403.

Mudd, G. M. (2010). The environmental sustainability of mining in Australia: Key mega-trends and looming constraints. *Resources Policy, 35*(2), 98–115.

Murguía, D. I., Bringezu, S., & Schaldach, R. (2016). Global direct pressures on biodiversity by large-scale metal mining: Spatial distribution and implications for conservation. *Journal of Environmental Management, 180*, 409–420.

Nagorski, S. A., McKinnon, T. E., & Moore, J. N. (2002). Seasonal and storm-scale variations in heavy metal concentrations of two mining-contaminated streams, Montana, USA. *Journal of the American Water Resources Association, 38*(1), 1363–1378.

Nordstrom, D. K. (2011). Hydrogeochemical processes governing the origin, transport and fate of major and trace elements from mine wastes and mineralized rock to surface waters. *Applied Geochemistry, 26*(11), 1777–1791.

Northey, S. A., Mudd, G. M., Saarivuori, E., Wessman-Jääskeläinen, H., & Haque, N. (2016). Water footprinting and mining: Where are the limitations and opportunities? *Journal of Cleaner Production, 135*, 1098–1116.

Northey, S. A., Mudd, G. M., Saarivuori, E., Wessman-Jääskeläinen, H., & Haque, N. (2018). Water footprinting and mining: Where are the limitations and opportunities? *Journal of Cleaner Production, 135*, 1098–1116.

Ongena, S., Bollen, T., & de Pooter, S. (2018). Mining waste management with blockchain technology. In *Proceedings of the APCOM mining conference. Santiago, Chile.*

Oyarzún, J., & Oyarzún, R. (2011). Sustainable development threats, inter-sector conflicts and environmental policy requirements in the arid, mining rich, northern Chile territory. *Sustainable Development, 19*(4), 263–274.

Palmer, M. A., Bernhardt, E. S., Schlesinger, W. H., Eshleman, K. N., Foufoula-Georgiou, E., Hendryx, M. S., Lemly, A. D., Likens, G. E., Loucks, O. L., Power, M. E., White, P. S., & Wilcock, P. R. (2010). Mountaintop mining consequences. *Science, 327*(5962), 148–149.

Patricelli, G. L., & Blickley, J. L. (2006). Avian communication in urban noise: Causes and consequences of vocal adjustment. *The Auk, 123*(3), 639–649.

Petavratzi, E., Kingman, S., & Lowndes, I. (2005). Particulates from mining operations: A review of sources, effects and regulations. *Minerals Engineering, 18*(12), 1183–1199.

Phillips, J. (2016). Climate change and surface mining: A review of environment-human interactions & their spatial dynamics. *Applied Geography, 74*, 95–108.

Priyadarshini, B., & Das, A. P. (2024). The adverse impact of synthetic microplastics released from disposable face masks pollutants on aquatic microbiota, aquatic species and humans: A review. *Environmental Science and Engineering, 2827*, 67–81. https://doi.org/10.1007/978-3-031-50840-0_5

Priyadarshini, S., Jagatee, S., & Das, A. P. (2024). Synthetic fabrics and microfiber pollution–an assessment of their global impact. *Environmental Science and Engineering, 2333*, 137–157. https://doi.org/10.1007/978-3-031-51792-1_8

Priyadarsini, S., & Das, A. P. (2025). Lithium bioleaching: A review on microbial-assisted sustainable technology for lithium bio-circularity. *Journal of Water Process Engineering, 69*, 106744. https://doi.org/10.1016/j.jwpe.2024.106744

Raiter, K. G., Possingham, H. P., Prober, S. M., & Hobbs, R. J. (2014). Under the radar: Mitigating enigmatic ecological impacts. *Trends in Ecology & Evolution, 29*(11), 635–644.

Ramirez-Llodra, E., Trannum, H. C., Evenset, A., Levin, L. A., Andersson, M., Finne, T. E., Hilario, A., Flem, B., Christensen, G., Schaanning, M., & Vanreusel, A. (2015). Submarine and deep-sea mine tailing placements: A review of current practices, environmental issues, natural analogs and knowledge gaps in Norway and internationally. *Marine Pollution Bulletin, 97*(1-2), 13–35.

Ranängen, H., & Lindman, Å. (2017). A path towards sustainability for the Nordic mining industry. *Journal of Cleaner Production, 151*, 43–52.

Rankine, R. M., Sivakugan, N., & Rankine, K. J. (2006). Geotechnical characteristics of mine tailings. In A. Phoon et al. (Eds.), *Proceedings of the international conference on geotechnical engineering*. Springer.

Ray, S. K., Dey, S., Das, A. P., & John, A. (2024). Microplastic waste management through greener technologies: A review. In *Sustainable management of urban plastic waste through circular economic approaches* (pp. 237–261). Springer. https://doi.org/10.1201/9781003465447-14

Reuter, M. A., van Schaik, A., Gutzmer, J., Bartie, N., & Abadías-Llamas, A. (2018). Challenges of the circular economy: A material, metallurgical, and product design perspective. *Annual Review of Materials Research, 49*, 253–274.

Roche, C., Thygesen, K., & Baker, E. (2017). *Mine tailings storage: Safety is no accident*. United Nations Environment Programme and GRID-Arendal.

Rooney, R. C., Bayley, S. E., & Schindler, D. W. (2012). Oil sands mining and reclamation cause massive loss of peatland and stored carbon. *Proceedings of the National Academy of Sciences, 109*(13), 4933–4937.

Sahu, A., Dey, S., Mallick, K., Priyadarsini, S., Behera, I. D., & Das, A. P. (2024). *A review on advanced technology for sustainable management of synthetic microplastic waste. Sustainable management of urban plastic waste through circular economic approaches* (pp. 262–279). Springer. https://doi.org/10.1201/9781003465447-15

Sahu, A., Mallick, K., & Das, A. P. (2024). Environmental impact of single-use synthetic face mask and its recycling: A sustainable approach. In *Impact of COVID-19 waste on environmental pollution and its sustainable management* (pp. 197–212). Springer. https://doi.org/10.1007/978-3-031-50840-0_11

Sánchez, F., & Hartlieb, P. (2020). Innovation in the mining industry: Technological trends and a case study of the challenges of disruptive innovation. *Mining, Metallurgy & Exploration, 37*, 1385–1399.

Sassoon, M. (2009). *Financial surety: Guidelines for the implementation of financial surety for mine closure. Extractive industries for development series #7*. World Bank.

Sheoran, V., Sheoran, A. S., & Poonia, P. (2010). Soil reclamation of abandoned mine land by revegetation: A review. *International Journal of Soil, Sediment and Water, 3*(2), 13.

Sheoran, V., Sheoran, A. S., & Poonia, P. (2016). Factors affecting phytoextraction: A review. *Pedosphere, 26*(2), 148–166.

Skousen, J., Zipper, C. E., Rose, A., Ziemkiewicz, P. F., Nairn, R., McDonald, L. M., & Kleinmann, R. L. (2017). Review of passive systems for acid mine drainage treatment. *Mine Water and the Environment, 36*(1), 133–153.

Sonter, L. J., Ali, S. H., & Watson, J. E. M. (2018). Mining and biodiversity: Key issues and research needs in conservation science. *Proceedings of the Royal Society B: Biological Sciences, 285*(1892), 20181926.

Sonter, L. J., Barrett, D. J., Soares-Filho, B. S., & Moran, C. J. (2014). Global demand for steel drives extensive land-use change in Brazil's iron quadrangle. *Global Environmental Change, 26*, 63–72.

Sonter, L. J., Herrera, D., Barrett, D. J., Galford, G. L., Moran, C. J., & Soares-Filho, B. S. (2017). Mining drives extensive deforestation in the Brazilian Amazon. *Nature Communications, 8*(1), 1013.

Swain, S., Purnima Sudha, M., Dey, S., Badu, Y., Bhoi, S. R., Alreshidi, M. A., Qasim, M. T., Nath, H., Ram, D. K., & Das, A. P. (2025). Sources of micro and nano plastic hydrocarbon pollutants in marine bodies and their treatment. *Environmental Science and Engineering, 218*, 45–67. https://doi.org/10.1007/978-3-031-81615-4_3

Taylor, M. P., Mould, D. J., Kesterton, R. G. H., & Flint, K. P. (1995). Environmental management of the Bougainville copper mine in Papua New Guinea. In M. J. Crozier, P. N. Preston, & D. S. Smith (Eds.), *Science and the environment* (pp. 543–586). Longman.

Thrush, S. F., & Dayton, P. K. (2002). Disturbance to marine benthic habitats by trawling and dredging: Implications for marine biodiversity. *Annual Review of Ecology and Systematics, 33*, 449–473.

Tordoff, G. M., Baker, A. J. M., & Willis, A. J. (2000). Current approaches to the revegetation and reclamation of metalliferous mine wastes. *Chemosphere, 41*(1–2), 219–228.

Virah-Sawmy, M., Ebeling, J., & Taplin, R. (2014). Mining and biodiversity offsets: A transparent and science-based approach to measure "no-net-loss". *Journal of Environmental Management, 143*, 61–70.

Whitmore, A. (2006). The emperor's new clothes: Sustainable mining? *Journal of Cleaner Production, 14*(3-4), 309–314.

Wickham, J., Wood, P. B., Nicholson, M. C., Jenkins, W., Druckenbrod, D., Suter, G. W., Strager, M. P., Mazzarella, C., Galloway, W., & Amos, J. (2013). The overlooked terrestrial impacts of mountaintop mining. *Bioscience, 63*(5), 335–348.

Wolfe, J. E., Vaughan, R. G., & Rice, M. A. (2000). Dredge mining in the United States: A review of techniques, impacts and regulations. In *Proceedings of the Western Dredging Association Annual Conference*. WEDA.

Wolkersdorfer, C. (2008). *Water management at abandoned flooded underground mines: Fundamentals, tracer tests, modelling, water treatment*. Springer.

Younger, P. L. (2001). Mine water pollution in Scotland: Nature, extent and preventative strategies. *Science of the Total Environment, 265*(1-3), 309–326.

Younger, P. L., Banwart, S. A., & Hedin, R. S. (2002). *Mine water: Hydrology, pollution, remediation*. Springer.

Younger, P. L., & Wolkersdorfer, C. (2004). Mining impacts on the fresh water environment: Technical and managerial guidelines for catchment scale management. *Mine Water and the Environment, 23*(S1), 2–80.

Zipper, C. E., Burger, J. A., Skousen, J. G., Angel, P. N., Barton, C. D., Davis, V., & Franklin, J. A. (2011). Restoring forests and associated ecosystem services on Appalachian coal surface mines. *Environmental Management, 47*(5), 751–765.

Health Hazards Related to Uranium Mining and Bioremediation as a Sustainable Approach

Anamika Basu

1 Introduction

Approximately 60,000 tons of uranium are produced worldwide at present, with five major nations—Kazakhstan (Bersimbaev & Bulgakova, 2015), Canada, Niger, Australia, and Namibia—producing 47,000 tons of uranium. In the Singhbhum Shear Zone in Jharkhand, Uranium Corporation of India Limited (UCIL) presently runs one open pit mine at Banduhurang and six underground uranium mines at Jaduguda (Panigrahi, Mishra, & Sahu, 2015), Turamdih, Bhatin, Bagjata (Sarangi et al., 2007), Narwapahar (Reddy et al., 2012), and Mohuldih. In addition to this, UCIL runs an underground uranium mine in the Andhra Pradesh state of Tummalapalle (Panigrahi et al., 2018) to supply the country's dedicated nuclear power program with uranium, which is in increasing demand.

The primary sources of radiological risks related to uranium mining are internal and exterior radiation exposures. Beginning with uranium-238 and ending with the stable isotope lead-206, the uranium disintegration series—also referred to as the uranium-radium series—is a chain of natural radioactive decay that passes through a number of radioactive intermediate isotopes. Thorium, uranium, protactinium, radium, polonium, radon, thallium, bismuth, and uranium isotopes are all included in the series. When ^{218}Po atoms form, they are very mobile and positively charged. They only need a few seconds to attach themselves to the surrounding aerosols. The alpha particles from radon's radioactive disintegration immediately affect delicate lung tissues when inhaled, causing harm that may eventually result in lung cancer (Sahu et al., 2016). However, the majority of radon is exhaled because it is a gas.

A. Basu (✉)
Department of Biochemistry, Gurudas College, Affiliated to University of Calcutta, Calcutta, India
e-mail: anamika.biochem@gurudas.education

© The Author(s), under exclusive license to Springer Nature Switzerland AG 2026

A. P. Das (ed.), *Bioremediation Technology for Sustainable Mining Waste Management*, Sustainable Environmental Waste Management Strategies,
https://doi.org/10.1007/978-3-032-03468-7_5

However, even after leaving the job, the heavy metal atoms that are left over from its decay accumulate (Chen & Harley, 2020) inside the respiratory system and continue to irradiate the lung tissues (Das & Ghosh, 2023a; Priyadarshini et al., 2024; Reddy Koteswara et al., 2025).

The harmful effects of uranium on human kidneys, lung epithelial cells, and osteocyte cells, including cell necrosis, as well as on many other organisms, such as plants, aquatic invertebrates and vertebrates, bacteria, and fungi, are particularly concerning. Although natural uranium has low radiotoxicity because of its presence of ^{238}U (>99%), but the chemotoxicity is prevalent due to the formation of uranyl ion in aerobic environment for all living organisms such as plants and animals (Das, Dey, Bala, et al., 2024; Mohanty, Dey, et al., 2024; Sahu, Dey, et al., 2024). This uranyl ion, acidic in nature, can bind with various oxygen donors such as carboxylate, phosphate, hydroxyl, and carbonate. This is actual cause of uranium toxicity in living systems. Inside the cell uranium increases the production of NO (nitric oxide) and H_2O_2 (hydrogen peroxide) as well as involves in the homeostasis mechanisms of iron, manganese, and phosphate (Sarthou et al., 2020).

The most stable uranium form, Uranyl ion (UO_2^{2+}) exists under physiological conditions, and the high toxicity of uranium may be due to the capacity of UO_2^{2+} to interact with biomolecules including proteins and nucleotides, changing or disrupting their natural activities (Lin, 2020). After binding with proteins, uranyl ion induces some conformational changes as well as disrupts biomolecular interactions such as protein–protein interactions, protein–ligand interactions and protein–DNA interactions. Under physiological condition, these interactions result in toxicity of uranium, by disrupting or altering the native structures of biomolecules such as proteins, DNA, or ligands. In general, uranyl (UO_2^{2+}) forms coordination complexes with four/five atoms from carboxylate groups of acidic amino acids such as glutamate, aspartate (Thuéry & Harrowfield, 2024). The nitrate and acetate groups of ligands (Cao et al., 2023) are participated with complex formation with uranyl ions, resulting in non-radioactive uranium toxicity in living organisms. These similar mechanisms are elucidated for bio-extraction method for uranium metals from environment (Vidaud et al., 2005).

Binding of blood proteins (such as hemoglobin, transferrin, serum albumin, etc.) (Sauge-Merle et al., 2017), bone growth related proteins (e.g., osteopontin, fetnin-A), and some intracellular proteins (such as cytochromes specifically, Cyt b_5/c, calcium binding proteins, calmodulin (Montavon et al., 2009), metallothionein, etc., with uranyl group, has been investigated thoroughly. Both albumin and transferrin can bind with hexavalent uranium U(VI) to form ternary complex in presence of carbonate as synergistic anion, in coordination sphere (Van Horn & Huang, 2006). Furthermore, transferrin protein containing two lobes (known as N and C lobes) binds with U (VI) through two tyrosine residues such as Tyr 188 and Tyr 95, which are also Fe (III) binding sites. UV absorption band at 242 nm confirms the interaction of uranium metal with the phenolic -OH group of tyrosine.

Upon uranyl ion binding, peroxidase activity of cytochrome Cyt c decreases and the protein–protein interaction between Cyt b_5 and Cyt c can be interfered. Similarly, ubiquitin, a small protein with 76 amino acids residues, degrades misfolded protein using ubiquitin proteosome system (UPS). The glutamate and aspartate residues (18 and 21, respectively) participate in coordination binding with UO_2^{2+}, results in dysfunction of UPS mechanisms in human lung cells (Dinocourt et al., 2015). Since, this uranyl ion can cross the blood–brain barrier (BBB) (Vellingiri, 2023), henceforth, uranium toxicity has strong relationship with occurrence of some neurodegenerative diseases such as Parkinson's disease, Alzheimer's disease, and prion diseases. After reaching brain, uranium causes neurobehavioral impairments by increasing locomotor neuron activity, decreasing memory, increasing anxiety and perturbing the sleep-wake cycle in nervous system (Das et al., 2023; Das, Dey, & Das, 2024; Das, Ray, Mohanty, et al., 2024; Dey, Tripathy, et al., 2023).

The Purbi Singhbhum district of Jharkhand is site to the underground uranium mine referred to as the Jaduguda uranium mine. A 5-m-diameter vertical shaft leads to the mine, which is at least 640 m below the surface of the earth. 99.28% of uranium ore that is mined is considered waste since the majority of the ore, which is composed of uranium-236 and uranium-238 along with a few other elements, is left behind because uranium-235 is the primary uranium isotope used in nuclear power plants. After being neutralized with lime, this waste—also known as tailings—is transported via pipes to a tailing pond (Panigrahi, Mishra, Sahu, & Bhowmik, 2015). This transportation is made possible by the supply of clean water (in pipelines) from decantation wells, which is subsequently transported to an effluent treatment facility for the removal of manganese and radium via a closed channel. After that, the solid tailings are kept in the tailing ponds.

Radiation from the tailing ponds has a negative impact on the approximately 35,000 residents who live within a 5-km radius of the mines (Sahu et al., 2014). Additionally, miscarriages and stillbirths are common in this area. Children were stillborn in the past. These days, kids pass just a few days after birth. The loss of fertility affects around one-third of the women in these places (Sonowal, 2020). Rare diseases are affecting even animals like cows and buffaloes. Serious environmental and health issues have been whipped up by the Jaduguda Uranium Mine in Jharkhand, India (Bose, 2010). These dangers are caused by exposure to radioactive materials (Jha et al., 2000), contamination of natural resources, and other hazards associated with mining (Sahu, Beg, et al., 2024).

Health risks to people who live close to uranium mines fall into a number of domains, including radiation exposure's negative consequences, heavy metal contamination of the soil and water, air pollution, mine workers' occupational risks, and environmental effects on people. Radioactive radon gas released during uranium mining and processing, tailings ponds used for storing radioactive slurry, and finally contaminated soil and dust particles are the causes of radiation exposure (Bala et al., 2024; Brugge & Buchner, 2011; Das, Behera, & Das, 2024; John & Das, 2024; Mohanty, Tripathy, et al., 2024; Raj & Das, 2023). Radiation exposure causes a number of ailments, such as a rise in blood, lung, and bone malignancies as well as chronic radiation syndrome, which manifests as weariness, immunological suppression, and neurological consequences. Genetic alterations lead to a higher rate of miscarriages, infertility, and congenital impairments (Srivastava et al., 2020).

Heavy metal pollution of the environment results from the leakage of hazardous heavy metals such as arsenic, lead, and cadmium into the soil and water during uranium mining (Pereira et al., 2014). Kidney function can be affected by long-term exposure to heavy metals. In particular, uranium bioaccumulation weakens bones and lead poisoning can induce neurological abnormalities, such as cognitive deficits and developmental problems in children. Tailings leakage is the term for the radioactive and chemical waste that frequently escapes into surrounding water bodies and groundwater from tailing ponds. Health problems including diarrhea and stomach ache are caused by this tainted water, which can lead to long-term digestive problems. Furthermore, eating tainted fish (Lourenço et al., 2017) or crops exposes people to more radiation and heavy metals, a process known as bioaccumulation of radioactive materials. Rashes and long-term skin issues are caused by direct contact with tainted water. Ecosystems are disturbed by mining operations, which lowers agricultural output. Radiation builds up in crops that are irrigated with tainted water, endangering the health of consumers (Dey, Das, et al., 2023; Ghosh, Sinha, et al., 2023; Jagatee et al., 2024; Kumar et al., 2023).

Lung fibrosis, lung cancer, and chronic bronchitis are all brought on by the fine radioactive dust and radon gas released into the atmosphere by mining operations. Due to a lack of proper safety precautions, miners are exposed to high quantities of dust and radiation, which increases their risk of developing silicosis, radiation sickness, and cancer (Au et al., 1998). There may be more infants born with birth defects and developmental impairments, according to anecdotal evidence (Priyadarshini & Das, 2024; Sahu, Mallick, & Das, 2024). Infertility, stillbirths, and miscarriages are said to be more common in the area (Carvalho, 2011). Mining operations inherently provide a significant safety risk to marine and terrestrial life, and most importantly, human health, due to the toxicity and radioactivity of U (National Research Council, 2011).

Different chemical forms of U, including UO_2, UO_3, UF_4, UF_6, UO_2F_2, U_3O_8, and $UO_2(NO_3)_2$, are exposed to mine workers. These personnel are also exposed to non-radiological risks (such as crystalline silica, acids, and noise) as well as radium, thorium, and RDPs (Canu et al., 2009). The most frequent exposure channel for the general public is ingestion through food or water, while the most frequent exposure pathway for plant personnel is inhalation of airborne U particles (Srivastava et al., 2020). Rather than radiation risks, the main factor influencing occupational exposure limits has been the chemotoxicity of U. Alpha particle exposure is increased by insoluble U particles because they activate macrophages in the lungs and are stored in the body for a longer period of time. Even though alpha particles cannot pass through the epidermis, they can enter the body through the mouth or nose (Rathod et al., 2023).

U has the highest solubility and is most likely to leak into surface and subsurface water in agricultural soil with high pH and redox potential. Numerous ligands, including phosphate, carbonate, nitrate, bacteria (Li et al., 2024), and humic compounds, can form complexes with U species. Therefore, uranium-contaminated soil can have a number of detrimental impacts on plant systems, including decreased seed germination rates, inhibitions of overall plant development, decreased photosynthesis rates, disruptions of plant nutrient intake, oxidative damage, and genotoxic effects (Chen et al., 2022). Most people concur that plants produce too many reactive oxygen species (ROS) as their first response to abiotic stress, particularly heavy metal stress (Das et al., 2011, 2015; Priyadarsini & Das, 2025).

The overproduction of ROS caused by U may lead to a variety of cellular damage, such as DNA and RNA mutilation, modifications to the cell cycle and division, and inhibition of the activity of various proteins (Chen et al., 2021; Gao et al., 2024). Plants have evolved several defensive mechanisms to lessen oxidative stress caused by ROS, such as U sequestration in root tissue, compartmentalization in the cell wall and vacuoles, and chelation by organic chemicals. The antioxidant enzyme system has been one of the most important defensive strategies as a secondary defense mechanism as it effectively scavenges the U-mediated overproduced ROS (Ma et al., 2020).

Uranium from soil penetrates into the plant roots and then accumulates in root cells, before translocation to the aerial parts of plants e.g. *Arabidopsis thaliana, Nicotina tabacum, Pisum sativum*, etc. Specifically in Arabidopsis (Doustaly et al., 2014), three transporter proteins viz., FIT1 (transcription factor), FRO2 (a ferric-chelate reductase), IRT1(a high-affinity Fe^{2+} transporter) are differentially expressed in presence of uranyl group (Berthet et al., 2018). IRT1, a Fe^{+2} transporter involves in uranium transport in plant. Here, uranium competing with iron, form complex with phosphate group. The molecular mechanism of uranium binding with metal transporters in plant is not elucidated thoroughly. Membrane transporter proteins regulate the permeability of uranium inside the plant cell. Translocation of uranium from root to shoot parts may be facilitated by the binding of UO_2 with lactate ion by symplastic movement (Wu et al., 2020). Transport of uranium via xylem occurs after crossing the Casparian strip cells by active transport (Pentyala & Eapen, 2020).

It is possible to draw the preliminary conclusion that long-term low-dose radiation exposure from internally deposited uranium may put workers at uranium processing and enrichment plants at risk for lung cancer, dementia, Alzheimer's disease, and circulatory system disorders. Radiation monitoring (by installing real-time radiation monitoring systems) (Lourenço et al., 2013), preventing tailing leakage (by sealing tailing ponds with impermeable lining), treating and recycling uranium-contaminated water, and aligning mining operations with International Atomic Energy Agency (IAEA) radiation protection guidelines can all help to reduce the health risks of uranium miners (Bjørklund et al., 2017).

2 Bioremediation of Waste from Uranium Mine

Only a small number of studies have been carried out in the field, despite reports of various methods for cleaning up U-contaminated soils. Because of their cheap cost, environmental safety, and in situ treatment capabilities, biological techniques (plants and microorganisms) for the removal of U from polluted settings have become more prominent among these remediation approaches. A viable and sustainable strategy for reducing the risks to the environment and human health presented by radioactive and heavy metal pollution is the bioremediation of uranium mine waste (Mkandawire, 2013). In order to neutralize, eliminate, or stabilize pollutants like uranium, radium, thorium, and other heavy metals in soil, microbes, plants, or their enzymes are used.

3 Methods for Bioremediation of Uranium Mine Waste

3.1 Microbial Bioremediation

Certain bacteria, such as Geobacter and Shewanella, have been shown to exhibit microbial reduction of uranium, which reduces the mobility and bioavailability of soluble uranium [U(VI)] by converting it to its insoluble form [U(IV)], which precipitates as uranium oxide (UO_2). During anaerobic respiration, these bacteria employ uranium as an electron acceptor. Uranium in tainted groundwater is efficiently immobilized by *Geobacter sulfurreducens*. Acidophilic bacteria such as *Acidithiobacillus ferrooxidans* may bioleach uranium by oxidizing sulfides, which makes recovery possible and waste reduction possible (Das & Ghosh, 2023a; Ghosh et al., 2016). The method's advantage is that it can transform solid waste into a soluble state that may be processed for recovery or safe disposal.

Dead microbial biomass (such as bacteria, fungus, or algae) or plant material can adsorb uranium ions from wastewater or soil utilizing ion exchange, complexation, or surface precipitation in the biosorption technique (Groudev et al., 2007), which is adsorption by biomaterials (Banerjee et al., 2022). *Sargassum* (seaweed) and *Aspergillus niger* (fungus) are two examples of efficient uranium biosorbents. Bioaugmentation technique speeds up the natural bioremediation process in highly polluted situations by introducing native or artificial microorganisms that can bioremediate uranium to contaminated places (Acharya, 2015, Dey, Tripathy, et al., 2023; Mohanty et al., 2016). *Pseudomonas aeruginosa* was shown to exhibit uranium biomineralization through regulated polyphosphate metabolism, whereby internal polyphosphate breakdown resulted in the release of phosphate, which precipitated uranium as uranyl phosphate on the cell surface (Newman-Portela et al., 2024).

The proper functioning of ecosystems and soils depends on soil microbes. The mobility and transmission of uranium are significantly influenced by bacterial interactions that change its speciation. Through a variety of mechanisms, including as sorption, reduction, precipitation, and mineralization (Chandwadkar et al., 2018), bacteria can immobilize uranium. Uranium reductive precipitation is mediated by a number of bacterial species in anoxic environments (Kulkarni et al., 2016). When oxygen is present, uranyl biosorption on chelating groups takes place on bacterial surfaces. By employing inorganic phosphate for bioprecipitation and biomineralization (Maleke et al., 2015), a range of bacterial strains can also reduce uranium mobility. Both processes can take place extracellularly or intracellularly and include phosphatase activity (Reninger et al., 2004). The sequestration of uranium may also be influenced by metabolites like siderophores. Each of these methods has been suggested for detoxifying uranium.

In addition, transporters like cation diffusion facilitator (CDF) and PIB-ATPases may have a role in uranium tolerance (Lebrazi & Fikri-Benbrahim, 2018). Some bacterial species can sequester uranium, making them interesting candidates for bioremediation procedures. Transporter protein for uranium transport present in different microorganisms is given in Table 1.

Microbacterium, a common genus of actinobacteria, has been researched for its uranium tolerance. The A9 strain of *Microbacterium oleivorans* was discovered from a polluted waste storage facility in Chernobyl (Fig. 1). A proteomics investigation revealed two proteins that are modified by uranyl exposure, namely UipA and UipB (Uranium-Induced Protein A and B). UipA protein (Gallois et al., 2022) is unique to uranium-tolerant strains (Ortet et al., 2019) and is rapidly accumulated inside the cells in response to uranyl exposure. UipA is a single-pass transmembrane protein found in the bacteria *Microbacterium oleivorans* A9 (Philippe et al., 2017). Its large C-terminal soluble domain has a nanomolar binding affinity for UO_2^{2+} and as well as Fe^{3+}. Different spectroscopic studies (Theodorakopoulos et al., 2015) revealed that the protein's mono and bidentate carboxylate groups interacted with both metals. Uranyl coordination requires carboxylate groups from Asp and Glu residues.

The crystal structure of UipA, solved in its apo state and bound to uranium, revealed two PepSY domains with uranium binding spots, located on a highly negatively charged face, and stabilized by a set of conserved residues. UipA proteins (Gallois et al., 2018) are predicted to be single-pass transmembrane proteins with a cytoplasmic N-terminal domain (38–75 aa in length), and an extracellular C-terminal domain of 185–186 residues. It has low pI values (3.86–3.99) due to their richness

Table 1 List of transport proteins for uranium transport in microorganisms

Entry name	Protein names	Gene names	Organism	Length
A0A0D0GNG7_9SPHI	Cation efflux protein cytoplasmic domain-containing protein	TH53_01180	*Pedobacter lusitanus*	470 AA
N0AVT6_9BACI	Cation diffusion facilitator family transporter	B1NLA3E_03700	*Bacillus* sp. *1NLA3E*	330 AA
A0A0D0G2N8_9SPHI	Cation diffusion facilitator family transporter	TH53_00125	*Pedobacter lusitanus*	325 AA
A0A411HE30_ARTSP	Cation transporter	czcD	*Arthrobacter* sp.	163 AA
N0AU14_9BACI	Cation efflux system protein	B1NLA3E_08525	*Bacillus* sp. *1NLA3E*	311 AA
A0A0D0GES1_9SPHI	Cation transporter	TH53_18540	*Pedobacter lusitanus*	296 AA

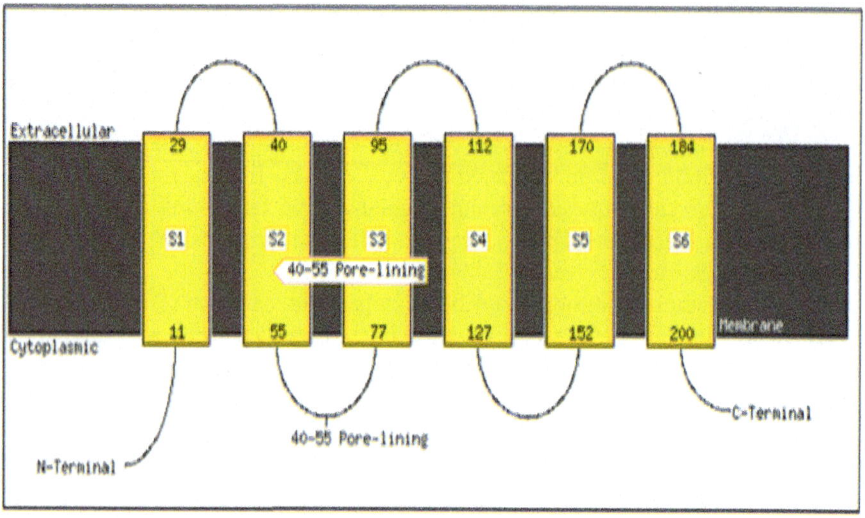

Fig. 1 Cation efflux protein cytoplasmic domain-containing protein from *Pedobacter lusitanus*

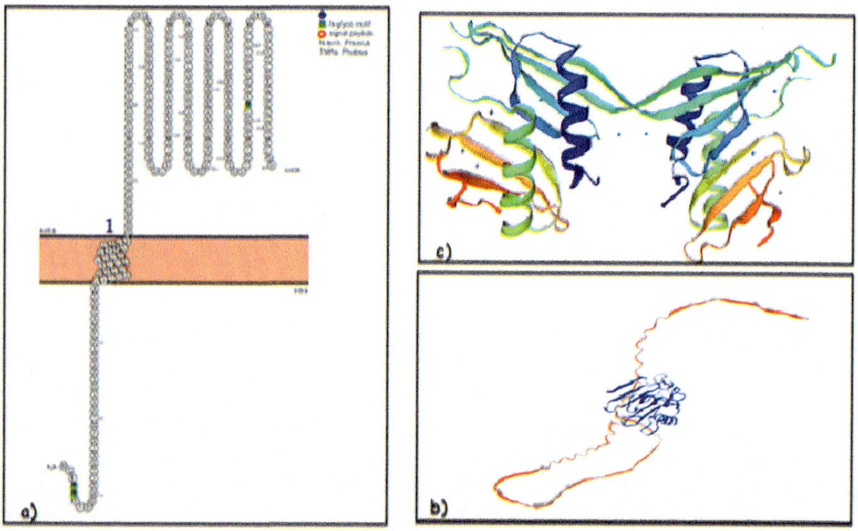

Fig. 2 Three-dimensional structure of UipA protein, (**a**) UipA as single-pass transmembrane protein, (**b**) molecular modeling structure of UipA with its disordered part structure, (**c**) dimeric structure of UpiA protein in presence of uranium

in negatively charged residues namely aspartate and glutamate, accounting for 18.6%–20.7% of the total amino acids residues in the protein (Fig. 2).

For drawing the structure of transmembrane protein, UipA, Protter, an open-source tool for interactive integration and visualization of annotated and predicted protein sequence features together with experimental proteomic evidence, is used (Omasits et al., 2014). By comparing the three-dimensional structures of apoprotein (PDB ID 7ATH) and uranium binding protein (PDB ID 7ATK), it can be concluded that the buried amino acid residues (positions 107, 176, 183, 193) in apoprotein structure become exposed to surface after uranium binding, due to conformational change in mettaloprotein.

Siderophores are actually organic compounds having low molecular weight, secreted by plants and microorganisms in specific abiotic condition such as low iron. These organic compounds having specific functional groups such as hydroxamates, carboxylates, catecholates, etc., help to make available iron from environment for plants and microorganisms. The application of siderophores in bioremediation of metals has been investigated due to their binding ability to various metals such as Mo, Mn, Ni, Co, and U, as chelating agents. At the siderophore-binding sites, iron can complete with free protons as well as various metal ions, for example, Cd^{+2}, PB^{+2}, Mn^{+3}, and U^{+4}. Examples of siderophores along with their chemical nature are enlisted in Table 2. Studies have indicated that uranium stress can induce siderophore production in some microorganisms, suggesting a potential mechanism for uranium sequestration.

Observations in cyanobacteria and other microbes suggest that siderophores might protect cells from uranium toxicity by binding and sequestering the metal, thereby reducing its availability for the cell. Some siderophores, like pyoverdines, have been shown to mobilize uranium, neptunium, and other metals from uranium mine waste (Ivanov et al., 2019). The ability of siderophores to interact with uranium and other metals has potential applications in bioremediation and environmental cleanup, including the treatment of uranium-contaminated sites. The cyanobacterium *Synechococcus elongatus* BDU 130911 (Rashmi et al., 2013) has been shown to produce siderophores under uranium stress, and the fungal siderophore ferrichrome has been used in bioleaching strategies to extract uranium and other metals (Fig. 3).

FhuA protein facilitates ligand-gated transport of ferrichrome-bound iron across *Escherichia coli* outer membranes (Locher et al., 1998).

Gene expression data for *Microbacterium oleivorans* Strain A9, a bacterium isolated from Chernobyl radionuclide-contaminated soil, shows the putative siderophore-binding lipoprotein YfiY gene with GI accession no. BWL13_00727 (PI 4.23) (Mol wt 30,140), is upregulated (T fold 1.300). The protein related with this gene is Iron-siderophore ABC transporter substrate-binding protein

Table 2 Siderophore production due to uranium stress

Origin of siderophore	Examples of siderophores	Chemical nature
Microbial	Enterobacteria	Catecholates
	Rhizobactin	Carboxylates
	Ferrioxamine B	Hydroxamates
Fungal	Ferrichrome (PDB ID 1BY5)	Hydroxamates
Phyto	Mugineic acid	

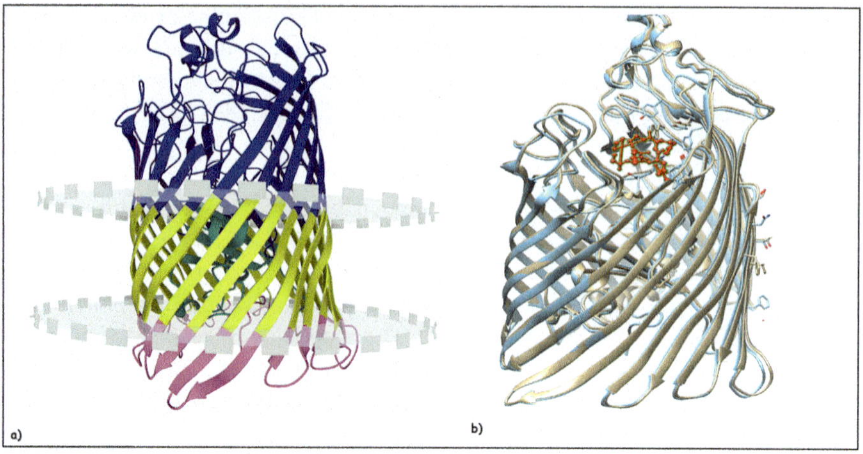

Fig. 3 (**a**) Ferric hydroxamate uptake protein (PDB ID 1BY5) in membrane bound structure. (**b**) Ferric hydroxamate uptake protein (PDB ID 1BY5) and protein (ferrichrome-iron receptor precursor (FHUA) (PDB ID 1BY3)) high lighting ferrichrome in red

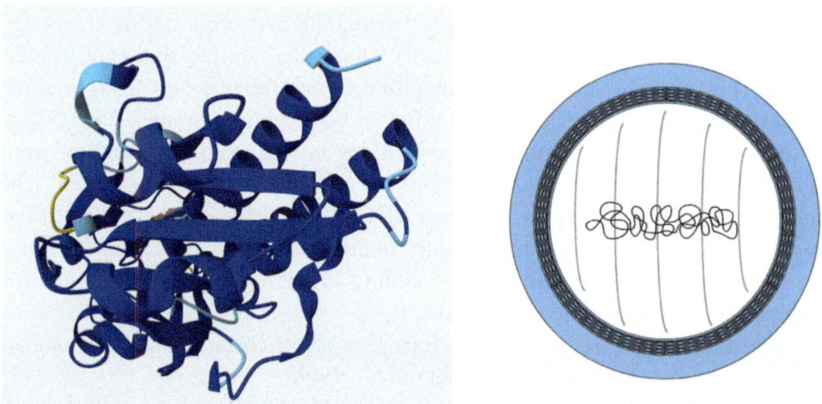

Fig. 4 Structure and location of Iron-siderophore ABC transporter substrate-binding protein (Microbacterium oleivorans)

(*Microbacterium oleivorans*) (UniProtKB/TrEMBL A0A031FUL0). It is present in outer membrane-bounded periplasmic space of bacterial cell (Fig. 4), belongs to the bacterial solute-binding protein 8 family. It protein binds organic molecules such as hydroxamate, dicitrate.

This protein is present in Gram-positive bacteria *Microbacterium oleivorans,* does not contain outer membrane. The iron-siderophore (i.e., hydroxamate, dicitrate) can be bound by the transporter protein. This protein is attached to the microbial cell membrane since no periplasmic space is present in the Gram-positive bacteria. Metals such as uranium, iron, and so on, can be transported toward the cytoplasm of bacteria with the help of the ABC transport system.

3.2 Phytoremediation

Because of their effective and ecologically acceptable capacity to absorb and accumulate U, plants are being recognized as a key tool for in situ remediation of U-contaminated soils. Both U(IV) and U(VI) are absorbed by plants through epidermal root cells, however, the exact chemical process is unclear. The majority of U cations, including UO_2^{2+}, are absorbed by roots mostly via the same ion channels or carriers as essential elements including calcium, iron, and magnesium. As analogs of bicarbonate or phosphate, anionic species of U, such as $UO_2(CO_3)_2^{2-}$ and $UO_2PO_4^-$, can cross cell membranes through anionic channels. Large, nonselective membrane anion channels are common.

Plants absorb uranium by passive absorption, such as assisted diffusion, or active transport, which involves energy expenditure. Uranium is mostly present in soil solution as UO_2^{2+} and $[UO_2OH]^+$. Through their roots, hyperaccumulator plants store uranium and other heavy elements in biomass. For instance, because they absorb and transfer large amounts of U to aerial parts, water lily, wild ramie, and mustard are potential U hyperaccumulators.

For possible detoxification of radionuclide uranium, from *Arabidopsis thaliana*, 38 uranyl binding proteins (UraBPs) have been identified from root and shoot parts of plant. Among them plasma membrane associated cation-binding protein 1 (PCaP1) (Vallet et al., 2023) can bind with U(VI), Cu (II), and Fe(III), by oligomer formation. Uranyl ion, a form of uranium causes toxic effect on plants. Research indicates that this protein plays an important role in managing uranium uptake from contaminated soil and potentially mitigating its harmful effects within the plant. PCaP1 proteins are proved to help plant sequestering uranium by binding to uranyl ions and thus preventing them reaching critical cellular components. The binding of uranyl to these proteins usually involves specific amino acid residues with oxygen-containing side chains, that can coordinate with the uranium ion.

4 Advantages of Bioremediation

Eco-friendly bioremediation minimizes secondary pollution compared to chemical or physical methods. It requires fewer resources and energy than traditional remediation methods. Moreover, bioremediation technique can be applied directly at the contamination site, reducing the need for excavation or transport. It can address multiple contaminants (e.g., uranium and heavy metals) at once.

5 Challenges in Bioremediation of Uranium Waste

Microorganisms and plants must survive and function in highly radioactive environments. Presence of mixed contaminants (e.g., uranium with arsenic or radium) complicates remediation. Bioremediation can take longer than chemical methods to show significant results. Effectiveness depends on factors like pH, temperature, and the presence of nutrients or competing ions.

Bioremediation efficiency can be enhanced by create genetically modified organisms (GMOs) with enhanced uranium tolerance and uptake capacity (Thakur & Kumar, 2024). Bioavailability of the technique can be increased by adding chelating agents like citric acid to increase uranium solubility and uptake. Combine bioremediation with other methods (e.g., phytoremediation + bioaugmentation) can be applied for improved results. Sensors and models can be used to monitor progress and optimize environmental conditions. In Savannah River Site (USA), *Geobacter* species are identified to immobilize uranium in groundwater, achieving significant reduction in contamination levels (Chang et al., 2014), (Scala et al., 2006). Similarly, in Chernobyl Exclusion Zone, Fungi like *Cladosporium* are shown to adsorb radioactive isotopes (Zhdanova et al., 2000), demonstrating potential for uranium waste treatment (Belozerskaya et al., 2010).

6 Conclusion

Uranium Corporation of India Limited (UCIL) operates several open pit and underground uranium mines in Jaduguda and other places in India. Toxicity related to uranium on the environment, can be divided into two types such as radiological and non-radiological. Different radioisotopes from uranium-radium radioactive series, especially radon gas are responsible for harmful effects on different living organisms such as plants, animals and microbes, known as radiological toxicity. Non-radiological toxicity or chemotoxicity of uranium on living systems is related with formation of uranyl ion in aerobic environment. The binding capacity of uranyl ion with proteins, nucleic acids, disrupts their structures, resulting their functional alteration in physiological condition.

Radiation emitted from waste part of uranium mine, causes huge impact on the health and environment of surrounding areas of uranium mines. Water bodies and ground water in those areas are contaminated with heavy metals and uranyl ion. This phenomenon affects the agricultural output and fish quality, which ultimately results in health problems to human populations in the 5 km radius of uranium mines. Overproduction of reactive oxygen species (ROS) due to uranium contamination and heavy metal poisoning, causes detrimental effects on human, plant and microbial activity.

Eco-friendly biological techniques, using plants and microorganisms for removal of uranium from soil, water, are cost effective. In order to neutralize, eliminate as well as stabilize the harmful ingredients like uranium, radium, thorium, and other uranium derivatives from soil, microbes bio, plants, various enzymes, and membrane bound transporter proteins can be utilized.

In case of microbial bioremediation, Geobacter and Shewanella bacteria precipitates uranium oxide from uranium waste. Similarly, seaweed and fungus Aspergillus species can act as uranium biosorbents from polluted environment. Pseudomonas species involve in biomineralization through polyphosphate metabolism pathway. Immobilization of uranium causes reduction in uranium mobility through bacterial surface. Furthermore, the sequestering of uranium by becteria has been highlighted using transporters like cation diffusion facilitator (CDF) and PIB-ATPases., In microbacterium genus, UipA transmembrane protein has been isolated from Chernobyl site, as uranium accumulator inside the cell. From gene expression data for *Microbacterium oleivoranus*, an upregulated protein with iron-siderophore ABC transporter substrate-binding property, has been identified, which facilitates transport of metal ions like uranium, iron toward the cytoplasm of bacteria as a method of bioremediation.

After absorption of uranium from soil, root, and shoot cells of plants can transfer uranium to aerial parts, known as hyperaccumulator plants. Similarly, 38 uranyl binding proteins, e.g., PCaP1, help for sequestering uranium from contaminated soil. Bioremediation methods have been applied in different uranium mines, showing significant results to protect the green planet from health hazards of uranium.

References

Acharya, C. (2015). Uranium bioremediation: Approaches and challenges. In *Environmental microbial biotechnology* (pp. 119–132). Springer.

Au, W. W., McConnell, M. A., Wilkinson, G. S., Ramanujam, V. M. S., & Alcock, N. (1998). Population monitoring: Experience with residents exposed to uranium mining/milling waste. *Mutation Research/Fundamental and Molecular Mechanisms of Mutagenesis, 405*(2), 237–245.

Bala, K., Anshul, Verma, S., Dogra, A., Das, A. P., Kumar, R., & Nandini. (2024). Sustainable management for urban plastic waste generation. In *Sustainable management of urban plastic waste through circular economic approaches* (pp. 211–236). Springer. https://doi.org/10.1201/9781003465447-13

Banerjee, S., Kundu, A., & Dhak, P. (2022). Bioremediation of uranium from waste effluents using novel biosorbents: A review. *Journal of Radioanalytical and Nuclear Chemistry, 331*(6), 2409–2435.

Belozerskaya, T., Aslanidi, K., Ivanova, A., Gessler, N., Egorova, A., Karpenko, Y., & Olishevskaya, S. (2010). Characteristics of extremophylic fungi from Chernobyl nuclear power plant. *Current Research Topics in Applied Microbiology and Microbial Biotechnology, 1*, 88–94.

Bersimbaev, R. I., & Bulgakova, O. (2015). The health effects of radon and uranium on the population of Kazakhstan. *Genes and Environment, 37*, 1–10.

Berthet, S., Villiers, F., Alban, C., Serre, N. B., Martin-Laffon, J., Figuet, S., & Bourguignon, J. (2018). Arabidopsis thaliana plants challenged with uranium reveal new insights into iron and phosphate homeostasis. *New Phytologist, 217*(2), 657–670.

Bjørklund, G., Christophersen, O. A., Chirumbolo, S., Selinus, O., & Aaseth, J. (2017). Recent aspects of uranium toxicology in medical geology. *Environmental Research, 156*, 526–533.

Bose, T. K. (2010). *Challenging the nuclear heart: The movement against uranium mining in Jharkhand*. Dissent, Self-Determination, and Resilience: Social Movements in India.

Brugge, D., & Buchner, V. (2011). Health effects of uranium: New research findings. *Reviews on Environmental Health, 26*(4), 231–249.

Canu, I. G., Paquet, F., Goldberg, M., Auriol, B., Bérard, P., Collomb, P., & Tirmarche, M. (2009). Comparative assessing for radiological, chemical, and physical exposures at the French uranium conversion plant: Is uranium the only stressor? *International Journal of Hygiene and Environmental Health, 212*(4), 398–413.

Cao, M., Peng, Q., Wang, Y., Luo, G., Feng, L., Zhao, S., & Wang, N. (2023). High-efficiency uranium extraction from seawater by low-cost natural protein hydrogel. *International Journal of Biological Macromolecules, 242*, 124792.

Carvalho, F. P. (2011). Environmental radioactive impact associated to uranium production. *American Journal of Environmental Sciences, 7*(6), 547.

Chandwadkar, P., Misra, H. S., & Acharya, C. (2018). Uranium biomineralization induced by a metal tolerant Serratia strain under acid, alkaline and irradiated conditions. *Metallomics, 10*(8), 1078–1088.

Chang, H. S., Buettner, S. W., Seaman, J. C., Jaffé, P. R., Koster van Groos, P. G., Li, D., & Kaplan, D. I. (2014). Uranium immobilization in an iron-rich rhizosphere of a native wetland plant from the Savannah River site under reducing conditions. *Environmental Science & Technology, 48*(16), 9270–9278.

Chen, J., & Harley, N. H. (2020). A review of radon equilibrium factors in underground mines, caves, and thermal spas. *Health Physics, 119*(3), 342–350.

Chen, L., Liu, J., Zhang, W., Zhou, J., Luo, D., & Li, Z. (2021). Uranium (U) source, speciation, uptake, toxicity and bioremediation strategies in soil-plant system: A review. *Journal of Hazardous Materials, 413*, 125319.

Chen, L., Wang, J., Beiyuan, J., Guo, X., Wu, H., & Fang, L. (2022). Environmental and health risk assessment of potentially toxic trace elements in soils near uranium (U) mines: A global meta-analysis. *Science of the Total Environment, 816*, 151556.

Das, A., Dey, S., Bala, K., Gnanasekaran, R., & Das, A. P. (2024). Microbial bioremediation technology for sustainable treatment and management of synthetic microfiber waste. In *Sustainable microbial technology for synthetic and cellulosic microfiber bioremediation* (pp. 221–238). Springer. https://doi.org/10.1007/978-3-031-62600-5_12

Das, A., Dey, S., & Das, A. P. (2024). Ecotoxicological impacts of synthetic microfiber pollutants and development of sustainable mitigation strategies. *Environmental Chemistry and Ecotoxicology*. https://doi.org/10.1016/J.ENCECO.2024.12.003

Das, A. P., Behera, I. D., & Das, N. P. (2024). *Renewable energy generation and value addition from environmental microfiber pollution through advanced greener solution* (p. 278). Springer.

Das, A., Ray, S. K., Mohanty, M., Mohanty, J., Dey, S., & Das, A. P. (2024). Ecotoxicity of microplastic wastes and their sustainable management: A review. *Environmental Chemistry and Ecotoxicology, 6*, 144–152. https://doi.org/10.1016/J.ENCECO.2024.05.003

Das, A. P., & Ghosh, S. (2023a). Manganese and its application. In *Manganese mining microorganisms* (pp. 29–40). Elsevier. https://doi.org/10.1016/B978-0-12-822147-1.00007-8

Das, A. P., Ghosh, S., Mohanty, S., & Sukla, L. B. (2015). Consequences of manganese compounds: A review. *Toxicological & Environmental Chemistry, 96*(7), 981–997. https://doi.org/10.1080/02772248.2015.1005428

Das, A. P., & Ghosh, S. (2023b). *Manganese mining microorganisms*. Elsevier.

Das, A. P., Mallavarapu, M., & Ghosh, S. (2023). Ecotoxicity of mining pollutants on the environment and their remediation. *Environmental Chemistry and Ecotoxicology, 5*, 165–167. https://doi.org/10.1016/J.ENCECO.2023.08.002

Das, A. P., Sukla, L. B., Pradhan, N., & Nayak, S. (2011). Manganese biomining: A review. *Bioresource Technology, 102*(16), 7381–7387. https://doi.org/10.1016/J.BIORTECH.2011.05.018

Dey, S., Das, A., Mallick, K., Sahu, A., & Das, A. P. (2023). *Environmental petroleum waste: Pollution, toxicity, sustainable remediation* (pp. 159–175). Springer. https://doi.org/10.1007/978-3-031-48220-5_7

Dey, S., Tripathy, B., Kumar, M. S., & Das, A. P. (2023). Ecotoxicological consequences of manganese mining pollutants and their biological remediation. *Environmental Chemistry and Ecotoxicology*. https://doi.org/10.1016/J.ENCECO.2023.01.001

Dinocourt, C., Legrand, M., Dublineau, I., & Lestaevel, P. (2015). The neurotoxicology of uranium. *Toxicology, 337*, 58–71.

Doustaly, F., Combes, F., Fiévet, J. B., Berthet, S., Hugouvieux, V., Bastien, O., & Bourguignon, J. (2014). Uranium perturbs signaling and iron uptake response in Arabidopsis thaliana roots. *Metallomics, 6*(4), 809–821.

Gallois, N., Alpha-Bazin, B., Bremond, N., Ortet, P., Barakat, M., Piette, L., & Chapon, V. (2022). Discovery and characterization of UipA, a uranium-and iron-binding PepSY protein involved in uranium tolerance by soil bacteria. *The ISME Journal, 16*(3), 705–716.

Gallois, N., Alpha-Bazin, B., Ortet, P., Barakat, M., Piette, L., Long, J., & Chapon, V. (2018). Proteogenomic insights into uranium tolerance of a Chernobyl's microbacterium bacterial isolate. *Journal of Proteomics, 177*, 148–157.

Gao, Y., Zhou, S., Yang, Z., Tang, Z., Su, Y., Duan, Y., & Wang, Y. (2024). Unveiling the role of uranium in enhancing the transformation of antibiotic resistance genes. *Journal of Hazardous Materials, 479*, 135624.

Ghosh, S., Mohanty, S., Nayak, S., Sukla, L. B., & Das, A. P. (2016). Molecular identification of indigenous manganese solubilising bacterial biodiversity from manganese mining deposits. *Journal of Basic Microbiology, 56*(3), 254–262. https://doi.org/10.1002/JOBM.201500477

Ghosh, S., Sinha, J. K., Ghosh, S., Vashisth, K., Han, S., & Bhaskar, R. (2023). Microplastics as an emerging threat to the global environment and human health. *Sustainability, 15*(14), 10821. https://doi.org/10.3390/SU151410821

Ghosh, S., Tripathy, B., Dey, S., & Das, A. P. (2023). Comparative investigation of fungal and bacterial manganese biomining mechanisms. In *Manganese mining microorganisms* (pp. 99–115). Elsevier. https://doi.org/10.1016/b978-0-12-822147-1.00016-9

Groudev, S. N., Georgiev, P. S., Spasova, I., & Nicolova, M. (2007). Bioremediation of acid mine drainage in an uranium deposit. *Advanced Materials Research, 20*, 248–257.

Ivanov, A. S., Parker, B. F., Zhang, Z., Aguila, B., Sun, Q., Ma, S., & Popovs, I. (2019). Siderophore-inspired chelator hijacks uranium from aqueous medium. *Nature Communications, 10*(1), 819.

Jagatee, S., Priyadarshini, S., Rath, C. C., & Das, A. P. (2024). Synthetic microfiber: An enduring environmental problem linked to sustainable development. *Environmental Science and Engineering, 2333*, 93–112. https://doi.org/10.1007/978-3-031-51792-1_6

Jha, S., Ghosh, D. K., & Mishra, U. C. (2000). Assessment of exposure of miners to the α-rays of long-lived radionuclides associated with respirable ore dust in the Jaduguda U mine. *Journal of Environmental Radioactivity, 48*(3), 317–326.

John, A., & Das, A. P. (2024). Government strategies for achieving Circular Economy for Plastic waste management in Southeast Asia. In *Sustainable management of urban plastic waste through circular economic approaches* (pp. 72–82). Routledge. https://doi.org/10.1201/9781003465447-6

Kulkarni, S., Misra, C. S., Gupta, A., Ballal, A., & Apte, S. K. (2016). Interaction of uranium with bacterial cell surfaces: Inferences from phosphatase-mediated uranium precipitation. *Applied and Environmental Microbiology, 82*(16), 4965–4974.

Kumar, A., Kumar, V., Saroop, S., Arsenov, D., Bali, S., Radziemska, M., & Bhardwaj, R. (2023). A comprehensive review of uranium in the terrestrial and aquatic environment: Bioavailability, immobilization, tolerance and remediation approaches. *Plant and Soil, 490*(1), 31–65.

Lebrazi, S., & Fikri-Benbrahim, K. (2018). Rhizobium-legume symbioses: Heavy metal effects and principal approaches for bioremediation of contaminated soil. In *Legumes for soil health and sustainable management* (pp. 205–233). Springer.

Li, Q., Xiong, Z., Xiang, P., Zhou, L., Zhang, T., Wu, Q., & Zhao, C. (2024). Effects of uranium mining on soil bacterial communities and functions in the Qinghai-Tibet plateau. *Chemosphere, 347*, 140715.

Lin, Y. W. (2020). Uranyl binding to proteins and structural-functional impacts. *Biomolecules, 10*(3), 457.

Locher, K. P., Rees, B., Koebnik, R., Mitschler, A., Moulinier, L., Rosenbusch, J. P., & Moras, D. (1998). Transmembrane signaling across the ligand-gated FhuA receptor: Crystal structures of free and ferrichrome-bound states reveal allosteric changes. *Cell, 95*(6), 771–778.

Lourenço, J., Marques, S., Carvalho, F. P., Oliveira, J., Malta, M., Santos, M., & Mendo, S. A. (2017). Uranium mining wastes: The use of the fish embryo acute toxicity test (FET) test to evaluate toxicity and risk of environmental discharge. *Science of the Total Environment, 605*, 391–404.

Lourenço, J., Pereira, R., Pinto, F., Caetano, T., Silva, A., Carvalheiro, T., & Mendo, S. (2013). Biomonitoring a human population inhabiting nearby a deactivated uranium mine. *Toxicology, 305*, 89–98.

Ma, M., Wang, R., Xu, L., Xu, M., & Liu, S. (2020). Emerging health risks and underlying toxicological mechanisms of uranium contamination: Lessons from the past two decades. *Environment International, 145*, 106107.

Maleke, M., Williams, P., Castillo, J., Botes, E., Ojo, A., DeFlaun, M., & van Heerden, E. (2015). Optimization of a bioremediation system of soluble uranium based on the biostimulation of an indigenous bacterial community. *Environmental Science and Pollution Research, 22*, 8442–8450.

Mkandawire, M. (2013). Biogeochemical behaviour and bioremediation of uranium in waters of abandoned mines. *Environmental Science and Pollution Research, 20*, 7740–7767.

Mohanty, J., Dey, S., Bala, K., & Das, A. P. (2024). Recent advances in microplastic pollution for its sustainable management. In *Sustainable management of urban plastic waste through circular economic approaches* (pp. 27–48). Routledge. https://doi.org/10.1201/9781003465447-3

Mohanty, M., Tripathy, B., Koteswara Reddy, G., & Das, A. P. (2024). Greener technologies for establishing restraint over microplastic pollution. In *Sustainable microbial technology for synthetic and cellulosic microfiber bioremediation* (pp. 239–258). Routledge. https://doi.org/10.1007/978-3-031-62600-5_13

Mohanty, S., Ghosh, S., Nayak, S., & Das, A. P. (2016). Isolation, identification and screening of manganese solubilizing fungi from low-grade manganese ore deposits. *Geomicrobiological Journal, 34*(4), 309–316. https://doi.org/10.1080/01490451.2016.1189016

Montavon, G., Apostolidis, C., Bruchertseifer, F., Repinc, U., & Morgenstern, A. (2009). Spectroscopic study of the interaction of U (VI) with transferrin and albumin for speciation of U (VI) under blood serum conditions. *Journal of Inorganic Biochemistry, 103*(12), 1609–1616.

National Research Council. (2011). Uranium mining in Virginia: Scientific, technical, environmental, human health and safety, and regulatory aspects of uranium mining and processing in Virginia.

Newman-Portela, A. M., Krawczyk-Bärsch, E., Lopez-Fernandez, M., Bok, F., Kassahun, A., Drobot, B., & Merroun, M. L. (2024). Biostimulation of indigenous microbes for uranium bioremediation in former U mine water: Multidisciplinary approach assessment. *Environmental Science and Pollution Research, 31*(5), 7227–7245.

Omasits, U., Ahrens, C. H., Müller, S., & Wollscheid, B. (2014). Protter: Interactive protein feature visualization and integration with experimental proteomic data. *Bioinformatics, 30*(6), 884–886.

Ortet, P., Gallois, N., Long, J., Zirah, S., Berthomieu, C., Armengaud, J., & Chapon, V. (2019). Complete genome sequences of four microbacterium strains isolated from metal-and radionuclide-rich soils. *Microbiology Resource Announcements, 8*(42), 10–1128.

Panigrahi, D. C., Mishra, D. P., & Sahu, P. (2015). Evaluation of inhalation exposure contributed by backfill mill tailings in underground uranium mine. *Environmental Earth Sciences, 74*, 4327–4334.

Panigrahi, D. C., Mishra, D. P., Sahu, P., & Bhowmik, S. C. (2015). Assessment of radiological parameters and radiation dose received by the miners in Jaduguda uranium mine, India. *Annals of Nuclear Energy, 78*, 33–39.

Panigrahi, D. C., Sahu, P., & Banerjee, M. (2018). Assessment to 222Rn and gamma exposure of the miners in Narwaphar underground uranium mine, India. *Radiation Physics and Chemistry, 151*, 225–231.

Pentyala, V. B., & Eapen, S. (2020). High efficiency phytoextraction of uranium using Vetiveria zizanioides L. Nash. *International Journal of Phytoremediation, 22*(11), 1137–1146.

Pereira, R., Barbosa, S., & Carvalho, F. P. (2014). Uranium mining in Portugal: A review of the environmental legacies of the largest mines and environmental and human health impacts. *Environmental Geochemistry and Health, 36*, 285–301.

Philippe, O., Gallois, N., Piette, L., Long, J., Berthomieu, C., Armengaud, J., & Virginie, C. (2017). Draft genome sequence of microbacterium oleivorans strain A9, a bacterium isolated from Chernobyl radionuclide-contaminated soil. *Microbiology Resource Announcements, 5*, 14.

Priyadarshini, B., & Das, A. P. (2024). The adverse impact of synthetic microplastics released from disposable face masks pollutants on aquatic microbiota, aquatic species and humans: A review. *Environmental Science and Engineering, 2827*, 67–81. https://doi.org/10.1007/978-3-031-50840-0_5

Priyadarshini, S., Jagatee, S., & Das, A. P. (2024). Synthetic fabrics and microfiber pollution–an assessment of their global impact. *Environmental Science and Engineering, 2333*, 137–157. https://doi.org/10.1007/978-3-031-51792-1_8

Priyadarsini, S., & Das, A. P. (2025). Lithium bioleaching: A review on microbial-assisted sustainable technology for lithium bio-circularity. *Journal of Water Process Engineering, 69*, 106744. https://doi.org/10.1016/j.jwpe.2024.106744

Raj, K., & Das, A. P. (2023). Lead pollution: Impact on environment and human health and approach for a sustainable solution. *Environmental Chemistry and Ecotoxicology, 5*, 79–85. https://doi.org/10.1016/J.ENCECO.2023.02.001

Rashmi, V., ShylajaNaciyar, M., Rajalakshmi, R., D'Souza, S. F., Prabaharan, D., & Uma, L. (2013). Siderophore mediated uranium sequestration by marine cyanobacterium Synechococcus elongatus BDU 130911. *Bioresource Technology, 130*, 204–210.

Rathod, A. M., Verpaele, S., Kelvin, M., & Leybourne, M. I. (2023). Uranium: An overview of physicochemical properties, exposure assessment methodologies, and health effects of environmental and occupational exposure. *Environmental Geochemistry and Health, 45*(5), 1183–1200.

Reddy, K., Gopal Reddy, C., Vidya Sagar, D., Yadagiri Reddy, P., & Rama Reddy, K. (2012). Environmental radioactivity studies in the proposed Lambapur and Peddagattu uranium mining areas of Andhra Pradesh, India. *Radiation Protection Dosimetry, 151*(2), 290–298.

Reddy Koteswara, G., Bhargavi, S., Yamuna, D., Anjani Devi, M., Jeevana Lakshmi Kumar, G., Manas, T., Adarsh, S., Bhavya Sai, V., Nath, H., Alreshidi, M. A., Togun, H., Ram, D. K., & Das, A. P. (2025). Recent advancements in the remediation of marine hydrocarbon and Plastic pollutants. *Environmental Science and Engineering, 218*, 127–142. https://doi.org/10.1007/978-3-031-81615-4_7

Renninger, N., Knopp, R., Nitsche, H., Clark, D. S., & Keasling, J. D. (2004). Uranyl precipitation by Pseudomonas aeruginosa via controlled polyphosphate metabolism. *Applied and Environmental Microbiology, 70*(12), 7404–7412.

Sahu, A., Dey, S., Mallick, K., Priyadarsini, S., Behera, I. D., & Das, A. P. (2024). A review on advanced technology for sustainable management of synthetic microplastic waste. In *Sustainable management of urban plastic waste through circular economic approaches* (pp. 262–279). Routledge. https://doi.org/10.1201/9781003465447-15

Sahu, A., Mallick, K., & Das, A. P. (2024). *Environmental impact of single-use synthetic face mask and its recycling: A sustainable approach* (pp. 197–212). Springer. https://doi.org/10.1007/978-3-031-50840-0

Sahu, P., Beg, I. A., Bhowmik, S. C., & Panigrahi, D. C. (2024). Evaluation of radiation dose associated with underground uranium mining activities in east Singhbhum, Jharkhand, India. *Radiation Physics and Chemistry, 223*, 111977.

Sahu, P., Mishra, D. P., Panigrahi, D. C., Jha, V., Patnaik, R. L., & Sethy, N. K. (2014). Radon emanation from backfilled mill tailings in underground uranium mine. *Journal of Environmental Radioactivity, 130*, 15–21.

Sahu, P., Panigrahi, D. C., & Mishra, D. P. (2016). A comprehensive review on sources of radon and factors affecting radon concentration in underground uranium mines. *Environmental Earth Sciences, 75*, 1–19.

Sarangi, A. K., Bhowmik, S. C., & Jha, V. N. (2007). Radiological impact assessment in bagjata uranium deposit: A case study. *Journal of Mines, Metals and Fuels, 55*(5), 127–133.

Sarthou, M. C. M., Revel, B. H., Villiers, F., Alban, C., Bonnot, T., Gigarel, O., & Bourguignon, J. (2020). Development of a metalloproteomic approach to analyse the response of Arabidopsis cells to uranium stress. *Metallomics, 12*(8), 1302–1313.

Sauge-Merle, S., Brulfert, F., Pardoux, R., Solari, P. L., Lemaire, D., Safi, S., & Berthomieu, C. (2017). Structural analysis of uranyl complexation by the EF-hand motif of calmodulin: Effect of phosphorylation. *Chemistry, 23*(61), 15505–15517.

Scala, D. J., Hacherl, E. L., Cowan, R., Young, L. Y., & Kosson, D. S. (2006). Characterization of Fe (III)-reducing enrichment cultures and isolation of Fe (III)-reducing bacteria from the Savannah River site, South Carolina. *Research in Microbiology, 157*(8), 772–783.

Sonowal, C. J. (2020). Uranium mining and public health issues: Reassessing the contention between scientific evidence and people's experience in Jadugoda, Jharkhand, India. *Journal of the Social Sciences, 48*(4), 425–437.

Srivastava, R. R., Pathak, P., & Perween, M. (2020). Environmental and health impact due to uranium mining. In *Uranium in plants and the environment* (pp. 69–89). Springer.

Thakur, A., & Kumar, A. (2024). Emerging paradigms into bioremediation approaches for nuclear contaminant removal: From challenge to solution. *Chemosphere, 352*, 141369.

Theodorakopoulos, N., Chapon, V., Coppin, F., Floriani, M., Vercouter, T., Sergeant, C., & Fevrier, L. (2015). Use of combined microscopic and spectroscopic techniques to reveal interactions between uranium and microbacterium sp. A9, a strain isolated from the Chernobyl exclusion zone. *Journal of Hazardous Materials, 285*, 285–293.

Thuéry, P., & Harrowfield, J. (2024). Bifunctional ligands in uranyl chemistry: Metalloligands and zwitterions. *Coordination Chemistry Reviews, 510*, 215821.

Vallet, A., Martin-Laffon, J., Favier, A., Revel, B., Bonnot, T., Vidaud, C., Armengaud, J., Gaillard, J. C., Delangle, P., Devime, F., Figuet, S., Serre, N. B. C., Erba, E. B., Brutscher, B., Ravanel, S., Bourguignon, J., & Alban, C. (2023). The plasma membrane-associated cation-binding protein PCaP1 of Arabidopsis thaliana is a uranyl-binding protein. *Journal of Hazardous Materials, 446*, 130668. https://doi.org/10.1016/j.jhazmat.2022.130668

Van Horn, J. D., & Huang, H. (2006). Uranium (VI) bio-coordination chemistry from biohemical, solution and protein structural data. *Coordination Chemistry Reviews, 250*(7-8), 765–775.

Vellingiri, B. (2023). A deeper understanding about the role of uranium toxicity in neurodegeneration. *Environmental Research, 233*, 116430.

Vidaud, C., Dedieu, A., Basset, C., Plantevin, S., Dany, I., Pible, O., & Quéméneur, E. (2005). Screening of human serum proteins for uranium binding. *Chemical Research in Toxicology, 18*(6), 946–953.

Wu, R., Fan, Y., Wu, Y., Zhou, S., Tang, S., Feng, X., & Xia, C. (2020). Insights into mechanism on organic acids assisted translocation of uranium in Brassica juncea var. foliosa by EXAFS. *Journal of Environmental Radioactivity, 218*, 106254.

Zhdanova, N. N., Zakharchenko, V. A., Vember, V. V., & Nakonechnaya, L. T. (2000). Fungi from Chernobyl: Mycobiota of the inner regions of the containment structures of the damaged nuclear reactor. *Mycological Research, 104*(12), 1421–1426.

Advancements and Prospects in Microbial Remediation of Mining Wastewater

Sweta Sinha, Ravi Ranjan, Sumira Malik, Gautam Sen, and Safwaan Hasan

1 Introduction

Environmental contamination presents one of the most formidable challenges of our time, with water pollution emerging as a particularly critical concern affecting ecosystems and public health worldwide (Sinha, 2022, 2024; Sinha et al., 2024). The accelerating pace of industrialization, urbanization, and agricultural intensification has introduced complex mixtures of contaminants into water systems, including pharmaceuticals, toxic metals, persistent organic pollutants, and pathogenic microorganisms. These pollutants, even at trace concentrations, pose significant risks to aquatic ecosystems and human well-being, necessitating effective remediation approaches (Malik et al., 2023; Saxena, 2025; Sinha, 2022, 2024).

Mining activities remain essential to global industrial development but produce contaminated wastewater requiring effective treatment prior to discharge. The distinctive characteristics of mining wastewater—including high metal concentrations, extreme pH values, and complex contaminant mixtures—present unique treatment challenges (Fernando et al., 2018). Among various treatment methodologies, microbial remediation has garnered substantial attention as a sustainable strategy for environmental restoration (Sharma et al., 2022; Sinha, 2018, 2023). This approach

S. Sinha (✉)
Amity Institute of Applied Sciences, Amity University Jharkhand, Ranchi, Jharkhand, India

R. Ranjan
Steel Authority of India (SAIL), R&D, Ranchi Division, Ranchi, Jharkhand, India

S. Malik
Amity Institute of Biotechnology, Amity University Jharkhand, Ranchi, Jharkhand, India

G. Sen · S. Hasan
Department of Chemistry, Birla Institute of Technology (BIT), Mesra, Ranchi, Jharkhand, India

© The Author(s), under exclusive license to Springer Nature Switzerland AG 2026

A. P. Das (ed.), *Bioremediation Technology for Sustainable Mining Waste Management*, Sustainable Environmental Waste Management Strategies, https://doi.org/10.1007/978-3-032-03468-7_6

harnesses the remarkable metabolic versatility of microorganisms to transform, degrade, or immobilize contaminants through natural biological processes. The intrinsic ability of certain microbial species to utilize environmental pollutants as carbon or energy sources represents a promising avenue for addressing diverse contamination scenarios. Recent advances in biotechnology and environmental microbiology have significantly enhanced our understanding of microbial remediation mechanisms. The application of next-generation sequencing and various omics technologies has illuminated previously uncharacterized microbial capabilities and metabolic pathways involved in pollutant degradation (Sharma et al., 2022). These molecular insights have enabled the development of more effective bioremediation strategies tailored to specific contaminants and environmental conditions.

This chapter examines contemporary developments in microbial remediation of contaminated waters, with particular emphasis on mining wastewater treatment. We explore the diverse mechanisms through which microorganisms interact with pollutants, the engineering of treatment systems to optimize microbial activity, and the challenges associated with implementing these technologies at field scale (Das et al., 2023, 2024; Das & Ghosh, 2023a). Additionally, we discuss emerging approaches that integrate microbial remediation with other treatment modalities to achieve more comprehensive contaminant removal and potential resource recovery, highlighting both current limitations and promising future directions in this evolving field.

2 Key Microbial Remediation Mechanisms

2.1 *Enzymatic Transformation*

Microbial enzymes capable of transforming toxic metal species into less harmful forms represent a sophisticated remediation mechanism with growing applications in mining wastewater treatment (Noman et al., 2022). Unlike biosorption or precipitation processes, enzymatic transformations often alter the chemical state of the contaminant, potentially changing properties including solubility, mobility, and toxicity.

Oxidoreductase enzymes play particularly important roles in metal transformation processes. These enzymes catalyze electron transfer reactions that reduce or oxidize metal species, frequently resulting in altered environmental behavior (Hay Mele et al., 2023). The metal transformations typically involve changes in oxidation state that affect the metal's bioavailability and toxicity profile.

Chromate reductases exemplify this category of enzymes, catalyzing the reduction of hexavalent chromium (Cr^{6+}) to trivalent chromium (Cr^{3+}). This transformation is environmentally significant because Cr^{6+} is highly soluble, mobile, and toxic, readily crossing biological membranes and causing DNA damage, while Cr^{3+} exhibits substantially lower solubility, mobility, and toxicity (Das & Ghosh, 2023a; Dey et al., 2023, 2024; Pradhan et al., 2023; Sundarraj et al., 2023). Several bacterial species including Pseudomonas, Escherichia, and Shewanella produce chromate reductases that effectively mediate this beneficial transformation. Recent structural characterization of these enzymes has revealed the presence of NADH-binding domains and specific metal-binding sites that facilitate electron transfer to chromate ions. Similarly, arsenate reductases convert arsenate (As^{5+}) to arsenite (As^{3+}). While arsenite exhibits higher acute toxicity than arsenate, this transformation paradoxically facilitates remediation by enabling more effective precipitation or volatilization processes to remove arsenic from water (Khanna et al., 2022). The ArsC family of reductases, widely distributed among environmental bacteria, mediates this reaction through a complex mechanism involving thiol chemistry and phosphorylation steps.

Recent enzyme engineering approaches have significantly enhanced the applicability of these biological catalysts for wastewater treatment. Directed evolution techniques have generated chromate reductase variants with improved stability at acidic pH values commonly encountered in mining effluents.

The integration of enzymatic transformation approaches with traditional physicochemical methods has demonstrated particular promise (Zha et al., 2024). For instance, systems combining microbial chromate reduction with subsequent chemical precipitation achieve chromium removal efficiencies exceeding 99.5%, substantially outperforming either approach independently. The coupling of arsenate reduction with specialized adsorbent materials similarly enhances overall treatment effectiveness.

2.2 *Biosorption and Bioaccumulation*

Biosorption and bioaccumulation represent critical microbial strategies for metal removal from mining wastewater. These processes involve complex interactions between microbial cells and metal contaminants, resulting in metal sequestration through passive and active mechanisms.

Biosorption occurs through physicochemical interactions between metal ions and microbial cell surfaces (Priya et al., 2022). The cellular envelope of microorganisms contains numerous functional groups including carboxyl, hydroxyl, phosphate, amino, and sulfhydryl moieties that serve as binding sites for metal cations. These interactions typically occur rapidly and independent of metabolism, making biosorption effective even with non-viable biomass. The efficiency of biosorption depends on factors including pH, temperature, metal concentration, biomass quantity, and contact time. Recent advances in understanding the molecular basis of biosorption have led to enhanced metal removal capabilities (Das & Ghosh, 2023b, 2023c, 2023d; Ghosh et al., 2023). The identification of specific binding proteins and polysaccharides has enabled targeted modifications to improve metal affinity. For instance, Bacillus subtilis strains modified to overexpress S-layer proteins have demonstrated 40%–50% improved cadmium binding compared to wild-type strains.

Similarly, Pseudomonas putida engineered to express additional carboxyl-rich extracellular polymeric substances has shown remarkable capacity for capturing copper and lead ions.

Bioaccumulation differs from biosorption in that it involves active cellular transport of metals across membranes and their subsequent intracellular sequestration. This metabolism-dependent process enables microorganisms to concentrate metals within their cytoplasm, often at levels significantly higher than environmental concentrations. Metal-binding proteins such as metallothionein's play central roles in this process by chelating metal ions and reducing their toxicity to cellular functions.

2.3 *Sulfate Reduction*

Sulfate-reducing bacteria (SRB) represent one of the most effective microbial groups for mining wastewater remediation due to their distinctive metabolic capabilities. These anaerobic organisms utilize sulfate as a terminal electron acceptor during respiration, producing hydrogen sulfide that subsequently reacts with dissolved metal ions to form insoluble metal sulfides. This precipitation mechanism effectively removes metals from solution while simultaneously generating alkalinity that helps neutralize the typically acidic pH of mining wastewater.

The biochemical pathway of sulfate reduction involves several key enzymes, with dissimilatory sulphite reductase (DSR) playing a central role in the final reduction step (Qian et al., 2019). The process can be represented by the simplified equation:

$$SO_4^{2-} + 2CH_2O \rightarrow H_2S + 2HCO_3^-$$

where CH_2O represents organic carbon sources essential for SRB metabolism. The generated bicarbonate ions (HCO_3^-) contribute to pH increase, while hydrogen sulfide reacts with dissolved metals (M^{2+}) to form highly insoluble precipitates:

$$M^{2+} + H_2S \rightarrow MS\downarrow + 2H^+$$

Recent advances in SRB-based remediation systems have addressed several historical limitations of this approach. Psychrotolerant SRB strains capable of maintaining significant metabolic activity at temperatures as low as 5 °C, extending the applicability of this treatment approach to colder climates where mining operations frequently occur (Kushkevych et al., 2021). Similarly, acid-tolerant SRB species and communities that can function at pH levels as low as 4.0 have been characterized, eliminating the need for extensive pre-treatment pH adjustment in many cases.

Innovations in organic carbon delivery systems have significantly improved treatment efficiency and operational sustainability. Traditional simple carbon sources like lactate and ethanol, while effective, often prove prohibitively expensive for large-scale applications. Novel approaches utilizing agricultural byproducts, lignocellulosic materials, and specialized slow-release carbon compounds have demonstrated comparable treatment efficacies at substantially reduced costs. Co-culture systems incorporating fermentative bacteria that break down complex organic substrates into simpler compounds readily utilized by SRB have shown particular promise for enhancing carbon utilization efficiency.

The development of advanced immobilization matrices represents another significant advancement in SRB-based treatment systems. These structures provide protective microenvironments that shield SRB from inhibitory conditions while maintaining high cell densities. Materials including calcium alginate beads, polyvinyl alcohol, and various porous ceramics have demonstrated the ability to sustain SRB activity under fluctuating conditions that would otherwise impair treatment performance. Long-term studies have documented stable sulfate reduction rates in such immobilized systems for periods exceeding 2 years without significant decline.

The integration of SRB with complementary microbial communities has emerged as a strategy for comprehensive treatment of complex mining wastewater (Table 1). Engineered consortia incorporating metal-resistant heterotrophs, acidophiles, and methanogens alongside SRB can address multiple contaminants simultaneously while improving overall system resilience. These synergistic communities more closely mimic natural attenuating processes and demonstrate enhanced adaptability to variable influent compositions commonly encountered in mining operations.

3 Treatment System Designs for Microbial Remediation of Mining Wastewater

3.1 Passive Treatment Systems

Passive treatment systems represent a cornerstone approach in mining wastewater remediation, combining ecological principles with engineering design to create sustainable, low-maintenance solutions (Table 2). These systems harness natural biogeochemical processes to treat contaminated water without requiring continuous energy input or chemical additions (Hou, 2019).

Constructed wetlands have evolved considerably from early designs that mimicked natural wetland systems. Contemporary engineered wetlands incorporate sophisticated substrate compositions that serve multiple functions simultaneously. Multi-layer substrates combining limestone aggregate, organic matter, and selected mineral additives achieved removal efficiencies of 98% for iron, 97% for aluminum, and 95% for copper in acid mine drainage (Broadhurst et al., 2022). These engineered substrates optimize hydraulic conductivity to prevent channeling and short-circuiting while providing optimal surface area for microbial colonization.

The development of layered treatment systems has addressed one of the primary limitations of earlier passive approaches: the inability to handle multiple contaminant types effectively. Modern vertical flow systems typically incorporate distinct

Table 1 Microbial remediation mechanisms for mining wastewater treatment

Mechanism	Description	Key elements	Examples	Advantages	Challenges
Enzymatic transformation	Enzymes convert toxic metals to less harmful forms.	Oxidoreductases (chromate, arsenate, mercury reductases)	Cr^{6+} to Cr^{3+}, As^{5+} to As^{3+}, Hg^{2+} to Hg^0	Alters metal properties, specific, integrates with other methods	Enzyme stability, cost, scaling
Biosorption and bioaccumulation	Microbes bind/absorb metals on cell surfaces or internally.	Cell surface groups, metallothioneins, active transport	Cadmium binding by *Bacillus subtilis*, copper/lead capture by *pseudomonas putida*, metal removal by mine isolates	Effective (biosorption), concentrates metals (bioaccumulation), diverse microbes	Affected by environment, requires molecular understanding
Sulfate reduction	Bacteria produce sulfide, precipitating metals.	Sulfate-reducing bacteria (SRB), dissimilatory sulfite reductase (DSR)	Precipitation of cu, Zn, Pb sulfides	Removes metals, neutralizes pH, works in some extreme conditions	Anaerobic, H_2S odor, carbon source optimization

Table 2 Treatment system designs for microbial remediation of mining wastewater

System type	Description	Key features	Examples	Advantages	Disadvantages
Passive	Low maintenance, uses natural processes	Constructed wetlands (layered, hybrid), local organic materials	Multi-layer wetlands, hybrid systems using waste	Sustainable, low cost	Larger area, slower, climate-sensitive
Active	Higher rate, controlled, more input	Membrane bioreactors (MBRs), fluidized beds, sequential batch reactors (SBRs), online monitoring	MBRs for high removal, fluidized beds for variable influent, SBRs for fluctuating waste	Faster, more control, smaller footprint potential	Higher cost, more complex operation

treatment zones, each targeting specific pollutants. For example, the upper aerobic layers facilitate iron oxidation and precipitation, while deeper anaerobic zones support sulfate reduction and metal sulfide formation. This stratified approach allows for comprehensive treatment of complex mining effluents containing both metals and acidity (Broadhurst et al., 2022).

The integration of vertical and horizontal flow components has created hybrid constructed wetlands with enhanced treatment capabilities. These systems direct water through a sequence of aerobic and anaerobic environments, optimizing different microbial communities and treatment mechanisms. Performance data from full-scale installations in former mining regions have demonstrated consistent metal removal efficiencies exceeding 90% for zinc, cadmium, and lead, while simultaneously increasing pH from 3.5 to circumneutral ranges and reducing sulfate concentrations by 60–80%.

3.2 Active Bioreactor Systems

Active bioreactor systems represent the technological counterpart to passive approaches, offering higher treatment rates and greater process control at the expense of increased operational requirements (Palladino et al., 2024). Recent innovations have significantly enhanced the performance, reliability, and applicability of these systems for mining wastewater treatment (Table 2).

Membrane bioreactors (MBRs) have transformed active biological treatment by integrating microbial processes with membrane filtration technology. These systems maintain high biomass concentrations, often 3–5 times greater than conventional suspended growth reactors, significantly increasing volumetric treatment capacity (Rahman et al., 2023). The ultrafiltration or microfiltration membranes retain microbial cells while allowing treated water to pass, eliminating biomass washout issues that plagued earlier designs. MBRs treating mining-influenced water achieved consistent removal efficiencies of >99% for copper, zinc, and nickel while operating at hydraulic retention times 50% shorter than conventional alternatives (Matebese et al., 2024).

Sequential batch reactors (SBRs) have gained prominence for their operational flexibility in addressing mining wastewaters with fluctuating compositions. By separating treatment processes temporally rather than spatially, these systems can adjust cycle times and operational parameters to optimize treatment for changing influent characteristics. This adaptability proves particularly valuable for mining operations where wastewater quality varies with extraction activities and seasonal influences (Mishra et al., 2022). Modern SBR designs incorporate automated control systems that adjust process parameters based on real-time monitoring data, maintaining optimal conditions despite influent variability.

Advances in process monitoring and control technologies have significantly enhanced active bioreactor performance. Online sensors measuring parameters including dissolved oxygen, oxidation-reduction potential, pH, and specific contaminant concentrations now provide continuous operational data. Integration of these sensors with sophisticated control algorithms enables automated adjustment of aeration rates, nutrient addition, and retention times to maintain optimal treatment conditions. These systems can detect and respond to perturbations before treatment performance declines, substantially improving operational stability and reliability.

4 Recent Technological Advances

4.1 Synthetic Biology and Genetic Engineering Applications

The application of synthetic biology and genetic engineering to mining wastewater treatment represents a frontier of innovation with significant potential for addressing recalcitrant contamination challenges. These approaches move beyond utilizing naturally occurring microbial capabilities to creating enhanced organisms with specifically designed remediation functions.

The CRISPR-Cas gene editing technology has revolutionized the precision and efficiency with which microbial genomes can be modified. This technology allows for targeted modifications that enhance metal tolerance and removal capabilities without introducing extraneous genetic material (Sahoo et al., 2022). Research teams have successfully employed CRISPR-Cas9 systems to modify metal transporter genes in organisms including Pseudomonas putida and Escherichia coli, enhancing their uptake capacity for metals including cadmium, arsenic, and mercury. A modified P. putida strain with a threefold increase in cadmium uptake capacity compared to wild-type populations through targeted enhancement of the cadA transport system.

Biofilm engineering represents another promising approach for improving treatment systems. Enhanced biofilm formation provides numerous advantages including increased resistance to toxic shock loads and improved long-term stability. Genetic modifications targeting exopolysaccharide production and cell surface adhesins have generated bacterial strains with superior attachment capabilities and denser biofilm structures (Das, 2022). These engineered biofilms demonstrate enhanced metal sorption capacities due to the increased binding sites within their matrix structure, while also exhibiting improved resistance to environmental stressors common in mining wastewater.

The development of microorganisms with multiple metal resistance mechanisms addresses the complex nature of mining effluents, which typically contain mixtures of contaminants rather than single substances. By incorporating complementary resistance and removal pathways, engineered organisms can simultaneously address multiple contaminants. For example, strains equipped with both mercury reduction and arsenate reduction operons can effectively treat waters containing both contaminants without requiring separate treatment processes. This integrated approach simplifies treatment system design while improving overall removal efficiencies.

Metabolic engineering has created organisms optimized for survival under the harsh conditions typical of mining wastewater. These modifications include enhanced acid tolerance, improved defense against oxidative stress, and optimized energy pathways for functioning in low-nutrient environments (Li & Wen, 2021). When deployed in treatment systems, these hardened strains maintain activity under conditions that would inhibit conventional microorganisms, extending the operational range of biological treatment approaches.

4.2 *Immobilization and Material Science Innovations in Microbial Remediation Systems.*

Recent advances in immobilization materials and techniques have transformed the performance and applicability of microbial remediation systems for mining wastewater treatment. These developments address historical challenges including biomass retention, operational stability, and cell viability under harsh conditions (Bouabidi et al., 2019).

Biodegradable polymers have emerged as versatile matrices for microbial immobilization, offering significant advantages over traditional materials. Recent formulations incorporating alginate, chitosan, and cellulose derivatives provide favorable mechanical properties while maintaining biocompatibility with remediation organisms. These materials create protective microenvironments that shield encapsulated cells from toxic shock loads commonly encountered in mining effluents. Sulfate-reducing bacteria immobilized in chitosan-alginate composite beads maintained 85% of their metabolic activity when exposed to copper concentrations that completely inhibited free cells, highlighting the protective effect of these matrices (Redwan, 2021).

Activated carbon-based immobilization systems combine the benefits of microbial activity with adsorptive properties, creating multifunctional treatment materials. These composites simultaneously support biofilm development while providing immediate sorption capacity for contaminant removal (Wang et al., 2024). The carbon substrate also serves as a buffer against concentration fluctuations, gradually releasing adsorbed contaminants for biological processing as external concentrations decrease. This integrated approach has proven particularly effective for treating intermittent high-concentration discharges typical of many mining operations.

Advances in ceramic materials have expanded the applicability of biological treatment to extreme conditions previously considered unsuitable for microbial remediation. High-porosity ceramic carriers with precisely controlled pore size distributions optimize microbial attachment while maintaining excellent hydraulic characteristics (Bartels, 2020). These materials withstand acidic conditions, high temperatures, and mechanical stress that would degrade conventional organic matrices. Long-term field trials have demonstrated stable biofilm development and consistent treatment performance on ceramic carriers despite pH fluctuations between 2.5 and 6.5, expanding the operational range of bioremediation systems.

The application of additive manufacturing technologies has revolutionized carrier design for immobilized cell systems. 3D-printed structures with optimized geometries enhance both hydraulic flow patterns and surface colonization characteristics. These precision-engineered carriers incorporate features at multiple scales—macro-channels for flow distribution, meso-scale structures for bacterial attachment, and micro-texturing for initial cell adhesion (Proaño-Peña, 2018). The ability to produce complex geometries impossible through conventional manufacturing techniques has led to significant improvements in mass transfer efficiency and biofilm development rates.

4.3 Integration with Emerging Technologies

The integration of microbial remediation with complementary technologies has created hybrid systems that exceed the capabilities of either approach independently. These integrated solutions address multiple aspects of mining wastewater treatment simultaneously, improving overall performance and expanding application ranges.

Bioelectrochemical systems represent a revolutionary approach that combines microbial metabolism with electrochemical processes. These systems utilize electroactive microorganisms capable of transferring electrons to or from solid electrodes, enabling simultaneous contaminant removal and electricity generation (Zheng et al., 2020). In microbial fuel cell configurations, metal reduction processes are coupled to electricity production, converting treatment operations from energy consumers to energy producers. Research installations have demonstrated successful treatment of acid mine drainage while generating power densities of 380–520 mW/m^2, offsetting operational energy requirements. Advanced electrode materials incorporating graphene, carbon nanotubes, and conductive polymers have significantly improved electron transfer efficiency between microorganisms and electrodes, enhancing both treatment rates and energy recovery (Sun et al., 2023).

Photobioreactor systems integrating microalgae with bacterial consortia offer comprehensive treatment capabilities through complementary metabolic processes. Microalgal photosynthesis generates oxygen that supports aerobic bacterial metabolism while simultaneously increasing pH through carbon dioxide consumption (Nagarajan et al., 2022). This pH elevation facilitates metal precipitation without chemical addition requirements. The microalgal biomass also provides significant heavy metal biosorption capacity through cell wall binding sites. Recent pilot-scale installations treating zinc mining effluents have achieved removal efficiencies exceeding 95% for zinc, 90% for cadmium, and 87% for manganese while producing valuable algal biomass that can be harvested for resource recovery.

Nanomaterial integration has enhanced microbial remediation through multiple mechanisms. Metallic and metal oxide nanoparticles serve as electron shuttles between microbial cells and insoluble substances, accelerating reduction reactions critical for metal immobilization. Carbon-based nanomaterials including graphene oxide and carbon nanotubes provide high-surface-area scaffolds for biofilm development while enhancing extracellular electron transfer (Wang et al., 2021). Nanoscale zero-valent iron particles combined with sulfate-reducing bacteria create synergistic treatment systems where chemical and biological reduction processes operate simultaneously. These integrated approaches have demonstrated treatment rates up to three times faster than conventional biological systems for selenate, chromate, and uranyl contaminants. Advanced oxidation process integration with biological treatment has proven particularly effective for mining wastewaters containing recalcitrant organic compounds alongside metal contamination (Nidheesh et al., 2022). Sequential systems applying controlled oxidation followed by biological treatment convert persistent organic pollutants into biodegradable intermediates that subsequent microbial processes can mineralize. This approach effectively addresses the mixed contaminant streams common in modern mining operations where both metals and process chemicals require treatment.

Real-time monitoring systems incorporating biosensor technologies have significantly improved treatment system control and performance verification (Hemdan et al., 2024). Whole-cell biosensors utilizing genetically modified microorganisms that produce measurable signals in response to specific contaminants provide continuous feedback on treatment progress. These systems enable adaptive control strategies that optimize operational parameters based on actual treatment performance rather than predetermined schedules, improving both efficiency and reliability.

The integration of these emerging technologies with established microbial remediation approaches represents a significant advancement in mining wastewater treatment capabilities (Fig. 1). By combining the selectivity and sustainability of biological processes with the speed and intensity of physicochemical methods, these hybrid systems address the complex challenges presented by modern mining operations while improving economic and environmental performance.

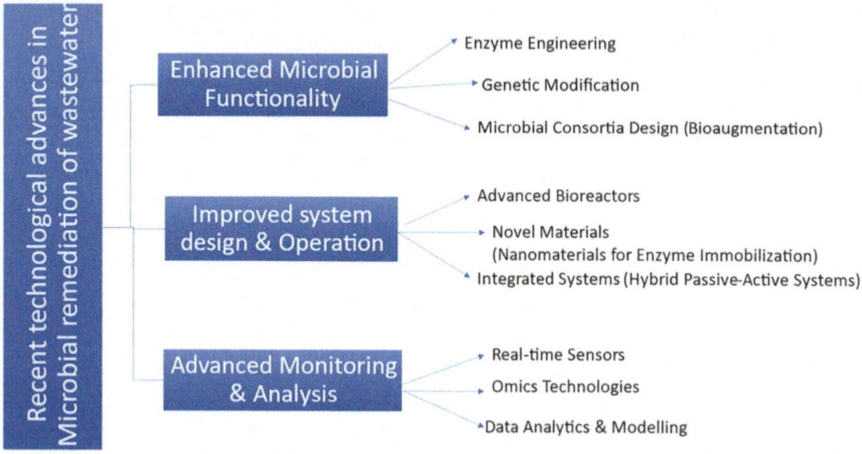

Fig. 1 Recent technological advances in microbial remediation

5 Challenges and Future Directions

Despite significant advances, several challenges hinder the widespread implementation of microbial remediation for mining wastewater. One major issue is scaling limitations, as translating laboratory success to field scale operations often encounters challenges related to system stability and performance consistency (Khanzada et al., 2023). Future research should focus on optimizing scale-up parameters and operating conditions. Additionally, extreme conditions pose a significant hurdle, as many mining operations occur in regions with harsh climates or produce wastewater with extreme properties that challenge microbial systems. Developing extremophile-based solutions offers promising directions. Another key challenge is process economics—while microbial remediation is generally more cost-effective than chemical treatments, initial capital requirements and operating costs still present barriers to implementation (Yadav et al., 2021). Conducting life cycle analyses that incorporate environmental benefits is essential for accurately assessing economic feasibility. Moreover, regulatory frameworks remain uncertain, with concerns about biological treatment systems and potential biosafety risks associated with engineered organisms. Addressing these concerns requires policy development and standardization.

Future research directions likely to drive significant advancements include optimizing microbial ecology by engineering complete microbial communities rather than focusing on individual strains, which appears crucial for robust treatment systems. Integrating resource recovery into treatment processes, such as selective metal recovery and the generation of valuable byproducts, can further enhance sustainability (Diaz-Elsayed et al., 2020). Additionally, the development of predictive models that incorporate microbial kinetics, fluid dynamics, and geochemistry will help optimize system design and operation. Finally, advancements in real-time monitoring, including the use of advanced sensor technologies for continuous assessment of microbial activity and treatment performance, will significantly improve process efficiency and reliability.

6 Conclusion

Microbial remediation of mining wastewater has advanced significantly, offering sustainable alternatives to conventional treatment approaches. The integration of biotechnology, materials science, and engineering principles continues to enhance system performance and applicability. While challenges remain, particularly in scaling and extreme condition operation, ongoing research promises to address these limitations. The evolution toward comprehensive resource recovery systems rather than simple treatment processes represents a paradigm shift with potential economic and environmental benefits. As mining activities continue globally, microbial remediation systems will play an increasingly important role in environmental protection and sustainable resource management.

References

Bartels, J. (2020). *Ceramic capillary membranes with tailored pore sizes and functionalizations for virus retention*. Universität Bremen.

Bouabidi, Z. B., El-Naas, M. H., & Zhang, Z. (2019). Immobilization of microbial cells for the biotreatment of wastewater: A review. *Environmental Chemistry Letters, 17*, 241–257.

Broadhurst, J., Petrik, L., Folifac, L., Ameh, A. E., Jason, J., Jehoma, A., & Gitari, W. (2022). *Integrating Nano iron production into the acid mine drainage neutralisation treatment process*.

Das, A., Dey, S., Bala, K., Gnanasekaran, R., & Das, A. P. (2024). Microbial bioremediation technology for sustainable treatment and management of synthetic microfiber waste. In *Sustainable microbial technology for synthetic and cellulosic microfiber bioremediation* (pp. 221–238). Springer. https://doi.org/10.1007/978-3-031-62600-5_12

Das, A. P., & Ghosh, S. (2023a). Evolution of biomining technology. In *Manganese mining microorganisms* (pp. 5–19). Elsevier. https://doi.org/10.1016/B978-0-12-822147-1.00012-1

Das, A. P., & Ghosh, S. (2023b). Manganese and its application. Manganese mining microorganisms. Elsevier, , 29–40. https://doi.org/10.1016/B978-0-12-822147-1.00007-8.

Das, A. P., & Ghosh, S. (2023c). Metagenomic insights into unculturable microbial diversity. In *Manganese mining microorganisms* (pp. 63–76). Elsevier. https://doi.org/10.1016/B978-0-12-822147-1.00002-9

Das, A. P., & Ghosh, S. (2023d). Microorganisms in manganese biomining processes. In *Manganese mining microorganisms* (pp. 55–61). Elsevier. https://doi.org/10.1016/B978-0-12-822147-1.00013-3

Das, A. P., Mallavarapu, M., & Ghosh, S. (2023). Ecotoxicity of mining pollutants on the environment and their remediation. *Environmental Chemistry and Ecotoxicology, 5*, 165–167. https://doi.org/10.1016/J.ENCECO.2023.08.002

Das, S. (2022). Genetic regulation, biosynthesis and applications of extracellular polysaccharides of the biofilm matrix of bacteria. *Carbohydrate Polymers, 291*, 119536.

Dey, S., Dash, A., & Das, A. P. (2024). Bioremediation as a sustainable and circular solution for mining waste management and bioenergy generation. In *Sustainable management of mining waste and tailings: A circular economy approach* (pp. 107–119). Routledge. https://doi.org/10.1201/9781003442455-6

Dey, S., Tripathy, B., Kumar, M. S., & Das, A. P. (2023). Ecotoxicological consequences of manganese mining pollutants and their biological remediation. *Environmental Chemistry and Ecotoxicology*. https://doi.org/10.1016/J.ENCECO.2023.01.001

Diaz-Elsayed, N., Rezaei, N., Ndiaye, A., & Zhang, Q. (2020). Trends in the environmental and economic sustainability of wastewater-based resource recovery: A review. *Journal of Cleaner Production, 265*, 121598.

Fernando, W. A. M., Ilankoon, I. M. S. K., Syed, T. H., & Yellishetty, M. (2018). Challenges and opportunities in the removal of sulphate ions in contaminated mine water: A review. *Minerals Engineering, 117*, 74–90.

Ghosh, S., Tripathy, B., Dey, S., & Das, A. P. (2023). Comparative investigation of fungal and bacterial manganese biomining mechanisms. In *Manganese mining microorganisms* (pp. 99–115). Elsevier. https://doi.org/10.1016/b978-0-12-822147-1.00016-9

Hay Mele, B., Monticelli, M., Leone, S., Bastoni, D., Barosa, B., Cascone, M., & Giovannelli, D. (2023). Oxidoreductases and metal cofactors in the functioning of the earth. *Essays in Biochemistry, 67*(4), 653–670.

Hemdan, M., Ali, M. A., Doghish, A. S., Mageed, S. S. A., Elazab, I. M., Khalil, M. M., & Amin, A. S. (2024). Innovations in biosensor technologies for healthcare diagnostics and therapeutic drug monitoring: Applications, recent progress, and future research challenges. *Sensors, 24*(16), 5143.

Hou, D. (Ed.). (2019). *Sustainable remediation of contaminated soil and groundwater: Materials, processes, and assessment*. Butterworth-Heinemann.

Khanna, K., Kohli, S. K., Kumar, P., Ohri, P., Bhardwaj, R., Alam, P., & Ahmad, P. (2022). Arsenic as hazardous pollutant: Perspectives on engineering remediation tools. *Science of the Total Environment, 838*, 155870.

Khanzada, A. K., Al-Hazmi, H. E., Śniatała, B., Joseph, T. M., Majtacz, J., Abdulrahman, S. A., & Mąkinia, J. (2023). Hydrochar-nanoparticle integration for arsenic removal from wastewater: Challenges, possible solutions, and future horizon. *Environmental Research, 238*, 117164.

Kushkevych, I., Kovářová, A., Dordevic, D., Gaine, J., Kollar, P., Vítězová, M., & Rittmann, S. K. M. (2021). Distribution of sulfate-reducing bacteria in the environment: Cryopreservation techniques and their potential storage application. *PRO, 9*(10), 1843.

Li, M., & Wen, J. (2021). Recent progress in the application of omics technologies in the study of bio-mining microorganisms from extreme environments. *Microbial Cell Factories, 20*(1), 178.

Malik, S., Bora, J., Nag, S., Sinha, S., Mondal, S., Rustagi, S., & Almutary, A. G. (2023). Fungal-based remediation in the treatment of anthropogenic activities and pharmaceutical-pollutant-contaminated wastewater. *Water, 15*(12), 2262.

Matebese, F., Mosai, A. K., Tutu, H., & Tshentu, Z. R. (2024). Mining wastewater treatment technologies and resource recovery techniques: A review. *Heliyon, 10*(3), e24730.

Mishra, S., Singh, R. P., Rout, P. K., & Das, A. P. (2022). Membrane bioreactor (MBR) as an advanced wastewater treatment technology for removal of synthetic microplastics. In *Development in wastewater treatment research and processes* (pp. 45–60). Elsevier. https://doi.org/10.1016/B978-0-323-85583-9.00022-3

Nagarajan, D., Lee, D. J., Varjani, S., Lam, S. S., Allakhverdiev, S. I., & Chang, J. S. (2022). Microalgae-based wastewater treatment–microalgae-bacteria consortia, multi-omics approaches and algal stress response. *Science of the Total Environment, 845*, 157110.

Nidheesh, P. V., Couras, C., Karim, A. V., & Nadais, H. (2022). A review of integrated advanced oxidation processes and biological processes for organic pollutant removal. *Chemical Engineering Communications, 209*(3), 390–432.

Noman, E. A., Al-Gheethi, A., Al-Sahari, M., Mohamed, R. M. S. R., Crane, R., Ab Aziz, N. A., & Govarthanan, M. (2022). Challenges and opportunities in the application of bioinspired engineered nanomaterials for the recovery of metal ions from mining industry wastewater. *Chemosphere, 308*, 136165.

Palladino, F., Marcelino, P. R. F., Schlogl, A. E., José, Á. H. M., Rodrigues, R. D. C. L. B., Fabrino, D. L., & Rosa, C. A. (2024). Bioreactors: Applications and innovations for a sustainable and healthy future—A critical review. *Applied Sciences, 14*(20), 9346.

Pradhan, G., Tripathy, B., Ram, D. K., Digal, A. K., & Das, A. P. (2023). Bauxite mining waste pollution and its sustainable management through bioremediation. *Geomicrobiology Journal.* https://doi.org/10.1080/01490451.2023.2235353

Priya, A. K., Gnanasekaran, L., Dutta, K., Rajendran, S., Balakrishnan, D., & Soto-Moscoso, M. (2022). Biosorption of heavy metals by microorganisms: Evaluation of different underlying mechanisms. *Chemosphere, 307*, 135957.

Proaño-Peña, G. F. (2018). *3D-printed custom substratum to accentuate fast functional responses from microbial colonization.*

Qian, Z., Tianwei, H., Mackey, H. R., van Loosdrecht, M. C., & Guanghao, C. (2019). Recent advances in dissimilatory sulfate reduction: From metabolic study to application. *Water Research, 150*, 162–181.

Rahman, T. U., Roy, H., Islam, M. R., Tahmid, M., Fariha, A., Mazumder, A., & Islam, M. S. (2023). The advancement in membrane bioreactor (MBR) technology toward sustainable industrial wastewater management. *Membranes, 13*(2), 181.

Redwan, A. M. (2021). *Semi-conductive materials for engineered bioremediation: From surface properties to microbial diversity* (doctoral dissertation).

Sahoo, S., Routray, S. P., Lenka, S., Bhuyan, R., & Mohanty, J. N. (2022). CRISPR/Cas-mediated functional gene editing for improvement in bioremediation: An emerging strategy. In *Omics insights in environmental bioremediation* (pp. 635–664). Springer.

Saxena, V. (2025). Water quality, air pollution, and climate change: Investigating the environmental impacts of industrialization and urbanization. *Water, Air, & Soil Pollution, 236*(2), 1–40.

Sharma, P., Singh, S. P., Iqbal, H. M., & Tong, Y. W. (2022). Omics approaches in bioremediation of environmental contaminants: An integrated approach for environmental safety and sustainability. *Environmental Research, 211*, 113102.

Sinha, S. (2018). Biodegradable superabsorbents: Methods of preparation and application—A review. In *Fundamental biomaterials: Polymers* (pp. 307–322). Elsevier.

Sinha, S. (2022). Textile wastewater clarification using milk based formulations as an effective dye absorbent and flocculant. *Materials Today: Proceedings, 67*, 1304–1309.

Sinha, S. (2023). Synthesis of biopolymer based superabsorbent: An eco-friendly approach towards future sustainability. In *Bio-based superabsorbents: Recent trends, types, applications and recycling* (pp. 29–49). Springer.

Sinha, S. (2024). Food security challenges and microplastics: A comprehensive review. In *Microplastics in African and Asian environments: The influencers, challenges, and solutions* (pp. 361–371). Springer.

Sinha, S., Singh, S., & Dey, K. P. (2024). Microplastics: A significant threat to freshwater ecosystems. In *Sustainable microbial technology for synthetic and cellulosic microfiber bioremediation* (pp. 91–105). Springer.

Sun, J., Rene, E. R., He, Y., Ma, W., Hu, Q., & Qiu, B. (2023). Carbon, iron, and polymer-based conductive materials for improving methane production in anaerobic wastewater treatment systems: A review on their direct interspecific electron transfer mechanism. *Fuel, 342*, 127703.

Sundarraj, S., Sudarmani, D. N. P., Samuel, P., & Sevarkodiyone, S. P. (2023). Bioremediation of hexavalent chromium by transformation of *Escherichia coli* DH5α with chromate reductase (ChrR) genes of *pseudomonas putida* isolated from tannery effluent. *Journal of Applied Microbiology, 134*(1), lxac019.

Wang, L., Cheng, D., Liu, X., & Ye, Y. (2024). Utilizing microorganisms immobilized on carbon-based materials for environmental remediation: A mini review. *Water Emerging Contaminants & Nanoplastics, 3*(3), 19.

Wang, R., Li, H., Sun, J., Zhang, L., Jiao, J., Wang, Q., & Liu, S. (2021). Nanomaterials facilitating microbial extracellular electron transfer at interfaces. *Advanced Materials, 33*(6), 2004051.

Yadav, G., Mishra, A., Ghosh, P., Sindhu, R., Vinayak, V., & Pugazhendhi, A. (2021). Technical, economic and environmental feasibility of resource recovery technologies from wastewater. *Science of the Total Environment, 796*, 149022.

Zha, S., Yu, A., Wang, Z., Shi, Q., Cheng, X., Liu, C., & Zhou, L. (2024). Microbial strategies for effective hexavalent chromium removal: A comprehensive review. *Chemical Engineering Journal, 489*, 151457.

Zheng, T., Li, J., Ji, Y., Zhang, W., Fang, Y., Xin, F., & Jiang, M. (2020). Progress and prospects of bioelectrochemical systems: Electron transfer and its applications in the microbial metabolism. *Frontiers in Bioengineering and Biotechnology, 8*, 10.

Advanced Physical and Biological Treatment Methods of Mining Pollutants

Kapilas Das, Devjani Das, Bikash Das, Sanjib Moulick, and Sanjoy Kumar Maji

1 Introduction

1.1 Mining and its Economic Significance

Mining is one of the greatest crucial industries worldwide. It delivers crucial resources used across different sectors, including construction, manufacturing, and energy generation, and high-tech industries. The extraction and processing of minerals and metals play a fundamental role in economic development by generating employment, contributing to national GDPs, and facilitating infrastructure growth (Chang et al., 2023).

Globally, mining supports steel manufacturing, electronics, and renewable energy industries. Iron ore, copper, aluminum, and lithium are critical resources for constructing buildings, developing electric vehicles, and producing consumer electronics (Bala et al., 2024; Das, Dey, & Das, 2024). The coal mining sector remains a dominant player in the energy industry, particularly in countries that rely heavily on fossil fuels. Similarly, rare earth elements, which are essential for producing semiconductors and batteries, are extracted from mining activities (Filho et al., 2023).

K. Das · B. Das · S. K. Maji (✉)
Department of Chemistry, SAS, KIIT University, Bhubaneswar, Odisha, India

D. Das · S. Moulick
School of Civil Engineering, KIIT University, Bhubaneswar, Odisha, India

© The Author(s), under exclusive license to Springer Nature Switzerland AG 2026

A. P. Das (ed.), *Bioremediation Technology for Sustainable Mining Waste Management*, Sustainable Environmental Waste Management Strategies,
https://doi.org/10.1007/978-3-032-03468-7_7

1.1.1 Contribution to Industrial Growth

Mining underpins industrial progress by ensuring a steady supply of essential materials. The steel industry, for example, depends on iron ore, which is mined and processed before being used in construction, automotive manufacturing, and machinery production. Similarly, aluminum, derived from bauxite mining, is widely used in packaging, transportation, and aeronautics. The electronics industry heavily relies on rare metals like lithium, cobalt, and tungsten, which are sourced through extensive mining operations (Das et al., 2025; Das, Dey, Bala, et al., 2024; Dey, Dash, & Das, 2024; Whitworth et al., 2022).

1.1.2 Economic Benefits and Job Creation

Mining significantly contributes to job creation, particularly in developing countries with vast mineral reserves. It offers employment opportunities in both direct and indirect sectors, including exploration, extraction, processing, and transportation. Additionally, mining operations stimulate local economies by supporting ancillary industries such as equipment manufacturing, logistics, and trade (Chandra et al., 2025). Mining also plays a vital role in international commerce. Many resource-rich countries, such as Australia, China, South Africa, and Canada, export large quantities of minerals to meet international demand. The revenue generated from these exports strengthens national economies, allowing for further investment in infrastructure, education, and healthcare (Kirsanov & Saaya, 2024).

1.1.3 Challenges in Balancing Economic Growth and Environmental Impact

Despite its economic significance, mining comes with substantial environmental costs. Extracting minerals often leads to deforestation, soil degradation, water pollution, and air contamination. Open-pit mining, for example, removes vast areas of vegetation and topsoil, leading to habitat destruction and loss of biodiversity. Moreover, the release heavy metals and toxic chemicals to environment poses severe risks to human health, ecosystems (Das, Behera, & Das, 2024; Das & Ghosh, 2023; Mitra et al., 2022).

Thus, while mining is indispensable for industrial progress, there is an immediate need to adopt responsible measures that minimize its environmental footprint (Rahnema et al., 2023). This requires stringent regulations, improved waste management strategies, and the integration of advanced pollution treatment technologies.

1.2 Environmental Challenges of Mining Activities

Mining activities are inherently disruptive to the environment. A process of removing minerals from the earth involves land clearing, drilling, blasting, and chemical treatment, all of which contribute to environmental degradation (John & Das, 2024; Mishra et al., 2024a, 2024b; Mohanty, Dey, et al., 2024). The main environmental concerns associated with mining include water pollution, soil contamination, air pollution, and habitat destruction (Dehkordi et al., 2024; Sharifi et al., 2023).

1.2.1 Water Pollution from Mining Waste

One of the most serious impacts of mining is water pollution. Mining operations produce significant amounts of wastewater that contain suspended solids, heavy metals, acids, and toxic chemicals. This polluted water often enters in to local rivers, lakes, and groundwater sources, leading to contamination.

- Acid mine drainage (AMD): Sulfide minerals in extracted rock interact with oxygen and water to form sulfuric acid, which leaches heavy metals like arsenic, lead, and cadmium. The resulting acidic water, as acid mine drainage, contaminates surface, groundwater, making it unsuitable for consumption and agricultural use (Ighalo et al., 2022).
- Heavy metal contamination: Mining processes release hazardous metals, including mercury, lead, and chromium, which accumulate in water bodies and bioaccumulate in aquatic life (Mohanty, Dey, et al., 2024; Mohanty, Tripathy, et al., 2024; Priyadarshini et al., 2024; Priyadarsini & Das, 2024, 2025). These metals pose serious health risks to humans who consume contaminated water or fish (Tolkou et al., 2023).

1.2.2 Soil Contamination and Land Degradation

Mining causes in large areas of land, loss of fertility, leading to soil erosion, and contamination by heavy metals (Haghighizadeh et al., 2024). Mine tailings, which are the residues left after mineral extraction, often contain toxic substances that seep into surrounding soil and render it unsuitable for agriculture.

- Deforestation and habitat destruction: Open-pit mining requires clearing of vast forested areas, resulting in the loss of natural habitats and biodiversity loss. The loss of vegetation also increases the likelihood of soil erosion and desertification (Aili et al., 2024).
- Toxic waste accumulation: Tailings and slag piles from mining operations contain high concentrations of heavy metals and other hazardous compounds. Without proper disposal measures, these wastes leach into the surrounding soil, affecting plant growth and entering the food chain (Talukder et al., 2024).

1.2.3 Air Pollution from Dust and Emissions

Mining activities emit large amounts of particulate matter, harmful chemicals, and greenhouse gases into the air. The combustion of fossil fuels in mining equipment contributes to carbon emissions, exacerbating climate change.

- Dust and particulate matter: Blasting, drilling, and transportation of minerals generate fine dust particles, which contribute to air contamination and lung-related illnesses in surrounding populations (Duarte et al., 2022).
- Toxic gas emissions: Smelting and refining processes release SO_2, (NOx), and volatile organic compounds (VOCs), leading to acid rain, respiratory issues, and environmental degradation.

1.2.4 Loss of Biodiversity and Ecological Imbalance

The destruction of ecosystems due to mining disrupts food chains and threatens wildlife populations. Many species lose their habitats due to deforestation, soil degradation, and water contamination (Ray et al., 2024; Reddy Koteswara et al., 2025; Sahu et al., 2024). In extreme cases, entire ecosystems collapse due to the loss of keystone species and the implementation of toxic substances (Ouma et al., 2022; Singh et al., 2023).

1.3 Need for Sustainable Mining Practices

Given the severe environmental challenges posed by mining, it is essential to adopt sustainable practices that minimize ecological damage while ensuring continued resource extraction. Sustainable mining involves using advanced treatment technologies, enforcing stricter regulations, and promoting environmentally responsible practices (Firoozi et al., 2024).

1.3.1 Adoption of Green Mining Technologies

Green mining technologies aim to reduce environmental foot of mining operations through improved efficiency and reduced waste generation (Zaman et al., 2024). Some key approaches include:

- Eco-friendly extraction methods: Techniques such as bioleaching, which uses bacteria to extract metals from ores, reduce the need for harsh chemical treatments (Nkuna et al., 2022).

- Zero liquid discharge (ZLD) Systems: Advanced wastewater treatment processes recycle water, reducing contamination and conserving water resources (Date et al., 2022).
- Renewable energy in mining: Using solar and wind energy for mining operations decreases dependency on fossil fuels and cuts down on greenhouse gas emissions (Abbaspour & Abbaspour, 2022).

1.3.2 Regulatory Measures and Environmental Policies

Governments and international organizations have implemented policies to ensure responsible mining practices. Some notable frameworks include:

- The Environmental Impact Assessment (EIA): Requires mining companies to evaluate and moderate potential environmental damage before starting operations (Amuah et al., 2023).
- Mine Rehabilitation Plans: Companies must restore mined areas by replanting vegetation and treating contaminated soil and water.
- Strict Water Quality Standards: Regulations impose limits on the discharge of pollutants into water bodies, reducing contamination risks (Kumar et al., 2024).

1.3.3 Community Involvement and Corporate Responsibility

Sustainable mining also involves engaging local communities in decision-making processes. Companies are increasingly adopting Corporate Social Responsibility (CSR) initiatives that prioritize environmental conservation, health and safety, and economic development in mining-affected regions.

- Local Community Participation: Encouraging dialogue between mining companies and indigenous/local populations to address concerns and ensure fair resource distribution.
- Rehabilitation Projects: Restoring landscapes affected by mining by planting trees, creating artificial wetlands, and implementing soil remediation techniques.

1.3.4 Integration of Biological and Physical Treatment Methods

Integration of physical and biological treatment techniques are found to be the viable solution for the managing of mining pollution (Table 1). By integrating advanced filtration, adsorption, and microbial degradation techniques, mining wastewater and solid waste can be treated efficiently, reducing environmental harm (Wibowo et al., 2024).

Table 1 Comparison of physical and biological treatment methods for mining pollutants

Treatment method	Mechanism	Advantages	Limitations	Reference
Sedimentation and filtration	Uses gravity and porous materials to remove solids	Effective for suspended solids and heavy metals	Ineffective for dissolved contaminants	Liu LianWen et al. (2018), Sukanya et al. (2022)
Membrane technologies (RO, UF, NF)	Uses semi-permeable membranes to separate pollutants	High purification efficiency, removes dissolved salts	Expensive, requires high energy	Mohammad et al. (2015), Van der Bruggen and Vandecasteele (2003)
Adsorption (activated carbon, biosorbents)	Adsorbs pollutants onto solid surfaces	Cost-effective, reusable adsorbents available	Saturation requires frequent replacement	Ali et al. (2012), Bhatnagar and Sillanpää (2010)
Electrochemical methods (EC, EO)	Uses electrical currents for coagulation or oxidation	Effective for metals, fast processing	High operational cost, sludge production	Panizza and Cerisola (2004), Vepsäläinen and Sillanpää (2020)
Microbial bioremediation	Uses bacteria and fungi to degrade pollutants	Eco-friendly, sustainable, cost-effective	Slow, requires specific environmental conditions	Azubuike et al. (2016), Gadd (2010)
Phytoremediation	Plants absorb or stabilize pollutants in soil and water	Improves ecosystem health, low cost	Time-consuming, dependent on plant survival	Ali et al. (2013), Salt et al. (1998)
Enzyme-based remediation	Uses enzymes to break down toxic compounds	High specificity, faster than microbial methods	Expensive, enzyme stability issues	Du et al. (2024), van Beilen and Li (2002)

2 Mining Pollution and Its Environmental Impacts

Mining is a major industrial activity that plays a substantial role in causing environmental pollution. While mining is essential for obtaining valuable minerals and metals, it also generates substantial amounts of waste that can have devastating effects on natural ecosystems, water bodies, and human health. The environmental impact of mining is due to the release of numerous pollutants, with heavy metals, acids, and sediments, which affect the surrounding air, water, and soil (Mardonova & Han, 2023). These pollutants continue in the environment for long periods and often spread beyond the immediate vicinity of mining sites, leading to widespread ecological disruption. The degree of pollution caused by mining depends on the types of minerals being extracted, the mining techniques used, and the effectiveness of pollution control measures in place. If left unchecked, mining pollution can severely affect biodiversity, degrade natural resources, and pose long-term health risks to human populations (Dey, Das, et al., 2023).

2.1 Sources of Mining Pollution

Mining pollution originates from multiple sources throughout the extraction, processing, and disposal stages. The three primary activities responsible for pollution include extraction, milling, and smelting. Each stage of mining releases different types of contaminants into the environment, affecting water quality, air purity, and soil composition (Haghighizadeh et al., 2024). The extraction process, which involves removing ore from underground or surface mines, is one of the most environmentally disruptive stages of mining. In open-pit mining, large amounts of rock and soil are removed to access mineral deposits, creating vast landscapes of waste material known as overburden. This overburden often contains traces of heavy metals, which can leach into groundwater or be carried into nearby rivers and lakes by rainwater. Underground mining also has its challenges, as the excavation of tunnels disturbs natural geological formations, increasing the risk of subsidence, water contamination, and acid mine drainage (AMD) (Bajrović et al., 2025).

Milling, or ore beneficiation, is another major source of pollution. This stage involves crushing and grinding the extracted ore to separate valuable minerals from unwanted materials. The process generates large amounts of fine matter, which can become airborne and pollute air. Additionally, the use of chemical reagents in flotation and leaching operations introduces toxic substances into wastewater, which may enter local water bodies if not properly treated. The waste material left behind after milling, known as tailings, is typically stored in tailings ponds. If these containment structures fail or are improperly managed, they can release toxic slurries containing heavy metals and chemicals into the environment.

Smelting and refining are the final stages of mineral processing, during which raw ore is heated to extract pure metals. These processes release large amounts of sulfur dioxide, carbon monoxide, and other harmful gases into the atmosphere, contributing to acid rain and respiratory health issues. Smelting also produces slag, a solid waste material that often contains residual heavy metals and other hazardous substances. If not properly disposed of, slag can leach contaminants into the soil and groundwater, further exacerbating mining-related pollution. The combined effect of these actions highlights the necessity for stringent environmental regulations and sustainable mining practices to mitigate pollution and minimize harm to ecosystems (Firoozi et al., 2024).

2.2 Types of Contaminants in Mining Waste

Mining waste contains a wide variety of contaminants that differ based on the specific mineral being mined and the methods used in processing. Some of the most hazardous pollutants released from mining operations include heavy metals, acid mine drainage, and sediments. These contaminants have long-term environmental and health implications, as they persist in the environment and accumulate in biological systems, leading to widespread ecological disruption (Whitworth et al., 2022).

2.2.1 Toxic Heavy Metals (Including Mercury, Cadmium, Arsenic, Lead, and Others)

The heavy metals are most dangerous pollutants associated with mining. These toxic elements are naturally present in ores and are released into the environment through excavation, milling, and waste disposal. Unlike heavy metals do not degrade over time; instead, they build up in soil, water, and living organisms, creating serious threats to both human health and wildlife (Upadhyay et al., 2023).

Arsenic, a common byproduct of gold and copper mining, is highly toxic and can cause severe health issues, including skin lesions, cardiovascular diseases, and various forms of cancer. It often enters water supplies through leaching from mine tailings and waste rock (Dey, Bissoyi, Koteswara Reddy, & Das, 2024; Dey, Dash, & Das, 2024; Dey, Tripathy, et al., 2023; Lafta et al., 2024). Lead, another prevalent mining pollutant, is particularly dangerous to children, as it affects brain development and cognitive function (Mandal et al., 2022). In areas with high lead contamination from mining activities, cases of developmental disorders and learning disabilities are significantly higher among exposed populations. Figure 1 indicates the primary conventional chemical and physical techniques employed for heavy metal removal, categorized under three main strategies. Table 2 offers a comparative evaluation of the various techniques utilized for removing heavy metals from wastewater.

Mercury, frequently used in small-scale gold mining, is highly volatile and easily enters the atmosphere, where it can be deposited into water bodies and transformed

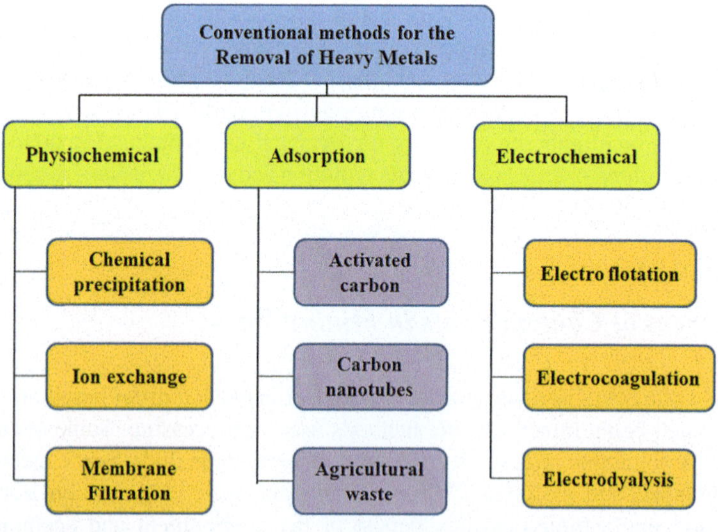

Fig. 1 Standard chemical and physical techniques for extracting heavy metals from wastewater

Table 2 Comparison of methods for heavy metal removal in wastewater treatment

Heavy metals	Techniques	C_o (mg/L)	pH	RE (%)	References
Pb	Adsorption Clinoptilolite	1036	4.0	55	Inglezakis et al. (2010), Kocasoy and Şahin (2007)
Ni	Clinoptilolite	25	7.0	93.4	Kędziora et al. (2014)
Zn	Clinoptilolite	65.4–654	5.0	>96	Tan et al. (2008)
Cd	Natural zeolite	1–50	5.5	99	Kędziora et al. (2014), Tan et al. (2008)
Hg and Cu	Activated carbon	6–7	6–7	Hg(II) >95% cu(II) >95%	Bhat et al. (2022), Kosa et al. (2012)
Zn	Precipitation CaO	32	9–10	>99.0	Mallampati et al. (2013)
Pb, Cu, Cr, Zn	CaO		7–11	>99.3	Benalia et al. (2022), Teh et al. (2016)
Pb, Zn, Cu	H_2S		3.0	>99, >94, >92	Fu and Wang (2011), Teh et al. (2016)
Zn, Pb, Cu	H_2O_2 and alum		7.6	82, 84, 86	Fu and Wang (2011)
Ni	Ion exchange Acidic cation exchange resin	1800, 2800	3–7	>97	Swelam et al. (2017)
Cu	Indion ^{225}H	50–150	6.3–6.5		Thakare and Jana (2015)
Zn, Cd	Ultrafiltration Amicone regenerated cellulose; SDS; MEUF	56.2, 32.7	–	99	Landaburu-Aguirre et al. (2009)
Ni, Cr (III), Cu	Carboxymethyl cellulose; polyether sulfone; PEUF	10	7	98, 99, 99	Barakat and Schmidt (2010)
Ni, Zn, Cu	Nanofiltration Membrane with metal complexing polymer	–	–	97.5	Mehdipour et al. (2015)
Ni, Zn	Reverse osmosis Pretreated aromatic polyamide (ES20) ultrafiltration membrane.	44–169	–	99.2%, 98.8%	Ipek (2005)
Fe, Zn	Coagulation/flocculation Polyaluminium chloride	25–45	8–9.3	92–98	Pang et al. (2011)
Co, Ni, Cu	Chitosan, montmorillonite, system	20–100	7	–	Assaad et al. (2007)
Cu, Cr, Zn	Sodium dodecyl sulphate	11.2	6.9	99	Shen et al. (2015)

into methyl mercury (Pant et al., 2024). This form of mercury is particularly toxic, as it bioaccumulates in aquatic organisms and moves up through food chain, leading to high levels of exposure in humans who consume contaminated fish. Cadmium, commonly found in zinc and phosphate mining, is another hazardous heavy metal that can cause kidney damage, bone fractures, and respiratory illnesses when ingested or inhaled (Charkiewicz et al., 2023). The widespread release of these heavy metals from mining operations highlights the critical necessity for efficient waste management and remediation techniques to prevent long-term environmental contamination.

2.2.2 Acid Mine Drainage (AMD)

Acid mine drainage (AMD), also known as acid rock drainage (ARD), is a pollution causing from chemical, physical, and biological interactions in active and abandoned mines (D'Sa, 2004; Ighalo et al., 2022; Lopez-Luna, 2008; Rajak, 2018). It has a very low pH (below 4), high sulfate ion levels, and elevated concentrations of dissolved metals like iron and zinc (Gundersen & Steinnes, 2003; Pohl, 2020; Sheoran & Sheoran, 2006; Shokes & Möller, 1999; Sun et al., 2017). AMD impacts the environment in four ways: chemical (increased acidity and metal levels), physical (reduced light in water bodies), biological (health and reproductive issues in organisms), and ecological (disrupted food chains and habitats) (Ngole-Jeme & Ndava, 2023). Figure 2 provides a detailed breakdown of these impacts.

Acid mine drainage is the most persistent and destructive forms of mining pollution (Simamora, 2024). It happens when sulfide minerals, like pyrite, coming into contact with air and water during mining activities. These minerals undergo oxidation, producing sulfuric acid, which lowers the pH of water bodies and accelerates the dissolution of heavy metals from surrounding rocks. The resulting acidic water is highly toxic and can persist for decades, continuously contaminating rivers, lakes, and groundwater.

AMD is particularly problematic in coal mining regions and areas with extensive metal ore extraction. Once initiated, it is challenging to stop, as the chemical

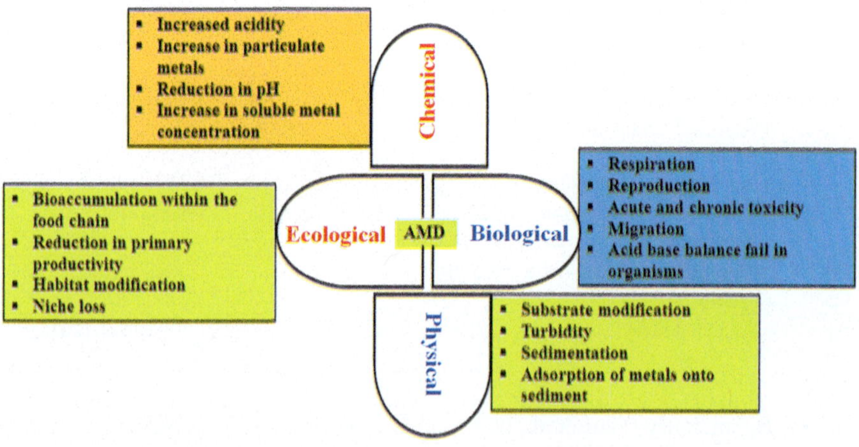

Fig. 2 Environmental effects of AMD

reactions responsible for its formation are self-sustaining. The low pH of AMD-contaminated water severely disrupts aquatic ecosystems, leading to a reduction in biodiversity, fish mortality, and the degradation of habitats. Additionally, AMD corrodes infrastructure, damages agricultural lands, and increases the costs of water treatment for local communities (Baloyi et al., 2023). Preventing AMD requires proper waste rock management, water treatment systems, and site rehabilitation measures, such as backfilling abandoned mines and using alkaline materials to neutralize acidity.

2.2.3 Sediments and Particulate Matter

Mining activities generate large quantities of sediments and particulate matter that contribute to water pollution and soil degradation. The excavation of minerals disturbs vast areas of land, loosening soil and increasing erosion (Pandey et al., 2024). When heavy rainfall occurs, loose sediments are washed into nearby rivers and lakes, leading to increased turbidity and sedimentation. This excess sedimentation reduces water quality by blocking sunlight, which affects photosynthetic aquatic plants and disrupts the natural balance of aquatic ecosystems.

Additionally, fine particulate matter from mining operations can become airborne, affecting air quality and human health. Dust generated from blasting, crushing, and transporting ores contains harmful compounds, including silica and heavy metals, which can cause respiratory diseases such as silicosis and lung cancer (Kasongo et al., 2024). Mining dust also settles on agricultural lands, reducing soil fertility and contaminating crops. The cumulative impact of sediment pollution and airborne particulate matter underscores the importance of erosion control measures, dust suppression techniques, and sustainable land management practices in mining regions.

2.3 *Impact on Water Resources and Ecosystems*

Mining pollution has severe consequences for water resources and ecosystems. The contamination of rivers, lakes, and groundwater sources disrupts aquatic life, reduces water quality, and alters hydrological cycles. Heavy metals released from mining sites accumulate in sediments, leading to long-term ecological damage. Acid mine drainage further exacerbates these effects by creating inhospitable conditions for fish and other aquatic organisms (Agarwal & Pandey, 2023). Biodiversity loss is a major concern in mining-affected areas, as species decline due to habitat destruction, water contamination, and reduced food availability. The introduction of toxic chemicals into the environment also leads to genetic mutations and reproductive failures in wildlife, further threatening ecosystem stability.

2.4 *Human Health Risks Associated with Mining Pollution*

Mining pollution posing significant threat to human health, especially in communities located near mining sites. Exposure to contaminated water sources can lead to neurological disorders, respiratory illnesses, and gastrointestinal diseases. Heavy metal poisoning from lead, mercury, and arsenic contamination has been linked to cognitive impairments, organ damage, and increased cancer rates. Poor air quality in mining regions contributes to higher incidences of asthma, bronchitis, and lung diseases. Additionally, the consumption of contaminated fish and crops exacerbates health risks, making mining pollution a global public health concern (Abera & Adimas, 2024).

3 Physical Treatment of Mining Pollutants

The physical treatment of mining pollutants involves techniques that aim to remove, separate, or immobilize contaminants in mining wastewater. These methods primarily rely on physical forces, such as filtration, adsorption, sedimentation, and electrochemical processes, to eliminate suspended particles, heavy metals, with other pollutants from wastewater prior to its release into the environment. Physical treatment technologies are widely used due to their effectiveness, cost-efficiency, and ability to be integrated with other remediation approaches, including biological and chemical treatments (Aparicio et al., 2022).

Physical techniques are crucial in minimizing the environmental effects of mining operations, as mining wastewater often contains high concentrations of heavy metals, sediments, and toxic chemicals (Noman et al., 2022). If left untreated, these contaminants can enter natural water bodies, which lead to ecological damage and health risks for nearby communities. Unlike biological treatment methods that rely on microorganisms for pollutant degradation, physical treatment methods focus on mechanical and electrochemical processes to achieve contaminant removal (Fitch et al., 2022). These methods are particularly useful for treating large volumes of wastewater with high pollutant loads. The following sections discuss various physical treatment techniques, including sedimentation, filtration, membrane separation, adsorption, and electrochemical methods.

3.1 *Overview of Physical Treatment Technologies*

Physical treatment technologies are essential for removing of pollutants from mining wastewater through mechanical separation, sedimentation, filtration, and adsorption (Jing et al., 2023). These methods are effective for removing suspended solids, dissolved heavy metals, and chemical residues, making wastewater less harmful before being discharged or further treatment. Unlike biological treatment methods, which rely on microbial activity for pollutant breakdown, physical treatment technologies focus on separating contaminants based on size, charge, or affinity for specific materials (Mohammadi et al., 2022).

Sedimentation and filtration are widely utilized to remove particulate matter and heavy metals from wastewater, preventing them from reaching natural water bodies. Membrane technologies such as reverse osmosis and ultrafiltration provide advanced purification by filtering out fine particles and dissolved pollutants. Adsorption techniques, including activated carbon adsorption and biosorbents, are employed to remove dissolved organic and inorganic contaminants (Tee et al., 2022). Electrochemical methods like electrocoagulation and electrochemical oxidation utilize electrical currents to break down and separate pollutants, offering an effective approach for treating highly contaminated wastewater. Each of the methods has its own benefits and limitations that depend on the specific characteristics of the wastewater being treated. The selection of technology is influenced by factors like pollutant concentration, wastewater volume, and implementation costs (Pistocchi et al., 2022). A combination of physical, chemical, and biological treatment methods is used to achieve optimal purification outcomes.

3.2 Sedimentation and Filtration

Sedimentation and filtration are fundamental physical treatment methods to remove suspended solids, heavy metals, and sediments from mining water. These techniques are based on the principle of gravity separation and mechanical filtration, ensuring that pollutants are effectively isolated from the water before further treatment or discharge (Anderson et al., 2022).

Sedimentation is a natural process where solid particles sink to the bottom of a treatment tank under the influence of gravity. This method is particularly use for removing large and heavy particles, such as rock fragments, soil, and metal hydroxides from wastewater. Sedimentation basins or clarifiers are commonly used in mining operations to allow particulates to settle over time. The clarified water is then collected from the top, while the settled sludge at the sediment is separated for additional processing or disposal. Filtration, on the other hand, is an active process that uses porous materials to trap and remove suspended particles from wastewater. Sand filters, gravel filters, and multimedia filters are commonly used in mining wastewater treatment plants to enhance water clarity and reduce pollutant levels (Etsuyankpa et al., 2024). These filtration systems work by forcing contaminated water through layers of porous media, where particulate matter is physically captured. Advanced filtration techniques, such as microfiltration and nanofiltration, provide even finer purification, eliminating both suspended solids and dissolved pollutants (Maroufi & Hajilary, 2023). Sedimentation and filtration are often used as preliminary treatment steps in mining wastewater treatment systems. They help reduce the load of contaminants entering downstream treatment processes, enhancing the effectiveness of water purification methods like adsorption and membrane filtration.

3.3 *Membrane Technologies*

3.3.1 Reverse Osmosis (RO)

Reverse osmosis (RO) is an effective membrane-based filtration technique used to remove dissolved heavy metals, salts, and other contaminants from mining wastewater (Xiang et al., 2022). This method uses a semi-permeable membrane that permits only water molecules to pass through while preventing dissolved contaminants from crossing. RO is particularly useful for treating wastewater with high doses of heavy metals, sulfates, and other dissolved substances that cannot be effectively removed through conventional filtration methods.

In a reverse osmosis (RO) system, wastewater is subjected to high pressure, driving it through the membrane. As a result, contaminants are left behind, and purified water is collected on the other side. RO systems are widely used in mining operations for water reclamation and reuse, as they produce high-quality effluent suitable for industrial applications or safe discharge into the environment (Date et al., 2022). However, RO requires significant energy input due to the high-pressure operation, and the concentrated waste stream, known as brine, must be properly managed to prevent secondary pollution.

3.3.2 Ultrafiltration and Nanofiltration

Ultrafiltration and nanofiltration are advanced membrane filtration techniques that provide effective separation of fine particles and metal ions from wastewater. Ultrafiltration has membranes with pore sizes ranging from 0.01 to 0.1 microns, making it capable of removing bacteria, viruses, and suspended solids. This method is commonly used as a pre-treatment step before reverse osmosis, ensuring that larger particles are removed to prevent membrane clogging (Covaliu-Mierlă et al., 2023). Nanofiltration, on the other hand, operates at a molecular level, removing dissolved salts, organic molecules, and specific metal ions from wastewater. It is particularly effective for the selective removal of divalent and multivalent ions, such as calcium, magnesium, and sulfate, while allowing some smaller molecules to pass through. Nanofiltration systems require lower pressure compared to RO, making them more energy-efficient while still providing excellent contaminant removal (Koli & Singh, 2023). Both ultrafiltration and nanofiltration plays crucial role in the efficiency of mining wastewater treatment by reducing pollutant concentrations and enabling water reuse in industrial processes.

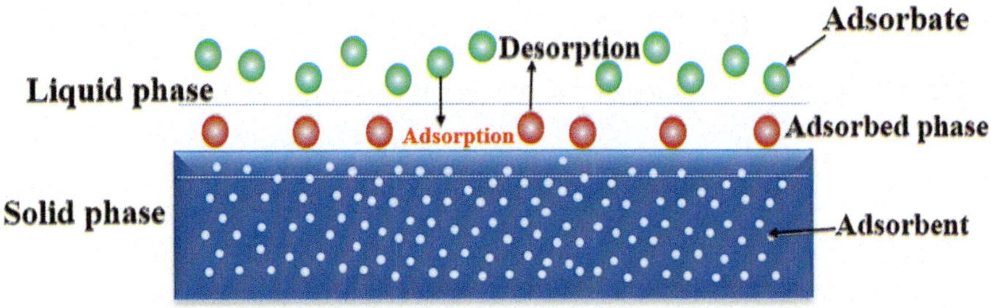

Fig. 3 The adsorption and desorption processes, involving the transition of molecules between the liquid phase and the solid surfaces of the used matrix

3.4 Adsorption Techniques

3.4.1 Adsorption Using Activated Carbon

Activated carbon adsorption is a widely used method for removing both organic and inorganic pollutants from mining wastewater. Activated carbon possesses a highly porous structure and a vast surface area, enabling it to efficiently capture pollutants through adsorption. This method (Fig. 3), is particularly useful for removing dissolved heavy metals, cyanide compounds, and organic pollutants such as hydrocarbons from wastewater (Hoang et al., 2022).

In the adsorption process, polluted water is filtered through an activated carbon bed, where contaminants stick to the surface of the carbon particles. The importance of this technique is depending on variables such as contact duration, the amount of carbon used, and the concentration of pollutants. Once the activated carbon reaches its adsorption capacity, it must be replaced or regenerated to maintain treatment efficiency (Wang et al., 2023). The adsorption and desorption processes involving the interaction between a liquid phase and a solid phase. The solid phase represents the adsorbent, a material that attracts and holds molecules (adsorbate) from the liquid phase on its surface. Adsorption is the process where these molecules move from the liquid phase and adhere to the solid surface. Conversely, desorption is the reverse process, where adsorbed molecules are released from the solid surface back into the liquid phase. These processes are crucial in various applications, such as purification and separation technologies (Wang et al., 2023). The diagram effectively captures the dynamic equilibrium between the liquid and solid phases, highlighting the movement of molecules during adsorption and desorption. This visualization aids in understanding the fundamental principles of surface chemistry and material interactions.

3.4.2 Biosorbents (Iron-Oxide Coated Materials, Clay, Etc.)

Biosorbents, such as iron-oxide coated materials and clay, are economical options for eliminating heavy metals from wastewater. Metal ions are highly attracted to these materials, allowing them to effectively capture and immobilize contaminants. Iron-oxide coated materials, for example, exhibit strong adsorption properties for arsenic, lead, and chromium, making them suitable for treating mining effluents (Raji et al., 2023).

3.5 Electrochemical Methods

3.5.1 Electrocoagulation

Electrocoagulation is an electrochemical method that applies electric current to clump together and eliminate metal pollutants from wastewater. The process involves passing an electric current through metal electrodes, which release charged particles that destabilize pollutants, leading them to aggregate and separate from the water. Electrocoagulation is particularly efficient to remove heavy metals, arsenic, and suspended solids from mining wastewater.

3.5.2 Electrochemical Oxidation

Electrochemical oxidation is another advanced treatment technique that uses electrical energy (Saravanan et al., 2022). This method is effective for degrading cyanide, organic compounds, and other persistent contaminants in mining wastewater. By applying an electric potential, pollutants are oxidized into non-toxic byproducts, making electrochemical oxidation a valuable tool for sustainable wastewater management.

4 Biological Treatment of Mining Pollutants

Biological treatment of mining pollutants is a sustainable and environmentally friendly approach to mitigating the adverse effects of mining waste. Unlike physical and chemical treatments that rely on filtration, adsorption, and oxidation, biological treatment methods use living organisms to degrade, transform, or restrain contaminants current in mining wastewater and polluted soils. This approach harnesses the natural ability of microorganisms, plants, and enzymes to break down or sequester toxic substances, thereby reducing their impact on the environment. Biological treatment techniques are particularly advantageous because they offer cost-effective and low-energy alternatives to conventional remediation strategies. They also generate minimal secondary pollution, making them a preferred choice for long-term ecological restoration of mining-affected areas. In recent years, extensive research has conducted to explore the potential of bioremediation, phytoremediation, and enzyme-based treatments in removing heavy metals, neutralizing acidic wastewater, and improving soil health in mining regions (Ojha et al., 2023). This section investigates into various biological treatment techniques, including microbial bioremediation, phytoremediation, and enzyme-based remediation, and their effectiveness in mitigating mining pollution.

4.1 Overview of Bioremediation

Bioremediation is an eco-friendly approach that employs biological agents like bacteria, fungi, and plants to break down, neutralize, or immobilize pollutants from mining activities. This technique is especially effective for addressing heavy metal contamination, acid mine drainage, and organic pollutants in areas impacted by mining. Bioremediation works by enhancing the metabolic activities of specific microorganisms or plants that either convert toxic compounds into less harmful forms or accumulate pollutants within their biomass for safe removal (Saeed et al., 2022).

Two main types of bioremediations: in situ and ex situ. In situ bioremediation is carried out at the contaminated site, reducing the need for excavation and transportation of polluted materials. This method is beneficial for large-scale mining sites where traditional remediation techniques may not be feasible. Ex situ bioremediation entails extracting contaminated soil or water and treating it under controlled conditions, allowing for greater optimization of biological processes as is described in Fig. 4.

The achievement of bioremediation depends on several factors, including the type and concentration of pollutants, environmental conditions (such as pH, temperature, and nutrient availability), and the presence of suitable microbial

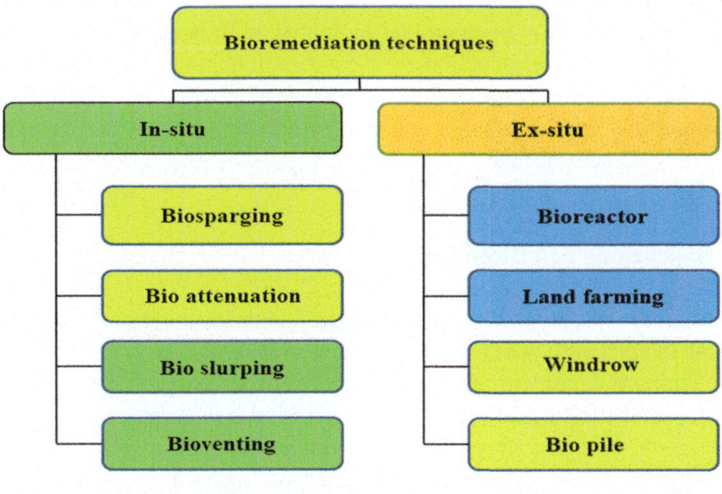

Fig. 4 Classification of bioremediation technologies

communities/population or plants. When appropriately implemented, bioremediation can significantly reduce pollutant levels, restore ecological balance, and promote long-term sustainability in mining-affected regions.

4.2 Microbial Bioremediation

Microbial bioremediation means to the use of bacteria and fungi to detoxify mining pollutants through natural metabolic processes. Microorganisms have advanced various mechanisms to survive in extreme environments, including the ability to absorb, transform, or sequester heavy metals and other toxic metals. These biological processes can be harnessed for the remediation of mining wastewater and contaminated soils.

4.2.1 Role of Bacteria and Fungi

Certain bacteria and fungi play a crucial role in microbial bioremediation due to their ability to metabolize toxic substances and facilitate heavy metal removal. Bacterial species like Pseudomonas and Bacillus are widely studied for their potential in bioremediation. These bacteria can resist heavy metal toxicity by transforming harmful metal ions into less bioavailable forms, thus reducing their environmental impact. Fungal species, including Aspergillus and Penicillium, are also known for their ability to degrade organic pollutants and accumulate heavy metals in their mycelial structures. Fungi produce organic acids that enhance metal solubility and promote metal precipitation, making them effective in the treatment of contaminated mine water. Additionally, fungi have extensive hyphal networks that enable them to stabilize pollutants in the soil, preventing their further spread into water bodies. The effectiveness of microbial bioremediation is often enhanced through bioaugmentation (the introduction of specific pollutant-degrading microbes into contaminated sites) or biostimulation (Chettri et al., 2024). These strategies have been successfully applied in numerous mining-affected areas to accelerate the degradation of toxic substances.

4.2.2 Bioleaching and Biosorption

Bioleaching and biosorption are two biological mechanisms through which microorganisms interact with heavy metals in mining waste.

Bioleaching is a process in which microorganisms extract metals from ores or tailings by producing organic acids and other bioactive compounds (Dong et al., 2022). This method is widely used in the mining industry to recover metals like gold, and uranium from low-grade ores. However, bioleaching can also be applied as a remediation strategy to reduce heavy metal concentrations in mine waste and

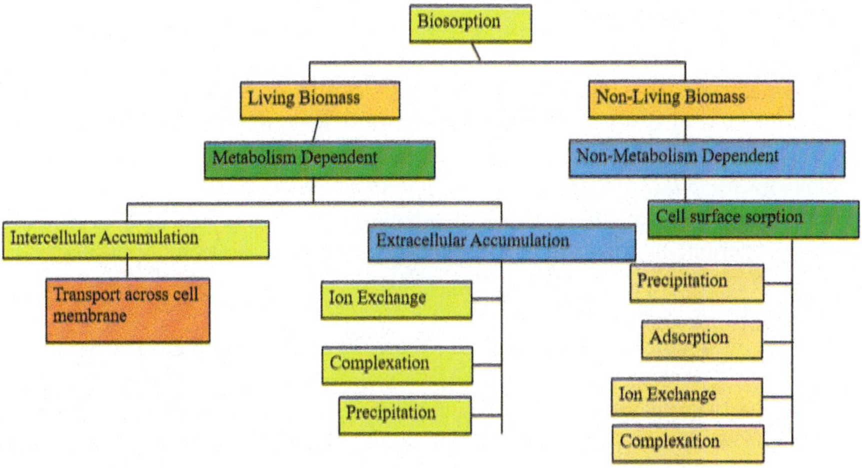

Fig. 5 Relationship between biosorption mechanisms, metabolism, and the effectiveness of metal removal

polluted water. Certain bacteria, such as Acid thiobacillusferroxidase and Leptospiralferroxidase, are commonly used in bioleaching due to their ability to oxidize sulfide minerals and mobilize metal ions.

Biosorption techniques are greatly affected with microorganism-driven processes and the processes involved in metal removal. Based on biomass activity, biosorption can either metabolism-dependent or independent. As illustrated in Fig. 5, metal removal takes place through (a) intracellular accumulation, (b) sorption or precipitation on cell surfaces, and (c) extracellular accumulation or precipitation. While non-metabolism-dependent processes involve rapid, reversible interactions with cell surface functional groups, intracellular bioaccumulation is metabolism-driven and occurs more slowly (Padhye et al., 2023).

Non-metabolism-dependent methods are quick and involve reversible physical and chemical interactions with metals and functional groups on the cell surface. In contrast, bioaccumulation is driven by metabolism and happens more slowly (Landaburu-Aguirre et al., 2009). Metal precipitation, when aided by microbial products, can be influenced by microbial metabolism or occur independently through straightforward chemical interactions (Abdelhamid et al., 2024). The mode and efficiency of metal binding are affected by the presence of other compounds and the condition of the biomass. In addition to the cell wall, metal binding can also occur on extracellular polysaccharides, varying based on whether the biomass is living or dead.

Biosorption, on the other hand, is a passive process in which microbial biomass binds and absorbs heavy metals from wastewater. This mechanism is highly effective for the removal of cadmium, lead, mercury, and other toxic metals from industrial and mining effluents. Biosorption occurs through various interactions, including ion exchange, complexation, and adsorption onto microbial cell walls. The use of microbial biomass as a biosorbent is gaining attention due to its low-cost and high efficiency in treating metal-contaminated water.

4.3 Phytoremediation

Phytoremediation is a plant-based approach to mining pollution treatment that involves the use of specific plant species to absorb, neutralize, or degrade pollutants from soil and water. This method is environmentally friendly, cost-effective, and suitable for large-scale remediation of contaminated landscapes. Plants used in phytoremediation are typically capable of accumulating high levels of heavy metals without experiencing toxicity, allowing for the gradual restoration of polluted sites.

4.3.1 Hyperaccumulator Plants (e.g., Water Hyacinth, Vetiver Grass)

Hyperaccumulator plants are a special group of plants that can absorb and store high concentrations of heavy metals in their tissues. These plants are particularly effective for remediating mining-affected environments, as they extract pollutants from contaminated soil and water through their root systems. Water hyacinth is widely recognized for its ability to absorb metals such as arsenic, lead, and mercury from polluted water bodies. It has an extensive root system that facilitates the uptake of contaminants, making it highly effective for water purification. Vetiver grass, another commonly used hyperaccumulator, is known for stabilizing soil and preventing the spread of heavy metal contamination in mining regions.

4.3.2 Phytostabilization and Phytoextraction

Phytostabilization and phytoextraction are two distinct phytoremediation techniques that address heavy metal remediation in different ways.

Phytostabilization involves the use of plants to fix in place pollutants in the soil, preventing their migration to nearby water bodies. Plants with deep root systems help stabilize heavy metals by binding them to the soil matrix, reducing the risk of groundwater contamination (Padhye et al., 2023).

Phytoextraction, on the other hand, is the process in which plants absorb heavy metals from the soil and store them in their biomass. Over time, these metals can be harvested and safely disposed of, gradually reducing pollutant levels in the affected area. Phytoextraction has been successfully applied in regions contaminated with lead, zinc, and cadmium, using plants such as Indian mustard and sunflowers.

4.4 Enzyme-Based Bioremediation

Enzyme-based bioremediation is an emerging field that focuses on using enzymes produced by microorganisms or plants to break down pollutants into less harmful substances. Enzymes such as peroxidases, oxidases, and hydrolases play a crucial role in degrading toxic chemicals, including cyanides, hydrocarbons, and persistent organic pollutants found in mining wastewater. Enzymatic treatments offer several advantages, including high specificity, rapid reaction rates, and minimal environmental impact. These enzymes can be immobilized onto solid supports to enhance their stability and reusability in large-scale remediation projects (Padhye et al., 2023).

5 Case Studies of Bioremediation in Mining Regions

Bioremediation has emerged as a crucial technique for addressing the environmental impacts of mining activities worldwide. Mining operations generate extensive pollution, including heavy metal contamination, acid mine drainage (AMD), and toxic effluents, all of which threaten local ecosystems and human health. Bioremediation strategies are the natural capabilities of microorganisms, plants, and biological processes to detoxify contaminated environments, providing a cost-effective alternative to conventional remediation techniques.

Across the world, bioremediation has been successfully implemented in mining-affected regions, helping to restore ecological balance and improve water quality. Countries with extensive mining industries, such as India and the USA, have adopted microbial bioremediation, phytoremediation, and enzyme-based remediation techniques to address the environmental challenges posed by coal and metal mining (Table 3). This section explores case studies of bioremediation initiatives in Indian mining areas, particularly in the coal-rich states of Jharkhand and Odisha, as well as global case studies, including acid mine drainage remediation efforts in the United States.

Table 3 Case studies of bioremediation in mining regions

Location	Issue	Bioremediation approach	Results	References
Jharkhand and Odisha, India	Coal mining pollution, acid mine drainage	Sulfate-reducing bacteria, phytoremediation with vetiver grass	Reduced sulfate levels, improved soil health	Singh et al. (2014)
Summitville mine, USA	Heavy metal contamination, acidic wastewater	Bioreactors with sulfate-reducing bacteria	Neutralized acidity, metal precipitation achieved	Neculita et al. (2007)
Appalachia, USA	Acid mine drainage from abandoned coal mines	Constructed wetlands with metal-absorbing plants	Lowered metal concentration in streams	Ziemkiewicz et al. (2021)
Iron Mountain mine, USA	Extreme acid mine drainage	Genetically modified bacteria for acid neutralization	Increased pH, reduced metal leaching	Rawlings (2005)

5.1 *Bioremediation Initiatives in Indian Mining Areas*

India is one of the largest producers of coal and other minerals, with extensive mining activities spread across various states. However, these operations have led to widespread environmental degradation, particularly in regions such as Jharkhand and Odisha, where coal mining has caused deforestation, soil erosion. The contamination of groundwater and surface water with heavy metals, arsenic, and acid mine drainage has severely impacted local communities, affecting agriculture and public health. Recognizing these challenges, Indian researchers and environmental agencies have implemented bioremediation techniques to restore mining-affected landscapes and reduce pollution levels.

5.1.1 Coal Mining Sites in Jharkhand and Odisha

Jharkhand and Odisha are home to some of India's largest coal reserves, with extensive open-cast and underground mining operations contributing significantly to the country's energy production. However, coal mining in these states has also resulted in severe environmental damage, including acid mine drainage and deforestation. The presence of sulfur-rich coal deposits leads to the generation of acid mine drainage, which lowers the pH of water bodies, making them uninhabitable for aquatic life and unfit for human consumption.

In recent years, microbial and phytoremediation strategies have been successfully implemented in several coal mining sites across Jharkhand and Odisha. Microbial remediation techniques, particularly the use of sulfate-reducing bacteria (SRB), have been applied to neutralize acid mine drainage and reduce heavy metal toxicity. These bacteria facilitate the conversion of sulfate into hydrogen sulfide, which precipitates heavy metals as insoluble sulfides, effectively removing them from contaminated water. In laboratory and field-scale trials, sulfate-reducing bacteria have demonstrated significant potential in improving water quality in abandoned coal mines, reducing sulfate concentrations and increasing the pH of acidic water bodies.

Phytoremediation has also been employed to restore degraded mining sites in Jharkhand and Odisha (Pandey et al., 2024). Native plant species such as Vetiver grass and Water hyacinth (Eichhornia crassipes) have been used to stabilize heavy metal-contaminated soil and extract toxic elements from polluted water sources. Vetiver grass, with its deep root system, effectively binds soil particles, preventing erosion and reducing the mobility of heavy metals. Meanwhile, water hyacinth is known for its high metal absorption capacity, making it an excellent candidate for phytoremediation in coal mining regions. Several pilot projects in Odisha's Talcher coalfields and Jharkhand's Jharia coalfields have demonstrated promising results in using phytoremediation to rehabilitate contaminated land. By combining microbial bioremediation with phytoremediation strategies, these initiatives have successfully reduced pollution levels, restored vegetation cover, and improved soil fertility, thereby contributing to the long-term sustainability of mining-affected landscapes. Rajasthan, although primarily known for its lignite mining rather than coal, faces similar environmental challenges due to mining activities. Lignite mines in regions like Barmer and Bikaner have reported significant groundwater contamination due to acid leachate and heavy metals. In response, bioremediation techniques, including the use of sulfate-reducing bacteria, have been tested to treat acidic drainage from lignite mines. Additionally, phytoremediation initiatives using salt-tolerant plant species such as Mesquite (Prosopis juliflora) and Indian mustard (Brassica juncea) have shown potential in stabilizing contaminated soils and absorbing heavy metals from mining-impacted areas. These efforts highlight the applicability of bioremediation strategies across diverse mining environments in India, paving the way for sustainable ecological restoration.

5.2 *Global Case Studies*

Bioremediation has been widely applied in mining-affected regions across the globe, particularly in countries with a long history of mining, such as the USA, Canada, Australia, and South Africa. These countries have faced significant challenges in managing acid mine drainage, heavy metal contamination, and soil degradation resulting from extensive mining operations. Several large-scale bioremediation initiatives have been applied to mitigate these environmental impacts, demonstrating the effectiveness of biological treatment methods in restoring polluted ecosystems.

5.2.1 Acid Mine Drainage Remediation in the USA

The USA has a vast mining industry, with thousands of abandoned mines contributing to environmental pollution across the country. One of the most severe issues associated with mining in the USA is AMD, which results from the oxidation of sulfide minerals, important to the formation of sulfuric acid and the leaching of heavy metals into nearby water bodies. This phenomenon has severely impacted rivers and lakes, particularly in coal mining regions such as West Virginia, Pennsylvania, and Colorado. To address the issue of AMD, the U.S. Environmental Protection Agency (EPA) and various research institutions have implemented bioremediation techniques involving sulfate-reducing bacteria (SRB). These bacteria are capable of reducing sulfate into hydrogen sulfide, which reacts with suspended heavy metals to form insoluble metal sulfides that precipitate out of the water. This process effectively removes contaminants such as iron, cadmium, and arsenic, thereby improving water quality in mining-affected rivers and lakes.

One of the most notable examples of AMD bioremediation is the remediation of the Summitville Mine in Colorado, a former gold mining site that experienced severe acid drainage contamination (Surimbayev et al., 2025). Bioremediation efforts at this site involved the construction of bioreactors, which provided optimal conditions for sulfate-reducing bacteria to thrive. These bioreactors contained organic substrates, such as wood chips and compost that served as a carbon source for the bacteria, enhancing their ability to neutralize acidity and remove heavy metals from the contaminated water. Over time, the bioremediation system successfully reduced acidity levels and improved water quality, demonstrating the potential of microbial treatment in AMD management.

Another significant example is the Pennsylvania Mine Drainage Remediation Project, which focused on rehabilitating abandoned coal mines in the Appalachian region. The initiative involved the use of wetland-based bioremediation systems, where natural and constructed wetlands were used to treat AMD through biological and chemical processes. Wetland plants, such as cattails (Typha spp.) and bulrushes (Scirpus spp.), played a crucial role in filtering out heavy metals, while microbial communities in the sediment facilitated the removal of sulfate and acidity. This natural treatment approach has been highly effective in restoring water quality in streams affected by mining pollution.

Furthermore, in the Iron Mountain Mine in California, one of the most acidic mine drainage sites in the world, researchers have tested the use of genetically engineered microbes capable of surviving in extreme acidic conditions to facilitate heavy metal removal. These engineered microbial strains exhibit enhanced sulfate reduction capabilities, providing a novel approach to treating AMD in highly contaminated environments. The successful implementation of bioremediation in the USA has set a precedent for other mining-affected countries to adopt biological treatment methods for acid mine drainage and heavy metal contamination. By utilizing sulfate-reducing bacteria, phytoremediation systems, and engineered bioreactors, these initiatives have contributed to the restoration of degraded landscapes and the protection of aquatic ecosystems from long-term pollution.

6 Advances in Biotechnological Approaches for Mining Waste Treatment

Mining waste is a major environmental concern due to the release of toxic heavy metals, acid mine drainage, and persistent pollutants into surrounding ecosystems. Traditional remediation techniques, including physical and chemical treatment, are often expensive, energy-intensive, and can generate secondary pollutants. To address these challenges, modern biotechnological approaches have been developed to enhance the efficiency of mining waste treatment.

Biotechnology provides innovative solutions for tackling mining pollution by leveraging the power of microorganisms, plants, and advanced materials such as nanoparticles (Eskandar, 2023). These approaches aim to improve heavy metal absorption, neutralize acidic contaminants, and accelerate the breakdown of harmful substances in mining wastewater and tailings. Among the most promising

Table 4 Advances in biotechnological treatment for mining waste

Technology	Description	Applications in mining waste treatment	Challenges	References
Genetic modification of microbes	Engineering bacteria for enhanced heavy metal absorption	Used for arsenic, lead, and mercury detoxification	Environmental concerns, regulatory approval	Das et al. (2008)
Nano-bioremediation	Nanoparticles aiding microbial and plant-based remediation	Iron-oxide nanoparticles remove heavy metals	High cost, potentialtoxicity	Braunschweig et al. (2013), Hemalatha et al. (2023)
Membrane bioreactors (MBRs)	Combines membrane filtration with biological treatment	Used for AMD and dissolved metal removal	Expensive setup, energy-intensive	Aslam et al. (2022)
Enzyme-based remediation	Uses enzymes to break down toxic compounds	Cyanide and hydrocarbon degradation	Enzyme stability and scalability	Mousavi et al. (2021)

developments in this field are the genetic modification of microbial strains and the integration of nanotechnology with bioremediation strategies. These cutting-edge techniques enhance the natural abilities of microbes and plants, making them more effective in detoxifying polluted environments. This section explores two significant advances in biotechnological mining waste treatment (Table 4). The first one is the genetic modification of microbial strains to enhance metal absorption and detoxification, and the second one is the application of nano-bioremediation strategies that use nanoparticles to improve pollutant removal efficiency.

6.1 *Genetic Modification of Microbial Strains*

Microorganisms play a crucial role in the natural detoxification of mining waste by transforming or immobilizing heavy metals and other toxic elements. However, many naturally occurring microbes have limited capacity for metal absorption and degradation, making bioremediation a slow and sometimes inefficient process. Advances in genetic engineering have allowed scientists to modify microbial strains to enhance their metal tolerance, bioaccumulation capacity, and metabolic activity, making them more efficient for use in mining waste treatment.

Genetic modification of microbes involves altering their DNA to introduce specific traits that improve their ability to interact with pollutants. One of the most effective genetic modifications is the insertion of metal-binding proteins or transport systems that increase the uptake and sequestration of heavy metals. For example, bacteria such as Pseudomonas putida and Escherichia coli have been genetically engineered to express metallothioneins, which are proteins with a high affinity for heavy metals like mercury, cadmium, and lead. These engineered bacteria can absorb and store large amounts of toxic metals, preventing them from leaching into the environment. Another approach in microbial genetic engineering is the enhancement of sulfate-reducing bacteria (SRB), which are commonly used in acid mine drainage (AMD) treatment. SRBs convert sulfate ions into hydrogen sulfide, effectively removing them from wastewater (Zhang et al., 2022). By introducing genes that increase sulfate reduction rates, researchers have developed SRB strains that can more efficiently neutralize acidic mine water and precipitate toxic metals.

In addition to bacteria, fungi have also been genetically modified for improved bioremediation performance. Aspergillus niger and Trametes versicolor, for example, have been engineered to produce higher levels of ligninolytic enzymes, which enhance their ability to degrade organic contaminants present in mining wastewater. These enzymes break down complex toxic compounds, reducing their persistence in the environment. Recent breakthroughs in synthetic biology have taken microbial engineering a step further by designing completely new metabolic pathways in microbes to improve their detoxification efficiency. Using CRISPR-Cas9 gene editing technology, scientists have been able to precisely modify bacterial genomes to optimize their resistance to harsh mining environments and increase their capacity to metabolize pollutants. These advancements hold great potential for developing highly efficient microbial consortia capable of remediating multiple contaminants simultaneously. Despite the promising potential of genetically engineered microbes in mining waste treatment, several challenges remain. The genetically modified organisms (GMOs) into natural ecosystems raise concerns about ecological balance and unintended genetic transfer to native microbial populations. Strict regulatory frameworks must be established to ensure that these engineered microbes do not disrupt local biodiversity or pose risks to human health. However, with proper containment strategies, genetically modified microbes represent a powerful for improving the efficiency of bioremediation in mining-affected areas.

6.2 Nano-Bioremediation Strategies

Nanotechnology has emerged as a revolutionary tool in environmental remediation, providing novel solutions for mining waste treatment. Nano-bioremediation is an advanced approach that integrates nanomaterials with microbial and plant-based bioremediation techniques to enhance the removal of heavy metals, organic pollutants, and other contaminants from mining-affected environments (Zhang et al., 2022). By utilizing nanoparticles, scientists have been able to significantly improve the efficiency, specificity, and adaptability of biological remediation systems. Nanoparticles are microscopic particles with a size range of 1–100 nm, possessing unique physicochemical properties that make them highly effective for pollutant removal. Their surface area-to-volume ratio, strong adsorption capacity, and ability to interact with metal ions at a molecular level allow them to act as powerful catalysts for contaminant degradation. In mining waste treatment, nanoparticles can be used in various ways to enhance bioremediation processes.

One of the most effective applications of nanotechnology in bioremediation is the use of iron-oxide nanoparticles (Fe_3O_4) for heavy metal removal. These nanoparticles have a high affinity for arsenic, lead, chromium, and mercury, binding to these metals and facilitating their separation from wastewater. When combined with bacterial or fungal bioremediation systems, iron-oxide nanoparticles improve the absorption efficiency of microorganisms by increasing the availability of metal ions. Additionally, they can be magnetically separated from treated water, making the recovery and reuse of adsorbed metals easier.

Another promising nanomaterial used in bioremediation is carbon-based nanoparticles, like graphene oxide and carbon nanotubes. These materials have exceptional adsorption properties and can remove a wide range of pollutants from mining wastewater, including organic contaminants and metal ions. Researchers have developed hybrid bioremediation systems in which bacteria or plants are coated with carbon-based nanoparticles to enhance their pollutant uptake capacity. These nanocomposites provide a protective barrier against toxic metals while increasing the efficiency of microbial and plant-based remediation.

Silver nanoparticles (AgNPs) have also been explored for their antimicrobial and catalytic properties in mining waste treatment. These nanoparticles can help control harmful microbial populations in contaminated sites while simultaneously promoting the growth of beneficial bacteria involved in bioremediation. When incorporated into biofilters or constructed wetlands, silver nanoparticles assist in the removal of organic pollutants and microbial toxins. In phytoremediation, nanoparticles can be used to improve the efficiency of plant uptake of heavy metals. Nano-chelating agents, such as nano-EDTA (ethylene diamine tetraacetate), help solubilize metal ions in the soil, making them more bioavailable for plant absorption. By increasing metal mobility, these nanoparticles enable hyperaccumulator plants to extract higher concentrations of toxic metals from contaminated sites. Additionally, nanoscale fertilizers can be applied to enhance plant growth and resistance to metal stress, ensuring the long-term success of phytoremediation strategies. While nano-bioremediation presents numerous advantages, challenges remain in its large-scale implementation. The environmental fate and potential toxicity of nanoparticles must be thoroughly assessed to ensure that their use does not lead to unintended ecological consequences. Research is ongoing to develop biodegradable nanoparticles that degrade after completing their remediation functions, reducing concerns about long-term nanoparticle accumulation in the environment.

Overall, the integration of nanotechnology with bioremediation offers a highly effective and innovative approach to mining waste treatment. By enhancing microbial and plant-based remediation strategies, nano-bioremediation has the potential to significantly improve the removal metals, acid mine drainage, and persistent pollutants from contaminated sites. Continued advancements in nanotechnology and biotechnology will further optimize these methods, paving the way for more sustainable and efficient mining waste management practices.

7 Future Directions and Challenges

Mining activities have long been associated with severe environmental impacts, particularly the contamination of soil, water, and air with hazardous pollutants, heavy metals, acid mine drainage acidic or basic gases (NO_x, SO_x, CO_2, NH_3, etc.), and toxic chemical residues (Mardonova & Han, 2023). While each of these methods has proven effective in certain contexts, they often exhibit limitations when applied independently. As a result, there is a growing need for an integrated approach that combines the strengths of physical, chemical, and biological treatment methods to maximize efficiency and sustainability. The future of mining waste treatment lies in the development of hybrid remediation technologies that integrate multiple treatment strategies to achieve comprehensive and long-term environmental restoration. These approaches aim to optimize pollutant removal, reduce treatment costs, minimize secondary pollution, and accelerate the rehabilitation of mining-affected ecosystems. However, despite their potential, several challenges must be addressed to ensure their successful implementation. This section explores the future directions in mining pollution remediation, focusing on the integration of physical and biological treatment methods, and discusses the key challenges that must be overcome to achieve sustainable mining waste management.

7.1 Integration of Physical and Biological Treatment Methods

The integration of physical and biological treatment methods represents a promising advancement in the field of mining waste remediation. While physical treatment techniques such as sedimentation, filtration, and membrane technologies are highly effective in removing suspended solids and heavy metals, they often fail to completely eliminate dissolved pollutants or degrade organic contaminants. On the other hand, biological treatments, including microbial bioremediation and phytoremediation, provide a sustainable and environmentally friendly approach to breaking down pollutants, but they can be slow and less effective for certain contaminants. By combining these approaches, a more robust and efficient remediation system can be developed, addressing the limitations of individual treatment methods.

7.1.1 Advantages of Hybrid Treatment Approaches

Hybrid remediation approaches offer several advantages over single-method treatments. First, they allow for a more comprehensive removal of pollutants, targeting both solid and dissolved contaminants. Physical treatment methods can serve as a pre-treatment step, reducing the pollutant load and preparing wastewater or contaminated soil for subsequent biological treatment. For example, sedimentation and filtration can be used to remove larger particles and heavy metals from mining wastewater, making it easier for microbial communities to degrade organic pollutants or neutralize toxic compounds. Another advantage of integrating physical and biological methods is the reduction of treatment time. Physical treatments provide immediate pollutant removal, while biological treatments ensure long-term degradation and stabilization of contaminants. This dual-action approach accelerates the remediation process, making it more suitable for large-scale applications in mining-affected regions. Furthermore, hybrid systems can be optimized to operate with minimal energy consumption and chemical inputs, making them more sustainable and cost-effective.

7.1.2 Examples of Hybrid Treatment Applications

Several case studies have demonstrated the effectiveness of hybrid treatment systems in mining waste remediation. One such example is the use of membrane bioreactors (MBRs) in treating mining wastewater. Membrane bioreactors combine membrane filtration with biological treatment, using microbial communities to degrade organic pollutants while filtering out suspended solids and heavy metals. This technology has been successfully implemented in several mining regions, improving wastewater quality and reducing the environmental impact of mining operations (Wibowo et al., 2024). Another notable example is the use of constructed wetlands in conjunction with sedimentation and adsorption techniques. When combined with physical treatment methods such as coagulation and activated carbon adsorption, these systems can effectively remove heavy metals, neutralize acidic mine drainage, and promote the biodegradation of organic pollutants. This approach has been successfully applied in coal mining regions, where acid mine drainage is a persistent issue.

Additionally, electrochemical-biological hybrid systems have gained attention in recent years. Electrocoagulation, a physical treatment method that uses electrical currents to destabilize and remove pollutants, has been integrated with microbial bioremediation to enhance heavy metal removal. The electrical current is not only aid in the coagulation of metal ions but also stimulate microbial activity, increasing the efficiency of metal detoxification (Wibowo et al., 2024). These systems have shown great promise in treating mining wastewater with high metal concentrations, offering a sustainable alternative to conventional chemical treatments.

7.1.3 Challenges in Implementing Hybrid Treatment Systems

Despite the numerous benefits of integrating physical and biological treatment methods in treating mining wastewater, several challenges must be addressed to facilitate their widespread adoption. One of the challenges is the high initial cost associated with establishing hybrid treatment facilities. Combining multiple treatment technologies requires significant investment in infrastructure, equipment, and operational maintenance. This financial barrier can be particularly limiting for mining companies operating in developing countries, where resources for environmental management may be scarce.

Another challenge is the technical complexity of hybrid treatment systems. The successful integration of physical and biological processes requires a deep understanding of their interactions and dependencies. For instance, biological treatments are highly sensitive to environmental conditions such as temperature, pH, and nutrient availability. If physical treatment methods alter these parameters unfavorably, microbial communities may struggle to survive and perform their detoxification functions. Therefore, extensive research and pilot-scale testing are necessary to optimize system performance and ensure compatibility between different treatment components. Additionally, the scalability of hybrid remediation approaches remains a concern. While laboratory-scale and small-scale pilot studies have demonstrated the effectiveness of integrated treatment methods, their application on a larger scale presents logistical and engineering challenges. Factors, such as variability in waste compositions, seasonal fluctuations in contaminant levels, and site-specific environmental conditions must be considered when designing and implementing the hybrid systems.

Regulatory and policy-related barriers also pose a significant challenge to the adoption of hybrid treatment technologies. Many countries lack clear guidelines and standardized protocols for integrating multiple remediation methods, leading to uncertainties in permitting and compliance. Mining companies may face difficulties in obtaining approval for hybrid treatment systems, particularly if regulatory agencies are unfamiliar with their technical and environmental implications. Strengthening environmental policies and providing incentives for sustainable remediation practices will be crucial in overcoming these regulatory hurdles.

7.1.4 Future Prospects for Hybrid Treatment Technologies

Looking ahead, the future of hybrid treatment technologies lies in the continued advancement of research, innovation, and collaboration between industries, academia, and government agencies. By analyzing vast amounts of environmental data, AI algorithms can predict pollutant behavior, optimize treatment processes, and improve system efficiency in real time.

Another promising development is the integration of nanotechnology with hybrid treatment systems. Nanomaterials, such as iron-oxide nanoparticles and graphene-based adsorbents, have shown exceptional potential for enhancing pollutant removal in both physical and biological treatment processes. When incorporated into hybrid systems, these nanomaterials can improve adsorption efficiency, facilitate microbial growth, and accelerate contaminant degradation. Furthermore, advancements in genetic engineering and synthetic biology hold great potential for improving the performance of biological treatment methods within hybrid systems. By engineering microorganisms with enhanced metal resistance and detoxification capabilities, researchers can create microbial consortia that are more resilient and efficient in remediating mining waste. These engineered microbes can be integrated with physical treatment methods to develop next-generation hybrid remediation technologies.

8 Conclusion

Mining activities have long been associated with significant environmental challenges, including soil degradation, water contamination, air pollution, and biodiversity loss. The extraction and processing of minerals generate large volumes of waste that, poses severe risks to ecosystems and human health. Heavy metals, acid mine drainage, and other toxic pollutants released from mining operations persist in the environment for decades, affecting water bodies, agricultural lands, and nearby communities. Addressing these challenges requires a comprehensive and sustainable approach to mining waste treatment.

Physical treatment methods, such as filtration, sedimentation, membrane technologies, and electrochemical techniques, play a crucial role in removing suspended solids, heavy metals, and other contaminants from mining wastewater. These methods offer immediate pollutant removal and are highly effective in the initial stages of treatment. However, they often fail to address dissolved pollutants and organic contaminants, making long-term remediation difficult. In contrast, biological treatment methods, including microbial bioremediation and phytoremediation, provide sustainable solutions by leveraging natural biological processes to degrade, immobilize, or extract pollutants. These approaches are environmentally friendly and promote ecosystem restoration but can be slow and highly dependent on site-specific conditions.

To enhance the effectiveness of mining waste treatment, hybrid remediation approaches have gained attention. The integration of physical and biological treatment methods provides a more comprehensive and efficient solution, combining the strengths of both techniques. Hybrid systems, such as membrane bioreactors, electrochemical-biological treatments, and constructed wetlands, have demonstrated significant success in improving pollutant removal while reducing secondary pollution. Furthermore, recent advancements in biotechnology, including genetically modified microorganisms and nano-bioremediation, have revolutionized the field of mining waste management. These cutting-edge techniques enhance the ability of microbes and plants to absorb, degrade, or neutralize contaminants, making bioremediation faster and more efficient.

Real-world case studies from India and the USA illustrate the successful application of bioremediation in mining regions. Microbial and phytoremediation strategies have been effectively implemented in coal mining areas of Jharkhand and Odisha, India, helping to reduce heavy metal contamination and restore soil fertility. Similarly, acid mine drainage remediation projects in the United States have utilized sulfate-reducing bacteria and constructed wetlands to neutralize acidic water and remove toxic metals. Despite these advancements, several challenges remain in achieving large-scale implementation of these remediation techniques.

High costs, technical complexities, regulatory barriers, and environmental uncertainties pose obstacles to widespread adoption. Effective mining waste management requires sustained investment in research, innovation, and policy interventions. Governments, industries, and researchers must collaborate to develop cost-effective, scalable, and environmentally responsible solutions. Strengthening environmental regulations, promoting cleaner mining technologies, and integrating sustainable waste treatment strategies will be critical in mitigating the long-term impacts of mining pollution.

As mining continues to expand globally, it is imperative to prioritize responsible mining practices and sustainable remediation strategies. The combination of advanced physical, biological, and biotechnological approaches provides a promising pathway toward reducing mining-related pollution and restoring degraded environments. By embracing these integrated treatment solutions, the mining industry can minimize its environmental footprint while ensuring the sustainable use of natural resources for future generations.

References

Abbaspour, H., & Abbaspour, F. (2022). Renewable energy systems in the mining industry: A literature review and research agenda. *International Journal of Renewable Energy Research, 12*(1), 569–580.

Abdelhamid, M. A., Khalifa, H. O., Yoon, H. J., Ki, M.-R., & Pack, S. P. (2024). Microbial immobilized enzyme biocatalysts for multipollutant mitigation: Harnessing nature's toolkit for environmental sustainability. *International Journal of Molecular Sciences, 25*(16), 8616.

Abera, B. D., & Adimas, M. A. (2024). Health benefits and health risks of contaminated fish consumption: Current research outputs, research approaches, and perspectives. *Heliyon, 10*(13), e09807.

Agarwal, C., & Pandey, A. K. (2023). Remediation and recycling of inorganic acids and their green alternatives for sustainable industrial chemical processes. *Environmental Science: Advances, 2*(10), 1306–1339.

Aili, A., Zhang, Y., Lin, T., Xu, H., Waheed, A., Zhao, W., Kuerban, A., Liu, K., & Dou, H. (2024). Optimizing vegetation restoration: A comprehensive index system for reclaiming abandoned mining areas in arid regions of China. *Biology, 14*(1), 23.

Ali, H., Khan, E., & Sajad, M. A. (2013). Phytoremediation of heavy metals—Concepts and applications. *Chemosphere, 91*(7), 869–881.

Ali, I., Asim, M., & Khan, T. A. (2012). Low cost adsorbents for the removal of organic pollutants from wastewater. *Journal of Environmental Management, 113*, 170–183.

Amuah, E. E. Y., Tetteh, I. K., Boadu, J. A., & Nandomah, S. (2023). Environmental impact assessment practices of the Federative Republic of Brazil: A comprehensive review. *Environmental Challenges, 13*, 100746.

Anderson, A., Anbarasu, A., Pasupuleti, R. R., Manigandan, S., Praveenkumar, T., & Kumar, J. A. (2022). Treatment of heavy metals containing wastewater using biodegradable adsorbents: A review of mechanism and future trends. *Chemosphere, 295*, 133724.

Aparicio, J. D., Raimondo, E. E., Saez, J. M., Costa-Gutierrez, S. B., Álvarez, A., Benimeli, C. S., & Polti, M. A. (2022). The current approach to soil remediation: A review of physicochemical and biological technologies, and the potential of their strategic combination. *Journal of Environmental Chemical Engineering, 10*(2), 107141.

Aslam, A., Khan, S. J., & Shahzad, H. M. A. (2022). Anaerobic membrane bioreactors (AnMBRs) for municipal wastewater treatment-potential benefits, constraints, and future perspectives: An updated review. *Science of the Total Environment, 802*, 149612.

Assaad, E., Azzouz, A., Nistor, D., Ursu, A., Sajin, T., Miron, D., Monette, F., Niquette, P., & Hausler, R. (2007). Metal removal through synergic coagulation–flocculation using an optimized chitosan–montmorillonite system. *Applied Clay Science, 37*(3–4), 258–274.

Azubuike, C. C., Chikere, C. B., & Okpokwasili, G. C. (2016). Bioremediation techniques–classification based on site of application: Principles, advantages, limitations and prospects. *World Journal of Microbiology and Biotechnology, 32*, 1–18.

Bajrović, H. B., Tomašević, V. P., & Raut, J. D. (2025). Analysis of the impact of mining activities on surface water quality–identification of key parameters through statistical and chemical analysis. *Polish Journal of Environmental Studies, 34*(3), 2567–2577.

Bala, K., Anshul, A., Verma, S., Dogra, A., Das, A. P., Kumar, R., & Nandini, A. (2024). Sustainable management for urban plastic waste generation. In *Sustainable management of urban plastic waste through circular economic approaches* (pp. 211–236). Routledge. https://doi.org/10.1201/9781003465447-13

Baloyi, J., Ramdhani, N., Mbhele, R., & Ramutshatsha-Makhwedzha, D. (2023). Recent progress on acid mine drainage technological trends in South Africa: Prevention, treatment, and resource recovery. *Water, 15*(19), 3453.

Barakat, M., & Schmidt, E. (2010). Polymer-enhanced ultrafiltration process for heavy metals removal from industrial wastewater. *Desalination, 256*(1–3), 90–93.

Benalia, M. C., Youcef, L., Bouaziz, M. G., Achour, S., & Menasra, H. (2022). Removal of heavy metals from industrial wastewater by chemical precipitation: Mechanisms and sludge characterization. *Arabian Journal for Science and Engineering, 47*(5), 5587–5599.

Bhat, S. A., Bashir, O., Haq, S. A. U., Amin, T., Rafiq, A., Ali, M., Américo-Pinheiro, J. H. P., & Sher, F. (2022). Phytoremediation of heavy metals in soil and water: An eco-friendly, sustainable and multidisciplinary approach. *Chemosphere, 303*, 134788.

Bhatnagar, A., & Sillanpää, M. (2010). Utilization of agro-industrial and municipal waste materials as potential adsorbents for water treatment—A review. *Chemical Engineering Journal, 157*(2–3), 277–296.

Braunschweig, J., Bosch, J., & Meckenstock, R. U. (2013). Iron oxide nanoparticles in geomicrobiology: From biogeochemistry to bioremediation. *New Biotechnology, 30*(6), 793–802.

Chandra, R., Pai, S., Nayak, S., & Devagudi, S. H. (2025). India's coal conundrum: Decarbonization amidst A developmental legacy. *Wiley Interdisciplinary Reviews: Climate Change, 16*(1), e928.

Chang, L., Taghizadeh-Hesary, F., & Mohsin, M. (2023). Role of mineral resources trade in renewable energy development. *Renewable and Sustainable Energy Reviews, 181*, 113321.

Charkiewicz, A. E., Omeljaniuk, W. J., Nowak, K., Garley, M., & Nikliński, J. (2023). Cadmium toxicity and health effects—A brief summary. *Molecules, 28*(18), 6620.

Chettri, D., Verma, A. K., & Verma, A. K. (2024). Bioaugmentation: An approach to biological treatment of pollutants. *Biodegradation, 35*(2), 117–135.

Covaliu-Mierlă, C. I., Păunescu, O., & Iovu, H. (2023). Recent advances in membranes used for nanofiltration to remove heavy metals from wastewater: A review. *Membranes, 13*(7), 643.

D'Sa, J. V. (2004). *Rapid removal of toxicity of acid mine drainage contaminated sediments in a clean stream: Relative contribution of acidity vs metals*. Ohio University.

Das, A., Dey, S., Bala, K., Gnanasekaran, R., & Das, A. P. (2024). Microbial bioremediation technology for sustainable treatment and management of synthetic microfiber waste. In *Sustainable microbial technology for synthetic and cellulosic microfiber bioremediation* (pp. 221–238). Springer. https://doi.org/10.1007/978-3-031-62600-5_12

Das, A., Dey, S., & Das, A. P. (2024). Ecotoxicological impacts of synthetic microfiber pollutants and development of sustainable mitigation strategies. *Environmental Chemistry and Ecotoxicology*. https://doi.org/10.1016/J.ENCECO.2024.12.003

Das, A., Dey, S., Gacem, A., Yadav, K. K., Nath, H., Ram, D. K., & Das, A. P. (2025). Revolutionizing solid hydrocarbon sourcing: Plastic's journey from waste to upcycled treasure. *Environmental Science and Engineering, 218*, 331–351. https://doi.org/10.1007/978-3-031-81615-4_14

Das, A. P., Behera, I. D., & Das, N. P. (2024). *Renewable energy generation and value addition from environmental microfiber pollution through advanced greener solution* (p. 278). Springer.

Das, A. P., & Ghosh, S. (2023). *Manganese mining microorganisms*. Elsevier.

Das, N., Vimala, R., & Karthika, P. (2008). *Biosorption of heavy metals–an overview*. IntechOpen.

Date, M., Patyal, V., Jaspal, D., Malviya, A., & Khare, K. (2022). Zero liquid discharge technology for recovery, reuse, and reclamation of wastewater: A critical review. *Journal of Water Process Engineering, 49*, 103129.

Dehkordi, M. M., Nodeh, Z. P., Dehkordi, K. S., Khorjestan, R. R., & Ghaffarzadeh, M. (2024). Soil, air, and water pollution from mining and industrial activities: Sources of pollution, environmental impacts, and prevention and control methods. *Results in Engineering, 23*, 102729.

Dey, S., Bissoyi, A., Koteswara Reddy, G., & Das, A. P. (2024). Synthetic microfibers: An urban emerging pollutant demanding for sustainable environmental solutions. In *Sustainable microbial technology for synthetic and cellulosic microfiber bioremediation* (pp. 107–121). Springer. https://doi.org/10.1007/978-3-031-62600-5_7

Dey, S., Das, A., Mallick, K., Sahu, A., & Das, A. P. (2023). *Environmental petroleum waste: Pollution, toxicity, sustainable remediation* (pp. 159–175). Springer. https://doi.org/10.1007/978-3-031-48220-5_7

Dey, S., Dash, A., & Das, A. P. (2024). Bioremediation as a sustainable and circular solution for mining waste management and bioenergy generation. In *Sustainable management of mining waste and tailings: A circular economy approach* (pp. 107–119). Springer. https://doi.org/10.1201/9781003442455-6

Dey, S., Tripathy, B., Kumar, M. S., & Das, A. P. (2023). Ecotoxicological consequences of manganese mining pollutants and their biological remediation. *Environmental Chemistry and Ecotoxicology*. https://doi.org/10.1016/J.ENCECO.2023.01.001

Dong, H., Huang, L., Zhao, L., Zeng, Q., Liu, X., Sheng, Y., Shi, L., Wu, G., Jiang, H., & Li, F. (2022). A critical review of mineral–microbe interaction and co-evolution: Mechanisms and applications. *National Science Review, 9*(10), 128.

Du, H., Wang, Z., Sun, Y., & Shah, K. J. (2024). An overview of the progress made in research on odor removal in water treatment plants. *Water, 16*(2), 280.

Duarte, J., Castelo Branco, J., Rodrigues, F., Vaz, M., & Santos Baptista, J. (2022). Occupational exposure to mineral dust in mining and earthmoving works: A scoping review. *Safety, 8*(1), 9.

Eskandar, K. (2023). Revolutionizing biotechnology and bioengineering: Unleashing the power of innovation. *Journal Of Applied Biology & Biotechnology, 10*(3), 81–88.

Etsuyankpa, M., Augustine, A., Musa, S., Mathew, J., Ismail, H., Salihu, A., & Mamman, A. (2024). An overview of wastewater characteristics, treatment and disposal: A review. *Journal of Applied Sciences and Environmental Management, 28*(5), 1553–1572.

Filho, W. L., Kotter, R., Özuyar, P. G., Abubakar, I. R., Eustachio, J. H. P. P., & Matandirotya, N. R. (2023). Understanding rare earth elements as critical raw materials. *Sustainability, 15*(3), 1919.

Firoozi, A. A., Tshambane, M., Firoozi, A. A., & Sheikh, S. M. (2024). Strategic load management: Enhancing eco-efficiency in mining operations through automated technologies. *Results in Engineering, 24*, 102890.

Fitch, A., Balderas-Hernandez, P., & Ibanez, J. G. (2022). Electrochemical technologies combined with physical, biological, and chemical processes for the treatment of pollutants and wastes: A review. *Journal of Environmental Chemical Engineering, 10*(3), 107810.

Fu, F., & Wang, Q. (2011). Removal of heavy metal ions from wastewaters: A review. *Journal of Environmental Management, 92*(3), 407–418.

Gadd, G. M. (2010). Metals, minerals and microbes: Geomicrobiology and bioremediation. *Microbiology, 156*(3), 609–643.

Gundersen, P., & Steinnes, E. (2003). Influence of pH and TOC concentration on cu, Zn, cd, and Al speciation in rivers. *Water Research, 37*(2), 307–318.

Haghighizadeh, A., Rajabi, O., Nezarat, A., Hajyani, Z., Haghmohammadi, M., Hedayatikhah, S., Asl, S. D., & Beni, A. A. (2024). Comprehensive analysis of heavy metal soil contamination in mining environments: Impacts, monitoring techniques, and remediation strategies. *Arabian Journal of Chemistry, 17*, 105777.

Hemalatha, M., Lakshmi, G. S., Megha, V., & Kotibagar, B. C. (2023). Applications of nanomaterials for water treatment: Current trends and future scope. In *Modern nanotechnology: Volume 1: Environmental sustainability and remediation* (pp. 145–175). Springer.

Hoang, A. T., Kumar, S., Lichtfouse, E., Cheng, C. K., Varma, R. S., Senthilkumar, N., Nguyen, P. Q. P., & Nguyen, X. P. (2022). Remediation of heavy metal polluted waters using activated carbon from lignocellulosic biomass: An update of recent trends. *Chemosphere, 302*, 134825.

Ighalo, J. O., Kurniawan, S. B., Iwuozor, K. O., Aniagor, C. O., Ajala, O. J., Oba, S. N., Iwuchukwu, F. U., Ahmadi, S., & Igwegbe, C. A. (2022). A review of treatment technologies for the mitigation of the toxic environmental effects of acid mine drainage (AMD). *Process Safety and Environmental Protection, 157*, 37–58.

Inglezakis, V., Stylianou, M., & Loizidou, M. (2010). Ion exchange and adsorption equilibrium studies on clinoptilolite, bentonite and vermiculite. *Journal of Physics and Chemistry of Solids, 71*(3), 279–284.

Ipek, U. (2005). Removal of Ni (II) and Zn (II) from an aqueous solutionby reverse osmosis. *Desalination, 174*(2), 161–169.

Jing, G., Meng, X., Sun, W., Kowalczuk, P. B., & Gao, Z. (2023). Recent advances in the treatment and recycling of mineral processing wastewater. *Environmental Science: Water Research & Technology, 9*(5), 1290–1304.

John, A., & Das, A. P. (2024). Government strategies for achieving Circular Economy for Plastic waste Management in Southeast Asia. In *Sustainable management of urban plastic waste through circular economic approaches* (pp. 72–82). Routledge. https://doi.org/10.1201/9781003465447-6

Kasongo, J., Alleman, L. Y., Kanda, J.-M., Kaniki, A., & Riffault, V. (2024). Metal-bearing airborne particles from mining activities: A review on their characteristics, impacts and research perspectives. *Science of the Total Environment, 951*, 175426.

Kędziora, K., Piasek, J., Szerement, J., & Ambrożewicz-Nita, A. (2014). Use of modified zeolite in environmental engineering. a review. *Вісник Національного університету Львівська політехніка Теорія і практика будівництва, 781*, 61–66.

Kirsanov, A., & Saaya, S. (2024). Global mining in the 21st century: An overview. *Journal of Geography and Cartography, 7*(1), 6090.

Kocasoy, G., & Şahin, V. (2007). Heavy metal removal from industrial wastewater by clinoptilolite. *Journal of Environmental Science and Health, Part A, 42*(14), 2139–2146.

Koli, M. M., & Singh, S. P. (2023). Surface-modified ultrafiltration and nanofiltration membranes for the selective removal of heavy metals and inorganic groundwater contaminants: A review. *Environmental Science: Water Research & Technology, 9*(11), 2803–2829.

Kosa, S. A., Al-Zhrani, G., & Salam, M. A. (2012). Removal of heavy metals from aqueous solutions by multi-walled carbon nanotubes modified with 8-hydroxyquinoline. *Chemical Engineering Journal, 181*, 159–168.

Kumar, R., Goyal, M. K., Surampalli, R. Y., & Zhang, T. C. (2024). River pollution in India: Exploring regulatory and remedial paths. *Clean Technologies and Environmental Policy, 26*(9), 2777–2799.

Lafta, M. H., Afra, A., Patra, I., Jalil, A. T., Mohammadi, M. J., Baqir Al-Dhalimy, A. M., Ziyadullaev, S., Kiani, F., Ekrami, H. A., & Asban, P. (2024). Toxic effects due to exposure heavy metals and increased health risk assessment (leukemia). *Reviews on Environmental Health, 39*(2), 351–362.

Landaburu-Aguirre, J., García, V., Pongrácz, E., & Keiski, R. L. (2009). The removal of zinc from synthetic wastewaters by micellar-enhanced ultrafiltration: Statistical design of experiments. *Desalination, 240*(1–3), 262–269.

Liu LianWen, L. L., Li Wei, L. W., Song WeiPing, S. W., & Guo MingXin, G. M. (2018). Remediation techniques for heavy metal-contaminated soils: Principles and applicability. *Science of The Total Environment, 633*, 206–219.

Lopez-Luna, E. L. (2008). *Contributions of iron (III) and sulfate-reducing bacteria to attenuation of an acid mine drainage site: Linking microcosm studies and geochemistry*. University of Massachusetts Amherst.

Mallampati, S. R., Mitoma, Y., Okuda, T., Sakita, S., & Kakeda, M. (2013). Total immobilization of soil heavy metals with nano-Fe/ca/CaO dispersion mixtures. *Environmental Chemistry Letters, 11*, 119–125.

Mandal, G. C., Mandal, A., & Chakraborty, A. (2022). The toxic effect of lead on human health: A review. *Human Biology and Public Health, 3*, 45.

Mardonova, M., & Han, Y.-S. (2023). Environmental, hydrological, and social impacts of coal and nonmetal minerals mining operations. *Journal of Environmental Management, 332*, 117387.

Maroufi, N., & Hajilary, N. (2023). Nanofiltration membranes types and application in water treatment: A review. *Sustainable Water Resources Management, 9*(5), 142.

Mehdipour, S., Vatanpour, V., & Kariminia, H.-R. (2015). Influence of ion interaction on lead removal by a polyamide nanofiltration membrane. *Desalination, 362*, 84–92.

Mishra, S., Dash, D., & Das, A. P. (2024a). Isolation and molecular characterization of polyester-degrading bacterial strains from municipal wastewater and their application in Microplastic degradation in contaminated wastewater. *Geomicrobiology Journal*. https://doi.org/10.1080/01490451.2024.2376101

Mishra, S., Dash, D., & Das, A. P. (2024b). Microbial degradation of polyester microfibers using indigenously isolated bacterial strain Exiguobacterium sp. *CLEAN, 52*(12), e202300343. https://doi.org/10.1002/CLEN.202300343

Mitra, S., Chakraborty, A. J., Tareq, A. M., Emran, T. B., Nainu, F., Khusro, A., Idris, A. M., Khandaker, M. U., Osman, H., & Alhumaydhi, F. A. (2022). Impact of heavy metals on the environment and human health: Novel therapeutic insights to counter the toxicity. *Journal of King Saud University-Science, 34*(3), 101865.

Mohammad, A. W., Teow, Y., Ang, W., Chung, Y., Oatley-Radcliffe, D., & Hilal, N. (2015). Nanofiltration membranes review: Recent advances and future prospects. *Desalination, 356*, 226–254.

Mohammadi, S. A., Najafi, H., Zolgharnian, S., Sharifian, S., & Asasian-Kolur, N. (2022). Biological oxidation methods for the removal of organic and inorganic contaminants from wastewater: A comprehensive review. *Science of the Total Environment, 843*, 157026.

Mohanty, J., Dey, S., Bala, K., & Das, A. P. (2024). Recent advances in microplastic pollution for its sustainable management. In *Sustainable management of urban plastic waste through circular economic approaches* (pp. 27–48). Routledge. https://doi.org/10.1201/9781003465447-3

Mohanty, M., Tripathy, B., Koteswara Reddy, G., & Das, A. P. (2024). Greener technologies for establishing restraint over microplastic pollution. In *Sustainable microbial technology for synthetic and cellulosic microfiber bioremediation* (pp. 239–258). Springer. https://doi.org/10.1007/978-3-031-62600-5_13

Mousavi, S. M., Hashemi, S. A., Iman Moezzi, S. M., Ravan, N., Gholami, A., Lai, C. W., Chiang, W.-H., Omidifar, N., Yousefi, K., & Behbudi, G. (2021). Recent advances in enzymes for the bioremediation of pollutants. *Biochemistry Research International, 2021*(1), 5599204.

Neculita, C. M., Zagury, G. J., & Bussière, B. (2007). Passive treatment of acid mine drainage in bioreactors using sulfate-reducing bacteria: Critical review and research needs. *Journal of Environmental Quality, 36*(1), 1–16.

Ngole-Jeme, V. M., & Ndava, J. (2023). The implications of AMD induced acidity, high metal concentrations and ochre precipitation on aquatic organisms. *Polish Journal of Environmental Studies, 32*(4), 2959.

Nkuna, R., Ijoma, G. N., Matambo, T. S., & Chimwani, N. (2022). Accessing metals from low-grade ores and the environmental impact considerations: A review of the perspectives of conventional versus bioleaching strategies. *Minerals, 12*(5), 506.

Noman, E. A., Al-Gheethi, A., Al-Sahari, M., Mohamed, R. M. S. R., Crane, R., Ab Aziz, N. A., & Govarthanan, M. (2022). Challenges and opportunities in the application of bioinspired engineered nanomaterials for the recovery of metal ions from mining industry wastewater. *Chemosphere, 308*, 136165.

Ojha, A., Jaiswal, S., Thakur, P., & Mishra, S. K. (2023). Bioremediation techniques for heavy metal and metalloid removal from polluted lands: A review. *International Journal of Environmental Science and Technology, 20*(9), 10591–10612.

Ouma, K. O., Shane, A., & Syampungani, S. (2022). Aquatic ecological risk of heavy-metal pollution associated with degraded mining landscapes of the southern Africa River basins: A review. *Minerals, 12*(2), 225.

Padhye, L. P., Srivastava, P., Jasemizad, T., Bolan, S., Hou, D., Shaheen, S. M., Rinklebe, J., O'Connor, D., Lamb, D., & Wang, H. (2023). Contaminant containment for sustainable remediation of persistent contaminants in soil and groundwater. *Journal of Hazardous Materials, 455*, 131575.

Pandey, S., Jha, A. K., & Singh, T. (2024). A sustainable approach for stabilization of coal mine overburden waste: A critical appraisal. *Indian Geotechnical Journal, 55*, 1301–1319.

Pang, F. M., Kumar, P., Teng, T. T., Omar, A. M., & Wasewar, K. L. (2011). Removal of lead, zinc and iron by coagulation–flocculation. *Journal of the Taiwan Institute of Chemical Engineers, 42*(5), 809–815.

Panizza, M., & Cerisola, G. (2004). Electrochemical oxidation as a final treatment of synthetic tannery wastewater. *Environmental Science & Technology, 38*(20), 5470–5475.

Pant, R., Mathpal, N., Chauhan, R., Singh, A., & Gupta, A. (2024). A review of mercury contamination in water and its impact on public health. In *Mercury toxicity mitigation: Sustainable nexus approach* (pp. 93–115). Springer.

Pistocchi, A., Andersen, H., Bertanza, G., Brander, A., Choubert, J.-M., Cimbritz, M., Drewes, J. E., Koehler, C., Krampe, J., & Launay, M. (2022). Treatment of micropollutants in wastewater: Balancing effectiveness, costs and implications. *Science of the Total Environment, 850*, 157593.

Pohl, A. (2020). Removal of heavy metal ions from water and wastewaters by sulfur-containing precipitation agents. *Water, Air, & Soil Pollution, 231*(10), 503.

Priyadarshini, S., Jagatee, S., & Das, A. P. (2024). Synthetic fabrics and microfiber pollution–an assessment of their global impact. *Environmental Science and Engineering, 2333*, 137–157. https://doi.org/10.1007/978-3-031-51792-1_8

Priyadarsini, S., & Das, A. P. (2024). Analytical and structural characterization of waste lithium-ion batteries for their effective recycling strategy. *Environmental Chemistry and Ecotoxicology*. https://doi.org/10.1016/j.enceco.2024.12.004

Priyadarsini, S., & Das, A. P. (2025). Lithium bioleaching: A review on microbial-assisted sustainable technology for lithium bio-circularity. *Journal of Water Process Engineering, 69*, 106744. https://doi.org/10.1016/j.jwpe.2024.106744

Rahnema, M., Amirmoeini, B., & Moradi Afrapoli, A. (2023). Incorporating environmental impacts into short-term mine planning: A literature survey. *Mining, 3*(1), 163–175.

Rajak, O. (2018). *Preparation of synthetic AMD (acid mine drainage) in laboratory and its treatment using biochar material (adsorption method)*. Indian Institute of Technology.

Raji, Z., Karim, A., Karam, A., & Khalloufi, S. (2023). Adsorption of heavy metals: Mechanisms, kinetics, and applications of various adsorbents in wastewater remediation—A review. *Waste, 1*(3), 775–805.

Rawlings, D. E. (2005). Characteristics and adaptability of iron-and sulfur-oxidizing microorganisms used for the recovery of metals from minerals and their concentrates. *Microbial Cell Factories, 4*, 1–15.

Ray, S. K., Dey, S., Das, A. P., & John, A. (2024). Microplastic waste management through greener technologies: A review. In *Sustainable management of urban plastic waste through circular economic approaches* (pp. 237–261). Routledge. https://doi.org/10.1201/9781003465447-14

Reddy Koteswara, G., Bhargavi, S., Yamuna, D., Anjani Devi, M., Jeevana Lakshmi Kumar, G., Manas, T., Adarsh, S., Bhavya Sai, V., Nath, H., Alreshidi, M. A., Togun, H., Ram, D. K., & Das, A. P. (2025). Recent advancements in the remediation of marine hydrocarbon and

Plastic pollutants. *Environmental Science and Engineering, 218*, 127–142. https://doi.org/10.1007/978-3-031-81615-4_7

Saeed, M. U., Hussain, N., Sumrin, A., Shahbaz, A., Noor, S., Bilal, M., Aleya, L., & Iqbal, H. M. (2022). Microbial bioremediation strategies with wastewater treatment potentialities–A review. *Science of the Total Environment, 818*, 151754.

Sahu, A., Dey, S., Mallick, K., Priyadarsini, S., Behera, I. D., & Das, A. P. (2024). A review on advanced technology for sustainable management of synthetic microplastic waste. In *Sustainable management of urban plastic waste through circular economic approaches* (pp. 262–279). Routledge. https://doi.org/10.1201/9781003465447-15

Salt, D. E., Smith, R., & Raskin, I. (1998). Phytoremediation. *Annual Review of Plant Biology, 49*(1), 643–668.

Saravanan, A., Deivayanai, V., Kumar, P. S., Rangasamy, G., Hemavathy, R., Harshana, T., Gayathri, N., & Alagumalai, K. (2022). A detailed review on advanced oxidation process in treatment of wastewater: Mechanism, challenges and future outlook. *Chemosphere, 308*, 136524.

Sharifi, S. A., Zaeimdar, M., Jozi, S. A., & Hejazi, R. (2023). Effects of soil, water and air pollution with heavy metal ions around lead and zinc mining and processing factories. *Water, Air, & Soil Pollution, 234*(12), 760.

Shen, L.-C., Nguyen, X.-T., & Hankins, N. P. (2015). Removal of heavy metal ions from dilute aqueous solutions by polymer–surfactant aggregates: A novel effluent treatment process. *Separation and Purification Technology, 152*, 101–107.

Sheoran, A., & Sheoran, V. (2006). Heavy metal removal mechanism of acid mine drainage in wetlands: A critical review. *Minerals Engineering, 19*(2), 105–116.

Shokes, T. E., & Möller, G. (1999). Removal of dissolved heavy metals from acid rock drainage using iron metal. *Environmental Science & Technology, 33*(2), 282–287.

Simamora, E. (2024). The literature review: Acid mine drainage remediation using alternative materials in the mining industry. In *Proceedings of the National Conference on electrical engineering, informatics, industrial technology, and creative media.*

Singh, R., Verma, A. K., & Prakash, S. (2023). The web of life: Role of pollution in biodiversity decline. *International Journal of Fauna and Biological Studies, 10*(3), 49–52.

Singh, R. B. P., Singh, A., & Choudhary, S. K. (2014). Impact of opencast coal mining on the quality of surfacewater, groundwater and vegetation: A case study in Simlong coalfield, Sahibganj, Jharkhand. *International Journal on Emerging Technologies, 5*(2), 95.

Sukanya, K., Sivarajasekar, N., & Saranya, K. (2022). Spent filter backwash water treatment by coagulation followed by ultrafiltration. In *Industrial wastewater treatment: Emerging technologies for sustainability* (pp. 27–40). Springer.

Sun, H., Shi, B., Yang, F., & Wang, D. (2017). Effects of sulfate on heavy metal release from iron corrosion scales in drinking water distribution system. *Water Research, 114*, 69–77.

Surimbayev, B., Bolotova, L., Akcil, A., Yessengarayev, Y., Kanaly, Y., Akzharkenov, M., Khumarbekuly, Y., Shalgymbayev, S., Zhumabai, S., & Mambetzhanova, A. (2025). Technology development of gold heap leaching in Kazakhstan: An overview. *Mineral Processing and Extractive Metallurgy Review*. https://doi.org/10.1080/08827508.2024.2448786

Swelam, A., Awad, M., & Saber, M. (2017). Removal of some toxic ions from industrial wastewater using batch technique by indion 225na resin. *Al-Azhar Bulletin of Science, 28*(1-A), 83–93.

Talukder, P., Ray, R., Sarkar, M., Das, A., & Chakraborty, S. (2024). Adverse effects of mining pollutants on terrestrial and aquatic environment and its remediation. *Environmental Quality Management, 33*(4), 595–610.

Tan, I., Ahmad, A., & Hameed, B. (2008). Optimization of preparation conditions for activated carbons from coconut husk using response surface methodology. *Chemical Engineering Journal, 137*(3), 462–470.

Tee, G. T., Gok, X. Y., & Yong, W. F. (2022). Adsorption of pollutants in wastewater via biosorbents, nanoparticles and magnetic biosorbents: A review. *Environmental Research, 212*, 113248.

Teh, C. Y., Budiman, P. M., Shak, K. P. Y., & Wu, T. Y. (2016). Recent advancement of coagulation–flocculation and its application in wastewater treatment. *Industrial & Engineering Chemistry Research, 55*(16), 4363–4389.

Thakare, Y. N., & Jana, A. K. (2015). Performance of high density ion exchange resin (INDION225H) for removal of cu (II) from waste water. *Journal of Environmental Chemical Engineering, 3*(2), 1393–1398.

Tolkou, A. K., Toubanaki, D. K., & Kyzas, G. Z. (2023). Detection of arsenic, chromium, cadmium, lead, and mercury in fish: Effects on the sustainable and healthy development of aquatic life and human consumers. *Sustainability, 15*(23), 16242.

Upadhyay, S. K., Rani, N., Kumar, V., Mythili, R., & Jain, D. (2023). A review on simultaneous heavy metal removal and organo-contaminants degradation by potential microbes: Current findings and future outlook. *Microbiological Research, 273*, 127419.

van Beilen, J. B., & Li, Z. (2002). Enzyme technology: An overview. *Current Opinion in Biotechnology, 13*(4), 338–344.

Van der Bruggen, B., & Vandecasteele, C. (2003). Removal of pollutants from surface water and groundwater by nanofiltration: Overview of possible applications in the drinking water industry. *Environmental Pollution, 122*(3), 435–445.

Vepsäläinen, M., & Sillanpää, M. (2020). Electrocoagulation in the treatment of industrial waters and wastewaters. In *Advanced water treatment* (pp. 1–78). Elsevier.

Wang, B., Lan, J., Bo, C., Gong, B., & Ou, J. (2023). Adsorption of heavy metal onto biomass-derived activated carbon. *RSC Advances, 13*(7), 4275–4302.

Whitworth, A. J., Vaughan, J., Southam, G., van der Ent, A., Nkrumah, P. N., Ma, X., & Parbhakar-Fox, A. (2022). Review on metal extraction technologies suitable for critical metal recovery from mining and processing wastes. *Minerals Engineering, 182*, 107537.

Wibowo, Y. G., Taher, T., Khairurrijal, K., Ramadan, B. S., Safitri, H., Sudibyo, S., Yuliansyah, A. T., & Petrus, H. T. B. M. (2024). Recent advances in the adsorptive removal of heavy metals from acid mine drainage by conventional and novel materials: A review. *Bioresource Technology Reports, 25*, 101797.

Xiang, H., Min, X., Tang, C.-J., Sillanpää, M., & Zhao, F. (2022). Recent advances in membrane filtration for heavy metal removal from wastewater: A mini review. *Journal of Water Process Engineering, 49*, 103023.

Zaman, Q. U., Zhao, Y., Zaman, S., Batool, K., & Nasir, R. (2024). Reviewing energy efficiency and environmental consciousness in the minerals industry amidst digital transition: A comprehensive review. *Resources Policy, 91*, 104851.

Zhang, Z., Zhang, C., Yang, Y., Zhang, Z., Tang, Y., Su, P., & Lin, Z. (2022). A review of sulfate-reducing bacteria: Metabolism, influencing factors and application in wastewater treatment. *Journal of Cleaner Production, 376*, 134109.

Ziemkiewicz, P. F., Skousen, J., White, K., Leavitt, B., & Stiles, J. (2021). *Guidelines for the design of abandoned mine land remediation and water treatment.*

Novel and Eco-Friendly Techniques for the Remediation of Mining Pollutants

Meysam Naseri and Rahele Khosravi Nessiani

1 Introduction

As a fundamental driver of worldwide economic growth, the extractive sector continues to supply critical mineral resources that underpin industrial expansion and innovation. As a critical sector supporting infrastructure, energy production, and manufacturing, mining operations contribute significantly to national economies, particularly in resource-rich regions. However, this vital industry faces growing scrutiny due to its substantial environmental footprint, particularly concerning waste generation and management (Agrawal et al., 2023; Harrami et al., 2024).

Contemporary mining processes generate diverse waste streams, including overburden, waste rock, tailings, slags, and acid mine drainage, each presenting unique environmental challenges (Fig. 1). Such waste streams frequently exhibit high levels of toxic elements, including transition metals, semi-metals, and persistent pollutants capable of long-term environmental accumulation (Das et al., 2025; Das & Ghosh, 2023a; Reddy Koteswara et al., 2025). Traditional disposal methods like landfill deposition have demonstrated significant ecological limitations, driving demand for innovative treatment approaches consistent with resource recovery paradigms (Ayangbenro et al., 2018; Huang et al., 2023; Park et al., 2019).

Recent advances in environmental biotechnology and materials science have yielded promising approaches for transforming mining waste from an environmental liability into potential resources. Phytoremediation has emerged as a particularly viable solution, leveraging specialized metal-accumulating plants to extract,

M. Naseri (✉)
Department of Mining Engineering, Amirkabir University of Technology, Tehran, Iran
e-mail: meysam.naseri@aut.ac.ir

R. K. Nessiani
Department of Chemical Engineering, Sahand University of Technology, Tabriz, Iran

© The Author(s), under exclusive license to Springer Nature Switzerland AG 2026

A. P. Das (ed.), *Bioremediation Technology for Sustainable Mining Waste Management*, Sustainable Environmental Waste Management Strategies,
https://doi.org/10.1007/978-3-032-03468-7_8

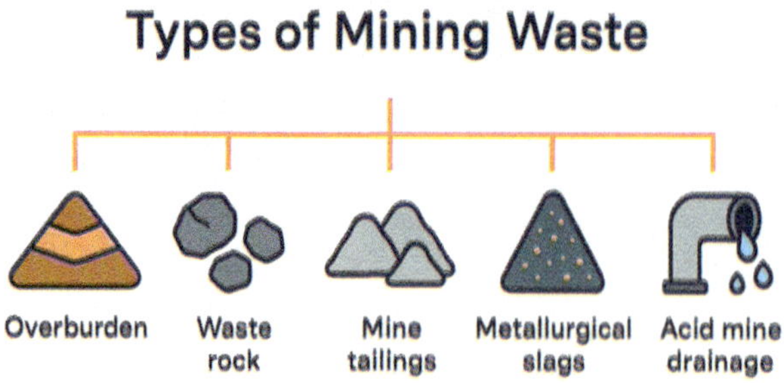

Fig. 1 Classification of mining waste byproducts

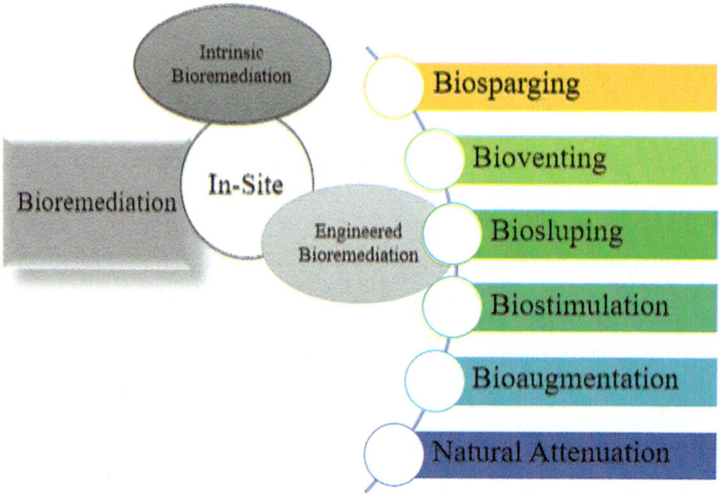

Fig. 2 Process step of bioremediation

stabilize, or degrade contaminants (Funari et al., 2023; Sarathchandra et al., 2023). Concurrently, microbial remediation techniques utilize extremophile microorganisms to neutralize acidic drainage and immobilize heavy metals (Fig. 2). These biological methods offer distinct advantages over traditional physical–chemical treatments, including lower energy requirements, reduced secondary pollution, and potential value recovery from waste streams (Thomas et al., 2022; Wilcox et al., 2024).

The integration of these nature-based solutions with emerging technologies such as nanotechnology and advanced materials recovery systems presents new opportunities for sustainable mine waste management (Mishra et al., 2024; Priyadarshini, Jagatee, & Das, 2024; Sahu et al., 2024). Nanoscale zero-valent iron particles demonstrate remarkable efficacy in heavy metal sequestration, while innovative hydrometallurgical processes enable economic recovery of residual metals from tailings. Furthermore, the repurposing of mineral wastes into construction materials or soil amendments exemplifies the practical application of circular economy concepts in mining operations (Li et al., 2022; Mahabub et al., 2023; Nancucheo & Johnson, 2020; Thomas et al., 2022; Wang et al., 2024).

This comprehensive review examines the multifaceted challenges of mining waste management through an interdisciplinary lens, evaluating both established and emerging remediation technologies. Special emphasis is placed on phytoremediation systems and their optimization through species selection, soil amendments, and agronomic practices. The analysis extends to regulatory frameworks and policy instruments that can incentivize sustainable practices while maintaining economic viability. By synthesizing current research findings with practical implementation case studies, this work aims to provide actionable insights for industry stakeholders, regulators, and researchers working toward environmentally responsible mineral resource development (Bala, Anshul, et al., 2024; Das, Dey, et al., 2024; Mohanty, Dey, et al., 2024).

The transition toward sustainable mining waste management requires balanced consideration of technical feasibility, economic viability, and environmental protection. This review contributes to that transition by critically assessing available technologies, identifying knowledge gaps, and proposing integrated solutions that reconcile resource extraction with ecosystem preservation. As global demand for minerals continues to rise alongside increasing environmental awareness, such holistic approaches will prove essential for ensuring the long-term sustainability of the mining sector.

2 Analysis of Mining Waste Management Options and Environmental Solutions

The mining industry, while economically vital, faces significant challenges in managing large volumes of hazardous waste. These byproducts contain toxic metallic and metalloid compounds that pose serious threats to ecosystems, human health, and biodiversity. In many mining regions worldwide, improper waste management practices have led to environmental degradation and public health crises (Tripathi et al., 2025).

Mineral extraction and processing operations generate substantial waste materials, including tailings, acid drainage, dust, and contaminated sludge. These pollutants not only compromise water and soil quality but also contribute to greenhouse gas emissions and acid rain. Consequently, developing sustainable waste management strategies has become an urgent priority for the sector (Faubert et al., 2017).

Ecosystem restoration represents a critical approach to mitigating mining impacts. Advanced rehabilitation techniques focus on soil remediation, vegetation recovery, and water quality improvement (Das, Behera, & Das, 2024; John & Das, 2024; Mohanty, Tripathy, et al., 2024). Cutting-edge solutions such as lime amendment for soil neutralization and phytoremediation using metal-accumulating plants show promise in reducing contamination and restoring ecological balance. However, successful implementation requires substantial investment and robust regulatory frameworks (Mensah, 2024).

Mining wastewater demands specialized treatment systems due to its complex mixture of toxic contaminants. Selecting appropriate treatment technologies depends on pollution characteristics, operational costs, and intended water reuse applications. Emerging solutions like constructed wetlands, membrane filtration, and biological treatment methods offer effective approaches for water purification (Abdullahi et al., 2023).

Mining activities face enduring ecological challenges from acidic drainage systems that develop when sulfide compounds oxidize. Effective management requires preventive measures (such as dry covers) combined with treatment options (like chemical neutralization). Developing cost-effective AMD treatment technologies, particularly for abandoned mines, represents a crucial research priority (Fig. 3) (Bala, Dogra, et al., 2024; Chang et al., 2024; Das, Dey, et al., 2024; Dey, Bissoyi, et al., 2024; Mohanty, Dey, et al., 2024; Ray et al., 2024).

Modern tailings disposal methods emphasize both environmental safety and resource recovery. Innovative approaches include (Xu et al., 2019):

- Metal extraction from tailings for economic benefit.
- Utilization in construction materials.
- Engineered dry stacking techniques.
- Secure underground storage solutions.

Tailings facility design must incorporate rigorous safety standards and site--specific geological considerations to prevent environmental disasters.

Fig. 3 Diagram of the process of acid mine drainage (AMD) formation

Transitioning to sustainable mining practices requires integration of three key elements: clean technologies, strong governance, and community engagement. Strategic investments in research, workforce training, and continuous monitoring systems are essential for this transformation. By adopting these comprehensive solutions, the mining industry can balance resource extraction with environmental protection and social responsibility (Xu et al., 2019).

This holistic approach demonstrates that responsible mineral development and environmental stewardship are not mutually exclusive, but rather complementary components of modern mining operations.

3 Microbial Remediation

Research indicates that microscopic organisms including bacteria, archaea, and fungi possess unique capabilities for interacting with heavy metal elements. Certain microorganisms can survive in metal-rich environments, while others have the ability to accumulate these elements within their cellular structures (Sun et al., 2024). Among bacteria, species such as Pseudomonas, Cupriavidus, and Bacillus demonstrate remarkable metal absorption capabilities (Das et al., 2015a; Dey, Dash, & Das, 2024; Paul et al., 2023; Pradhan et al., 2023). Extremophilic archaea, due to their high tolerance to elevated metal concentrations, serve as excellent candidates for bioremediation in harsh environmental conditions (Azam et al., 2025). Among fungi, Aspergillus, Trichoderma, and Penicillium show significant potential for mycoremediation applications due to their ability to accumulate metals within their mycelial networks. The metal resistance mechanisms in these organisms can be broadly categorized into two types (Cecchi et al., 2017):

1. Active processes: Involve specialized transport systems that pump metal ions out of cells.
2. Passive processes: Rely on metal binding to surface molecules such as lipopolysaccharides and exopolysaccharides in cell walls.

Certain microbial species demonstrate remarkable capabilities to convert toxic metallic species into less mobile forms through biomineralization processes, resulting in stable sulfate, oxide, or carbonate precipitates. Practical applications of these capabilities in environmental remediation have been well documented (Cánovas et al., 2023). For instance, specific bacteria have been employed at mercury-contaminated sites to convert the element into less toxic forms. Fungi such as Aspergillus niger have proven effective in cleaning soils contaminated with chromium, lead, and cadmium. In abandoned mine restoration projects, communities of metal-resistant microbes have been utilized to reduce toxic element concentrations and rehabilitate ecosystems (Liaquat et al., 2020; Sun et al., 2020).

These findings suggest that harnessing the natural capabilities of microorganisms provides an effective strategy for addressing heavy metal pollution challenges (Bhattacharjee et al., 2021; Das, Ray, et al., 2024; Dey et al., 2023). These methods not only demonstrate high efficiency but also offer significantly greater environmental compatibility compared to conventional chemical approaches (Thai et al., 2023).

Empirical evidence shows that indigenous microflora in mining environments can catabolize various persistent organic pollutants via multiple biochemical mechanisms. These contaminants include petroleum hydrocarbons, polycyclic aromatic compounds, and various other toxic organic substances (Das, Mallavarapu, & Ghosh, 2023; Das, Mohanty, et al., 2023). The underlying mechanism of this degradation process relies on specialized enzymatic activities and unique metabolic pathways unique to these microorganisms (Almutairi, 2024).

In the degradation process, bacteria typically initiate the breakdown. They employ specialized enzymes to transform complex molecular structures into simpler intermediate compounds. Subsequently, fungi utilize these intermediate products as nutrient sources to complete the degradation cycle. This biological synergy between bacterial and fungal communities significantly enhances the overall efficiency of the remediation system (Ufarté et al., 2018).

Microbial enzymes operate through various mechanisms including oxygen removal, hydrogen addition, and aromatic ring cleavage. Certain bacterial species such as Pseudomonas and Mycobacterium exhibit exceptional degradation capabilities for aromatic compounds due to their robust enzymatic systems. The final products of these degradation processes are typically harmless substances like water and carbon dioxide (Yang et al., 2018).

The advancement of modern biotechnologies has opened new horizons in this field. Through genetic engineering techniques, researchers have successfully developed modified microbial strains capable of degrading resistant compounds like chlorinated substances. These scientific achievements represent significant progress toward developing targeted and efficient bioremediation methods (Bhatt et al., 2021).

These findings indicate that harnessing the natural capabilities of microorganisms, combined with advanced technologies, can provide effective solutions for addressing environmental challenges caused by mining activities. Such approaches not only demonstrate high efficiency but also prove to be economically viable and fully compatible with environmental conservation efforts (Clagnan et al., 2024).

Scientific research demonstrates that microorganisms possess remarkable abilities to counteract a wide range of toxic substances. These microscopic organisms employ sophisticated mechanisms to transform hazardous materials into less harmful compounds (Yamaguchi & Miyazaki, 2024).

Specialized bacteria, such as certain Rhodococcus species, can break down complex chlorinated compounds. This process occurs through the production of specific enzymes that alter the molecular structure of toxic substances. Microscopic algae also play a significant role in aquatic environment purification by absorbing and storing dangerous materials (Gałązka & Jankiewicz, 2022).

Recent studies have revealed that some microorganisms can produce nanoparticles with unique properties. These particles can transform hazardous chemicals into harmless compounds by modifying their chemical structure. For instance, the bacterium Bacillus licheniformis can produce silver nanoparticles with antimicrobial properties (Dalvand et al., 2018).

3.1 Innovative Approaches in Bioremediation of Mining Waste

Recent advances in biotechnology have opened new horizons for mining waste management. Gene-editing systems like CRISPR-Cas9 have revolutionized microbial engineering, enabling precise genetic modifications that enhance pollutant degradation efficiency. Laboratory studies demonstrate that genetically optimized strains can achieve up to 70% greater effectiveness in breaking down toxic compounds compared to wild-type microorganisms (Dixit et al., 2025).

Biomineralization processes offer nature-inspired solutions for heavy metal management. Experimental research confirms that specialized bacterial strains can extract valuable metals like copper and cobalt from mining effluents with recovery rates reaching 85%. This dual-function approach not only cleanses the environment but also facilitates resource recovery (Diallo et al., 2015).

The integration of multi-scale data with deep learning algorithms has transformed predictive capabilities for biological systems (Das & Ghosh, 2023a, 2023b, 2023c; Ghosh & Das, 2018). Cutting-edge research indicates that AI-powered predictive models can forecast bioremediation outcomes with 92% accuracy, significantly improving process optimization (Wang, Guo, et al., 2025).

Understanding complex microbial interactions is crucial for designing effective remediation systems. Recent field studies reveal that engineered microbial consortia can perform remediation tasks 40% faster than natural communities, highlighting the potential of targeted community design (Tsoi et al., 2019).

Modern approaches emphasize combining biological methods with land rehabilitation techniques. Controlled experiments demonstrate that simultaneous application of microbial inoculation and native plant cultivation can reduce mine site restoration time by up to 60% (Wang, Meng, et al., 2025).

Despite significant progress, obstacles such as process scalability and biosafety considerations require special attention. Recent pilot projects show that integrated approaches can reduce remediation costs by 35% while increasing system efficiency by 25%. These developments indicate that the future of mining waste management lies in smart utilization of biological potential and advanced technologies (Liu et al., 2024). However, ultimate success depends on interdisciplinary collaboration and comprehensive attention to technical, economic, and social dimensions of this challenge. The field stands at a critical juncture where responsible innovation must balance environmental benefits with ethical implementation.

3.2 Key Factors Influencing Microbial Performance in Mine Waste Remediation

Mining environments present unique biological challenges that require careful consideration for effective bioremediation strategies. Three fundamental microbial degradation mechanisms play critical roles: complexolysis, where microorganisms secrete organic compounds like citrate and oxalate to release metals from mineral structures without direct cell contact; redox reactions that alter metal oxidation states to modify solubility and toxicity, proving valuable for precious metal recovery; and acidolysis, in which sulfur-oxidizing bacteria generate sulfuric acid to dissolve ore matrices, widely applied in bioleaching operations (Dong et al., 2023; Jones & Santini, 2023) (Fig. 4).

The success of these processes heavily depends on environmental parameters, particularly pH levels ranging from extremely acidic (pH 2) to alkaline (pH 11) conditions, with acidophiles thriving in low-pH environments and alkaliphiles dominating basic conditions (Dhakar & Pandey, 2016). Temperature significantly impacts reaction kinetics, with each 10 °C increase typically doubling or tripling metabolic rates, though temperatures exceeding 60 °C are only tolerable by thermophilic species (Das et al., 2011; Ghosh, Mohanty, Nayak, et al., 2016; Mohanty et al., 2016; Vandekerckhove et al., 2018). Nutrient availability follows the ideal 100:10:1 carbon-nitrogen-phosphorus ratio, with phosphorus deficiency often being the most limiting factor that can be addressed using carbon sources like beet molasses (Yang et al., 2024) (Fig. 5).

However, significant challenges persist, including high salinity (above 35 g/L) that disrupts enzymatic activity, the toxicity of heavy metals like cadmium and mercury even at low concentrations, and the recalcitrance of complex organic

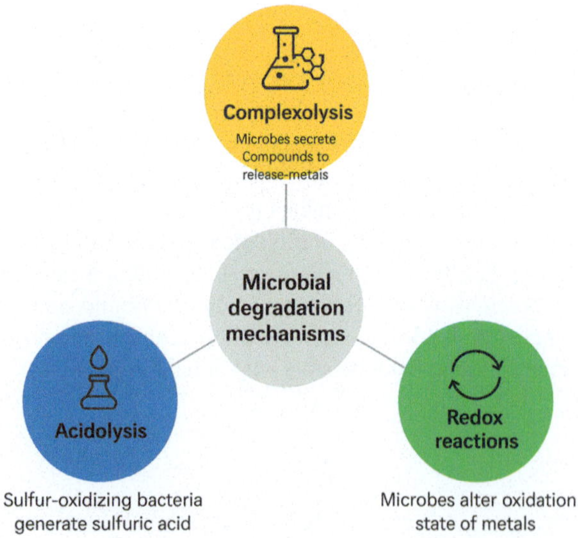

Fig. 4 Microbial degradation mechanisms play critical roles

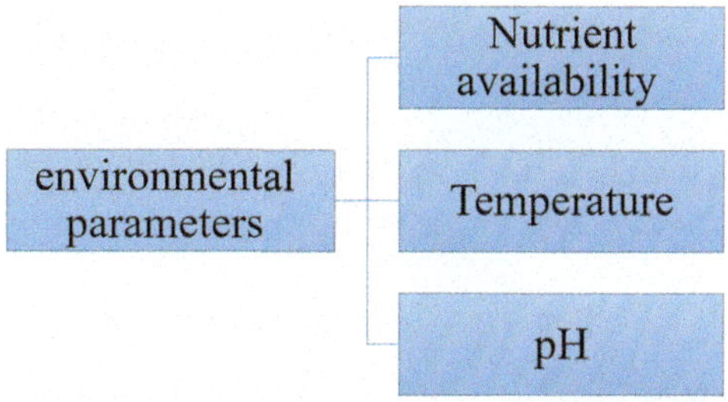

Fig. 5 Processes heavily depends on environmental parameters

compounds such as aromatic hydrocarbons. Practical solutions involve employing microbial consortia rather than single species, optimizing environmental conditions through aeration and temperature control, adding soil amendments like compost or biochar, and implementing hybrid chemical-biological approaches (Wawra et al., 2018). This comprehensive understanding of microbial ecology and site-specific conditions forms the foundation for successful bioremediation projects in mining environments, where proper management of these factors can achieve high-efficiency cleanup while recovering valuable resources. The integration of advanced monitoring technologies with these biological processes promises to further enhance remediation outcomes in these complex ecosystems (Haldar & Ghosh, 2020).

3.3 Optimization of Bioremediation Processes in Mining Waste Management

The harsh and complex conditions of mining environments create unique challenges for microbial activity. To overcome these limitations, researchers have developed advanced solutions focusing on three key areas: intelligent bioreactor design, selection of appropriate operational strategies, and implementation of precise monitoring systems (Ahmed & Geed, 2025) (Fig. 6).

In bioreactor design, the selection of reactor type is based on waste characteristics and remediation process requirements. Aerobic reactors provide ideal conditions for organic compound-degrading and metal-oxidizing microbes by supplying sufficient oxygen. Conversely, anaerobic reactors create oxygen-free environments optimal for metal-reducing and chlorinated compound-degrading microorganisms. A recent innovation in this field is the development of hybrid reactors that simultaneously establish aerobic and anaerobic zones, enabling concurrent treatment of various pollutants (Shi et al., 2018).

For operational strategies, two main approaches are employed: continuous and batch systems. Continuous systems maintain more stable environmental conditions through constant material flow but require precise control. Batch systems allow more accurate parameter adjustment but may show variable efficiency. The choice between these methods depends on factors such as waste volume, pollutant type, and process control capabilities (Sanchez-Ramos et al., 2023).

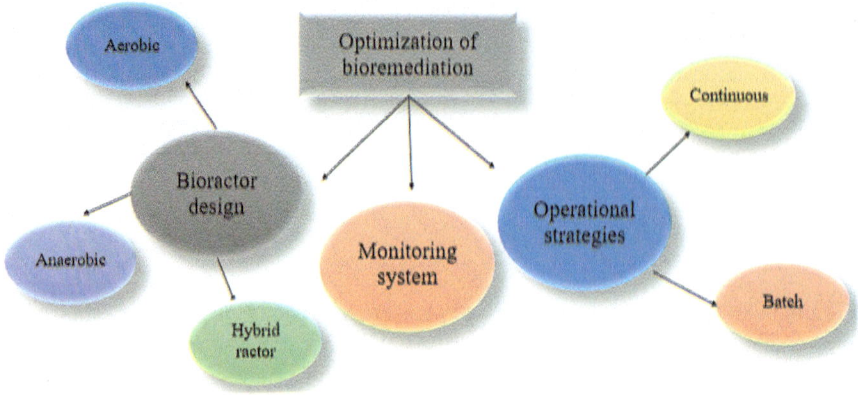

Fig. 6 Optimization of bioremediation processes in mining waste management

The most critical component of these systems is their intelligent monitoring and control mechanisms. Real-time measurement of key parameters like pH, temperature, and dissolved oxygen concentration enables optimal condition adjustment. Advanced control systems using smart algorithms can automatically regulate environmental conditions for maximum microbial activity. These technologies not only enhance remediation efficiency but also improve resource management and reduce operational costs (Anbarasu et al., 2024).

The integration of these three key elements—bioreactor design, operational strategy, and monitoring system—creates a comprehensive framework for sustainable mining waste management. This holistic approach not only mitigates environmental challenges in mining but also facilitates the recovery of valuable resources from waste. Further development of these technologies requires interdisciplinary collaboration among microbiologists, chemical engineers, and environmental scientists. This cooperative effort will drive innovation in bioremediation techniques, making them more effective and adaptable to diverse mining waste scenarios while maintaining economic viability and environmental safety standards. The continuous refinement of these systems promises to revolutionize mining waste treatment, offering scalable solutions that balance industrial needs with ecological preservation.

3.4 *Successful Applications of Microorganisms in Mine Site Remediation*

Field studies from various parts of the world have demonstrated the critical role microorganisms can play in cleaning up contaminated mining areas. At the former Berkeley copper mine in the USA, researchers successfully transformed highly acidic water ($pH < 3$) containing dangerous metal concentrations into near-natural conditions using specialized bacteria. These microorganisms reduced soluble metal concentrations by up to 90% through metal sulfide production (Tuovinen et al., 2018).

The industrial-scale Talvivaara mine project in Finland employed biological methods for nickel extraction from low-grade ores. The designed system achieved 25% higher extraction efficiency while reducing costs by 40% compared to conventional methods, simultaneously minimizing environmental pollution (Riekkola-Vanhanen, 2013).

The integrated Wheal Jane project in England successfully combined native plants for soil stabilization with metal-degrading microbes, achieving regulatory-compliant levels of lead and cadmium concentrations within 5 years. Similarly, the Dexing copper mine in China developed acid-resistant microbial technology that reduced sulfuric acid consumption by 60%, increased copper recovery by 35%, and halved solid waste production (Krümmelbein et al., 2012).

Key findings from these projects indicate that complete remediation typically requires 3–10 years, with biological methods costing 30%–50% less than conventional approaches. Microbial systems demonstrate up to 85% efficiency in acidic environments, with combined biological and technological solutions yielding superior results. These practical examples prove that nature-based solutions can effectively restore environments while remaining economically viable, inspiring many countries to reconsider their mine waste management strategies. The success of these initiatives highlights how biological approaches can achieve both environmental protection and cost-effectiveness, representing a paradigm shift in sustainable mining practices worldwide (Anawar, 2015).

3.5 *Innovative Approaches to Mining Waste Management Using Biotechnologies*

The mining industry, as a cornerstone of economic development, has consistently faced the challenge of managing hazardous waste. In recent years, microorganism-based biotechnologies have brought about a significant transformation in this field. These methods utilize natural decomposition and transformation mechanisms to provide sustainable and environmentally friendly solutions (Navina et al., 2024).

The most notable advantage of these technologies lies in their economic efficiency. Unlike traditional methods that require substantial investment in equipment and chemicals, biological systems leverage naturally occurring microorganisms to significantly reduce operational costs. Furthermore, these approaches minimize harm to surrounding ecosystems and often enable the recovery of valuable metals (Berens et al., 2024).

Operational flexibility stands as another prominent feature of this technology. In situ methods facilitate cleanup without the need to relocate contaminated materials, which is particularly crucial for large mines and hard-to-access areas. Conversely, ex situ methods allow for more precise environmental control. This flexibility enables environmental engineers to design customized solutions for each mining site (Burchart-Korol et al., 2016).

Despite these advantages, several implementation barriers remain. Microbial systems exhibit pronounced susceptibility to operational parameters including temperature fluctuations, pH variations, and ionic strength, demanding rigorous process optimization. Furthermore, the relatively prolonged treatment duration of biological systems compared to physicochemical alternatives may restrict their application in time-sensitive scenarios. Addressing these constraints will require investments in dedicated infrastructure and workforce capacity building (Burchart-Korol et al., 2016).

The future of this technology depends on applied research and the development of hybrid methods. Combining microbiological knowledge with emerging technologies like nanotechnology and artificial intelligence could substantially enhance the efficiency of biological systems. Investment in pilot projects and the establishment of national standards are also essential for advancing this technology domestically.

Ultimately, despite existing challenges, biotechnologies have the potential to become the gold standard for mining waste management. These methods not only offer cost-effective solutions but also align perfectly with sustainable development principles. Success in this endeavor requires close collaboration among universities, research centers, and the mining industry. This interdisciplinary approach will be crucial for developing comprehensive solutions that address both environmental concerns and industrial needs, paving the way for more sustainable mining practices worldwide (Das et al., 2015b; Ghosh & Das, 2017; Ghosh, Mohanty, Akcil, et al., 2016; Ghosh, Tripathy, et al., 2023). The integration of advanced monitoring systems and adaptive management strategies will further enhance the reliability and effectiveness of these biological treatment methods in diverse mining environments.

3.6 Challenges and Constraints of Biological Remediation Approaches for Mine Waste Treatment

Despite their promise for environmental restoration, bioremediation techniques encounter notable operational constraints. A primary limitation stems from the inherently slower kinetics of biological degradation processes when compared to conventional physicochemical treatment methods. These temporal constraints become especially acute under suboptimal conditions, including (Newsome & Falagán, 2021; Rosanti et al., 2020; Samarasekere, 2024):

- Extreme pH fluctuations (either highly acidic or alkaline regimes).
- Elevated salinity concentrations exceeding microbial tolerance thresholds.
- Insufficient nutrient availability for microbial metabolism.

Such environmental stressors can substantially prolong remediation timelines, potentially rendering biological approaches inadequate for time-sensitive decontamination requirements or stringent regulatory compliance deadlines.

The inherent complexity of microbial communities and their behaviors in different environments presents another challenge for implementing these methods. Precise monitoring of microorganism activity requires advanced and costly molecular techniques that may not be economically viable for many projects. Moreover, maintaining optimal conditions for microbial activity at large scales in constantly changing natural environments appears difficult and sometimes impossible (Newsome & Falagán, 2021; Rosanti et al., 2020; Samarasekere, 2024).

Significant regulatory and social barriers also exist. Regulatory bodies often lack sufficient experience in evaluating and permitting large-scale bioremediation projects. Additionally, public concerns about using genetically modified or non-native organisms for environmental cleanup may create resistance to these technologies. This highlights the crucial role of communication planning and public education to increase awareness and social acceptance (Newsome & Falagán, 2021; Rosanti et al., 2020; Samarasekere, 2024).

Despite these challenges, bioremediation methods can be applied to various types of mining waste. For mineral tailings containing heavy metals, metal-resistant microorganisms and bioleaching processes can prove effective. Acidic wastewater can be treated using acidophilic and metal-reducing microbes. Microorganisms can also transform hazardous metals from smelting and refining waste into more stable forms like metal sulfides. Even in leaching piles, microorganisms can be used to accelerate metal extraction and reduce pollutants (Newsome & Falagán, 2021; Rosanti et al., 2020; Samarasekere, 2024).

Ultimately, despite all limitations, the advantages of bioremediation methods—including relatively low costs, environmental compatibility, and potential for long-term use—make them an attractive option for mining waste management. However, overcoming existing challenges will require further research, development of advanced monitoring technologies, and careful planning to secure social and regulatory support. The future of these technologies depends on solving these problems and finding innovative solutions to enhance their efficiency. Continued progress in microbial research and biotechnology will be essential to address current limitations and unlock the full potential of bioremediation for sustainable mining practices.

4 Phytoremediation in Environmental Pollution Management Mining

Plants possess unique capabilities in addressing chemical pollution. Research has demonstrated that many plant species can absorb toxic organic compounds without sustaining damage, transforming them into less hazardous substances. This distinctive characteristic makes plants an ideal solution for remediating contaminated environments (Mandal et al., 2025).

The phytoremediation process operates through several specialized mechanisms that leverage plants' natural biological processes. These include phytoextraction, where hyperaccumulator plants absorb and concentrate heavy metals in their tissues; rhizofiltration, using aquatic plant roots to filter pollutants from water; and phytostabilization, where root secretions immobilize contaminants in soil (Ghosh, Sinha, et al., 2023; Mallick et al., 2024; Priyadarshini & Das, 2024; Priyadarshini, Santhosh Kumar, & Das, 2024). Additional methods involve phytovolatilization for converting pollutants to gaseous forms and phytodegradation through enzymatic breakdown of toxic compounds (de Oliveira et al., 2023) (Fig. 7).

This green technology offers multiple environmental and economic benefits compared to conventional remediation methods. Its cost-effectiveness makes large-scale implementation feasible, while its ecological compatibility ensures minimal environmental disruption. The approach simultaneously facilitates ecosystem restoration through revegetation and presents opportunities for valuable metal recovery from harvested plants. Particularly for extensive but moderately contaminated sites, phytoremediation provides a sustainable alternative to energy-intensive traditional methods (Zhakypbek et al., 2024).

Despite its potential, phytoremediation faces several implementation barriers. The process requires careful species selection tailored to specific contaminants and

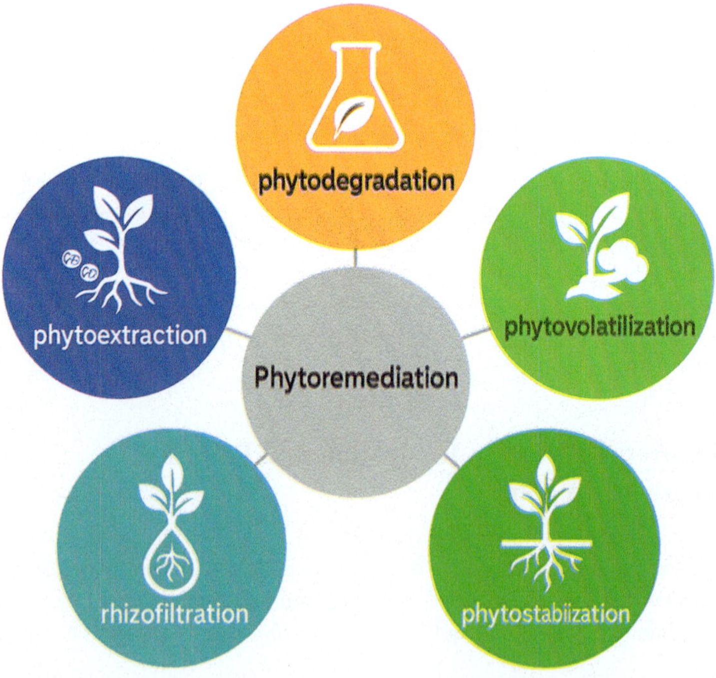

Fig. 7 Phytoremediation process

site conditions, and typically operates on longer timescales than chemical treatments. Root depth limitations and post-harvest management of contaminated biomass present additional considerations. Ongoing advances in plant genetic engineering promise to enhance treatment efficiency and expand applicability to more challenging pollution scenarios (Zhakypbek et al., 2024).

4.1 Phytoremediation Applications for Metal Mine Waste Management

Phytoremediation has emerged as an ecologically sustainable approach for mitigating inorganic pollutants, leveraging the inherent metal-accumulating capabilities of specialized vegetation. This technique utilizes vascular plants to extract, sequester, or detoxify contaminants through natural physiological processes (Abdullah et al., 2020). Compared to conventional methods, plant-based remediation offers dual advantages of being economically viable while simultaneously rehabilitating degraded ecosystems and enhancing the visual landscape of disturbed mining sites. Its applications span diverse pollution scenarios, including mine tailings, industrial effluents, and agricultural runoff, making it a versatile solution for environmental restoration (Raklami et al., 2022; Yu et al., 2022).

A distinct class of vegetation exhibits extraordinary metal-accumulation capacities, with tissue concentrations exceeding typical plants by two orders of magnitude. Notable examples include (Table 1):

- Thlaspi caerulescens (Zn/Cd accumulator; leaf concentrations ≤30,000 mg/kg Zn).
- Arabidopsis halleri (Pb/Cd accumulator).
- Pteris vittata (As accumulator).

These species employ specialized biochemical adaptations for metal uptake and storage, including:

1. Root exudation of metal-chelating compounds.
2. Enhanced transmembrane metal transport.
3. Vacuolar sequestration in aerial tissues.

Such physiological mechanisms enable effective translocation and compartmentalization of toxic elements, making hyperaccumulators invaluable for targeted decontamination strategies (Gallo et al., 2022; Mishra et al., 2017; Rosenfeld et al., 2018).

The selection of metal-tolerant plant species represents a critical factor in successfully rehabilitating degraded mine soils. Extensive research has demonstrated that various plant species, including woody legumes like Albizia and grasses such as Vetiver (Vetiveria zizanioides), exhibit remarkable capabilities to thrive while accumulating heavy metals in their tissues. Field studies at lead-zinc mines in

Table 1 Studies phytoremediation in mine

Mining site name	Waste treatment type	Source phytoremediation	Contaminant removal	Removal (%)	References
Nampundwe Mine	Mine tailings	Sunflower (Helianthus annuus L.)	Heavy metals (Cu, Zn and Fe)	53.7	Kaonda and Chileshe (2023)
Jiahang, Beijing, China Mine	Alum mine soils	Saccharomyces cerevisiae	Heavy metals (Pb, Cd and Cu)	90	Wang, Xie, et al. (2025)
Pasir Sakti District, East Lampung Regency, Indonesia	Sand mining waters	Azolla pinnata	Iron (Fe)	98.1	Hasani et al. (2021)
China Mine	Soil chromium	Pteris vittata	Cr(VI)	33.4	Zheng et al. (2024)
Gold Mining Activities Colombia	Soils contaminated by potentially toxic elements	Clidemia sericea D	Heavy metals (Pb, Cd and Hg)	95	Durante--Yánez et al. (2022)
gold mine in Muara Bungo, Jambi, Indonesia	Gold mine tailings	Arbuscular mycorrhizal	Heavy metals and their effect on the growth of Pennisetum purpureum cv.	95	Putra et al. (2022)
Romania Mine	Soils contaminated	Silphium perfoliatum L	Heavy metals (Pb, Cu, Zn and Cr)	99.98	Nescu et al. (2022)
Chahar Gonbad copper mining Iran, Kerman	Soils contaminated	Euphorbia gedrosiaca, Scariola orientalis and Eremurus persicus	Heavy metals (Cu, Zn, and Fe)	94.98	Mahdavian et al. (2022)

Southern China, for instance, revealed that combining organic and chemical fertilizers significantly enhances plant survival and growth rates (Mu et al., 2019).

International research across mining regions in Greece, Italy, and India has identified specialized plant species with exceptional metal accumulation capacities. Notable examples include Minuartia cf. bulgarica, which can concentrate lead up to 1175 mg/kg, and Scleranthus perennis, accumulating copper at 205 mg/kg. These investigations also revealed that calcium-to-magnesium absorption ratios in plants typically far exceed those found in surrounding soils (Wan et al., 2024).

Field observations at copper mining sites like Rakha Mines in India documented metal accumulation exceeding 1000 mg/kg copper in root tissues of species such as Ammannia baccifera. Such plants employ effective metal stabilization strategies by sequestering contaminants primarily in their root systems (Punia et al., 2017).

The application of chemical agents like EDTA can dramatically improve phytoremediation outcomes. Studies demonstrate this approach may increase lead uptake by 60%, though response varies among species—with maize and poplar showing distinct performance differences depending on soil pH and contamination levels (Hamidpour et al., 2024).

Operational projects utilizing Vetiver and various Rumex species have yielded promising results in soils contaminated with multiple heavy metals. While these plants differ in their metal distribution patterns, they collectively offer viable options for both phytoextraction and phytostabilization approaches (Mu et al., 2019).

Scientific evidence confirms that selecting appropriate plant species based on site-specific metal profiles and environmental conditions provides an effective waste management solution. When combined with supplemental techniques like chemical amendments and organic fertilizers, the remediation potential increases substantially. However, the need for further large-scale optimization research remains, particularly regarding implementation protocols and long-term efficacy monitoring. These nature-based solutions present sustainable alternatives for addressing the global challenge of mine waste rehabilitation while supporting ecosystem recovery.

5 Nanotechnology in Mineral Pollution Management

Nanotechnology, as one of the new technologies, plays an important role in the management of mineral pollution. This technology, using nanoscale materials (1–100 nm), enables the effective removal of pollutants from polluted environments. Due to high specific surface area, high chemical activity and ability to modify the surface, nanomaterials have high efficiency in absorbing, breaking down and stabilizing pollutants. Nanotechnology can not only reduce mineral pollution, but also help to recover valuable resources from waste. For example, nanoparticles can adsorb heavy metals from water and soil and convert them into recoverable forms (Mabudi et al., 2025; Vidu et al., 2020) (Table 2).

Zero-valent iron nanoparticles (nZVI) are recognized as one of the most widely used nanomaterials in environmental remediation. These nanostructures can degrade various organic and inorganic pollutants due to their strong reducing properties. Their mechanism of action is based on electron release, converting pollutants into less hazardous forms. For example, these nanoparticles can transform toxic hexavalent chromium (Cr^{6+}) into less toxic trivalent chromium (Cr^{3+}). They also exhibit the ability to adsorb and stabilize heavy metals such as lead, cadmium, and arsenic. Their nanoscale size and high specific surface area enhance their penetration into contaminated areas, making them highly effective for water and soil purification (Dalal & Reddy, 2019; Li et al., 2019; Ou et al., 2020).

Nanosorbents and nanocatalysts are other widely used nanomaterials in environmental cleanup. Nanosorbents, with their high specific surface area and active functional groups, demonstrate exceptional adsorption capacity for pollutants. Carbon-based nanosorbents, such as carbon nanotubes and graphene oxide, can effectively adsorb heavy metals and toxic organic compounds from water and soil. On the other hand, nanocatalysts utilize oxidation and reduction processes to break down pollutants. For instance, titanium dioxide (TiO_2)-based nanocatalysts can

Table 2 Studies nanotechnology

Type of NPs	Contaminant removed	Contact time (min)	Removal efficiency (%)	Size particle (nm)	pH	References
ZnO	Cu(II)	5	98.4	<100	4.8	Leiva et al. (2021)
IL/ZnO	Nickel	90	92.5	–	6	Panneerselvam et al. (2020)
ZnO	Nickel	120	81	–	6	Panneerselvam et al. (2020)
ZnO	Arsenic(III)	60	99.90	25–30	7	Cheraghi et al. (2024)
ZnO	Arsenic(III)	40	90.3	25–30	7	Chandan et al. (2023)
ZnO	As(III)	40	93	62.03 ± 4.06	4	Morales et al. (2023)
ZnO	As(V)	40	95	62.03 ± 4.06	4	Morales et al. (2023)
ZnO	Copper	15	>95	43.3	4	Primo et al. (2020)
Fe_3O_4@EPS	PO_4^{2-}	46,800	91	10–20	7	Govarthanan et al. (2020)
KP-FeNPs	Cr(VI)	7200	99.1	20–90	5	Rong et al. (2020)
nZVI-NPs	Lead	86,400	79.33	5–10	4–8	Shad et al. (2020)
Fe-NPs	Chromium	16,200	90	10.6	4.7	Pan et al. (2020)
Calcined-Fe/Pd NPs	As (III)	1800	100	30–60	6.2	Lin et al. (2021)
nFeCu-CH	Cr(VI)	1800	97.1	125	5	Wang et al. (2021)
NiO– Fe_3O_4	Pb (II)	1800	100	25–35	2–7	Alswat et al. (2023)
NiO– Fe_3O_4	Hg (II)	1800	99.71	25–35	2–7	Alswat et al. (2023)
Fe_3O_4	Hg(II)	3600	99.2	NCs:190 NPs:<10	7	Venkateswarlu et al. (2022)
Fe/Mn oxides	As(III)	129,600	98	1–20	4	Pranudta et al. (2020)
AgNPs/GO	Fe(III)	1800	60–98	33.64–74.87	6	El Shahawy et al. (2022)
AgNPs/GO	Cr(VI)	4800	60–98	33.64–74.87	6	El Shahawy et al. (2022)
TiO_2	Cd2+	24 h	80	4.40	7	Banivaheb et al. (2021)
Fe/Mn oxides	As(III)	36 h	98	1-20	4	Pranudta et al. (2020)

Table 2 (continued)

Type of NPs	Contaminant removed	Contact time (min)	Removal efficiency (%)	Size particle (nm)	pH	References
MNp-FeMn	As(III)	12 h	71	8.72	6.50	Nikić et al. (2024)
MNp-FeMn	As(V)	12 h	72	8.72	6.50	Nikić et al. (2024)
MNp	As(V)	12 h	54	8.19	6.11	Nikić et al. (2024)
MNp	As(III)	12 h	63	8.19	6.11	Nikić et al. (2024)
$pSiO_2$-NH_2	Cr(VI)	2 h	97	80–200	1	Jadhav et al. (2021)

degrade toxic organic compounds into water and carbon dioxide under ultraviolet (UV) light irradiation (Campos-Delgado & Mendoza, 2023; Kong et al., 2022; Taka et al., 2017).

Mining pollution management has been made possible through nanotechnology. This technology can remove heavy metals from water and soil while decomposing chemical pollutants. Heavy metals are among the most significant contaminants resulting from mining activities, and nanotechnology employs nanoparticles and nanosorbents to eliminate them from the environment. For example, zero-valent iron nanoparticles can adsorb and stabilize lead, cadmium, and arsenic in water and soil. Additionally, carbon-based nanosorbents, such as carbon nanotubes, can remove up to 90% of lead from contaminated water. These methods are not only highly efficient but also enable the recovery of valuable metals (Kheskwani & Ahammed, 2023; Tuovinen et al., 2018).

Chemical pollutants, including cyanide, phenol, and polycyclic aromatic hydrocarbons (PAHs), also pose environmental challenges in mining areas. Nanotechnology addresses these pollutants using nanocatalysts. For instance, TiO_2-based nanocatalysts can break down toxic organic compounds into water and carbon dioxide. Similarly, zero-valent iron nanoparticles can convert cyanide into less toxic ammonia and formate. These approaches not only reduce chemical pollution but also contribute to the restoration of contaminated environments (Chen et al., 2024).

Despite their numerous advantages, nanotechnology faces challenges and limitations. Environmental and safety concerns are among the most critical issues. Due to their small size and high chemical reactivity, nanomaterials may adversely affect human health and ecosystems. For example, nanoparticles can enter the body through inhalation or skin contact, potentially causing inflammation, oxidative stress, and DNA damage. Furthermore, nanomaterials may accumulate in the environment, impacting aquatic and terrestrial ecosystems. Thus, further research is needed to assess nanomaterial risks and develop safer application methods (Mutalik et al., 2024).

The cost of producing and applying nanomaterials presents another challenge. Manufacturing nanoparticles and nanosorbents requires advanced technologies and expensive raw materials. Large-scale deployment of nanomaterials may also incur high costs. For instance, producing zero-valent iron nanoparticles via chemical methods is costly. Therefore, developing more cost-effective production and application techniques is essential (Xu et al., 2025).

6 Electrochemical Methods for Mineral Pollution Management

Electrochemical methods such as electrocoagulation and electrodialysis are known as effective and low-cost methods for mineral pollution management. These methods are based on the use of electric current to remove pollutants from polluted environments (Hansen et al., 2023). The mechanisms of pollutant removal in electrochemical methods include complex processes that take place at the surface of electrodes and solution. In electrocoagulation, by applying electric current, metal ions are released from anode electrodes and react with pollutants. These ions cause the formation of solid flocs that absorb and separate pollutants from the environment. For example, aluminum or iron ions can adsorb heavy metals and toxic organic matter (Xu et al., 2017). In electrodialysis, by applying an electric current, polluting ions are separated from the environment through selected membranes. These membranes allow only certain ions to pass through and cause the selective removal of pollutants (Veerman et al., 2023).

Electrochemical methods have several advantages that make them an attractive method for mineral pollution management. These advantages include high efficiency, usability at different scales, compatibility with the environment, and cost reduction. For example, electrocoagulation can remove up to 95% of heavy metals from mineral effluents. Also, these methods require less chemicals and can be operated continuously and automatically. These characteristics make electrochemical methods to be recognized as an effective alternative to traditional methods of environmental purification (de Vargas Brião et al., 2024; Fan et al., 2020; Shahedi et al., 2023).

Electrochemical methods are widely used in mineral pollution management. These methods can be used to treat mineral wastes and recover valuable metals. Mineral wastewater contains significant amounts of heavy metals and toxic chemicals that can harm the environment. Electrochemical methods such as electrocoagulation and electrodialysis remove these pollutants from wastewater. For example, electrocoagulation can remove heavy metals such as lead, cadmium, and arsenic from mineral effluents. This method absorbs and separates pollutants from the environment by producing solid flocs. Electrodialysis also separates pollutant ions from wastewater using selective membranes. These methods are not only effective, but can also help recover valuable metals (de Vargas Brião et al., 2024; Fan et al., 2020; Hansen et al., 2023).

Recovery of precious metals from mineral wastes is one of the most important applications of electrochemical methods. These methods can extract metals such as copper, gold, and silver from wastewater. For example, electrocoagulation can absorb copper ions from mineral wastewater and recover them in the form of copper metal. Also, electrodialysis can separate gold and silver ions from wastewater. These methods not only reduce mineral pollution, but also help recover valuable resources (Cheng et al., 2025).

Despite many advantages, electrochemical methods also face challenges and limitations that should be considered. One of the major challenges of electrochemical methods is their high energy consumption. Applying electric current to remove contaminants requires significant energy, which can increase operating costs. For example, electrocoagulation needs a lot of energy to remove heavy metals from mineral wastewater. Therefore, there is a need to develop low-consumption methods and use renewable energy sources. Electrochemical methods require advanced and expensive equipment. For example, electrodialysis requires selective membranes and sophisticated control systems. Also, maintenance and repair of this equipment can be expensive. Therefore, there is a need to develop more cost-effective technologies and improve equipment design (AlJaberi et al., 2023; Zhu & Jassby, 2019).

7 Integration of Multiple Methods for Sustainable Management of Mining Pollution

The remediation of contaminated mining areas requires intelligent and comprehensive solutions capable of addressing the complexities of these environments. In recent years, the combination of biological methods with other technologies has emerged as an effective strategy. These integrated approaches significantly enhance remediation efficiency by simultaneously leveraging the advantages of multiple techniques (Unimke et al., 2024).

One particularly interesting method combines bioremediation with electrokinetics. This technique accelerates the movement of contaminants toward degrading microorganisms by establishing an electric field in polluted soil. Not only does this approach increase remediation speed, but it also proves highly effective for heavy metals like lead and cadmium. Research indicates this method can improve remediation efficiency by up to 40% compared to conventional approaches (He et al., 2018).

Another promising strategy integrates phytoremediation with microbial bioremediation. In this system, plants act as natural filters by absorbing contaminants from soil and transferring them to above-ground parts. Simultaneously, bacteria in the root zone secrete specialized enzymes to break down remaining pollutants. This natural collaboration can reduce the time required for complete remediation by half (Raklami et al., 2022).

Nanotechnology has also revolutionized bioremediation. Nanoparticles enhance microbial activity by increasing contact surfaces and improving nutrient transfer. For instance, zero-valent iron nanoparticles not only serve as nutrient carriers but also actively participate in degradation processes. This technology has shown particularly remarkable results in heavily contaminated areas (Lacalle et al., 2018; Nessiani et al., 2025).

Among the most fascinating innovations is the use of microbial fuel cells. These intelligent systems both generate energy through contaminant degradation and utilize this energy to enhance the remediation process. This self-sustaining cycle represents an ideal solution for remote mining sites with limited access to power sources (Lovecchio et al., 2024).

These integrated approaches demonstrate that the future of mining waste management lies in the intelligent combination of natural capabilities with advanced technologies. With continued research and development of these methods, we can look forward to a time when the mining industry not only avoids environmental damage but actively contributes to ecosystem restoration. The success of these projects worldwide heralds a cleaner, more sustainable future for mineral industries.

8 Circular Economy in Mining Waste Management

The mining industry stands at the threshold of a fundamental transformation, with the circular economy at its core. This new paradigm focuses on three key pillars: first, the gradual elimination of waste and pollutants; second, maintaining material value at its highest level through recirculation; and third, the regeneration of natural ecosystems.

In recent years, policymakers have evolved toward smarter mechanisms for managing mining waste. This shift extends beyond mere regulation to emphasize public education and environmental awareness. However, a significant gap remains between waste generation volumes and sustainable management capacity, necessitating policy reforms.

A crucial aspect of applying circular economy principles to mining is redefining "waste" as a "potential resourcee" This mindset shift paves the way for technological innovations in recovering valuable materials from mining residues and byproducts. For instance, transforming waste into construction materials or extracting precious metals from slag represents tangible manifestations of this transformation.

The primary challenge lies in the global disparity of adoption. While advanced economies have made substantial progress, many developing nations remain in early transition stages. This imbalance calls for international knowledge-sharing platforms and technology transfer initiatives.

Success hinges on aligning industry, academia, and governance. Such collaboration can spawn innovative business models where waste management becomes a value-creation opportunity rather than a cost center. In these models, outsourcing recycling processes may lead to specialized, efficient ecosystems.

Realizing this vision requires reimagining the mining value chain—from exploration to production and post-consumer management. Digital technologies like industrial IoT and artificial intelligence can serve as key enablers in this transition.

Ultimately, the shift toward circular mining isn't merely a technical choice but demands organizational culture transformation. This change will only prove sustainable when it balances economic objectives, environmental requirements, and social responsibilities—creating shared value for all stakeholders in the mining ecosystem.

The path forward demands nothing less than reinventing mining's fundamental business logic, where every byproduct becomes an input, every operation regenerates ecosystems, and every stakeholder participates in creating circular value networks. This systemic approach promises to reconcile resource extraction with planetary boundaries while securing the industry's social license to operate in an increasingly sustainability-conscious world.

8.1 Innovative Approaches to Mining Waste Management

The mining industry faces significant challenges in managing mineral tailings, the residual materials left after valuable mineral extraction. Recent research has demonstrated promising methods to transform these waste products into valuable construction resources, offering both environmental and economic benefits.

One particularly effective application involves repurposing mining tailings for construction materials. These mineral byproducts show excellent potential for manufacturing bricks, building blocks, paving stones, and even concrete mixtures. This approach simultaneously addresses waste accumulation issues while reducing pressure on natural resource extraction. Studies confirm that mineral tailings can meet quality standards for various construction applications.

A more advanced solution involves converting mining waste into geopolymers through alkali activation. These innovative materials not only solve storage challenges but also significantly lower the carbon footprint of construction activities. The resulting geopolymers have proven effective in concrete production, mortar mixtures, and other building applications.

Backfilling mine cavities with processed tailings represent another practical waste management strategy. This technique, known as mine backfilling, provides multiple advantages: enhanced underground mine safety, reduced landfill requirements, and prevention of ground subsidence. The process involves mixing tailings with binding agents before injection into abandoned mine workings.

Mining rock waste also offers valuable applications in civil engineering projects. These materials, typically stockpiled near open-pit mines, can serve as road base materials and construction aggregates. Research indicates many mining rocks meet technical specifications for infrastructure development.

However, utilizing mining byproducts requires thorough environmental assessments. Critical evaluation areas include permeability testing, toxicological analysis, and groundwater quality monitoring. Process optimization for waste conversion remains a key research challenge in this field.

These innovative approaches demonstrate how creative thinking can transform mining waste from an environmental liability into an economic asset. Successful implementation requires collaboration between mining experts, civil engineers, and materials scientists. With targeted investment in applied research, the mining sector can evolve to not only minimize environmental impact but actively contribute to ecosystem restoration. This transition represents a sustainable path forward for responsible resource extraction and waste management practices (Table 3).

8.2 Transforming the Mining Industry Through Circular Economy Principles

The mining sector currently faces significant challenges in managing its waste streams. In this context, the circular economy has emerged as an innovative solution that can transform mining byproducts into valuable resources. One particularly promising application involves repurposing mining tailings for construction materials, simultaneously reducing demand for virgin natural resources while optimizing production costs. Research indicates this approach could generate up to 40% cost savings in construction projects.

Concurrently, the recovery of valuable metals from mining waste has attracted considerable research attention. Advanced hydrometallurgical technologies now enable up to 85% recovery rates for metals like iron, titanium, and aluminum from

Table 3 Application involves repurposing mining

Type waste	Recovery	References
Gold mine waste rock	Construction materials	Taha et al. (2021), Benarchid et al. (2018)
Mine tailings	Construction materials	Argane et al. (2015)
Tailings (copper mine)	Construction materials	Thomas et al. (2013)
Tailings phosphate mine waste rock	Construction materials	Ince (2019), El Machi et al. (2021), Yang et al. (2014)
Fe tailings >80%	Geopolymers	Loutou et al. (2019)
Phosphate mining waste	Geopolymers	Loutou et al. (2019)
Mine tailings	Geopolymers	Taha et al. (2017)
Phosphate sludge >99%	Geopolymers	Ettoumi et al. (2021)
Coal dust/ powder	Geopolymers	Luo et al. (2020)

these waste streams. These processes prove economically viable while substantially reducing the environmental footprint of mining operations.

However, implementing these solutions presents several challenges. Significant initial capital investment requirements must be addressed, alongside necessary revisions to construction standards and regulations. Furthermore, building public trust in the quality and safety of products derived from mining waste demands careful planning and transparent communication.

Successful implementations in leading countries demonstrate the potential of this approach when properly managed. Several European mines adopting circular economy principles have achieved up to 60% reduction in natural resource consumption while simultaneously increasing profitability by 25%. These successes confirm that circular economy practices represent not just an environmental imperative but also a strategic economic opportunity for the mining industry.

The transition requires coordinated efforts across three key areas: policy incentives from governments, targeted R&D investments from industry, and comprehensive public education initiatives. When properly implemented, this transformation can create a sustainable mining sector that balances economic growth with environmental responsibility, establishing new standards for resource efficiency in the "twenty-first" century.

9 Conclusion and Future Perspectives

This comprehensive study has systematically evaluated innovative approaches for sustainable mining pollution management through an interdisciplinary lens. Our analysis reveals that integrated biological-phytoremediation systems demonstrate remarkable efficacy in heavy metal sequestration, with field studies documenting 70%–95% contaminant removal rates when properly optimized. The emerging nanotechnology applications show particular promise, with engineered nanomaterials achieving 30%–40% greater remediation efficiency than conventional methods while enabling valuable resource recovery. Circular economy principles have proven transformative, with case studies demonstrating 25%–30% cost reductions through innovative waste valorization into construction materials and industrial catalysts.

Several critical challenges must be addressed to advance these solutions. The inherent time requirements of biological systems (typically 3–5 years for full remediation) remain a significant implementation barrier, while nanotechnology applications face cost-prohibitive scaling challenges ($50–100/m^3 treatment costs). Safety concerns regarding nanoparticle mobility in ecosystems and potential bioaccumulation pathways require further toxicological assessment. Standardization gaps in monitoring protocols and performance metrics create uncertainty in comparing technology effectiveness across different mining contexts.

Strategic recommendations emerge from this analysis. First, developing AI-optimized hybrid systems that combine biological and nanotechnological approaches could potentially reduce treatment timelines by 40% while maintaining cost-effectiveness. Second, establishing international certification standards for remediation technologies would enable more reliable performance benchmarking. Third, creating incentive structures for circular economy adoption through tax benefits or green certification programs could accelerate industry uptake. Fourth, investing in advanced workforce training programs will be essential to build technical capacity for these sophisticated solutions.

The path forward requires coordinated action across multiple fronts. Research institutions should prioritize developing next-generation bioengineered microbes with enhanced metal affinity and faster metabolic rates. Industry partners need to pilot integrated treatment systems at demonstration scales to validate economic viability. Policymakers must create regulatory frameworks that encourage innovation while ensuring environmental safety. Financial institutions can play a crucial role by developing specialized green financing instruments for sustainable mining projects.

Ultimately, the transition to sustainable mining pollution management represents both a formidable challenge and tremendous opportunity. By leveraging cutting-edge science, smart policy design, and collaborative governance models, we can transform mining waste from an environmental liability into a source of value. This paradigm shift will not only mitigate ecological damage but also create new economic opportunities in the growing green technology sector. The solutions outlined here provide a roadmap for achieving environmentally responsible mineral resource development that meets both present needs and future sustainability goals.

References

Abdullah, S. R. S., Al-Baldawi, I. A., Almansoory, A. F., Purwanti, I. F., Al-Sbani, N. H., & Sharuddin, S. S. N. (2020). Plant-assisted remediation of hydrocarbons in water and soil: Application, mechanisms, challenges and opportunities. *Chemosphere, 247*, 125932. https://doi.org/10.1016/j.chemosphere.2020.125932

Abdullahi, M., Stead, I., Bennett, S., Orozco, R., Abdallah, M. A. E., Jabbari, S., & Orsini, L. (2023). Harnessing water fleas for water reclamation: A nature-based tertiary wastewater treatment technology. *Science of the Total Environment, 905*, 167224. https://doi.org/10.1016/j.scitotenv.2023.167224

Agrawal, R., Bhagia, S., Satlewal, A., & Ragauskas, A. J. (2023). Urban mining from biomass, brine, sewage sludge, phosphogypsum and e-waste for reducing the environmental pollution: Current status of availability, potential, and technologies with a focus on LCA and TEA. *Environmental Research, 224*, 115523. https://doi.org/10.1016/j.envres.2023.115523

Ahmed, A., & Geed, S. R. (2025). Sustainable refinery waste management through biotechnological interventions: Health impacts, historical successes, and emerging solutions. *Environmental Research, 270*, 120967. https://doi.org/10.1016/j.envres.2025.120967

AlJaberi, F. Y., Ahmed, S. A., Makki, H. F., Naje, A. S., Zwain, H. M., Salman, A. D., & Nguyen, D. D. (2023). Recent advances and applicable flexibility potential of electrochemical processes for wastewater treatment. *Science of the Total Environment, 867*, 161361. https://doi.org/10.1016/j.scitotenv.2022.161361

Almutairi, H. H. (2024). Microbial communities in petroleum refinery effluents and their complex functions. *Saudi Journal of Biological Sciences, 31*, 104008. https://doi.org/10.1016/j.sjbs.2024.104008

Alswat, A. A., Ashmali, A. M., Alqasmi, T. M., Alhassani, H. R., & Alshorifi, F. T. (2023). Role of nanohybrid NiO–Fe3O4 in enhancing the adsorptive performance of activated carbon synthesized from Yemeni-Khat leave in removal of Pb (II) and hg (II) from aquatic systems. *Heliyon, 9*(3), e14301. https://doi.org/10.1016/j.heliyon.2023.e14301

Anawar, H. M. (2015). Sustainable rehabilitation of mining waste and acid mine drainage using geochemistry, mine type, mineralogy, texture, ore extraction and climate knowledge. *Journal of Environmental Management, 158*, 111–121. https://doi.org/10.1016/j.jenvman.2015.04.045

Anbarasu, K., Thanigaivel, S., Sathishkumar, K., Alam, M. M., Al-Sehemi, A. G., & Devarajan, Y. (2024). Harnessing artificial intelligence for sustainable bioenergy: Revolutionizing optimization, waste reduction, and environmental sustainability. *Bioresource Technology, 418*, 131893. https://doi.org/10.1016/j.biortech.2024.131893

Argane, R., Benzaazoua, M., Hakkou, R., & Bouamrane, A. (2015). Reuse of base-metal tailings as aggregates for rendering mortars: Assessment of immobilization performances and environmental behavior. *Construction and Building Materials, 96*, 296–306. https://doi.org/10.1016/j.conbuildmat.2015.08.029

Ayangbenro, A. S., Olanrewaju, O. S., & Babalola, O. O. (2018). Sulfate-reducing bacteria as an effective tool for sustainable acid mine bioremediation. *Frontiers in Microbiology, 9*, 1986. https://doi.org/10.3389/fmicb.2018.01986

Azam, T., Dai, X., Chen, X., Ali, I., Chen, S., Noor, F., & Haider, S. Z. (2025). Comparative transcriptomic and physiological analysis of extremophilic and non-extremophilic fungi in bioremediation of cadmium (cd) and strontium (Sr). *Environmental Pollution, 367*, 125678. https://doi.org/10.1016/j.envpol.2025.125678

Bala, K., Anshul, A., Verma, S., Dogra, A., Das, A. P., Kumar, R., & Nandini, A. (2024). Sustainable management for urban plastic waste generation. In *Sustainable management of urban plastic waste through circular economic approaches* (pp. 211–236). Routledge. https://doi.org/10.1201/9781003465447-13

Bala, K., Dogra, A., Lata, P., & Das, A. P. (2024). Discussing sources and biological-chemical remediation approaches for microplastic pollution. In *Sustainable microbial technology for synthetic and cellulosic microfiber bioremediation* (pp. 37–59). Springer. https://doi.org/10.1007/978-3-031-62600-5_3

Banivaheb, S., Dan, S., Hashemipour, H., & Kalantari, M. (2021). Synthesis of modified chitosan TiO2 and SiO2 hydrogel nanocomposites for cadmium removal. *Journal of Saudi Chemical Society, 25*(8), 101283. https://doi.org/10.1016/j.jscs.2021.101283

Benarchid, Y., Taha, Y., Argane, R., & Benzaazoua, M. (2018). Application of Quebec recycling guidelines to assess the use feasibility of waste rocks as construction aggregates. *Resources Policy, 59*, 68–76. https://doi.org/10.1016/j.resourpol.2018.01.004

Berens, M. J., Deen, T. W., & Chun, C. L. (2024). Bioelectrochemical reactor to manage anthropogenic sulfate pollution for freshwater ecosystems: Mathematical modeling and experimental validation. *Chemosphere, 357*, 142054. https://doi.org/10.1016/j.chemosphere.2024.142054

Bhatt, P., Gangola, S., Bhandari, G., Zhang, W., Maithani, D., Mishra, S., & Chen, S. (2021). New insights into the degradation of synthetic pollutants in contaminated environments. *Chemosphere, 268*, 128827. https://doi.org/10.1016/j.chemosphere.2020.128827

Bhattacharjee, J., Mishra, S., & Das, A. P. (2021). Recent advances in sensor-based detection of toxic dyes for bioremediation application: A review. *Applied Biochemistry and Biotechnology, 2021*, 1–20. https://doi.org/10.1007/S12010-021-03767-7

Burchart-Korol, D., Fugiel, A., Czaplicka-Kolarz, K., & Turek, M. (2016). Model of environmental life cycle assessment for coal mining operations. *Science of the Total Environment, 562*, 61–72. https://doi.org/10.1016/j.scitotenv.2016.03.202

Campos-Delgado, J., & Mendoza, M. E. (2023). Ternary graphene oxide and titania nanoparticles-based nanocomposites for dye photocatalytic degradation: A review. *Materials, 17*(1), 135. https://doi.org/10.3390/ma17010135

Cánovas, C. R., González, R. M., Vieira, B. J., Waerenborgh, J. C., Marques, R., Macías, F., & Prudencio, M. I. (2023). Metal mobility and bioaccessibility from cyanide leaching heaps in a historical mine site. *Journal of Hazardous Materials, 448*, 130948. https://doi.org/10.1016/j.jhazmat.2023.130948

Cecchi, G., Marescotti, P., Di Piazza, S., & Zotti, M. (2017). Native fungi as metal remediators: Silver myco-accumulation from metal contaminated waste-rock dumps (Libiola mine, Italy). *Journal of Environmental Science and Health, Part B, 52*(3), 191–195. https://doi.org/10.1080/03601234.2017.1261549

Chandan, A. K., Mallika, G. N., & Narsaiah, T. B. (2023). A green approach to arsenic removal using ZnO nanoparticles synthesized from acacia catechu leaf extract. *Materials Today: Proceedings, 72*, 110–119. https://doi.org/10.1016/j.matpr.2022.06.199

Chang, W., Ke, X., Wang, W., & Liu, P. (2024). Identifying sources of acid mine drainage and major hydrogeochemical processes in abandoned mine adits (Southeast Shaanxi, China). *Environmental Geochemistry and Health, 46*(2), 60. https://doi.org/10.1007/s10653-024-01858-y

Chen, A., Wang, H., Zhan, X., Gong, K., Xie, W., Liang, W., & Peng, C. (2024). Applications and synergistic degradation mechanisms of nZVI-modified biochar for the remediation of organic polluted soil and water: A review. *Science of the Total Environment, 911*, 168548. https://doi.org/10.1016/j.scitotenv.2023.168548

Cheng, Y. Z., Bao, X., Jiang, D., Ji, W., Yang, D. H., Ding, X., & Han, B. H. (2025). Light-promoted extraction of precious metals using a porphyrin-integrated one-dimensional covalent organic framework. *Angewandte Chemie International Edition, 64*(2), e202414943. https://doi.org/10.1002/anie.202414943

Cheraghi, R., Abrishamkar, M., Jahromi, H. J., & Hoseini, F. (2024). Synthesized polyetheretherketone/polyvinylalcohol nanocomposite modified with zinc oxide nanoparticles: As an effective adsorbent for removal of arsenic (III) ion from wastewater. *Desalination and Water Treatment, 317*, 100008. https://doi.org/10.1016/j.dwt.2024.100008

Clagnan, E., Costanzo, M., Visca, A., Di Gregorio, L., Tabacchioni, S., Colantoni, E., & Bevivino, A. (2024). Culturomics-and metagenomics-based insights into the soil microbiome preservation and application for sustainable agriculture. *Frontiers in Microbiology, 15*, 1473666. https://doi.org/10.3389/fmicb.2024.1473666

Dalal, U., & Reddy, S. N. (2019). A novel nano zero-valent iron biomaterial for chromium (Cr6+ to Cr3+) reduction. *Environmental Science and Pollution Research, 26*, 10631–10640. https://doi.org/10.1007/s11356-019-04528-0

Dalvand, L. F., Hosseini, F., Dehaghi, S. M., & Torbati, E. S. (2018). Inhibitory effect of bismuth oxide nanoparticles produced by bacillus licheniformis on methicillin-resistant Staphylococcus aureus strains (MRSA). *Iranian Journal of Biotechnology, 16*(4), e2102. https://doi.org/10.21859/ijb.2102

Das, A., Dey, S., Bala, K., Gnanasekaran, R., & Das, A. P. (2024). Microbial bioremediation technology for sustainable treatment and management of synthetic microfiber waste. In *Sustainable microbial technology for synthetic and cellulosic microfiber bioremediation* (pp. 221–238). Springer. https://doi.org/10.1007/978-3-031-62600-5_12

Das, A. P., Behera, I. D., & Das, N. P. (2024). *Renewable energy generation and value addition from environmental microfiber pollution through advanced greener solution* (p. 278). Springer.

Das, A., Dey, S., Gacem, A., Yadav, K. K., Nath, H., Ram, D. K., & Das, A. P. (2025). Revolutionizing solid hydrocarbon sourcing: Plastic's journey from waste to upcycled treasure. *Environmental Science and Engineering, 218*, 331–351. https://doi.org/10.1007/978-3-031-81615-4_14

Das, A., Ray, S. K., Mohanty, M., Mohanty, J., Dey, S., & Das, A. P. (2024). Ecotoxicity of microplastic wastes and their sustainable management: A review. *Environmental Chemistry and Ecotoxicology, 6*, 144–152. https://doi.org/10.1016/J.ENCECO.2024.05.003

Das, A. P., & Ghosh, S. (2023a). Manganese and its application. In *Manganese mining microorganisms* (pp. 29–40). Elsevier. https://doi.org/10.1016/B978-0-12-822147-1.00007-8

Das, A. P., & Ghosh, S. (2023b). Mechanism of manganese bioleaching. In *Manganese mining microorganisms* (pp. 87–98). Elsevier. https://doi.org/10.1016/B978-0-12-822147-1.00014-5

Das, A. P., & Ghosh, S. (2023c). Microorganisms in manganese biomining processes. In *Manganese mining microorganisms* (pp. 55–61). Elsevier. https://doi.org/10.1016/B978-0-12-822147-1.00013-3

Das, A. P., Ghosh, S., Mohanty, S., & Sukla, L. B. (2015a). Advances in manganese pollution and its bioremediation. In L. B. Sukla, N. Pradhan, S. Panda, & B. K. Mishra (Eds.), *Environmental microbe biotechnology* (1st ed., pp. 313–328). Springer. https://doi.org/10.1007/978-3-319-19018-1_16

Das, A. P., Ghosh, S., Mohanty, S., & Sukla, L. B. (2015b). Consequences of manganese compounds: A review. *Toxicological & Environmental Chemistry, 96*(7), 981–997. https://doi.org/10.1080/02772248.2015.1005428

Das, A. P., Mallavarapu, M., & Ghosh, S. (2023). Ecotoxicity of mining pollutants on the environment and their remediation. *Environmental Chemistry and Ecotoxicology, 5*, 165–167. https://doi.org/10.1016/J.ENCECO.2023.08.002

Das, A. P., Mohanty, A. K., van Hullebusch, E. D., Figueiredo, G., Das, A. P., Mohanty, A. K., van Hullebusch, E. D., & Figueiredo, G. (2023). Bioremediation as an emerging technology for the removal of synthetic microplastic pollutants from marine ecosystem. *MarPB, 194*, 115297. https://doi.org/10.1016/J.MARPOLBUL.2023.115297

Das, A. P., Sukla, L. B., Pradhan, N., & Nayak, S. (2011). Manganese biomining: A review. *Bioresource Technology, 102*(16), 7381–7387. https://doi.org/10.1016/J.BIORTECH.2011.05.018

de Oliveira, E. C. M., Pires, L. P., Santos, V. S. V., Caixeta, E. S., Bravo, J. V. M., & Pereira, B. B. (2023). Phytoremediation, bioaccessibility and ecotoxicological risk assessment of arsenic in a gold mining area. *Chemosphere, 319*, 138030. https://doi.org/10.1016/j.chemosphere.2023.138030

de Vargas Brião, G., da Costa, T. B., Antonelli, R., & Costa, J. M. (2024). Electrochemical processes for the treatment of contaminant-rich wastewater: A comprehensive review. *Chemosphere, 355*, 141884. https://doi.org/10.1016/j.chemosphere.2024.141884

Dey, S., Bissoyi, A., Koteswara Reddy, G., & Das, A. P. (2024). Synthetic microfibers: An urban emerging pollutant demanding for sustainable environmental solutions. In *Sustainable microbial technology for synthetic and cellulosic microfiber bioremediation* (pp. 107–121). Springer. https://doi.org/10.1007/978-3-031-62600-5_7

Dey, S., Dash, A., & Das, A. P. (2024). Bioremediation as a sustainable and circular solution for mining waste management and Bioenergy generation. In *Sustainable management of mining waste and tailings: A circular economy approach* (pp. 107–119). Routledge. https://doi.org/10.1201/9781003442455-6

Dey, S., Tripathy, B., Kumar, M. S., & Das, A. P. (2023). Ecotoxicological consequences of manganese mining pollutants and their biological remediation. *Environmental Chemistry and Ecotoxicology*. https://doi.org/10.1016/J.ENCECO.2023.01.001

Dhakar, K., & Pandey, A. (2016). Wide pH range tolerance in extremophiles: Towards understanding an important phenomenon for future biotechnology. *Applied Microbiology and Biotechnology, 100*, 2499–2510. https://doi.org/10.1007/s00253-016-7285-2

Diallo, M. S., Kotte, M. R., & Cho, M. (2015). Mining critical metals and elements from seawater: Opportunities and challenges. *Environmental Science & Technology, 49*(16), 9390–9399. https://doi.org/10.1021/acs.est.5b00463

Dixit, Y., Yadav, P., Asnani, H., & Sharma, A. K. (2025). CRISPR/Cas9-engineering for increased Amylolytic potential of microbes for sustainable wastewater treatment: A review. *Current Microbiology, 82*(1), 44. https://doi.org/10.1007/s00284-024-04024-w

Dong, Y., Zan, J., & Lin, H. (2023). Bioleaching of heavy metals from metal tailings utilizing bacteria and fungi: Mechanisms, strengthen measures, and development prospect. *Journal of Environmental Management, 344*, 118511. https://doi.org/10.1016/j.jenvman.2023.118511

Durante-Yánez, E. V., Martínez-Macea, M. A., Enamorado-Montes, G., Combatt Caballero, E., & Marrugo-Negrete, J. (2022). Phytoremediation of soils contaminated with heavy metals from gold mining activities using Clidemia sericea D. Don. *Plants, 11*(5), 597. https://doi.org/10.3390/plants11050597

El Machi, A., Mabroum, S., Taha, Y., Tagnit-Hamou, A., Benzaazoua, M., & Hakkou, R. (2021). Valorization of phosphate mine waste rocks as aggregates for concrete. *Materials Today: Proceedings, 37*, 3840–3846. https://doi.org/10.1016/j.matpr.2020.08.404

El Shahawy, A., Mubarak, M. F., El Shafie, M., & Abdulla, H. M. (2022). Fe (III) and Cr (VI) ions' removal using AgNPs/GO/chitosan nanocomposite as an adsorbent for wastewater treatment. *RSC Advances, 12*(27), 17065–17084. https://doi.org/10.1039/D2RA01612E

Ettoumi, M., Jouini, M., Neculita, C. M., Bouhlel, S., Coudert, L., Taha, Y., & Benzaazoua, M. (2021). Characterization of phosphate processing sludge from Tunisian mining basin and its potential valorization in fired bricks making. *Journal of Cleaner Production, 284*, 124750. https://doi.org/10.1016/j.jclepro.2020.124750

Fan, T., Deng, W., Feng, X., Pan, F., & Li, Y. (2020). An integrated electrocoagulation–electrocatalysis water treatment process using stainless steel cathodes coated with ultrathin TiO2 nanofilms. *Chemosphere, 254*, 126776. https://doi.org/10.1016/j.chemosphere.2020.126776

Faubert, P., Durocher, S., Bertrand, N., Ouimet, R., Rochette, P., Tremblay, P., & Villeneuve, C. (2017). Greenhouse gas emissions after application of landfilled paper mill sludge for land reclamation of a nonacidic mine tailings site. *Journal of Environmental Quality, 46*(5), 950–960. https://doi.org/10.2134/jeq2017.03.0119

Funari, V., Toller, S., Vitale, L., Santos, R. M., & Gomes, H. I. (2023). Urban mining of municipal solid waste incineration (MSWI) residues with emphasis on bioleaching technologies: A critical review. *Environmental Science and Pollution Research, 30*(21), 59128–59150. https://doi.org/10.1007/s11356-023-26790-z

Gałązka, A., & Jankiewicz, U. (2022). Endocrine disrupting compounds (nonylphenol and bisphenol A)–sources, harmfulness and laccase-assisted degradation in the aquatic environment. *Microorganisms, 10*(11), 2236. https://doi.org/10.3390/microorganisms10112236

Gallo, V., Serianni, V. M., Imperiale, D., Zappettini, A., Villani, M., Marmiroli, M., & Marmiroli, N. (2022). Protein analysis of A. halleri and N. caerulescens hyperaccumulators when exposed to nano and ionic forms of CD and Zn. *Nanomaterials, 12*(23), 4236. https://doi.org/10.3390/nano12234236

Ghosh, S., & Das, A. P. (2017). Bioleaching of manganese from mining waste residues using Acinetobacter sp. *Geology, Ecology, and Landscapes, 1*(2), 77–83. https://doi.org/10.1080/24749508.2017.1332847

Ghosh, S., & Das, A. P. (2018). Metagenomic insights into the microbial diversity in manganese-contaminated mine tailings and their role in biogeochemical cycling of manganese. *Scientific Reports, 8*(1), 1–12. https://doi.org/10.1038/s41598-018-26311-w

Ghosh, S., Mohanty, S., Akcil, A., Sukla, L. B., & Das, A. P. (2016). A greener approach for resource recycling: Manganese bioleaching. *Chemosphere, 154*, 628–639. https://doi.org/10.1016/J.CHEMOSPHERE.2016.04.028

Ghosh, S., Mohanty, S., Nayak, S., Sukla, L. B., & Das, A. P. (2016). Molecular identification of indigenous manganese solubilising bacterial biodiversity from manganese mining deposits. *Journal of Basic Microbiology, 56*(3), 254–262. https://doi.org/10.1002/JOBM.201500477

Ghosh, S., Sinha, J. K., Ghosh, S., Vashisth, K., Han, S., & Bhaskar, R. (2023). Microplastics as an emerging threat to the global environment and human health. *Sustainability, 15*(14), 10821. https://doi.org/10.3390/SU151410821

Ghosh, S., Tripathy, B., Dey, S., & Das, A. P. (2023). Comparative investigation of fungal and bacterial manganese biomining mechanisms. In *Manganese mining microorganisms* (pp. 99–115). Elsevier. https://doi.org/10.1016/b978-0-12-822147-1.00016-9

Govarthanan, M., Jeon, C. H., Jeon, Y. H., Kwon, J. H., Bae, H., & Kim, W. (2020). Non-toxic nano approach for wastewater treatment using Chlorella vulgaris exopolysaccharides immobilized in iron-magnetic nanoparticles. *International Journal of Biological Macromolecules, 162*, 1241–1249. https://doi.org/10.1016/j.ijbiomac.2020.06.227

Haldar, S., & Ghosh, A. (2020). Microbial and plant-assisted heavy metal remediation in aquatic ecosystems: A comprehensive review. *3 Biotech, 10*(5), 205. https://doi.org/10.1007/s13205-020-02195-4

Hamidpour, M., Sadeghi, R., Abbaszadeh-Dahaji, P., Alaei, H., Shafigh, M., Omidvari, M., & Kariman, K. (2024). The effects of EDTA and Trichoderma species on growth and cu uptake of maize (Zea mays) plants grown in a cu-contaminated soil. *Environmental Geochemistry and Health, 46*(10), 372. https://doi.org/10.1007/s10653-024-02159-0

Hansen, H. K., Gutiérrez, C., Leiva Gonzalez, J., Lazo, A., Hansen, M. E., Lazo, P., & Ortiz, R. (2023). Combined electrodialysis and electrocoagulation as treatment for industrial wastewater containing arsenic and copper. *Membranes, 13*(3), 264. https://doi.org/10.3390/membranes13030264

Harrami, M., Ez-Zaki, H., Fami, N. E., Khachani, N., & Diouri, A. (2024). Recovery of mining and Agri-food wastes in fired materials: A case study of the Moroccan industry. *Environmental Science and Pollution Research, 31*(51), 60852–60868. https://doi.org/10.1007/s11356-024-35256-9

Hasani, Q., Pratiwi, N. T. M., Effendi, H., Wardiatno, Y., Raja Guk Guk, J. A., Maharani, H. W., & Rahman, M. (2021). Azolla pinnata as phytoremediation agent of iron (Fe) in ex sand mining waters. *Chiang Mai University Journal of Natural Sciences, 20*(1), e2021017.

He, J., He, C., Chen, X., Liang, X., Huang, T., Yang, X., & Shang, H. (2018). Comparative study of remediation of Cr (VI)-contaminated soil using electrokinetics combined with bioremediation. *Environmental Science and Pollution Research, 25*, 17682–17689. https://doi.org/10.1007/s11356-018-1741-8

Huang, J., Lin, Y., Chen, Y., Xie, X., Cui, X., & Li, C. (2023). Study on tailings covering system constructed by geological polymerization of mine waste. Part 2: Simulation of oxidation inhibition of overburden on acid-producing tailings. *Bulletin of Environmental Contamination and Toxicology, 110*(1), 25. https://doi.org/10.1007/s00128-022-03679-0

Ince, C. (2019). Reusing gold-mine tailings in cement mortars: Mechanical properties and socio-economic developments for the Lefke-Xeros area of Cyprus. *Journal of Cleaner Production, 238*, 117871. https://doi.org/10.1016/j.jclepro.2019.117871

Jadhav, S. A., Garud, H. B., Thoravat, S. S., Patil, V. S., Shinde, P. S., Burungale, S. H., & Patil, P. S. (2021). Synthesis and testing of functional mesoporous silica nanoparticles for removal of Cr (VI) ions from water. *Biointerface Research in Applied Chemistry, 11*, 8599–8607. https://doi.org/10.33263/BRIAC112.85998607

John, A., & Das, A. P. (2024). Government strategies for achieving circular economy for plastic waste management in Southeast Asia. In *Sustainable management of urban plastic waste through circular economic approaches* (pp. 72–82). Routledge. https://doi.org/10.1201/9781003465447-6

Jones, S., & Santini, J. M. (2023). Mechanisms of bioleaching: Iron and sulfur oxidation by acidophilic microorganisms. *Essays in Biochemistry, 67*(4), 685–699. https://doi.org/10.1042/EBC20220257

Kaonda, M. K. M., & Chileshe, K. (2023). Assessment of sunflower (Helianthus annuus L.) for phytoremediation of heavy metal polluted mine tailings—A case study of Nampundwe mine tailings dam, Zambia. *Journal of Environmental Protection, 14*(7), 481–492.

Kheskwani, U., & Ahammed, M. M. (2023). Removal of water pollutants using plant-based nanoscale zero-valent iron: A review. *Water Science & Technology, 88*(5), 1207–1231. https://doi.org/10.2166/wst.2023.270

Kong, E. D. H., Chau, J. H. F., Lai, C. W., Khe, C. S., Sharma, G., Kumar, A., & Sanjay, M. R. (2022). GO/TiO2-related nanocomposites as photocatalysts for pollutant removal in wastewater treatment. *Nanomaterials, 12*(19), 3536. https://doi.org/10.3390/nano12193536

Krümmelbein, J., Bens, O., Raab, T., & Anne Naeth, M. (2012). A history of lignite coal mining and reclamation practices in Lusatia, Eastern Germany. *Canadian Journal of Soil Science, 92*(1), 53–66. https://doi.org/10.4141/cjss2010-063

Lacalle, R. G., Gómez-Sagasti, M. T., Artetxe, U., Garbisu, C., & Becerril, J. M. (2018). Brassica napus has a key role in the recovery of the health of soils contaminated with metals and diesel by rhizoremediation. *Science of the Total Environment, 618*, 347–356. https://doi.org/10.1016/j.scitotenv.2017.10.334

Leiva, E., Tapia, C., & Rodríguez, C. (2021). Highly efficient removal of cu (II) ions from acidic aqueous solution using ZnO nanoparticles as nano-adsorbents. *Water, 13*(21), 2960. https://doi.org/10.3390/w13212960

Li, S., Yu, L., Jiang, W., Yu, H., & Wang, X. (2022). The recent progress China has made in green mine construction, part I: Mining groundwater pollution and sustainable mining. *International Journal of Environmental Research and Public Health, 19*(9), 5673. https://doi.org/10.3390/ijerph19095673

Li, Z., Xu, S., Xiao, G., Qian, L., & Song, Y. (2019). Removal of hexavalent chromium from groundwater using sodium alginate dispersed nano zero-valent iron. *Journal of Environmental Management, 244*, 33–39. https://doi.org/10.1016/j.jenvman.2019.04.130

Liaquat, F., Munis, M. F. H., Haroon, U., Arif, S., Saqib, S., Zaman, W., & Liu, Q. (2020). Evaluation of metal tolerance of fungal strains isolated from contaminated mining soil of Nanjing, China. *Biology, 9*(12), 469. https://doi.org/10.3390/biology9120469

Lin, Y., Jin, X., Khan, N. I., Owens, G., & Chen, Z. (2021). Efficient removal of as (III) by calcined green synthesized bimetallic Fe/Pd nanoparticles based on adsorption and oxidation. *Journal of Cleaner Production, 286*, 124987. https://doi.org/10.1016/j.jclepro.2020.124987

Liu, N., Zhao, J., Du, J., Hou, C., Zhou, X., Chen, J., & Zhang, Y. (2024). Non-phytoremediation and phytoremediation technologies of integrated remediation for water and soil heavy metal pollution: A comprehensive review. *Science of the Total Environment, 948*, 174237. https://doi.org/10.1016/j.scitotenv.2024.174237

Loutou, M., Misrar, W., Koudad, M., Mansori, M., Grase, L., Favotto, C., & Hakkou, R. (2019). Phosphate mine tailing recycling in membrane filter manufacturing: Microstructure and filtration suitability. *Minerals, 9*(5), 318. https://doi.org/10.3390/min9050318

Lovecchio, N., Giuseppetti, R., Bertuccini, L., Columba-Cabezas, S., Di Meo, V., Figliomeni, M., & D'Ugo, E. (2024). Hydrocarbonoclastic biofilm-based microbial fuel cells: Exploiting biofilms at water-oil Interface for renewable energy and wastewater remediation. *Biosensors, 14*(10), 484. https://doi.org/10.3390/bios14100484

Luo, L., Li, K., Fu, W., Liu, C., & Yang, S. (2020). Preparation, characteristics and mechanisms of the composite sintered bricks produced from shale, sewage sludge, coal gangue powder and iron ore tailings. *Construction and Building Materials, 232*, 117250. https://doi.org/10.1016/j.conbuildmat.2019.117250

Mabudi, A., Naseri, M., Zamzami, S. M., & Khosravi Nessiani, R. (2025). Enhanced wastewater treatment using metal-based nanoparticles: A comprehensive study. *International Journal of Mining and Geo-Engineering, 59*, 97–113. https://doi.org/10.22059/IJMGE.2025.386830.595211

Mahabub, M. S., Alahi, F., & Al Imran, M. (2023). Unlocking the potential of microbes: Biocementation technology for mine tailings restoration—A comprehensive review. *Environmental Science and Pollution Research, 30*(40), 91676–91709. https://doi.org/10.1007/s11356-023-28937-4

Mahdavian, K., Asadigerkan, S., Sangtarash, M. H., & Nasibi, F. (2022). Phytoextraction and phytostabilization of copper, zinc, and iron by growing plants in Chahar Gonbad copper mining area, Iran. *Proceedings of the National Academy of Sciences, India Section B: Biological Sciences, 92*(2), 319–327. https://doi.org/10.1007/s40011-021-01345-9

Mallick, K., Sahu, S., Sahu, A., Dey, S., & Das, A. P. (2024). A critical review of marine microfiber pollution routes, toxicity, and its sustainable remediation. *Environmental Science and Engineering, 2333*, 189–211. https://doi.org/10.1007/978-3-031-51792-1_10

Mandal, R. R., Bashir, Z., & Raj, D. (2025). Microbe-assisted phytoremediation for sustainable management of heavy metal in wastewater-A green approach to escalate the remediation of heavy metals. *Journal of Environmental Management, 375*, 124199. https://doi.org/10.1016/j.jenvman.2025.124199

Mensah, A. K. (2024). Soil amendment-assisted phytoremediation with ryegrass offers a promising approach to mitigate environmental health concerns. *International Journal of Phytoremediation, 26*(13), 2216–2233. https://doi.org/10.1080/15226514.2024.2380039

Mishra, S., Dash, D., & Das, A. P. (2024). Microbial degradation of polyester microfibers using indigenously isolated bacterial strain Exiguobacterium sp. *CLEAN, 52*(12), e202300343. https://doi.org/10.1002/CLEN.202300343

Mishra, S., Mishra, A., & Küpper, H. (2017). Protein biochemistry and expression regulation of cadmium/zinc pumping ATPases in the hyperaccumulator plants Arabidopsis halleri and Noccaea caerulescens. *Frontiers in Plant Science, 8*, 835. https://doi.org/10.3389/fpls.2017.00835

Mohanty, J., Dey, S., Bala, K., & Das, A. P. (2024). Recent advances in microplastic pollution for its sustainable management. In *Sustainable management of urban plastic waste through circular economic approaches* (pp. 27–48). Routledge. https://doi.org/10.1201/9781003465447-3

Mohanty, M., Tripathy, B., Koteswara Reddy, G., & Das, A. P. (2024). Greener technologies for establishing restraint over microplastic pollution. In *Sustainable microbial technology for synthetic and cellulosic microfiber bioremediation* (pp. 239–258). Springer. https://doi.org/10.1007/978-3-031-62600-5_13

Mohanty, S., Ghosh, S., Nayak, S., & Das, A. P. (2016). Isolation, identification and screening of manganese solubilizing fungi from low-grade manganese ore deposits. *Geomicrobiological Journal, 34*(4), 309–316. https://doi.org/10.1080/01490451.2016.1189016

Morales, H. M., Torreblanca, G., Mar, A., Alcoutlabi, M., Eubanks, T. M., Plata, E., & Parsons, J. G. (2023). Investigation of the thermodynamics for the removal of as (III) and as (V) from water using synthesized ZnO nanoparticles and the effects of pH, temperature, and time. *Applied Sciences, 13*(18), 10525. https://doi.org/10.3390/app131810525

Mu, J., Hu, Z., Xie, Z., Huang, L., & Holm, P. E. (2019). Influence of CaO-activated silicon-based slag amendment on the growth and heavy metal uptake of vetiver grass (Vetiveria zizanioides) grown in multi-metal-contaminated soils. *Environmental Science and Pollution Research, 26*, 32243–32254. https://doi.org/10.1007/s11356-019-06429-8

Mutalik, C., Nivedita, Sneka, C., Krisnawati, D. I., Yougbaré, S., Hsu, C. C., & Kuo, T. R. (2024). Zebrafish insights into nanomaterial toxicity: A focused exploration on metallic, metal oxide, semiconductor, and mixed-metal nanoparticles. *International Journal of Molecular Sciences, 25*(3), 1926. https://doi.org/10.3390/ijms25031926

Nancucheo, I., & Johnson, D. B. (2020). Characteristics of an iron-reducing, moderately acidophilic Actinobacterium isolated from pyritic mine waste, and its potential role in mitigating mineral dissolution in mineral tailings deposits. *Microorganisms, 8*(7), 990. https://doi.org/10.3390/microorganisms8070990

Navina, B. K., Velmurugan, N. K., Kumar, P. S., Rangasamy, G., Palanivelu, J., Thamarai, P., & Shakoor, A. (2024). Fungal bioremediation approaches for the removal of toxic pollutants: Mechanistic understanding for biorefinery applications. *Chemosphere, 350*, 141123. https://doi.org/10.1016/j.chemosphere.2024.141123

Nescu, V., Ciulca, S., Sumalan, R. M., Berbecea, A., Velicevici, G., Negrea, P., & Sumalan, R. L. (2022). Physiological aspects of absorption, translocation, and accumulation of heavy metals in Silphium perfoliatum L. plants grown in a mining-contaminated soil. *Minerals, 12*(3), 334. https://doi.org/10.3390/min12030334

Nessiani, R. K., Naseri, M., Erfani, H., Khodayari, S., Shangdiar, S., & Amesho, K. T. (2025). Utilizing biosorbents for sustainable mercury removal from aqueous solutions: Efficiency and practical applications. *Desalination and Water Treatment, 322*, 101123. https://doi.org/10.1016/j.dwt.2025.101123

Newsome, L., & Falagán, C. (2021). The microbiology of metal mine waste: Bioremediation applications and implications for planetary health. *GeoHealth, 5*(10), e2020GH000380. https://doi.org/10.1029/2020GH000380

Nikić, J., Watson, M., Jokić Govedarica, J., Vujić, M., Pešić, J., Rončević, S., & Agbaba, J. (2024). Adsorption performance of Fe–Mn polymer nanocomposites for arsenic removal: Insights from kinetic and isotherm models. *Materials, 17*(20), 5089. https://doi.org/10.3390/ma17205089

Ou, J. H., Sheu, Y. T., Tsang, D. C., Sun, Y. J., & Kao, C. M. (2020). Application of iron/aluminum bimetallic nanoparticle system for chromium-contaminated groundwater remediation. *Chemosphere, 256*, 127158. https://doi.org/10.1016/j.chemosphere.2020.127158

Pan, Z., Lin, Y., Sarkar, B., Owens, G., & Chen, Z. (2020). Green synthesis of iron nanoparticles using red peanut skin extract: Synthesis mechanism, characterization and effect of conditions on chromium removal. *Journal of Colloid and Interface Science, 558*, 106–114. https://doi.org/10.1016/j.jcis.2019.09.106

Panneerselvam, A., Rajadurai, V., & Anguraj, B. L. (2020). Removal of nickel from aqueous solution using synthesized IL/ZnO NPs. *Environmental Science and Pollution Research, 27*(24), 29791–29803. https://doi.org/10.1007/s11356-019-07425-8

Park, I., Tabelin, C. B., Jeon, S., Li, X., Seno, K., Ito, M., & Hiroyoshi, N. (2019). A review of recent strategies for acid mine drainage prevention and mine tailings recycling. *Chemosphere, 219*, 588–606. https://doi.org/10.1016/j.chemosphere.2018.11.053

Paul, A., Dey, S., Ram, D. K., & Das, A. P. (2023). Hexavalent chromium pollution and its sustainable management through bioremediation. *Geomicrobiology Journal*. https://doi.org/10.1080/01490451.2023.2218377

Pradhan, G., Tripathy, B., Ram, D. K., Digal, A. K., & Das, A. P. (2023). Bauxite mining waste pollution and its sustainable management through bioremediation. *Geomicrobiology Journal*. https://doi.org/10.1080/01490451.2023.2235353

Pranudta, A., Klysubun, W., El-Moselhy, M. M., & Padungthon, S. (2020). Synthesis optimization and X-ray absorption spectroscopy investigation of polymeric anion exchanger supported binary Fe/Mn oxides nanoparticles for enhanced as (III) removal. *Reactive and Functional Polymers, 147*, 104441. https://doi.org/10.1016/j.reactfunctpolym.2019.104441

Primo, J. D. O., Bittencourt, C., Acosta, S., Sierra-Castillo, A., Colomer, J. F., Jaerger, S., & Anaissi, F. J. (2020). Synthesis of zinc oxide nanoparticles by ecofriendly routes: Adsorbent for copper removal from wastewater. *Frontiers in Chemistry, 8*, 571790. https://doi.org/10.3389/fchem.2020.571790

Priyadarshini, B., & Das, A. P. (2024). The adverse impact of synthetic microplastics released from disposable face masks pollutants on aquatic microbiota, aquatic species and humans: A review. *Environmental Science and Engineering, 2827*, 67–81. https://doi.org/10.1007/978-3-031-50840-0_5

Priyadarshini, B., Santhosh Kumar, M., & Das, A. P. (2024). Microbial degradation of synthetic plastics released from COVID-19 generated single use disposable face masks. *Environmental Science and Engineering, 2827*, 1–16. https://doi.org/10.1007/978-3-031-50840-0_1

Priyadarshini, S., Jagatee, S., & Das, A. P. (2024). Synthetic fabrics and microfiber pollution–an assessment of their global impact. *Environmental Science and Engineering, 2333*, 137–157. https://doi.org/10.1007/978-3-031-51792-1_8

Punia, A., Siddaiah, N. S., & Singh, S. K. (2017). Source and assessment of metal pollution at Khetri copper mine tailings and neighboring soils, Rajasthan, India. *Bulletin of Environmental Contamination and Toxicology, 99*, 633–641. https://doi.org/10.1007/s00128-017-2175-6

Putra, B., Warly, L., Evitayani, E., & Utama, B. P. (2022). The role of arbuscular mycorrhizal fungi in phytoremediation of heavy metals and their effect on the growth of Pennisetum purpureum cv. Mott on gold mine tailings in Muara Bungo, Jambi, Indonesia. *Biodiversitas Journal of Biological Diversity, 23*(1), 51. https://doi.org/10.13057/biodiv/d230151

Raklami, A., Meddich, A., Oufdou, K., & Baslam, M. (2022). Plants—Microorganisms-based bioremediation for heavy metal cleanup: Recent developments, phytoremediation techniques, regulation mechanisms, and molecular responses. *International Journal of Molecular Sciences, 23*(9), 5031. https://doi.org/10.3390/ijms23095031

Ray, S. K., Dey, S., Das, A. P., & John, A. (2024). Microplastic waste management through greener technologies: A review. In *Sustainable management of urban plastic waste through circular economic approaches* (pp. 237–261). Routledge. https://doi.org/10.1201/9781003465447-14

Reddy Koteswara, G., Bhargavi, S., Yamuna, D., Anjani Devi, M., Jeevana Lakshmi Kumar, G., Manas, T., Adarsh, S., Bhavya Sai, V., Nath, H., Alreshidi, M. A., Togun, H., Ram, D. K., & Das, A. P. (2025). Recent advancements in the remediation of marine hydrocarbon and Plastic pollutants. *Environmental Science and Engineering, 218*, 127–142. https://doi.org/10.1007/978-3-031-81615-4_7

Riekkola-Vanhanen, M. (2013). Talvivaara mining company–from a project to a mine. *Minerals Engineering, 48*, 2–9. https://doi.org/10.1016/j.mineng.2013.04.018

Rong, K., Wang, J., Zhang, Z., & Zhang, J. (2020). Green synthesis of iron nanoparticles using Korla fragrant pear peel extracts for the removal of aqueous Cr (VI). *Ecological Engineering, 149*, 105793. https://doi.org/10.1016/j.ecoleng.2020.105793

Rosanti, D., Wibowo, Y. G., Safri, M., Maryani, A. T., & Ramadhan, B. S. (2020). Bioremediations technologies on wastewater treatment: Opportunities, challenges and economic perspective. *Sainmatika: Jurnal Ilmiah Matematika dan Ilmu Pengetahuan Alam, 17*(2), 142–156.

Rosenfeld, C. E., Chaney, R. L., & Martinez, C. E. (2018). Soil geochemical factors regulate cd accumulation by metal hyperaccumulating Noccaea caerulescens (J. Presl & C. Presl) FK Mey in field-contaminated soils. *Science of the Total Environment, 616*, 279–287. https://doi.org/10.1016/j.scitotenv.2017.11.016

Sahu, A., Dey, S., Mallick, K., Priyadarsini, S., Behera, I. D., & Das, A. P. (2024). A review on advanced technology for sustainable management of synthetic microplastic waste. In *Sustainable management of urban plastic waste through circular economic approaches* (pp. 262–279). Routledge. https://doi.org/10.1201/9781003465447-15

Samarasekere, P. W. (2024). Microbial remediation technologies for mining waste management. In *Harnessing microbial potential for multifarious applications* (pp. 31–67). Springer. https://doi.org/10.1007/978-981-97-1152-9_3

Sanchez-Ramos, D., Garrido, F. L. B., Hernández, I. A., Romero, L. R., Camacho, J. V., & Fernández-Morales, F. J. (2023). Sustainable use of wastes as reactive material in permeable reactive barrier for remediation of acid mine drainage: Batch and continuous studies. *Journal of Environmental Management, 345*, 118765. https://doi.org/10.1016/j.jenvman.2023.118765

Sarathchandra, S. S., Rengel, Z., & Solaiman, Z. M. (2023). A review on remediation of iron ore mine tailings via organic amendments coupled with phytoremediation. *Plants, 12*(9), 1871. https://doi.org/10.3390/plants12091871

Shad, S., Belinga-Desaunay-Nault, M. F. A., Bashir, N., & Lynch, I. (2020). Removal of contaminants from canal water using microwave synthesized zero valent iron nanoparticles. *Environmental Science: Water Research & Technology, 6*(11), 3057–3065. https://doi.org/10.1039/D0EW00157K

Shahedi, A., Darban, A. K., Jamshidi-Zanjani, A., & Homaee, M. (2023). An overview of the application of electrocoagulation for mine wastewater treatment. *Environmental Monitoring and Assessment, 195*(4), 522. https://doi.org/10.1007/s10661-023-11044-9

Shi, X., Guo, X., Zuo, J., Wang, Y., & Zhang, M. (2018). A comparative study of thermophilic and mesophilic anaerobic co-digestion of food waste and wheat straw: Process stability and microbial community structure shifts. *Waste Management, 75*, 261–269. https://doi.org/10.1016/j.wasman.2018.02.004

Sun, J., He, X., Al-Tohamy, R., & Ali, S. S. (2024). Potential applications of extremophilic bacteria in the bioremediation of extreme environments contaminated with heavy metals. *Journal of Environmental Management, 352*, 120081. https://doi.org/10.1016/j.jenvman.2024.120081

Sun, Z., Hu, Y., & Cheng, H. (2020). Public health risk of toxic metal (loid) pollution to the population living near an abandoned small-scale polymetallic mine. *Science of the Total Environment, 718*, 137434. https://doi.org/10.1016/j.scitotenv.2020.137434

Taha, Y., Benarchid, Y., & Benzaazoua, M. (2021). Environmental behavior of waste rocks based concrete: Leaching performance assessment. *Resources Policy, 74*, 101419. https://doi.org/10.1016/j.resourpol.2019.101419

Taha, Y., Benzaazoua, M., Mansori, M., & Hakkou, R. (2017). Recycling feasibility of glass wastes and calamine processing tailings in fired bricks making. *Waste and Biomass Valorization, 8*, 1479–1489. https://doi.org/10.1007/s12649-016-9657-3

Taka, A. L., Pillay, K., & Mbianda, X. Y. (2017). Nanosponge cyclodextrin polyurethanes and their modification with nanomaterials for the removal of pollutants from waste water: A review. *Carbohydrate Polymers, 159*, 94–107. https://doi.org/10.1016/j.carbpol.2016.12.027

Thai, T. D., Lim, W., & Na, D. (2023). Synthetic bacteria for the detection and bioremediation of heavy metals. *Frontiers in Bioengineering and Biotechnology, 11*, 1178680. https://doi.org/10.3389/fbioe.2023.1178680

Thomas, B. S., Damare, A., & Gupta, R. C. (2013). Strength and durability characteristics of copper tailing concrete. *Construction and Building Materials, 48*, 894–900. https://doi.org/10.1016/j.conbuildmat.2013.07.075

Thomas, G., Sheridan, C., & Holm, P. E. (2022). A critical review of phytoremediation for acid mine drainage-impacted environments. *Science of the Total Environment, 811*, 152230. https://doi.org/10.1016/j.scitotenv.2021.152230

Tripathi, M., Singh, P., Pathak, S., Manimekalai, R., Garg, D., & Dashora, K. (2025). Strategies for the remediation of micro-and Nanoplastics from contaminated food and water: Advancements and challenges. *Journal of Xenobiotics, 15*(1), 30. https://doi.org/10.3390/jox15010030

Tsoi, R., Dai, Z., & You, L. (2019). Emerging strategies for engineering microbial communities. *Biotechnology Advances, 37*(6), 107372. https://doi.org/10.1016/j.biotechadv.2019.03.011

Tuovinen, H., Pelkonen, M., Lempinen, J., Pohjolainen, E., Read, D., Solatie, D., & Lehto, J. (2018). Behaviour of metals during bioheap leaching at the talvivaara mine, Finland. *Geosciences, 8*(2), 66. https://doi.org/10.3390/geosciences8020066

Ufarté, L., Potocki-Veronese, G., Cecchini, D., Tauzin, A. S., Rizzo, A., Morgavi, D. P., & Laville, E. (2018). Highly promiscuous oxidases discovered in the bovine rumen microbiome. *Frontiers in Microbiology, 9*, 861. https://doi.org/10.3389/fmicb.2018.00861

Unimke, A. A., Okezie, O., Mohammed, S. A. E., Mmuoegbulam, A. O., Abdullahi, S., Ofon, U. A., & Ayanda, O. S. (2024). Microbe–plant–nanoparticle interactions: Role in bioremediation of petroleum hydrocarbons. *Water Science & Technology, 90*(10), 2870–2893. https://doi.org/10.2166/wst.2024.362

Vandekerckhove, T. G., De Mulder, C., Boon, N., & Vlaeminck, S. E. (2018). Temperature impact on sludge yield, settleability and kinetics of three heterotrophic conversions corroborates the prospect of thermophilic biological nitrogen removal. *Bioresource Technology, 269*, 104–112. https://doi.org/10.1016/j.biortech.2018.08.012

Veerman, J., Gómez-Coma, L., Ortiz, A., & Ortiz, I. (2023). Resistance of ion exchange membranes in aqueous mixtures of monovalent and divalent ions and the effect on reverse electrodialysis. *Membranes, 13*(3), 322. https://doi.org/10.3390/membranes13030322

Venkateswarlu, S., Yoon, M., & Kim, M. J. (2022). An environmentally benign synthesis of Fe3O4 nanoparticles to Fe3O4 nanoclusters: Rapid separation and removal of Hg (II) from an aqueous medium. *Chemosphere, 286*, 131673. https://doi.org/10.1016/j.chemosphere.2021.131673

Vidu, R., Matei, E., Predescu, A. M., Alhalaili, B., Pantilimon, C., Tarcea, C., & Predescu, C. (2020). Removal of heavy metals from wastewaters: A challenge from current treatment methods to nanotechnology applications. *Toxics, 8*(4), 101. https://doi.org/10.3390/toxics8040101

Wan, Y., Peng, M., & Wang, Y. P. (2024). Assessment of heavy metal concentrations in roadside soils and plants around the Dexing copper mine: Implications for environmental management and remediation. *Environmental Monitoring and Assessment, 196*(3), 251. https://doi.org/10.1007/s10661-017-6409-1

Wang, B., Guo, Y., Li, X., Dong, C., Sha, H., Li, H., & Li, T. (2025). Phytostabilization potential and microbial response to the reclamation of native Cynodon dactylon in spoil heaps from a multiple-metal mining site in Southwest China. *Journal of Environmental Management, 378*, 124758. https://doi.org/10.1016/j.jenvman.2025.124758

Wang, K., Meng, J., Wang, Z., Zhao, K., & Liu, B. (2025). Robust adaptive optimization for sustainable water demand prediction in water distribution systems. *Scientific Reports, 15*(1), 4039. https://doi.org/10.1038/s41598-025-88628-7

Wang, P., Li, J., Hu, Y., & Cheng, H. (2024). Environmental performance of unfired bricks produced from co-disposal of mine tailings and municipal solid waste incineration fly ash based on comprehensive leaching tests. *Environmental Pollution, 347*, 123795. https://doi.org/10.1016/j.envpol.2024.123795

Wang, W., Xie, L., Zhao, L., & Yu, Q. (2025). An engineered yeast expressing an artificial heavy metal-binding protein enhances the phytoremediation of alum mine soils. *Microorganisms, 13*(3), 612. https://doi.org/10.3390/microorganisms13030612

Wang, Y., Lin, N., Gong, Y., Wang, R., & Zhang, X. (2021). Cu–Fe embedded cross-linked 3D hydrogel for enhanced reductive removal of Cr (VI): Characterization, performance, and mechanisms. *Chemosphere, 280*, 130663. https://doi.org/10.1016/j.chemosphere.2021.130663

Wawra, A., Friesl-Hanl, W., Jäger, A., Puschenreiter, M., Soja, G., Reichenauer, T., & Watzinger, A. (2018). Investigations of microbial degradation of polycyclic aromatic hydrocarbons based on 13 C-labeled phenanthrene in a soil co-contaminated with trace elements using a plant assisted approach. *Environmental Science and Pollution Research, 25*, 6364–6377. https://doi.org/10.1007/s11356-017-0941-y

Wilcox, S. M., Mulligan, C. N., & Neculita, C. M. (2024). Microbially induced calcium carbonate precipitation as a bioremediation technique for mining waste. *Toxics, 12*(2), 107. https://doi.org/10.3390/toxics12020107

Xu, D. M., Zhan, C. L., Liu, H. X., & Lin, H. Z. (2019). A critical review on environmental implications, recycling strategies, and ecological remediation for mine tailings. *Environmental Science and Pollution Research, 26*, 35657–35669. https://doi.org/10.1007/s11356-019-06555-3

Xu, L., Cao, G., Xu, X., Liu, S., Duan, Z., He, C., & Huang, Q. (2017). Simultaneous removal of cadmium, zinc and manganese using electrocoagulation: Influence of operating parameters and electrolyte nature. *Journal of Environmental Management, 204*, 394–403. https://doi.org/10.1016/j.jenvman.2017.09.020

Xu, Y., Li, Y., Li, R., Xu, H., Ouyang, S., & Yuan, H. (2025). Cost-effective synthesis of Fe5C2 catalyst from nanosized zero-valent iron to achieve efficient photothermocatalytic CO hydrogenation to light olefins. *Advanced Science, 12*(4), https://doi.org/10.1002/advs.2024102152410215.

Yamaguchi, H., & Miyazaki, M. (2024). Bioremediation of hazardous pollutants using enzyme-immobilized reactors. *Molecules, 29*(9), 2021.https://doi.org/10.3390/molecules29092021

Yang, C., Cui, C., Qin, J., & Cui, X. (2014). Characteristics of the fired bricks with low-silicon iron tailings. *Construction and Building Materials, 70*, 36–42. https://doi.org/10.1016/j.conbuildmat.2014.07.075

Yang, C., Yue, F., Cui, Y., Xu, Y., Shan, Y., Liu, B., & Lü, X. (2018). Biodegradation of lignin by pseudomonas sp. Q18 and the characterization of a novel bacterial DyP-type peroxidase. *Journal of Industrial Microbiology and Biotechnology, 45*(10), 913–927. https://doi.org/10.1007/s10295-018-2064-y

Yang, R., Ma, J., Wang, Z., Du, Y., Tian, S., Fan, G., & Teng, C. (2024). The identification of a strain for the biological purification of soy molasses to produce functional soy oligosaccharides and optimize purification conditions. *Food, 13*(2), 296. https://doi.org/10.3390/foods13020296

Yu, H., Li, S., Yu, L., & Wang, X. (2022). The recent progress China has made in green mine construction, part II: Typical examples of green mines. *International Journal of Environmental Research and Public Health, 19*(13), 8166. https://doi.org/10.3390/ijerph19138166

Zhakypbek, Y., Kossalbayev, B. D., Belkozhayev, A. M., Murat, T., Tursbekov, S., Abdalimov, E., & Allakhverdiev, S. I. (2024). Reducing heavy metal contamination in soil and water using phytoremediation. *Plants, 13*(11), 1534. https://doi.org/10.3390/plants13111534

Zheng, Y., Li, Y., Guo, S., Yu, J., Chi, R., & Xiao, C. (2024). Enhanced phytoremediation of hexavalent chromium contamination in phosphate mining wasteland by a phosphate solubilizing bacterium. *Bioremediation Journal, 28*(1), 121–132. https://doi.org/10.1080/10889868.2022.2136135

Zhu, X., & Jassby, D. (2019). Electroactive membranes for water treatment: Enhanced treatment functionalities, energy considerations, and future challenges. *Accounts of Chemical Research, 52*(5), 1177–1186. https://doi.org/10.1021/acs.accounts.8b00558

Bioremediation of Mining Waste Contaminated Water

Debasis Behera, Simran SinghDeo, Swagat K. Mohapatra, and Jagnyaseni Tripathy

1 Introduction

1.1 Brief Overview on Mining

Mining activities are integral to global industrial and technological advancements, providing essential resources like coal, metals, and rare earth elements (Fig. 1). However, modern mining practices pose significant risk to our health and sustenance caused by deforestation, environmental pollution, water contamination, and greenhouse gas emissions. Often mining activities release pollutants such as heavy metals, acidic compounds, and hazardous chemicals which seep into nearby water sources thereby causing various water body contaminations. The water contaminated with mining waste poses significant adverse impact on human health, livestock, farming and aquatic ecosystems, etc. There have been several reports worldwide where these contaminated waste waters have evidently posed serious threat to health and living, and largely disrupted the ecosystem (Das et al., 2025; Priyadarsini & Das, 2024, 2025). The acidic water with heavy metals, released from the abandoned mines containing oxidized sulfide minerals in the Appalachian coalfields have adversely affected the local ecosystems and drinking supplies (Clark et al., 2018). In India's Jharia coalfield prolonged mining has led to high arsenic, lead and mercury levels in soil and water, threatening local communities and

D. Behera · S. SinghDeo · J. Tripathy (✉)
Department of Physics, School of Applied Sciences, KIIT Deemed to be University, Bhubaneswar, Odisha, India
e-mail: jtripathyfpy@kiit.ac.in

S. K. Mohapatra
Department of Chemistry, ICT-IOCB, Bhubaneswar, Odisha, India
e-mail: sk.mohapatra@iocb.ictmumbai.edu.in

© The Author(s), under exclusive license to Springer Nature Switzerland AG 2026

A. P. Das (ed.), *Bioremediation Technology for Sustainable Mining Waste Management*, Sustainable Environmental Waste Management Strategies,
https://doi.org/10.1007/978-3-032-03468-7_9

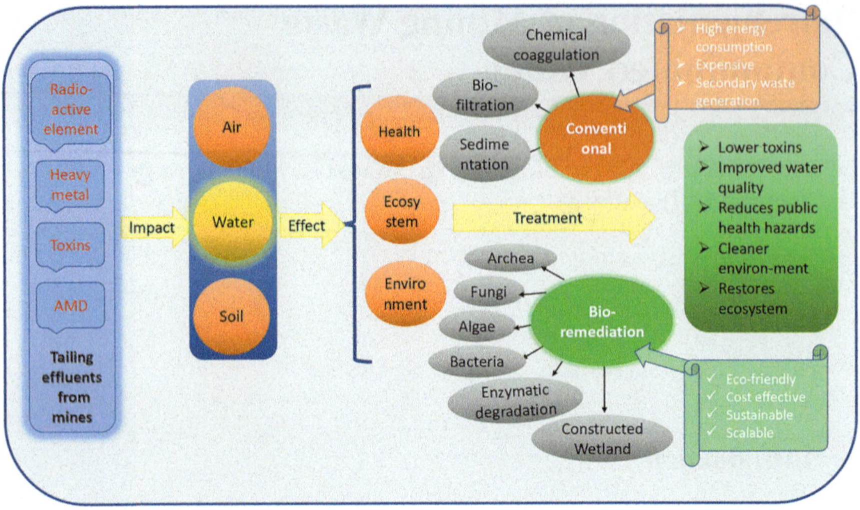

Fig. 1 Schematic presentation of mining waste water treatment

agriculture (Raj et al., 2017). Similarly, coal mining in Mpumalanga, South Africa has contaminated groundwater with sulfates and trace metals, raising concerns over environmental and health risks (McCarthy, 2011).

1.2 Limitations of Mining

Although mining offers numerous/countless advantages, it also has several limitations such as excessive exploitation of resources, land degradation, and high operational costs. Mineral reserves are finite and excessive extraction accelerates resource exhaustion. Additionally mining leads to soil erosion and loss of biodiversity thereby disrupting the ecosystems. One of the major concerns of mining is generation of large volumes of heavy metal, sulfate or otherwise contaminated toxic waste water, which adversely affect health and environment. Different kinds of mining operations are associated with release of different kind of pollutants and toxic substances. Mining waste water often results from the acid mine drainage (AMD) and tailings effluents leading to severe water pollution if not properly treated (Johnson & Hallberg, 2005). Surface mining methods such as open-pit mining expose sulfide minerals to air and water resulting in AMD which generates acidic waste water containing heavy metals. Similarly, placer mining can introduce mercury and other toxic substances into water systems (Johnson & Hallberg, 2005; Lottermoser, 2010; Ugya et al., 2018). For instance, discharging of untreated tailings from Ok Tedi mine into the river system caused widespread ecological damage in the Papua New Guinea (Kirsch, 2007). Similarly, in the USA, AMD from abandoned coal mines in the Appalachian region has led to the acidification of freshwater sources harming aquatic life and reducing water usability. These examples highlight that treating mining waste water pollution is essential to reduce its adverse impacts on both environment and public health, emphasizing the importance of adopting sustainable and efficient treatment processes.

1.3 Major Mining Waste, the Environmental and Societal Challenges

Among the various types of hazardous waste generated by mining contaminated water poses greatest threat due to its ability to spread pollutants over vast distances, impacting both surface and groundwater sources. Unlike solid waste which remains localized, mining waste water can infiltrate drinking water supplies, disrupt aquatic ecosystems and persist in the environment for decades. The uncontrolled discharge of contaminated water intensifies environmental degradation making it a critical issue that demands immediate attention (Das et al., 2011; Das & Ghosh, 2023a, 2023d; Ghosh & Das, 2018). Mining waste water a by-product of mineral extraction contains toxic metals, acidic compounds (AMD) and other pollutants. AMD is a major issue, leading to severe contamination of rivers, groundwater and ecosystems. Notable cases include the degradation of Tasmania's King River due to mining effluents and the 2014 Mount Polley disaster in Canada which released vast amounts of toxic waste water affecting local need for improved communities and aquatic life (Byrne et al., 2018; Nascimento et al., 2023). These incidents highlight the urgent waste water treatment technologies.

1.4 Need for Sustainable Waste Water Treatment and Bioremediation

Untreated mining effluents contain hazardous substances that degrade water quality and disrupt ecosystem. Effective waste water treatment is key to mitigating environmental and health risks associated with mining. Several sustainable methods such as phytoremediation, constructed wetlands, adsorption techniques etc. have been in use for removing pollutants from mining waste water. While, each one of these methods has its own advantages and limitations, microbial mediated bioremediation stands out from the rest of the techniques owing to its low energy demands, minimal chemical input and ability to naturally restore ecosystem (Ahmad et al., 2015). Bioremediation have emerged as a promising strategy for detoxifying mining waste water, reducing its ecological impact and preventing the bioaccumulation of toxic elements in food chains (Pande et al., 2022).

2 Types of Mining Waste Water Contaminants

Mining operations introduce various contaminants into water bodies, depending on both the mineral ore involved in a particular mining activity and the mining process adopted such as open-pit or surface mining, etc. The contaminants found in the mining waste water encompasse a wide range and type of materials from heavy metals to toxic elements as listed in Table 1.

2.1 *Heavy Metals*

Heavy metals are a major pollutant in mining waste water, posing serious environmental and health hazards owing to their toxicity and bioaccumulation potential. Mining activities, viz. ore extraction, processing and tailings disposal, release significant amounts of heavy metals, viz. arsenic (As), lead (Pb), cadmium (Cd), chromium (Cr), and mercury (Hg) into surrounding water bodies (Tchounwou et al., 2012). These metals contaminate groundwater and surface water affecting aquatic life and the food chain, leading to severe ecological imbalances (Singh et al., 2022). Unlike organic pollutants, heavy metals do not degrade over time and require specialized treatment methods for their removal (Das, Mallavarapu, & Ghosh, 2023; Das & Mishra, 2008; Das, Ray, et al., 2024). Conventional physicochemical treatments as chemical precipitation and ion exchange are often expensive and generate secondary waste (Fu & Wang, 2011). This necessitates the exploration of eco-friendly and economic alternatives such as bioremediation, which utilizes microorganisms, plants, and bioadsorbents to immobilize or remove heavy metals from mining waste water (Gadd, 2010).

Table 1 Sources and elements present in different kind of mining waste water

Type of mine	Primary contaminants in waste water	Examples of elements/chemicals
Gold mines	Heavy metals, cyanide	Arsenic, lead, mercury, cyanide
Uranium mines	Radioactive elements, heavy metals	Uranium, radium, thorium
Copper mines	Heavy metals, acid mine drainage (AMD)	Copper, lead, zinc, sulfates
Lead-zinc mines	Heavy metals, acid mine drainage	Lead, zinc, cadmium, sulfates
Coal mines	Acid mine drainage, heavy metals	Iron, aluminum, manganese, sulfates
Bauxite (Aluminum) mines	Heavy metals, suspended solids	Aluminum, iron, arsenic
Nickel mines	Heavy metals, acidic effluents	Nickel, cobalt, copper
Platinum and palladium mines	Heavy metals, cyanide	Platinum, palladium, chromium, cyanide
Phosphate mines	Radioactive elements, fluoride	Uranium, radium, fluoride
Silver mines	Heavy metals, cyanide	Silver, lead, zinc, cyanide

2.2 Radioactive Elements

Radioactive elements in mining waste water originate primarily from uranium, thorium, and other naturally occurring radioactive materials (NORM) present in ore deposits. Mining and processing activities particularly in uranium and rare earth element extraction, release radionuclides such as uranium-238 (U-238), radium-226 (Ra-226), and thorium-232 (Th-232) into water bodies leading to environmental contamination and health hazards (Al-Hwaiti & Al-Khashman, 2015). These radioactive contaminants pose serious risks owing to their long half-lives, bioaccumulation potential and radiotoxicity, which can cause genetic mutations, cancer, and organ damage upon prolonged exposure (Dey et al., 2023; Ma et al., 2023; Mishra et al., 2009; Paul et al., 2023). Conventional remediation techniques, for instance, chemical precipitation, ion exchange and membrane filtration, though effective, often generate secondary radioactive waste, and are expensive (Pabby et al., 2022). Bioremediation on the other hand offers a more sustainable and eco-friendly alternative employing microorganisms and biological materials to immobilize or transform radioactive elements, reducing their mobility and bioavailability in aquatic systems. Certain bacteria and fungi, such as *Pseudomonas* spp. and *Aspergillus* spp., have demonstrated the ability to adsorb or reduce radionuclide toxicity, making bioremediation a promising approach for managing radioactive pollution in mining waste water (Fomina & Gadd, 2014; Francis, 1990).

2.3 Acid Mine Drainage (AMD)

Acid mine drainage (AMD) is often associated with mining activities and causes significant environmental impact, Especially in metal sulfide mines where the oxidation of minerals, for example, pyrite (FeS_2) leads to the formation of sulfuric acid, and dissolved heavy metals are released into surrounding water bodies (Ríos et al., 2008). This process lowers the pH of water, increasing its acidity and enhancing the mobility of toxic metals like iron (Fe), copper (Cu), lead (Pb), and arsenic (As), which pose severe threats to aquatic ecosystems and human health (Gibert et al., 2004). Conventional treatment methods such as chemical neutralization and lime precipitation are commonly used to manage AMD; however, they can be costly and generate secondary sludge waste (Sheoran et al., 2010). However, bioremediation offers a sustainable alternative by employing sulfate-reducing bacteria (SRB) and other microorganisms to neutralize acidity and precipitate metals as stable sulfides, thereby, reducing toxicity and environmental impact. Microbial approaches including constructed wetlands and bioreactors have shown promising results in AMD treatment, making bioremediation an environmental-friendly and more economical alternative to traditional waste water treatment methods.

3 Cyanide and Other Toxic Chemicals

Cyanide and other toxic chemicals are commonly found in mining waste water, particularly from gold and silver extraction processes, where cyanide is used as a lixiviant to dissolve precious metals (Akcil, 2003). While effective, cyanide poses severe threat to environmental and health owing to its high toxicity, persistence and ability to form harmful by-products such as cyanate and thiocyanate (Dash et al., 2009). Additionally, mining waste water contains other hazardous chemicals such as ammonia, nitrates, arsenic, and organic solvents, which can contaminate surface and groundwater, endangering aquatic life and human populations (Mudhoo et al., 2011). Compared to the conventional water treatment methods, bioremediation involving cyanide-degrading bacteria such as *Pseudomonas* and *Bacillus* species offer a rather less expensive, eco-friendly approach by enzymatically breaking down cyanide into less toxic compounds like ammonia and carbon dioxide. Moreover, microbial consortia and constructed wetlands have shown promise in degrading other mining-related toxicants.

3.1 Conventional Methods for Waste Water Treatment Solution

Conventional waste water treatment utilizes various techniques including sedimentation, biological processes, and chemical coagulation to achieve efficient water purification. Waste water management through conventional treatment methods comprises of three stages of treatment: primary, secondary, and tertiary. The primary stage employs sedimentation or similar physical processes where gravity causes large solids or heavier suspended particles in waste water to settle at the bottom, facilitating their removal. The secondary stage often utilizes biological treatment techniques such as activated sludge systems and biofilters where microorganisms decompose organic pollutants, thereby reducing contamination and enhancing water quality (Muga & Mihelcic, 2008; Robles-González et al., 2008). The tertiary stage incorporates advanced treatment methods including chemical coagulation process to eliminate any unwanted fine particles and other dissolved substances from the mining waste water by promoting the aggregation of impurities into larger clusters for easier separation (Spellman, 2003). These combined processes enhance waste water purification, ensuring it meets safety standards for further use.

3.2 Limitations of Conventional Waste Water Treatment Methods

Conventional waste water treatment methods are undeniably largely effective and play a pivotal role in establishing modern waste water treatment systems and help uphold environmental regulations. However, they also present several challenges affecting their long-term sustainability and efficiency. One major drawback is their high energy demand, particularly due to the use of mechanical aeration in secondary treatment processes and the reliance on fossil fuels for energy-intensive operations (Spellman, 2003). Additionally, these methods often require significant amounts of chemicals for coagulation, flocculation, and disinfection, raising concerns about the environmental and health risks posed by residual chemical by-products (Dey et al., 2024; Ghosh et al., 2023; Mohanty, Dey, et al., 2024; Ray et al., 2024). Another critical issue is their limited ability to eliminate emerging contaminants such as pharmaceuticals, pesticides, endocrine-disrupting compounds, microplastics, and heavy metals. Moreover, conventional systems produce secondary pollutants, including excess sludge that requires proper disposal, as well as potentially hazardous by-products from chlorination (Verlicchi et al., 2012; Kümmerer, 2009). The high costs associated with constructing and maintaining these treatment facilities further pose economic challenges, particularly, in resource-limited regions. Lastly, the substantial land area required for these systems makes them less feasible in densely populated urban settings.

4 Introduction to Bioremediation: An Eco-friendly, Cost-Effective Alternative Approach

Although widely utilized, conventional remediation techniques have several drawbacks as discussed earlier. Their implementation on a large scale is often costly, involves intricate procedures, poses environmental risks, and often disturbs natural ecosystems, leading to ecological imbalances (Chatterjee, 2008). Bioremediation, on the other hand, offers a sustainable and greener alternative, harnessing biological agents to eliminate or transform contaminants (Das et al., 2015; Das, Dey, et al., 2024; Das, Mohanty, et al., 2023; Mohanty, Tripathy, et al., 2024). This method utilizes microorganisms, plants, fungi, and their enzymes to degrade, transform or neutralize pollutants converting them into less harmful substances. Unlike conventional physical and chemical treatments which are rather energy consuming and can generate secondary waste, bioremediation aligns well with natural ecological processes making it an environmental-friendly alternative (Cunningham & Ow, 1996). With a growing emphasis on sustainable environmental management in the recent years, bioremediation techniques have garnered huge attention among researchers globally. This has since emerged particularly effective in treating pollutants such as hydrocarbons, heavy metals, pesticides and organic solvents. The adaptability of bioremediation to various environmental conditions enhances its potential as both a large-scale and localized waste water remediation strategy (Suresh Kumar et al., 2015). Furthermore, by utilizing the metabolic capabilities of native microorganisms, bioremediation reduces reliance on external chemical treatments, making it a cost-efficient solution (Vidali, 2001). Table 2 presents a brief comparison between the conventional and bioremediation methods for waste water treatment considering five key points, that is, efficacy, energy usage, environmental impact, costing, and scalability.

Table 2 Comparison between conventional waste water treatment methods and bioremediation techniques

Criteria of comparison	Conventional treatment method	Bioremediation
Efficiency in heavy metal removal	Moderate	High
Energy requirement	High (mechanical and chemical processes)	Low (microbial/metabolic action)
Environmental impact	Generate secondary pollutants	Eco-friendly with minimal waste
Cost involved	High (due to chemical usage)	Low (natural processes)
Scalability	Requires large infrastructure	Can be applied in natural settings

4.1 Principles of Bioremediation

Bacteria and fungi help with the reduction of pollutants into less harmful substances thus playing a key role in bioremediation, while certain plant species engage in phytoremediation by absorbing or stabilizing toxins in the environment. Enzymes derived from these organisms further enhance the degradation process. Bioremediation can be realized in situ or ex situ depending on whether the contaminants are treated at their original location or contaminated materials are extracted for treatment. The overall treatment process can be seen as a combination of several functions viz., biodegradation, bioaccumulation, and biotransformation. Biodegradation allows pollutants to be broken down into simpler compounds while bioaccumulation lets the pollutants to be stored within biological systems, and biotransformation converts harmful substances into less toxic forms (Vidali, 2001). Bioremediation is widely applied in environmental restoration including oil spill cleanup using hydrocarbon-degrading bacteria, removal of heavy metal pollutant, and industrial waste water treatment.

4.2 Types of bioremediations

Bioremediation can be of several different types. It can be categorized as in situ or ex situ bioremediation depending on where the treatment is carried out. Similarly, depending on the mediating biological agents involved viz., bacteria, fungi, plant etc. it can be termed as aerobic, anaerobic, algal, or constructed wetland bioremediation etc.

4.2.1 In Situ Bioremediation

In in situ bioremediation, contaminants are treated directly at the contamination site without requiring excavation or removal of polluted materials. This method is environmentally friendly as it minimizes disturbances to the natural landscape and reduces the need for handling or transporting contaminants (Kuppusamy et al., 2016). Common applications include bioventing which introduces air into the soil to enhance microbial activity for organic pollutant degradation, and biosparging where air and nutrients are injected into groundwater to stimulate microbial growth and pollutant breakdown. Although in situ bioremediation offers significant advantages, it can be slower than other remediation methods and presents challenges in regulating various environmental factors such as oxygen levels, pH and temperature quintessential for effective microbial activity (Azubuike et al., 2016; Vidali, 2001).

4.2.2 Ex Situ Bioremediation

In ex situ bioremediation, contaminated material is extracted from its original location prior to their treatment either on-site or at a separate facility under controlled conditions. This approach provides greater control over key treatment factors which include temperature, nutrient availability and oxygen levels, leading to enhanced pollutant breakdown. Common techniques include biopiles, where contaminated soil is placed in aerated piles to stimulate microbial activity, and slurry-phase bioreactors which suspend pollutants in a liquid medium to maximize interaction with microorganisms. These controlled settings enable more predictable and efficient remediation, making ex situ bioremediation a preferred option for sites requiring swift cleanup or where in situ methods are impractical (Das & Chandran, 2011; Mary Kensa, 2011; Vidali, 2001).

4.2.3 Aerobic Bioremediation

Aerobic bioremediation relies on oxygen-dependent microorganisms to break down various pollutants including hydrocarbons, pesticides, and organic solvents. These microbes metabolize contaminants to produce simpler, less toxic substances such as carbon dioxide and water. This approach is particularly effective for treating petroleum-based pollutants and other organic compounds. Key techniques include bioventing which increases microbial activity by supplying oxygen to contaminated soil, biosparging where air is injected into groundwater to stimulate aerobic degradation, and biofilters which use microorganisms within a filtration medium to remove airborne contaminant (Azubuike et al., 2016; Vidali, 2001). Due to its efficiency in degrading a wide range of pollutants under oxygen-rich conditions, aerobic bioremediation is widely implemented in environmental cleanup efforts.

4.2.4 Anaerobic Bioremediation

Anaerobic bioremediation utilizes microorganisms that thrive in oxygen-deprived conditions to break down or transform contaminants into less toxic forms. This approach is particularly effective for treating chlorinated compounds, heavy metals, and certain pesticides as the pollutant which are resistant to aerobic degradation. Reductive dechlorination, a well-known process, involves microorganisms using chlorinated solvents as electron acceptors, leading to their breakdown into harmless byproducts. This method is commonly used for groundwater remediation, particularly involving chlorinated solvents like trichloroethylene (TCE) (He et al., 2003; Lovley & Coates, 2000) as the pollutant. Anaerobic bioremediation provides an efficient solution for addressing contamination in subsurface environments where oxygen levels are insufficient.

4.2.5 Phytoremediation

Phytoremediation is a green and cost-effective approach at remediation of contaminated soils and water. This technique leverages the natural abilities of plants to absorb, accumulate, and detoxify pollutants including heavy metals, pesticides, solvents, and crude oil derivatives (Marmiroli & Monciardini, 1999). The complete process of phytoremediation comprises of phytoextraction, phytodegradation, phytostabilization, and phytovolatilization. Phytoextraction is where contaminants are absorbed through roots and store those in above-ground tissues; phytodegradation involves the breakdown of organic pollutants within plant tissues; phytostabilization lessens mobility of the contaminants in the environment; phytovolatilization lets plants to uptake pollutants, modify and eventually release the modified pollutants into the atmosphere (Nemati et al., 2024). The phytoremediation efficacy depends on plant species, pollutant type, and environmental conditions (Lone et al., 2008). While, it offers numerous advantages, including soil preservation and ecosystem restoration, challenges remain in optimizing plant selection and understanding plant–microbe interactions to enhance remediation efficiency. Ongoing research aims to address these challenges and expand the practical applications of phytoremediation in diverse environmental contexts (Tiodar et al., 2021).

4.2.6 Algal Remediation

Algae are increasingly recognized for their potential in bioremediation. Algal biomass can adsorb metals and degrade organic pollutants. Cyanobacteria and green algae have shown promising results in treating AMD and heavy metal-contaminated water (Ciani & Adessi, 2023).

4.2.7 Constructed Wetlands

Constructed wetlands are the engineered ecosystems mimicing natural wetlands. These systems combine physical, chemical, and biological processes to remove contaminants. They are particularly effective for treating AMD and heavy metal pollution (Kadlec & Wallace, 2008).

4.2.8 Enzymatic Bioremediation

Enzymes derived from microorganisms or plants can catalyze the breakdown of complex pollutants. For example, peroxidases and laccases are effective in degrading organic compounds in mining waste water (Smyth et al., 2016).

4.3 Bioremediation Strategies for Mining Waste Water Treatment

Bacteria, fungi, algae, and other microorganisms have pivotal role in the bioremediation of mining waste contaminants, especially in treating waste water. These microbial bioremediation techniques leverage natural metabolic processes of the microorganisms to treat the waste water contaminated by heavy metals, sulfates, and toxic compounds posing significant environmental risks. These microbes bioremediate and treat the contaminated water by transforming, immobilizing, or degrading toxic substances in it and reducing these harmful pollutants into less toxic or inert forms. For instance, sulfate-reducing bacteria (SRB) precipitate heavy metals as insoluble sulfides, while certain algae absorb and accumulate metals through biosorption. These microorganisms thrive in tailored bioreactors or natural wetlands, where they degrade organic contaminants and neutralize AMD (Sheoran et al., 2010). The strategy to bioremediation of waste water can generally be realized in four sequential steps as shown below comprising of microbial breakdown of contaminants, adsorption and bioaccumulation, followed by enzymatic degradation into inert forms, and finally purification (Fig. 2).

Desulfovibrio spp., a type of SRB which convert sulfate into sulfide, are widely used in the treatment of AMD (Johnson & Hallberg, 2003). Other widely utilized bacteria are *Pseudomonas*, *Bacillus subtilis*, *Acinetobacter*, *Acidithiobacillus ferrooxidans*, and *Leptospirillum ferrooxidans*, commonly employed in bioleaching process for metal recovery. *Acidithiobacillus ferrooxidans*, for instance, plays a key role in detoxification of mining effluents by oxidizing sulfide minerals in the contaminated water and facilitating the removal of heavy metals such as iron and arsenic (Johnson & Hallberg, 2003). Reportedly, research on chalcopyrite ore in Zambia demonstrated that bioleaching using native *A. ferrooxidans* strains significantly enhanced copper extraction, underscoring its potential in mitigating copper contamination in mining tailings (Nangali et al., 2023). Fungi, including *Aspergillus* and *Peniciium* species, contribute to bioremediation through bioaccumulation and biosorption, effectively binding metals like lead, cadmium, and arsenic onto their cell walls (Gadd, 2010). Algae, such as *Chlorella vulgaris* and *Spirulina platensis,* are used for metal sequestration and cyanide degradation. These algae facilitate the oxidation and breakdown of organic pollutants by absorbing heavy metals while also producing oxygen (Diaconu et al., 2023). Meanwhile, archaea, known for their resilience in extreme environments facilitate bioremediation by metabolizing toxic

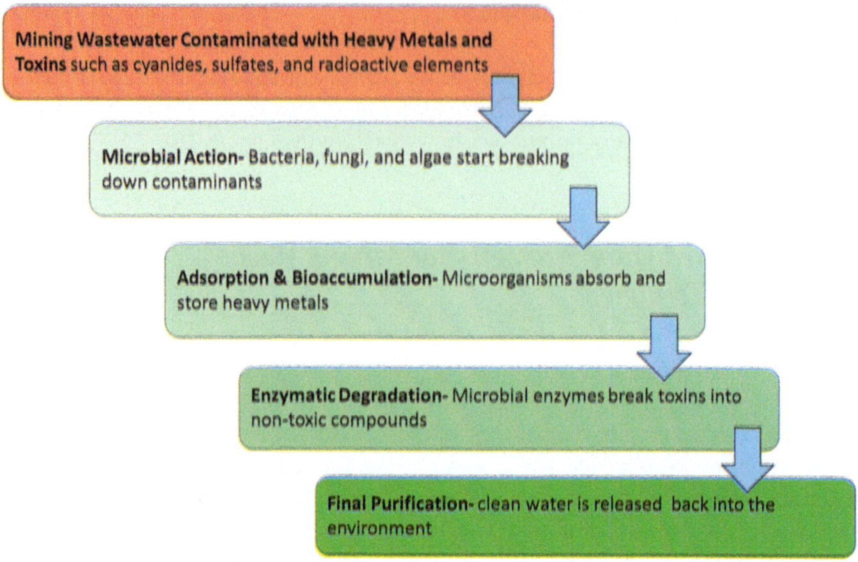

Fig. 2 Process chart for bioremediation

compounds like cyanide and heavy metals through unique biochemical pathways. Extremophilic archaea like *Ferroplasma acidarmanus*, which thrives in highly acidic conditions, contributes significantly to bioleaching in gold and copper mining. Studies have shown that *F. acidarmanus* is particularly effective in bioleaching processes, especially in environments characterized by extreme acidity and elevated heavy metal concentrations (Dopson et al., 2004). *Pseudomonas* species are well-known for degrading organic pollutants and facilitate metal precipitation. *Bacillus subtilis* contributes by producing biosurfactants that enhance the bioavailability of contaminants thus easier microbial degradation (Das & Ghosh, 2022, 2023b, 2023c). The synergistic interactions among these microorganisms enhance the efficiency of contaminant removal making microbial bioremediation a promising and sustainable approach for treating mining waste water. Table 3 enlists several such microorganisms widely used for bioremediation of mining waste water along with the respective bioremediation mechanism. By optimizing pH, temperature, and oxygen levels, which are key to a conducive environmental condition, the efficiency of microbial bioremediation can be significantly enhanced. (Gadd, 2010) This sustainable approach not only reduces the ecological footprint of mining operations but also promotes the recovery of valuable metals from waste water contributing to resource conservation and environmental protection.

Table 3 Microorganism and mechanism used to remediate different kinds of element present in mining waste water

Mining waste type	Contaminants	Microorganism used	Bioremediation mechanism	References
Acid mine drainage (AMD)	Sulfates, iron, aluminum, manganese, heavy metals	Acidithiobacillus ferrooxidans, Leptospirillum ferrooxidans	Oxidation of iron and sulfur compounds to reduce acidity and precipitate metals	Johnson and Hallberg (2003)
Tailings	Arsenic, cadmium, lead, zinc	Lysinibacillus sphaericus	Biosorption and bioaccumulation of heavy metals	Garbisu and Alkorta (2001)
Slag and smelter waste	Lead, arsenic, copper, cadmium	Acidithiobacillus thiooxidans, fusarium oxysporum	Bioleaching to solubilize and recover metals	Hocheng et al. (2014)
Waste rock	Heavy metals, radionuclides	Pseudomonas putida, Bacillus subtilis	Biosorption and enzymatic transformation of contaminants	Ni et al. (2012)
Processing chemicals	Cyanide, mercury	Pseudomonas fluorescens, bacillus megaterium	Enzymatic degradation of cyanide and reduction of mercury toxicity	Dash et al. (2009)
Sediments and sludges	Chromium, nickel, zinc, copper	Geobacter sulfurreducens, Shewanella oneidensis	Reduction of metal ions to less soluble and less toxic forms	Call and Logan (2011)
Coal combustion residues	Selenium, arsenic, mercury	Enterobacter cloacae, pseudomonas stutzeri	Bioprecipitation and reduction of soluble metals to insoluble forms	Butterfield et al. (2016)
Oil and hydrocarbon wastes	Hydrocarbons, polycyclic aromatic hydrocarbons (PAHs)	Alcanivorax borkumensis, mycobacterium	Metabolism and breakdown of hydrocarbon pollutants	Kadri et al. (2018)
Radionuclide contaminants	Uranium, thorium, radium	Deinococcus radiodurans, Geobacter metallireducens	Reduction and immobilization of radioactive elements	Prakash et al. (2013)

5 Case Studies of Successful Bioremediation Projects Applied to Mining Waste Water

Bioremediation has been effectively employed in various mining regions to treat contaminated waste water, utilizing the metabolic capabilities of specific microorganisms to neutralize pollutants.

Leviathan Mine, California, USA: At the Leviathan Mine, a former open-pit sulfur mine, a compost-free SRB was implemented to address acid rock drainage. This system utilized SRB, such as *Desulfovibrio* species, which convert sulfate to sulfide, precipitating heavy metals from the water. The bioreactor successfully reduced metal concentrations, improving the quality of water released into nearby creeks (Doshi, 2006). Copper Mining Effluents in Chile: In Chile, a leading copper-producing country, innovative bioremediation strategies have been explored to treat mining waste water. One study investigated the use of biosorption processes employing sustainable biosorbents to recover copper from ammoniacal waste streams. Materials such as pinecone biomass demonstrated high efficiency in selectively adsorbing copper ions, offering a cost-effective and environmental-friendly solution for waste water treatment in the mining industry (Nicomel et al., 2022). Gold Mining in South Africa: South Africa's gold mining industry has faced challenges with cyanide-containing waste waters. Research has identified microorganisms capable of detoxifying elevated concentrations of cyanide. For instance, the alkaliphilic bacterium *Pseudomonas pseudoalcaligenes* CECT5344 can degrade cyanide in the presence of heavy metals, showcasing its potential for bioremediation applications in gold mining effluents (Roldán et al., 2021). Ridgeway Mine, South Carolina, USA: Following the cessation of operations at Ridgeway Mine in 1999, reclamation efforts focused on managing pyrite-bearing tailings, which pose a risk of acid rock drainage when exposed to water and oxygen. The strategy involved creating an impoundment to isolate the tailings from fresh water and atmospheric oxygen. By submerging the tailings in stagnant water already saturated with sulfides, further oxidation and leaching were prevented. This approach aimed to establish a meromictic lake with distinct, non-mixing water layers, effectively containing potential contaminants (Location & Stabilization, 2002). Lemna minor Application in waste water Treatment: *Lemna minor*, commonly known as duckweed, has been utilized in waste water treatment due to its rapid growth and ease of harvesting. In North Dakota, USA, the Devils Lake waste water treatment facility employs *Lemna minor* and other aquatic plants to treat municipal and industrial waste water. The plants absorb contaminants, and after a growth period, they are harvested for use as soil amendments, compost material, or livestock feed. However, in regions contaminated with heavy metals from activities like mining, the harvested biomass must be disposed of properly to prevent environmental harm (Rahmani & Sternberg, 1999). Rampgill Mine, United Kingdom: At the Rampgill Mine, a pilot-scale vertical flow wetland was constructed to treat mine water with elevated zinc (Zn) and sulfate levels. The system utilized limestone gravel and an organic substrate comprising compost, woodchips, and dried activated sewage sludge to promote SRB. This setup achieved a mean total Zn removal efficiency of 68%, with sulfate concentrations also decreasing. Solid phase analysis confirmed the precipitation of ZnS, with up to 12,000 mg/kg Zn in the solid phase. The study highlighted the system's effectiveness and the role of temperature in Zn removal rates (Newsome & Falagán, 2021). Force Crag Mine, United Kingdom: The Force Crag Mine implemented a full-scale passive treatment system to address drainage containing Zn, lead (Pb), and cadmium (Cd). The system comprised two vertical flow ponds acting as downflow bioreactors filled with limestone and organic substrate. Within the first three months, the system removed 97% of Zn from the mine waters, reducing concentrations from

3.7 mg/L to 0.12 mg/L. Sulfate levels also decreased significantly. Despite initial challenges with elevated ammonium and organic content, the river returned to acceptable levels within four months. Long-term monitoring is ongoing to assess the system's efficacy amid variable rainfall (Newsome & Falagán, 2021). Mosaboni Copper Mines, India: In Eastern India, a pilot-scale study at the Mosaboni copper mines involved exposing 14 tons of copper tailings to natural weathering and plant colonization over one year. The research focused on determining the mineralogical and geochemical characteristics of the tailings and assessing the bioaccumulation patterns of metals in naturally colonizing vegetation. Findings indicated that specific plant species could survive in the contaminated environment and accumulate heavy metals, suggesting potential for phytoremediation strategies in copper mine waste management (Maiti, 2010). Cho Don Pb–Zn Mine, Vietnam: A pilot-scale experiment at the Cho Don Pb–Zn mine in Vietnam combined adsorption with a constructed wetland. Over four months, this system treated real waste water, achieving average removal efficiencies of 80.3% for arsenic (As), 96.9% for manganese (Mn), 79.6% for cadmium (Cd), 52.9% for zinc (Zn), and 38.7% for lead (Pb). The effluent concentrations met the established industrial waste water standards, demonstrating the feasibility of this combined approach for treating mining waste water (Nguyen et al., 2019). Hidden Valley Gold Mine, Papua New Guinea: At the Hidden Valley Gold Mine, a physicochemical assessment of a waste rock dump was conducted to evaluate its suitability for bioremediation. The study involved analyzing the concentrations of heavy metals and nutrients to understand the site's potential for supporting vegetation. Findings suggested that certain forest species could be introduced to the area to enhance natural attenuation processes, thereby reducing the mobility of heavy metals and improving the site's ecological health (Lewis et al., 2020). Paya Indah Wetland, Malaysia: In Malaysia, the Paya Indah Wetland, a former mining lake, was assessed for water quality and decolorization potential using a bioreactor system inoculated with the fungus Ganoderma lucidum. The study found that the lake water had a low pH and high iron content, with an intense dark brownish color. The bioreactor successfully reduced more than 90% of the water's color demonstrating the potential of *G. lucidum* in treating contaminated lake water (Hanafiah et al., 2024). These case studies illustrate diverse bioremediation techniques tailored to specific environmental challenges in mining regions, highlighting the adaptability and effectiveness of biological approaches in mitigating mining-related water contamination.

6 Challenges and Limitations of Bioremediation in Mining Waste Water

Implementing bioremediation for mining waste water though advantageous from many aspects presents several technical and operational challenges. One primary concern is the variability in waste water composition, which can affect microbial efficacy due to fluctuations in contaminant types and concentrations. Additionally, presence of toxins and extreme pH levels can inhibit microbial activity, complicating the bioremediation process. Environmental conditions such as temperature, pH, and nutrient availability must be meticulously maintained for achieving microbial viability and performance (Rosanti et al., 2020) thus scalability from successful laboratory experiments to real world industrial applications is often a challenging process. One of the limitations in using bioremediation techniques for mining waste water treatment is the associated health and safety concerns, which necessitates careful navigation of complex regulatory structures and maintaining substantial safety considerations. Regulatory bodies frequently mandate comprehensive risk analyses to establish site-specific remediation standards for area demographics, land usage patterns, and contaminant bioavailability. These multi-variate analyses play a vital role in ensuring that cleanup efforts align with the objectives of safeguarding human health and preserving ecological balance. Safety issues also stem from the potential discharge of secondary pollutants during bioremediation processes, which may introduce additional environmental and public health hazards. Additionally, the deployment of genetically modified microorganisms (GMMs) to boost bioremediation efficiency is subject to rigorous regulations due to potential ecological ramifications and public perception challenges. The safe and effective implementation of bioremediation technologies demands a cross-disciplinary approach incorporating scientific research with robust regulatory oversight to protect both environmental and human health (Naismith, 2021). Evaluating the economic feasibility of bioremediation for mining waste water obviates a comprehensive cost-benefit analysis considering both direct expenses and potential financial gains. Direct costs encompass capital investments in infrastructure, operational expenditures for maintaining optimal microbial activity, and expenses related to monitoring and compliance with environmental regulations. For instance, the levelized cost of mining waste water treatment can vary significantly based on the specific contaminants and technologies employed, with studies indicating a range from \$1.90 to \$24.30/m^3, depending on the waste water type and treatment method (Able et al., 2025). However, these costs can be offset by the recovery of valuable minerals from the treated effluent and the revenue thus generated (Fabregat & Pagán, 2024). Additionally, implementing bioremediation can mitigate long-term liabilities associated with environmental contamination, potentially avoiding costly future remediation efforts and legal penalties. Therefore, a detailed economic assessment tailored to the specific context of the mining operation is essential to determine the viability and sustainability of bioremediation strategies.

7 Future Prospects and Research Directions

Future research in bioremediation of mining waste water focuses on enhancing microbial efficiency, integrating cutting-edge technologies and ensuring large-scale feasibility. Advances in microbial engineering such as genetically modifying bacteria and fungi aim to improve their ability to tolerate heavy metals, and degrade pollutants more effectively. For instance, engineered strains of *Pseudomonas* and *Bacillus* have demonstrated enhanced capabilities in breaking down contaminants under extreme conditions (Yang et al., 2024). The combination of phytoremediation with microbial-assisted strategies is also being explored for sustainable and long-term heavy metal remediation in mining-affected areas (Ali et al., 2013). Moreover, focus must be on the development of cost-effective, field-deployable bioreactor systems capable of treating high contaminant loads under diverse environmental conditions (Glick, 2010). Future studies must also assess the long-term ecological impacts of bioremediation, and address regulatory challenges to ensure its practical implementation on an industrial scale. Collaborative efforts between scientists, environmental agencies and mining industries will be essential in advancing bioremediation technologies, and achieving sustainable waste water treatment solutions.

8 Conclusion

Bioremediation with its efficacy in the removal of mining waste contaminants, cost-effectiveness, and minimal disruption to ecosystems makes for a rising alternative to the conventional water treatment methods. Bioremediation provides a sustainable and effective alternative to conventional mining waste water treatment but challenges such as environmental variability and scalability remain. Advancements in microbial engineering, nanotechnology, and bioaugmentation hold promise for further enhancing its effectiveness. Future of bioremediation in treating mining waste water is full of scope and opportunity to explore wider application encompassing safe removal of many more harmful contaminants while focusing on optimizing bioremediation strategies to improve feasibility for large-scale applications in mining-affected areas.

References

Able, C., Schmitt, T., & Siefert, N. (2025). Baseline cost analysis of energy wastewater treatment with preliminary feasibility analysis of critical mineral recovery. *Minerals*. https://doi.org/10.3390/min15030213

Ahmad, A., Mohd-Setapar, S. H., Chuong, C. S., Khatoon, A., Wani, W. A., Kumar, R., & Rafatullah, M. (2015). Recent advances in new generation dye removal technologies: Novel search for approaches to reprocess wastewater. *RSC Advances, 5*(39), 30801–30818. https://doi.org/10.1039/C4RA16959J

Akcil, A. (2003). *Turkish gold mining and monitoring in Ovacik gold-silver mine*. Tailings and Mine Waste.

Al-Hwaiti, M., & Al-Khashman, O. (2015). Health risk assessment of heavy metals contamination in tomato and green pepper plants grown in soils amended with phosphogypsum waste materials. *Environmental Geochemistry and Health, 37*(2), 287–304. https://doi.org/10.1007/s10653-014-9646-z

Ali, H., Khan, E., & Sajad, M. A. (2013). Phytoremediation of heavy metals—Concepts and applications. *Chemosphere, 91*(7), 869–881. https://doi.org/10.1016/j.chemosphere.2013.01.075

Azubuike, C. C., Chikere, C. B., & Okpokwasili, G. C. (2016). Bioremediation techniques–classification based on site of application: Principles, advantages, limitations and prospects. *World Journal of Microbiology and Biotechnology, 32*(11), 180. https://doi.org/10.1007/s11274-016-2137-x

Butterfield, C. N., Lee, S.-W., & Tebo, B. M. (2016). The role of bacterial spores in metal cycling and their potential application in metal contaminant bioremediation. *Microbiology Spectrum, 4*(2), 18. https://doi.org/10.1128/microbiolspec.TBS-0018-2013

Byrne, P., Hudson-Edwards, K. A., Bird, G., Macklin, M. G., Brewer, P. A., Williams, R. D., & Jamieson, H. E. (2018). Water quality impacts and river system recovery following the 2014 mount Polley mine tailings dam spill, British Columbia, Canada. *Applied Geochemistry, 91*, 64–74. https://doi.org/10.1016/j.apgeochem.2018.01.012

Call, D. F., & Logan, B. E. (2011). Lactate oxidation coupled to iron or electrode reduction by Geobacter sulfurreducens PCA. *Applied and Environmental Microbiology, 77*(24), 8791–8794. https://doi.org/10.1128/AEM.06434-11

Chatterjee, S. (2008). Bioremediation: A tool for cleaning polluted environments. *Journal of Applied Biosciences, 11*, 594–601.

Ciani, M., & Adessi, A. (2023). Cyanoremediation and phyconanotechnology: Cyanobacteria for metal biosorption toward a circular economy. *Frontiers in Microbiology, 14*(May), 1–9. https://doi.org/10.3389/fmicb.2023.1166612

Clark, E. V., Zipper, C. E., Daniels, W. L., & Keefe, M. J. (2018). Appalachian coal mine spoil elemental release patterns and depletion. *Applied Geochemistry, 98*, 109–120. https://doi.org/10.1016/j.apgeochem.2018.09.016

Cunningham, S. D., & Ow, D. W. (1996). Promises and prospects of phytoremediation. *Plant Physiology, 110*(3), 715–719. https://doi.org/10.1104/pp.110.3.715

Das, A., Dey, S., Bala, K., Gnanasekaran, R., & Das, A. P. (2024). Microbial bioremediation technology for sustainable treatment and management of synthetic microfiber waste. In *Sustainable microbial technology for synthetic and cellulosic microfiber bioremediation* (pp. 221–238). Springer. https://doi.org/10.1007/978-3-031-62600-5_12

Das, A., Dey, S., Gacem, A., Yadav, K. K., Nath, H., Ram, D. K., & Das, A. P. (2025). Revolutionizing solid hydrocarbon sourcing: Plastic's journey from waste to upcycled treasure. *Environmental Science and Engineering, 218*, 331–351. https://doi.org/10.1007/978-3-031-81615-4_14

Das, A., & Ghosh, S. (2022). Role of microorganisms in extenuation of mining and industrial wastes. *Geomicrobiology Journal, 39*, 173–175. https://doi.org/10.1080/01490451.2022.2038953

Das, A., Ray, S. K., Mohanty, M., Mohanty, J., Dey, S., & Das, A. P. (2024). Ecotoxicity of microplastic wastes and their sustainable management: A review. *Environmental Chemistry and Ecotoxicology, 6*, 144–152. https://doi.org/10.1016/J.ENCECO.2024.05.003

Das, A. P., & Ghosh, S. (2023a). Manganese and its application. In *Manganese mining microorganisms* (pp. 29–40). Elsevier. https://doi.org/10.1016/B978-0-12-822147-1.00007-8

Das, A. P., & Ghosh, S. (2023b). Mechanism of manganese bioleaching. In *Manganese mining microorganisms* (pp. 87–98). Elsevier. https://doi.org/10.1016/B978-0-12-822147-1.00014-5

Das, A. P., & Ghosh, S. (2023c). Metagenomic insights into unculturable microbial diversity. In *Manganese mining microorganisms* (pp. 63–76). Elsevier. https://doi.org/10.1016/B978-0-12-822147-1.00002-9

Das, A. P., & Ghosh, S. (2023d). Microorganisms in manganese biomining processes. In *Manganese mining microorganisms* (pp. 55–61). Elsevier. https://doi.org/10.1016/B978-0-12-822147-1.00013-3

Das, A. P., Ghosh, S., Mohanty, S., & Sukla, L. B. (2015). Advances in manganese pollution and its bioremediation. In L. B. Sukla, N. Pradhan, S. Panda, & B. K. Mishra (Eds.), *Environmental microbe biotechnology* (1st ed., pp. 313–328). Springer. https://doi.org/10.1007/978-3-319-19018-1_16

Das, A. P., Mallavarapu, M., & Ghosh, S. (2023). Ecotoxicity of mining pollutants on the environment and their remediation. *Environmental Chemistry and Ecotoxicology, 5*, 165–167. https://doi.org/10.1016/J.ENCECO.2023.08.002

Das, A. P., & Mishra, S. (2008). Hexavalent chromium (VI): Health hazards and environmental pollutant. *Journal of Environment & Development, 2*(3), 386–392.

Das, A. P., Mohanty, A. K., van Hullebusch, E. D., Figueiredo, G., Das, A. P., Mohanty, A. K., van Hullebusch, E. D., & Figueiredo, G. (2023). Bioremediation as an emerging technology for the removal of synthetic microplastic pollutants from marine ecosystem. *MarPB, 194*, 115297. https://doi.org/10.1016/J.MARPOLBUL.2023.115297

Das, A. P., Sukla, L. B., Pradhan, N., & Nayak, S. (2011). Manganese biomining: A review. *Bioresource Technology, 102*(16), 7381–7387. https://doi.org/10.1016/J.BIORTECH.2011.05.018

Das, N., & Chandran, P. (2011). Microbial degradation of petroleum hydrocarbon contaminants: An overview. *Biotechnology Research International, 2011*, 1–13. https://doi.org/10.4061/2011/941810

Dash, R. R., Gaur, A., & Balomajumder, C. (2009). Cyanide in industrial wastewaters and its removal: A review on biotreatment. *Journal of Hazardous Materials, 163*(1), 1–11. https://doi.org/10.1016/j.jhazmat.2008.06.051

Dey, S., Bissoyi, A., Koteswara Reddy, G., & Das, A. P. (2024). Synthetic microfibers: An urban emerging pollutant demanding for sustainable environmental solutions. In *Sustainable microbial technology for synthetic and cellulosic microfiber bioremediation* (pp. 107–121). Springer. https://doi.org/10.1007/978-3-031-62600-5_7

Dey, S., Tripathy, B., Kumar, M. S., & Das, A. P. (2023). Ecotoxicological consequences of manganese mining pollutants and their biological remediation. *Environmental Chemistry and Ecotoxicology*. https://doi.org/10.1016/J.ENCECO.2023.01.001

Diaconu, M., Soreanu, G., Balan, C. D., Buciscanu, I. I., Maier, V., & Cretescu, I. (2023). Study of Spirulina platensis (Arthrospira) development under the heavy metals influence, as a potential promoter of wastewater remediation. *Water, 15*(22), 3962. https://doi.org/10.3390/w15223962

Dopson, M., Baker-Austin, C., Hind, A., Bowman, J. P., & Bond, P. L. (2004). Characterization of ferroplasma isolates and Ferroplasma acidarmanus sp. nov., extreme acidophiles from acid mine drainage and industrial bioleaching environments. *Applied and Environmental Microbiology, 70*(4), 2079–2088. https://doi.org/10.1128/AEM.70.4.2079-2088.2004

Doshi, S. M. (2006). *Bioremediation of acid mine drainage using sulfate-reducing bacteria prepared by* (pp. 1–72). U.S. Environmental Protection Agency.

Fabregat, V., & Pagán, J. M. (2024). Technical–economic feasibility of a new method of adsorbent materials and advanced oxidation techniques to remove emerging pollutants in treated wastewater. *Water, 16*(6), 814. https://doi.org/10.3390/w16060814

Fomina, M., & Gadd, G. M. (2014). Biosorption: Current perspectives on concept, definition and application. *Bioresource Technology, 160*, 3–14. https://doi.org/10.1016/j.biortech.2013.12.102

Francis, A. J. (1990). Microbial dissolution and stabilization of toxic metals and radionuclides in mixed wastes. *Experientia, 46*(8), 840–851. https://doi.org/10.1007/BF01935535

Fu, F., & Wang, Q. (2011). Removal of heavy metal ions from wastewaters: A review. *Journal of Environmental Management, 92*(3), 407–418. https://doi.org/10.1016/j.jenvman.2010.11.011

Gadd, G. M. (2010). Metals, minerals and microbes: Geomicrobiology and bioremediation. *Microbiology, 156*(3), 609–643. https://doi.org/10.1099/mic.0.037143-0

Garbisu, C., & Alkorta, I. (2001). Phytoextraction: A cost-effective plant-based technology for the removal of metals from the environment. *Bioresource Technology, 77*(3), 229–236. https://doi.org/10.1016/S0960-8524(00)00108-5

Ghosh, S., & Das, A. P. (2018). Metagenomic insights into the microbial diversity in manganese-contaminated mine tailings and their role in biogeochemical cycling of manganese. *Scientific Reports, 8*(1), 1–12. https://doi.org/10.1038/s41598-018-26311-w

Ghosh, S., Tripathy, B., Dey, S., & Das, A. P. (2023). Comparative investigation of fungal and bacterial manganese biomining mechanisms. In *Manganese mining microorganisms* (pp. 99–115). Elsevier. https://doi.org/10.1016/b978-0-12-822147-1.00016-9

Gibert, O., de Pablo, J., Luis Cortina, J., & Ayora, C. (2004). Chemical characterisation of natural organic substrates for biological mitigation of acid mine drainage. *Water Research, 38*(19), 4186–4196. https://doi.org/10.1016/j.watres.2004.06.023

Glick, B. R. (2010). Using soil bacteria to facilitate phytoremediation. *Biotechnology Advances, 28*(3), 367–374. https://doi.org/10.1016/j.biotechadv.2010.02.001

Hanafiah, Z. M., Azmi, A. R., Wan-Mohtar, W. A. A. Q. I., Olivito, F., Golemme, G., Ilham, Z., Jamaludin, A. A., Razali, N., Halim-Lim, S. A., & Wan Mohtar, W. H. M. (2024). Water quality assessment and decolourisation of contaminated ex-mining lake water using bioreactor dye-eating fungus (BioDeF) system: A real case study. *Toxics, 12*(1), 1–17. https://doi.org/10.3390/toxics12010060

He, J., Ritalahti, K. M., Yang, K.-L., Koenigsberg, S. S., & Löffler, F. E. (2003). Detoxification of vinyl chloride to ethene coupled to growth of an anaerobic bacterium. *Nature, 424*(6944), 62–65. https://doi.org/10.1038/nature01717

Hocheng, H., Su, C., & Jadhav, U. U. (2014). Bioleaching of metals from steel slag by Acidithiobacillus thiooxidans culture supernatant. *Chemosphere, 117*, 652–657. https://doi.org/10.1016/j.chemosphere.2014.09.089

Johnson, D. B., & Hallberg, K. B. (2003). The microbiology of acidic mine waters. *Research in Microbiology, 154*(7), 466–473. https://doi.org/10.1016/S0923-2508(03)00114-1

Johnson, D. B., & Hallberg, K. B. (2005). Acid mine drainage remediation options: A review. *Science of the Total Environment, 338*(1–2), 3–14. https://doi.org/10.1016/j.scitotenv.2004.09.002

Kadlec, R. H., & Wallace, S. (2008). *Treatment wetlands*. CRC Press. https://doi.org/10.1201/9781420012514

Kadri, T., Magdouli, S., Rouissi, T., & Brar, S. K. (2018). Ex-situ biodegradation of petroleum hydrocarbons using Alcanivorax borkumensis enzymes. *Biochemical Engineering Journal, 132*, 279–287. https://doi.org/10.1016/j.bej.2018.01.014

Kirsch, S. (2007). Indigenous movements and the risks of counterglobalization: Tracking the campaign against Papua New Guinea's Ok Tedi mine. *American Ethnologist, 34*(2), 303–321. https://doi.org/10.1525/ae.2007.34.2.303

Kümmerer, K. (2009). Antibiotics in the aquatic environment – A review – Part I. *Chemosphere, 75*(4), 417–434. https://doi.org/10.1016/j.chemosphere.2008.11.086

Kuppusamy, S., Palanisami, T., Megharaj, M., Venkateswarlu, K., & Naidu, R. (2016). *In-situ remediation approaches for the management of contaminated sites: A comprehensive overview* (pp. 1–115). Springer. https://doi.org/10.1007/978-3-319-20013-2_1

Lewis, L., Hossain, M., & Rao, B. K. R. (2020). Physicochemical assessment of a waste rock dump in the hidden valley gold mine: A case study for bioremediation in Papua New Guinea. *Environmental Earth Sciences, 79*(5), 1–11. https://doi.org/10.1007/s12665-020-8841-8

Location, M., & Stabilization, S. (2002). The Kennecott ridgeway mine. November 1999.

Lone, M. I., He, Z. L., Stoffella, P. J., & Yang, X. E. (2008). Phytoremediation of heavy metal polluted soils and water: Progresses and perspectives. *Journal of Zhejiang University: Science B, 9*(3), 210–220. https://doi.org/10.1631/jzus.B0710633

Lottermoser, B. G. (2010). *Mine wastes: Characterization, treatment and environmental impacts* (3rd ed.). Springer. https://doi.org/10.1007/978-3-642-12419-8

Lovley, D. R., & Coates, J. D. (2000). Novel forms of anaerobic respiration of environmental relevance. *Current Opinion in Microbiology, 3*(3), 252–256. https://doi.org/10.1016/S1369-5274(00)00085-0

Ma, H., Shen, M., Tong, Y., & Wang, X. (2023). Radioactive wastewater treatment technologies: A review. *Molecules, 28*(4), 1935. https://doi.org/10.3390/molecules28041935

Maiti, S. K. (2010). Bioremediation of copper mine waste: A case study from Mosaboni copper mines, eastern India. *International Journal of Environment and Pollution, 43*(1-3), 78–89. https://doi.org/10.1504/IJEP.2010.035915

Marmiroli, N., & Monciardini, P. (1999). PHYTONET: A thematic network devoted to the scientific, economical, environmental, and social aspects of phytoremediation. *International Journal of Phytoremediation, 1*(2), 125–138. https://doi.org/10.1080/15226519908500011

Mary Kensa, V. (2011). Bioremediation - An overview. *Journal of Industrial Pollution Control, 27*(2), 161–168.
McCarthy, T. S. (2011). The impact of acid mine drainage in South Africa. *South African Journal of Science, 107*(5/6), 712. https://doi.org/10.4102/sajs.v107i5/6.712
Mishra, S., Das, A. P., & Seragadam, P. (2009). Microbial remediation of hexavalent chromium from chromite contaminated mines of Sukinda Valley, Orissa (India). *Journal of Environment & Development, 2009*, 1122–1127.
Mohanty, J., Dey, S., Bala, K., & Das, A. P. (2024). Recent advances in microplastic pollution for its sustainable management. In *Sustainable management of urban plastic waste through circular economic approaches* (pp. 27–48). Routledge. https://doi.org/10.1201/9781003465447-3
Mohanty, M., Tripathy, B., Koteswara Reddy, G., & Das, A. P. (2024). Greener technologies for establishing restraint over microplastic pollution. In *Sustainable microbial technology for synthetic and cellulosic microfiber bioremediation* (pp. 239–258). Springer. https://doi.org/10.1007/978-3-031-62600-5_13
Mudhoo, A., Sharma, S. K., Garg, V. K., & Tseng, C.-H. (2011). Arsenic: An overview of applications, health, and environmental concerns and removal processes. *Critical Reviews in Environmental Science and Technology, 41*(5), 435–519. https://doi.org/10.1080/10643380902945771
Muga, H. E., & Mihelcic, J. R. (2008). Sustainability of wastewater treatment technologies. *Journal of Environmental Management, 88*(3), 437–447. https://doi.org/10.1016/j.jenvman.2007.03.008
Naismith, L. K. (2021). Bioremediation: Breaking Down the regulations of genetically modified microorganisms. *Texas A & M Law Review, 8*(3), 607.
Nangali, D. J., Siame, J., Mzula, A., Musango, L., Mugala, A., Mukosha, L., Ngulube, R., & Dusengemungu, L. (2023). Bioleaching of chalcopyrite using native Acidithiobacillus ferrooxidans isolated in a mining area in Kitwe Zambia. *Scientific African, 22*, e01882. https://doi.org/10.1016/j.sciaf.2023.e01882
Nascimento, S. C., Cooke, D. R., Townsend, A. T., Davidson, G., Parbhakar-Fox, A., Cracknell, M. J., & Miller, C. B. (2023). Long-term impact of historical mining on water quality at mount Lyell, Western Tasmania, Australia. *Mine Water and the Environment, 42*(3), 399–417. https://doi.org/10.1007/s10230-023-00943-5
Nemati, B., Baneshi, M. M., Akbari, H., Dehghani, R., & Mostafaii, G. (2024). Phytoremediation of pollutants in oil-contaminated soils by Alhagi camelorum: Evaluation and modeling. *Scientific Reports, 14*(1), 1–12. https://doi.org/10.1038/s41598-024-56214-y
Newsome, L., & Falagán, C. (2021). The microbiology of metal mine waste: Bioremediation applications and implications for planetary health. *GeoHealth, 5*(10), 1–53. https://doi.org/10.1029/2020GH000380
Nguyen, H. T. H., Nguyen, B. Q., Duong, T. T., Bui, A. T. K., Nguyen, H. T. A., Cao, H. T., Mai, N. T., Nguyen, K. M., Pham, T. T., & Kim, K. W. (2019). Pilot-scale removal of arsenic and heavy metals from mining wastewater using adsorption combined with constructed wetland. *Minerals, 9*(6), 379. https://doi.org/10.3390/min9060379
Ni, H., Xiong, Z., Ye, T., Zhang, Z., Ma, X., & Li, L. (2012). Biosorption of copper(II) from aqueous solutions using volcanic rock matrix-immobilized pseudomonas putida cells with surface-displayed cyanobacterial metallothioneins. *Chemical Engineering Journal, 204*, 264–271. https://doi.org/10.1016/j.cej.2012.05.029
Nicomel, N. R., Otero-Gonzalez, L., Williamson, A., Ok, Y. S., Van Der Voort, P., Hennebel, T., & Du Laing, G. (2022). Selective copper recovery from ammoniacal waste streams using a systematic biosorption process. *Chemosphere, 286*, 131935. https://doi.org/10.1016/j.chemosphere.2021.131935
Pabby, A. K., Swain, B., Sonar, N. L., Mittal, V. K., Valsala, T. P., Ramsubramanian, S., Sathe, D. B., Bhatt, R. B., & Pradhan, S. (2022). Radioactive waste processing using membranes: State of the art technology, challenges and perspectives. *Separation & Purification Reviews, 51*(2), 143–173. https://doi.org/10.1080/15422119.2021.1878221

Pande, V., Pandey, S. C., Sati, D., Bhatt, P., & Samant, M. (2022). Microbial interventions in bioremediation of heavy metal contaminants in agroecosystem. *Frontiers in Microbiology, 13*. https://doi.org/10.3389/fmicb.2022.824084

Paul, A., Dey, S., Ram, D. K., & Das, A. P. (2023). Hexavalent chromium pollution and its sustainable management through bioremediation. *Geomicrobiology Journal*. https://doi.org/10.1080/01490451.2023.2218377

Prakash, D., Gabani, P., Chandel, A. K., Ronen, Z., & Singh, O. V. (2013). Bioremediation: A genuine technology to remediate radionuclides from the environment. *Microbial Biotechnology, 6*(4), 349–360. https://doi.org/10.1111/1751-7915.12059

Priyadarsini, S., & Das, A. P. (2024). Analytical and structural characterization of waste lithium-ion batteries for their effective recycling strategy. *Environmental Chemistry and Ecotoxicology*. https://doi.org/10.1016/j.enceco.2024.12.004

Priyadarsini, S., & Das, A. P. (2025). Lithium bioleaching: A review on microbial-assisted sustainable technology for lithium bio-circularity. *Journal of Water Process Engineering, 69*, 106744. https://doi.org/10.1016/j.jwpe.2024.106744

Rahmani, G. N. H., & Sternberg, S. P. K. (1999). Bioremoval of lead from water using Lemna minor. *Bioresource Technology, 70*(3), 225–230. https://doi.org/10.1016/S0960-8524(99)00050-4

Raj, D., Chowdhury, A., & Maiti, S. K. (2017). Ecological risk assessment of mercury and other heavy metals in soils of coal mining area: A case study from the eastern part of a Jharia coal field, India. *Human and Ecological Risk Assessment: An International Journal, 23*(4), 767–787. https://doi.org/10.1080/10807039.2016.1278519

Ray, S. K., Dey, S., Das, A. P., & John, A. (2024). Microplastic waste management through greener technologies: A review. In *Sustainable management of urban plastic waste through circular economic approaches* (pp. 237–261). Springer. https://doi.org/10.1201/9781003465447-14

Ríos, J. J., Rosales, M. A., Blasco, B., Cervilla, L. M., Romero, L., & Ruiz, J. M. (2008). Biofortification of se and induction of the antioxidant capacity in lettuce plants. *Scientia Horticulturae, 116*(3), 248–255. https://doi.org/10.1016/j.scienta.2008.01.008

Robles-González, I. V., Fava, F., & Poggi-Varaldo, H. M. (2008). A review on slurry bioreactors for bioremediation of soils and sediments. *Microbial Cell Factories, 7*(1), 5. https://doi.org/10.1186/1475-2859-7-5

Roldán, M. D., Olaya-Abril, A., Sáez, L. P., Cabello, P., Luque-Almagro, V. M., & Moreno-Vivián, C. (2021). Bioremediation of cyanide-containing wastes. *EMBO Reports, 22*(11), 1–5. https://doi.org/10.15252/embr.202153720

Rosanti, D., Wibowo, Y. G., Safri, M., Maryani, A. T., & Ramadhan, B. S. (2020). Bioremediations technologies on wastewater treatment: Opportunities, challenges and economic perspective. *Sainmatika: Jurnal Ilmiah Matematika Dan Ilmu Pengetahuan Alam, 17*(2), 142. https://doi.org/10.31851/sainmatika.v17i2.5085

Sheoran, A. S., Sheoran, V., & Choudhary, R. P. (2010). Bioremediation of acid-rock drainage by sulphate-reducing prokaryotes: A review. *Minerals Engineering, 23*(14), 1073–1100. https://doi.org/10.1016/j.mineng.2010.07.001

Singh, E., Kumar, A., Mishra, R., & Kumar, S. (2022). Solid waste management during COVID-19 pandemic: Recovery techniques and responses. *Chemosphere, 288*, 132451. https://doi.org/10.1016/j.chemosphere.2021.132451

Smyth, S. J., Phillips, P. W. B., & Kerr, W. A. (2016). EU failing FAO challenge to improve global food security. *Trends in Biotechnology, 34*(7), 521–523. https://doi.org/10.1016/j.tibtech.2016.04.003

Spellman, F. R. (2003). *Handbook of water and wastewater treatment plant operations*. CRC Press. https://doi.org/10.1201/9780203489833

Suresh Kumar, K., Dahms, H.-U., Won, E.-J., Lee, J.-S., & Shin, K.-H. (2015). Microalgae—A promising tool for heavy metal remediation. *Ecotoxicology and Environmental Safety, 113*, 329–352. https://doi.org/10.1016/j.ecoenv.2014.12.019

Tchounwou, P. B., Yedjou, C. G., Patlolla, A. K., & Sutton, D. J. (2012). Heavy metal toxicity and the environment. *Experientia Supplementum, 101*, 133–164. https://doi.org/10.1007/978-3-7643-8340-4_6

Tiodar, E. D., Văcar, C. L., & Podar, D. (2021). Phytoremediation and microorganisms-assisted phytoremediation of mercury-contaminated soils: Challenges and perspectives. *International Journal of Environmental Research and Public Health, 18*(5), 1–38. https://doi.org/10.3390/ijerph18052435

Ugya, A. Y., Ajibade, F. O., & Ajibade, T. F. (2018). Water pollution resulting from mining activity: An overview. In *Proceedings of the 2018 annual conference of the School of Engineering & Engineering Technology (SEET)*. The Federal University of Technology.

Verlicchi, P., Al Aukidy, M., & Zambello, E. (2012). Occurrence of pharmaceutical compounds in urban waste water: Removal, mass load and environmental risk after a secondary treatment—A review. *Science of The Total Environment*, 429, 123–155. https://doi.org/10.1016/j.scitotenv.2012.04.028

Vidali, M. (2001). Bioremediation. An overview. *Pure and Applied Chemistry, 73*(7), 1163–1172. https://doi.org/10.1351/pac200173071163

Yang, P., Condrich, A., Lu, L., Scranton, S., Hebner, C., Sheykhhasan, M., & Ali, M. A. (2024). Genetic engineering in bacteria, fungi, and oomycetes, taking advantage of CRISPR. *DNA, 4*(4), 427–454. https://doi.org/10.3390/dna4040030

Bioremediation of Mining Pollutants: Eco-Innovative Approaches Involved in Sustainable Development

Amisha Mohanty, Ankita Agrawal, and Amiya Kumar Patel

1 Introduction

Mining activities are cornerstone in advancing the global economy and serve as the primary source for extracting and refining valuable resources (Dehkordi et al., 2024). Growing population and rapid industrial expansion has resulted in the excessive utilization of natural resources through mining activities (Mondal & Palit, 2022). It is a complex and multi-staged process involving exploration, extraction, processing and reclamation. Mining industries extract and process three types of minerals such as metallic minerals (aluminium, iron, lead, copper, zinc, gold, nickel), non-metallic minerals (limestone, gravel, clay, phosphate rock) and energy minerals (oil, gas, coal, oil shale) (Jakovljević et al., 2021). In addition, mining activities generate substantial amount of wastes that disrupt the ecosystem (Onifade et al., 2024). Mine wastes are the unwanted and unprofitable byproducts left behind after extraction of economically valuable ores, which includes overburden material, waste rocks, tailings and chemical residues, which must be strategically treated for effective environmental sustainability (Bala et al., 2024; Das, Dey, & Das, 2024; Hassan et al., 2024).

Global mining activities generate more than 100 billion tons of solid waste annually, which require effective management techniques (El Machi et al., 2024). Mining waste generated can be categorized into three categories: (a) solid waste that comprises of dust, waste rock, sludge, slurries and slag, (b) liquid wastes such as effluents and waste water, (c) along with harmful gaseous emission (Parida et al., 2024). In addition, the chemical processing generates substantial amounts of acid wastes including nitric, sulphuric and halogenic acids that undergo oxidative dissolution

A. Mohanty · A. Agrawal · A. K. Patel (✉)
Department of Biotechnology and Bioinformatics, Sambalpur University, Sambalpur, Odisha, India

© The Author(s), under exclusive license to Springer Nature Switzerland AG 2026

A. P. Das (ed.), *Bioremediation Technology for Sustainable Mining Waste Management*, Sustainable Environmental Waste Management Strategies, https://doi.org/10.1007/978-3-032-03468-7_10

and leaching to nearby water bodies producing acid mine drainage. To avoid the direct exposure of mining wastes to nearby forests, residents and water bodies, the wastes are dumped in tailing ponds following strict waste control treatment measures (Hudson-Edwards & Dold, 2015). They are characterized based on their hazardous properties including reactivity, toxicity, flammability and corrosivity (Mishra et al., 2019) that pose adverse environmental impacts. The volume of mining waste varies significantly and influenced by several factors such as location of mining site, extraction process implemented, economic value of ore, current market demand for metals and minerals. In particular, lower-grade ore generate substantially more waste than higher-grade ores (Norgate & Jahanshahi, 2010). Covering millions of hectares worldwide, these waste sites contain harmful heavy metals such as arsenic, cadmium, lead, mercury and metalloids, which leach into nearby soil and water bodies. Over time, such contamination leads to cascade of ecological impacts, reduce soil quality status, inhibit vegetation development, microbial growth and proliferation, threaten public health and ecosystem (Das et al., 2025; Das, Dey, Bala, et al., 2024; Das, Dey, & Das, 2024; Rodriguez-Franco & Page-Dumroese, 2021).

Establishing an equilibrium between economic development and ecological consequences has always remained challenging and matter of concern. Significant challenges due to mining activities require actions in the national and international regulatory framework to oversee proper waste management strategies and discover innovative pollution degrading techniques for minimizing the adverse environmental consequences (Das & Ghosh, 2023; John & Das, 2024). The standard of living is directly linked to the health of the ecosystem; hence, the extent and nature of mining pollutants demands stringent approaches for waste management (Firoozi et al., 2024). Traditional mining pollutant remediation techniques like adsorption, chemical oxidation, pyrolysis, flocculation, landfilling, and incineration are expensive and tedious with limited efficacy (Hassan et al., 2024), whereas biological remediation techniques are ecofriendly, sustainable and hence alternatively preferred for the breakdown of mining contaminants (Irshad et al., 2021; Tyagi & Kumar, 2021). Restoration of mine effluents by bioremediation is an efficient, economic and eco-reliable method essential for ecosystem sustainability and ecological stability. Besides, the microbial based bioremediation and phytoremediation has been extensively studied in metal-contaminated mining sites (Kour et al., 2022). Furthermore, the sequence-based approaches including DNA and RNA extracted from mine tailings provide critical support in designing effective restoration strategies for heavy metal-contaminated site.

Among the methods implemented for the bioremediation of heavy metal pollutants and acid mine drainage, the microbial based remediation strategy has been considered as the most powerful approach to treat mining pollutants (Mishra et al., 2024a, 2024b; Mohanty, Dey, et al., 2024). Moreover, the method is regarded as the gold standard technique since microbes can adapt to extreme environmental conditions with varied metabolic activity for processing mining waste (Zhou et al., 2023). This chapter explored different bioremediation approaches used to detoxify huge overburden of mining pollutants responsible for negative environmental impacts. It also delves into the identification of potent microbial strains involved in bioremediation along with their mechanism of action. Besides, the present study emphasized innovative and advanced techniques utilized in bioremediation approaches such as genetic engineering, advances in bioinformatics tools, omics technology, microarray and nano-bioremediation strategies for the effective bioremediation. In addition, the discovery of novel waste management approaches at the regulatory level has been discussed. Further, the advantages and limitations of different technological interventions implemented for effective bioremediation of the contaminated mine site have been discussed. The innovations towards research advancements in bioremediation, future perspectives and recommendations to harness full potential of technologies in the field of bioremediation are discussed for ecosystem sustainability.

2 Impacts of Mining Waste

The exploitation of mineral resources has been practiced with huge environmental concern (da Silva-Rêgo et al., 2022; Parida et al., 2024). The indiscriminate mining activities have catastrophic environmental impacts, including soil contamination, landscape alterations, substantial depletion of vegetation, loss of biodiversity, and aquatic ecosystem pollution. Moreover, the mining disrupts entire ecosystems, impacting global ecology (Worlanyo & Jiangfeng, 2021). It also contributes to surface and groundwater contamination, climatic changes, and prolonged health issues in humans and animals.

2.1 Environmental Impacts

Mining activities have a perilous mark on the environment, primarily by polluting the soil. The process typically begins with deforestation at the mining site, further exacerbating ecological damage (Worlanyo & Jiangfeng, 2021). The mining industry necessitates the removal of vegetation, which not only serves as the gateway for easy erosion of topsoil but also alters the landscape, significantly disrupting the soil subsystem and habitat (Cooke et al., 2024). Besides, the mining activities generate a substantial number of wastes, including solid waste, slurry, and heavy metal contaminants that require proper disposal. The clearance of huge land areas for disposal of mining waste exposes soil to erosion especially during heavy rainfall, which washes away loose particles into nearby water bodies (Candeias et al., 2018) causing various mine-related damage including formation of huge sinkholes and contamination of soil, surface and ground water with chemicals involved in mining operations (Parida et al., 2024).

2.2 Impacts on Vegetation Growth

Heavy metal accumulation causes the displacement and loss of plant species that are highly responsive to environmental changes. The mine-pollutant contaminated soil in different mining sites causes disequilibrated physiological, biochemical, and metabolic pathways that inhibit root growth and penetration, leading to poor plant development (Kumar, Malik, et al., 2023; Liu et al., 2019). Hence, oxidative stress condition arises, leading to the release of reactive oxygen species (ROS), which can disturb cellular functions by causing dismantling of the cell membrane, ion leakage, lipid peroxidation, and DNA strand breakage (Mansoor et al., 2023). Long-term exposure of toxic metals including zinc, nickel, chromium, cadmium and mercury led to necrosis, chlorosis, clogging of stomata, stunted root and shoot growth, reduced chlorophyll production and disruption in enzyme mediated functions leading to decreased plant productivity and inhibit vegetation development (Kalaivanan & Ganeshamurthy, 2016; Kumar, Malik, et al., 2023; Mohanty, Dey, et al., 2024; Mohanty, Tripathy, et al., 2024; Yu et al., 2023).

2.3 Impacts on Water Quality

Mining operations generate large quantities of solid wastes in the form of overburden, which are inappropriately disposed into environment. Overburden dumping causes leaching of the toxic trace elements that degrade surface and ground water quality (Alao, 2023). Accumulation of mine waste sediments in waterways disrupt aquatic ecosystem and raises the risk of localized flooding due to the filling of riverbeds leading to reduction in water flow capacity (Ciszewski & Grygar, 2016). Besides, the formation of acid mine drainage is the most pervasive consequence of mining. Acid runoff occurs when minerals exposed due to mining activities react with atmospheric oxygen leading to the oxidation and release of sulphuric acid. AMD when introduced to the nearby water bodies dramatically deteriorate water quality, endanger aquatic organisms and render water unfit for use (Jain & Das, 2017; Priyadarshini et al., 2024; Priyadarsini & Das, 2025).

2.4 Impacts on Air Quality

Particulate matter from mine tailing dumps and stockpiles contain toxic heavy metals that are easily dispersed by wind (Paluchamy et al., 2021). The emission of particulate matter (PM) during hauling, drilling, dumping and waste handling is a potent air pollutant that affects the human respiratory tract, skin and eyes (Worlanyo & Jiangfeng, 2021). During exploration, construction and operational phases of mining, the toxic air pollutants such as CO, fugitive dust, particulate matter, heavy metals, hydrocarbon fume, nitrogen oxides and SO_2 are released, which leads to local climatic changes through acid rain formation. Besides, the overburden blasting causes both air and noise shock that are classified as air impacts.

2.5 Health Impacts on Human

Mining activities pose significant present as well as future risks to nearby residents. According to Lancet Commission on Pollution and Health, the mining pollutants are leading cause of illness and mortality (Entwistle et al., 2019). Beyond these immediate hazards, the prolonged exposure to mining pollutants represents critical health risks due to inhalation, ingestion and skin exposure (Olafisoye et al., 2013). Respiratory issues such as, chronic obstructive pulmonary disease (COPD), emphysema, black lung disease, chronic bronchitis, asthma, pneumoconiosis and silicosis have been diagnosed among the workers (Karatela et al., 2022). Besides, the occurrence of hypertension and cardiovascular disorders such as myocardial infarction, arrhythmias, heart failure and stroke are prevalent (Martinelli et al., 2013). The mining dust with toxic gases and improper management of wastes lead to permanent health hazards in workers and resident community (Mohsin et al., 2021). Several studies have indicated lung cancer as one of the prominent causes of mortality among miners and nearby residents (Gautam et al., 2018; Ray et al., 2024; Reddy Koteswara et al., 2025). Prolonged exposer and iron accumulation in brain cells retard its functioning leading to neurodegenerative illness like multiple sclerosis and Parkinson's disorder. The buildup of heavy metals such as Cr and Hg can lead to cognitive impairment, renal failure, plumbism, anaemia and skin irritation (Nouri et al., 2017). In addition, exposure to particulate matter even for instances results in autonomic nervous dysfunction and reduced heart rate. Studies reported that higher concentration and prolong exposure to particulate matter and radionuclides lead to cytogenetic instability leading to epigenetic alterations such as DNA methylation inducing carcinogenesis (Gasparotto & Martinello, 2021). Besides, the mining pollutants affect both terrestrial and aquatic organisms. Terrestrial animals are severely impacted due to their direct exposure to mine wastes through the inhalation of contaminated air, consumption of polluted water along with tainted plants and insects (Rathod et al., 2025; Sahu et al., 2024). Once pollutants enter the animal body, they accumulate in different body tissues. Excessive accumulation triggers oxidative stress by generating large amount of free radicals that lead to liver damage, neurological disorders and carcinogenic effects (Talukder et al., 2024). Additionally, mine discharge can have sub-lethal impacts on marine life causing immediate fish mortality. To improve planetary health, it is important to mitigate the harmful impacts of mining activities. This demands holistic and multidisciplinary approach that integrates microbiological, mineralogical and geochemical considerations (Newsome & Falagán, 2021).

3 Conventional Remediation Method for Mining Pollutants

Various physico-chemical methods including stabilization, electrolysis, vitrification, ion exchange, immobilization, precipitation and evapotranspiration have been employed for remediation of mining pollutants. Such approaches are ineffective and expensive, which involve the logistic challenges and generate significant amount of toxic by-products (El Rasafi et al., 2021). Their application alters the chemical, biological and physical aspects of soil leading to generation of several secondary pollutants. Keeping in view, several ecofriendly approaches have been reported to be used for bioremediation of mining pollutants. Due to the limitations of conventional physico-chemical approaches, there is an urgent need for implementation of remediation techniques that are ecologically feasible and effectively reduce toxicity of mining pollutants (Azubuike et al., 2016). Thus, the biological approaches have been applied extensively and considered as the most adequate and ecologically efficient methods employed for removing environmental pollutants with least side-effects (Tyagi & Kumar, 2021).

4 Postulates of Bioremediation

Bioremediation is an integrated method involving living organisms such as microbes, green plants, microalgae, microbial enzymes and metabolites used to eliminate, breakdown, transform, mineralize and detoxify the toxic mining pollutants into less toxic form to restore the ecosystem to its pre-disturbed state (Agrawal et al., 2024; Sharma, Goutam, et al., 2021). This method is economically-feasible and ecologically safe to detoxify the harmful mining pollutants such as distinct metalloids and heavy metals to restore contamination site into its pre-pollution state (Raklami et al., 2022). Microbes are best applied as bioremediation reagents due to presence of enzymes that readily degrade complex pollutant into innocuous form (Agrawal et al., 2024). Many enzymes act on wide range of pollutants due to non-specific substrate binding ability that allow effective bioremediation. Microbes utilize oxygen and available pollutants as nutrients to undergo numerous metabolic processes and mortify pollutants. Bioremediation comprises fundamental research involving enzyme kinetics, co-metabolism, bio-treatment, biotransformation, environmental attenuation, biogeochemical and fate modelling (Dey, Das, et al., 2023; Dey, Dash, Ray, et al., 2024; Tyagi & Kumar, 2021). Environmental factors such as pollutant types, its solubility, its bioavailability to plants and microbes, adequate microbial growth condition play vital role influencing the effective bioremediation of mining pollutants (Bala et al., 2022). The bioremediation at the disturbed ecosystem occurs naturally is termed as 'natural attenuation' or 'intrinsic bioremediation', addition of organic amendments in mining site to improve bioavailability of pollutant is termed as stimulated bioremediation.

4.1 Microbe Based Bioremediation

The microbes are prevalent everywhere where they use variety of substrates to produce carbon and energy even in disturbed ecosystems. The ability of microbes to sequester, precipitate, biosorb and degrade the pollutants depend upon their varied metabolic activities and microbial mediate enzyme catalysis (Ibrahim et al., 2021). Besides, the microbes utilize toxic mining pollutants as substrates under both anaerobic and aerobic conditions forming H_2O, CO_2, metabolites and microbial biomass, which are further degraded by microbial colonies through mineralization (Saravanan, Kumar, Karishma, et al., 2021). Additionally, the microbes contain several genes encoding heavy metal resistant transporters and proteins located in the plasmids and transposons that carry out metal uptake from the surrounding medium (Ramakrishnan et al., 2011). The potent microbial population can either be indigenous to the mining site or amended experimentally for effective degradation (da Silva et al., 2022), which play key role in ecosystem function (Table 1).

Table 1 Microbes undergoing effective bioremediation of several mining pollutants

Mine pollutants	Microorganisms	Category	References
Arsenic	*Saccharomyces cerevisiae* *Micrococcus* sp. *Shigella* sp. *Rhizopus* sp. *Aspergillus* sp.	Fungi Bacteria Bacteria Fungi Fungi	Mitra et al. (2017) Bhakat et al. (2019) Román-Ponce et al. (2018) Flores-Amaro et al. (2024) Srivastava et al. (2011)
Chromium	*Pseudomonas aeruginosa* *Thermus* sp. *Enterobacter cloacae* *Bacillus subtilis* *Escherichia coli* *Aspergillus fumigatus*	Bacteria Bacteria Bacteria Bacteria Bacteria Fungi	Mala et al. (2020) Malaviya and Singh (2016) Pattnaik et al. (2022) Upadhyay et al. (2017) Samuel et al. (2013) Yin et al. (2019)
Arsenic	*Chlamydomonas reinhardtii* *Microcystis aeruginosa* *Trichoderma* sp.	Algae Algae Fungi	Li et al. (2024) Deng et al. (2020) Kidwai et al. (2022)
Copper, cobalt	*Penicillium* sp. *Trichoderma* sp. *Aspergillus* sp. *Chlorococcum humicola* *Bacillus licheniformis*	Fungi Fungi Fungi Algae Bacteria	Dusengemungu et al. (2020) Kidwai et al. (2022) Dusengemungu et al. (2020) Sarma et al. (2024) Sarwan (2022)
Lead	*Arthrospira platensis* *Klebsiella* sp. *Saccharomyces cerevisiae* *Aspergillus niger* *Microbacterium* sp.	Bacteria Bacteria Fungi Fungi Bacteria	Balaji et al. (2014) Oladipo et al. (2018) Duc et al. (2021) Yin et al. (2019) Ashrafi et al. (2022)

(continued)

Table 1 (continued)

Mine pollutants	Microorganisms	Category	References
Hydrocarbon	*Pseudoalteromonas* sp. *Rhodovibrio* sp. *Spingomonas* sp. *Bacillus licheniformis* *Fusarium* sp. *Candida* sp.	Bacteria Bacteria Bacteria Bacteria Fungi Fungi	Dell'Anno et al. (2020) Rana et al. (2023) Bastida et al. (2016) Eskandari et al. (2017) Okoye et al. (2019) Rana et al. (2023)
Cu (II) oxide	*Rhodobacter blasticus*	Bacteria	Upadhyay et al. (2023)
Mercury	*Nitrococcus mobilis* *Spirogyra hyaline* *Clamydomonas reinhardtii* *Pseudomonas* sp. *Staphylococcus* sp. *Cunninghamella elegans* *Brevundimonas vesicularis*	Bacteria Algae Algae Bacteria Bacteria Fungi Bacteria	Chasanah et al. (2018) Abioye et al. (2020) Balzano et al. (2020) Duc et al. (2021) Duc et al. (2021) Torimiro et al. (2021) Chasanah et al. (2018)
Uranium	*Pseudomonas aeruginosa* *Aeromonas* sp.	Bacteria Bacteria	Choudhary and Sar (2011) Banala et al. (2021)
Lead	*Spirogyra* sp. *Bacillus cereus* *Bacillus licheniformis*	Algae Bacteria Bacteria	Husain et al. (2022) Anusha and Natarajan (2020) Wen et al. (2018)
Endosulfans	*Staphyllococcus* sp. *Bacillus subtilis*	Bacteria Bacteria	Widyati et al. (2024) Singh et al. (2019)
Polycyclic aromatic hydrocarbon (PAHs)	*Myceliophthora thermophila* *S. maltophilia* *Acinetobacter* sp. *Penicillium chrysogenum* *Penicillium oxalicum* *Ascomycota* sp.	Fungi Fungi Bacteria Fungi Fungi Fungi	Mohapatra and Phale (2021) Arulazhagan et al. (2017) Basu et al. (2018) Leitão (2009) Zehra et al. (2018) Aranda et al. (2017)
Cadmium	*Chlamydomonas reinhardtii* *Serratia liquefaciens* *Aspergillus* sp. *Cladosporium* sp. *Rhodopseudomonas palustris* *Klebsiella pneumoniae*	Algae Bacteria Fungi Fungi Bacteria Bacteria	Ibuot et al. (2020) Kumar et al. (2019) Oladipo et al. (2016) Dusengemungu et al. (2020) Ma et al. (2023) Kumar et al. (2019)
Zinc	*Fucus vesiculosus* *Thiobacillus* sp.	Algae Bacteria	Brinza et al. (2020) Mathiyazhagan and Natarajan (2011)
Phenanthrene	*Phomopsis liquidambari* *Candida viswanathii*	Fungi Fungi	Fu et al. (2018) Khanna et al. (2023)
Phenol	*Bacillus subtilis* *Alcaligenes odorans* *Pseudomonas aeruginosa*	Bacteria Bacteria Bacteria	Gaur et al. (2018) Singh et al. (2013) Wang et al. (2011)
Nickel	*Desmodesmus* sp.	Algae	Yin et al. (2019)

Bacteria are ideally used for bioremediation since they grow and survive in diverse micro-environment (Kaushik et al., 2021). The predominance of negative functional groups such as carbonyl, hydroxyl and phosphate in the biomolecules of microbial cell membrane bind with toxic metal ions rapidly and accelerate bioremediation. The fungi and algae function as the promising biocatalysts that detoxify harmful mining pollutants by valance transformation, accumulation and precipitation inside the spores and mycelium (Goutam et al., 2021). Two principal mechanisms involved in microbial-based degradation of mining pollutants include mobilization and immobilization. The mobilization involves bioleaching, bioaugmentation, biostimulation and enzymatic mediated redox reactions, whereas the immobilization involves biosorption, bioaccumulation, complexation and precipitation. Bioleaching is the process of employing acidophilic microbes including iron and sulphur oxidizing microbial strain such as *Acidithiobacillus thiooxidans* for solubilization of heavy metals present in sulphur or iron ore (Sajjad et al., 2019). Bioaugmentation is the addition of specific combination of microbes to the pollution site to hasten biodegradative capacity of existing microbial populations (Agrawal et al., 2024). It is a fast, effective and economic method of bioremediation of mining pollutants. Biostimulation is the addition of nutrients (phosphorus, nitrogen), enzymes, biosurfactants, metabolites, growth promoting factors, electron donors and acceptors to enhance microbial activity and accelerate bioremediation process in polluted habitat (Patel et al., 2022). Biostimulation process is often preferred over bioaugmentation since the native microbes tend to be more competitive than the introduced species (Ayilara & Babalola, 2023). Such approaches preserve microbial diversity in the environment. Besides, it leads to eutrophication due to the excess of available nutrients (Nwankwegu et al., 2022). Enzymatic oxidation is the process where mining pollutants are transformed to reduced oxidation state in the presence of specialized enzymes (Saravanan, Kumar, Karishma, et al., 2021). Oxidoreductase is the microbial enzymes that convert mining pollutants such as heavy metal into high molecular weight compounds after degradation. Laccase is oxidoreductase enzyme produced in *Pcynoporus* sp. is useful in degradation of heavy metals in mine wastes (Singh, Bilal, et al., 2021). Enzymatic reduction is the process carried out by facultative and obligate anaerobes for the conversion of mining pollutants to higher oxidized state compounds (Tiwary et al., 2022), which is efficient for bioremediation of dibenzofurans, PAHs and hydrocarbons. Biosorption involves the adsorption of heavy metals in mine wastes through chelation, complexation, electrostatic interaction, proton and ion displacement (Priyadarshanee & Das, 2021). Bioaccumulation refers to the hyperaccumulation of toxic pollutants within microbial intracellular membrane. Pollutants such as heavy metals cross the intracellular membrane through ion pumps, carrier mediated transport or protein channels. Complexation refers to the formation of inorganic metal complexes through interaction with higher molecular weight ligands, siderophores, metal binding agents and organic acids like citric acids (Newsome & Falagán, 2021). Microbes undergo several multifaceted and comprehensive pathways for bioremediation of mining pollutants for ecosystem stability (Dey, Bissoyi, Koteswara Reddy, & Das, 2024; Dey, Tripathy, et al., 2023).

4.2 Phytoremediation

Plants play pivotal role in mitigating environmental pollutants. Phytoremediation is an emerging, sustainable and economical method of restoring environment, which leverages the capacity of plants to absorb and neutralize broad array of mining contaminants (Zine et al., 2024). However, negative ecological factors such as hyper acidity, heavy metal toxicity, and nutrient deficiency allow rare plant species to survive. Certain plant species possess the ability to degrade, detoxify, immobilize or metabolize an array of mining pollutants (Dada et al., 2015). Ideal phytoremediating plants are woody, heavy metal and contaminant resistant, easily cultivable, fast growing, high biomass yielding capacity, having high absorption capacity, tolerant to climate and adverse physio-chemical conditions and non-attractive to herbivores (Gajić et al., 2018; Kafle et al., 2022), which are highly effective for remediation of mining pollutants as compared to other conventional methods.

The process of phytoremediation is economical and ecofriendly approach with long-term applications, which aims in improving soil quality and productivity of contaminated mine pollutants by enhancing soil stability, biomass accumulation and carbon sequestration (Liu et al., 2023). Biomass generated is recycled by photoautotrophic bacteria in an ecofriendly manner with no secondary contamination. The primary phytoremediation approaches mostly include phytoextraction, phytostabilization, phytovolatilization, phytodegradation and phytostimulation. Phytoextraction, also referred as phytosequestration involves absorption of mining pollutants or heavy metal ions through the plant root system, which leads to hyperaccumulation or compartmentalization of metal ions in aerial plant tissues (Mahar et al., 2016). To prevent the reintroduction of pollutants into the soil, these accumulated pollutants must be harvested before leaf abscission or plant senescence (Ali et al., 2013). Studies revealed the potency of *Rhizophora mangle, Viola baoshaenis, Spartima maritima, Sedum alfredii* and *Rumex crispus* are effective for phytoextraction of metal ions (Pd, As, Zn, Cd), radionucleotides and hydrocarbons across the mining sites (Mesa et al., 2015). Phytostabilization involves the indigenous or plantation of heavy metal tolerant plant species that interchelate metals ions to their root system by absorption, precipitation and sequestration forming a stable plant mass that reduces the heavy metal ion mobility thereby preventing their vertical and lateral migration (Verâne et al., 2020). Subsequently, the plants transform toxic chemical compounds into innocuous form mediated by various chemical reactions. Several plant species such as *Cinnamomum camphora*, *Loropetalum chinense*, *Osmanthus fragrans*, *Acanthus ilicifolius*, *Festuca arvernensis*, *Anthyllis vulneraria* have been reported for extensive phytostabilization ability (Chaudhary & Sindhu, 2024; Deepika Haritash, 2023; Zeng et al., 2018). Phytovolatilization is an approach by which the plants detoxify mine waste contaminants into volatile gaseous compounds that undergo transpiration via the air exposed parts of plant (Sharma et al., 2024). This process involves absorption of heavy metals by plant root system, their transformation to ionic form and then their emission as volatile gases. Studies have identified *Brassica* sp. and *Luffa acutangular* as effective plant species involved in phytovolatilization of mercury and selenium as well as volatile hydrocarbon (Ghosh et al., 2021). Furthermore, phytodegradation is the approach involving enzyme catalysed decomposition of organic compounds by plant root system followed by their absorption and degradation. *Elodea canadensis*, *Mirabilis jalapa*, and *Pueraria* sp. have been reported to degrade nitrates, chlorinated hydrocarbons and explosive wastes in mining sites (Gill et al., 2023).

Altogether, the contaminated infertile land is transformed into cultivable form through the mechanism termed as 'phytotransformation'. Xenobiotic compounds present in mining sites can be efficiently phytotransformed by planting *Cannas* sp. in the mining sites (Singh et al., 2023). Phytostimulation involves the enhancement of microbial flora within the rhizosphere to boost the capacity of plants to degrade contaminants. It has significant efficacy in remediating hydrocarbon pollutants. With emerging technologies and the implementation of gene editing and genetic engineering, several plant species can be modified for detoxifying the vast range of mining pollutants (Kumar, Shinde, et al., 2023). Moreover, the metal transporter, chelator, phytochelatin and metallothionein specific genes are engineered in indigenous plant species to improve the metal uptake, transport and sequestration (Yadav et al., 2010).

4.3 Integrated Plant–Microbe Approach

Integrated plant–microbe interaction used for the detoxification of mining pollutants is a promising technological advancement due to its significantly higher efficacy compared to the individual microbe-based or plant-based method. It involves synergistic interactions between plant root exudates and soil rhizospheric microbes (Agarwal et al., 2020). Microbial species that establish strong mutualistic relationship with host plants are effective for soil restoration. Microbes are not only beneficial in detoxifying pollutants but also enhance plant growth thereby improving overall soil quality. Besides, the occurrence of heavy metal tolerant plant species and microbial species across mining sites can accelerate the rate of bioremediation by improving nutrient uptake, absorption and metal bioleaching (Dong et al., 2024). There are two aspects of integrated plant-microbes interactions such as (a) microbes assist plants in reducing heavy metal toxicity through sequestering metal ions into cell wall components, altering metabolic pathways to block metal uptake, lowering the intracellular metal ion concentration through precise efflux systems, which ultimately transform the toxic metal ions into less hazardous form; (b) plants play vital role in enhancing nutrient availability and soil organic matter content so that microbes can thrive and accelerate the process of reclamation effectively (Agarwal et al., 2020).

Plant root system acts as the chemical factory that secrete compounds like phenolic compounds, amino acids, simple sugars, organic acids, aliphatic and aromatic compounds, protein and polygalatic acids (Ite & Ibok, 2019), which promote survivability and mediate various underground mutualistic association among indigenous microbial populations present in mining sites such as mycorrhizae, rhizobia, endophytes and plant growth promoting bacteria (PGPB) (Ma et al., 2011). Microbes either minimize the cytotoxicity of mining pollutants or increase their bioavailability to plants through absorption. Such synergism between plant and microbe accelerates the potency of phytoremediation (Saharan et al., 2023). Several reports elucidated elimination of PAHs in *Festuca arundinacea* by inoculating with rhizobacterial species (Gayathiri et al., 2022). The presence of PGPB promotes root growth, improves plant metal tolerance, enhances uptake and translocation of metal and nutrients, shields plants from phytopathogens. They secrete various chelators, polymeric compounds and hormones with diverse functions including indole-3-acetic acid, polysaccharides, coelichelin and azotobactin, which together mediate decline in soil acidity and enhance metal bioavailability. Chelators act as the metal binding ligands, which improve metal bioavailability, metal uptake capacity, translocation via the root-shoot system and intracellular heavy metal accumulation (Lopez et al., 2022). Several PGPB such as *Azotobacter*, *Agrobacterium*, *Micrococcus*, *Flavobacterium*, *Chromobacterium* and *Erwinia* have been extensively applied in bioremediation of mining pollutants (Chauhan et al., 2021). Association of leguminous plants with PGPB has been the most successful approach for remediation of highly metal-contaminated mining sites (Saha et al., 2021). Efficient microbial and plant combinations should be developed with enhanced efficiency, adaptability and remediation capacity in disturbed ecosystems such as mining site (Fig. 1).

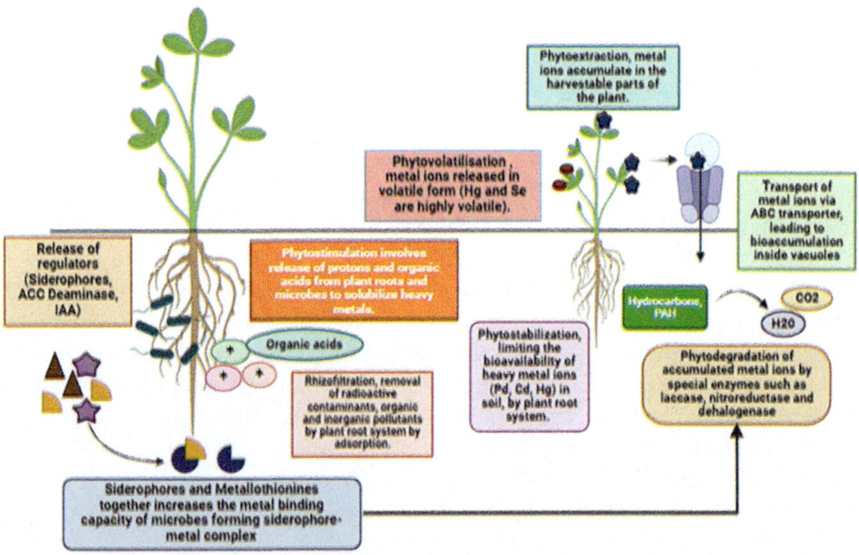

Fig. 1 Plant-mediated and integrated plant–microbe-mediated phytotransformation of mining pollutants into less toxic innocuous form highlighting the role of chelators, siderophores in metal binding and absorption. Mechanisms involved in bioremediation of toxic heavy metal ions, PAHs and hydrocarbon are highlighted

5 Emerging Techniques Applied for Bioremediation of Mining Pollutants

According to Environment Protection Agency (EPA), the successful implementation of emerging technologies including genetic engineering, recombinant DNA technology, metagenomics, metatranscriptomics, metaproteomics and metabolomics techniques, advanced bioinformatics tools, microarray technique and application of nanotechnology have been significantly provided deep insights into the efficiency of bioremediation techniques.

5.1 Genetic Engineering and r-DNA Technology

Unlike conventional microbial based approaches that treat single soil contaminant, lack specificity and efficacy, the genetic engineering ensures high efficacy with low potential risks (Azad et al., 2014). Genetic engineering through the recombinant DNA technology plays vital role in discovery of novel restriction enzymes and DNA ligases that allow the insertion of foreign genes from same or other species into genome of indigenous microbes to develop proficient microbial strain for effective bioremediation of mine wastes (Sharma et al., 2022). Genetically modified microbes degrade hazardous pollutants into innocuous products through the altered co-metabolism. The genetically engineered microbes are considered as artificially engineered microbes carrying novel heterogeneous genes from other microbes to acquire new traits and altered metabolic pathways for biodegradation of contaminants (Sharma, Kumar, & Pandey, 2021). Genetic engineering facilitates the de novo designing and artificial modification of microbial strains, discovering novel genes, enhancing expression of functional genes and degradation pathways. Amongst the bioremediation agents, bacteria play vital role in remediation of soil pollutants through mobilization, transformation and detoxification.

Recent developments in systemic biology have encouraged the discovery of various gene editing techniques like TALEN, CRISPR/Cas and ZFN supported by multi-omics techniques for predicting specific genes and proteins engaged in bioremediation of pollutants (Chakraborty et al., 2022). Genetic modified *Rhodopseudomonas* sp. revealed three times greater mercury binding capacity as compared to wild strain in mining site (Jia et al., 2022). Wild *Bacillus subtilis* that lack methylation and volatilization, on genetic engineering with S-adenosylmethionine methyltransferase gene of *Thermus thermophilus* enhanced methylation of inorganic As and other heavy metals in contaminated sites (Ghosh & Sarkar, 2023). Various genetically modified *Pseudomonas* strains were reported to degrade metal pollutants. *Pseudomonas aeruginosa* isolated from roots of *Phragmites australis* showed high efficiency for PAHs degradation including naphthalene, pyrene, phenanthrene and n-alkanes (Wu et al., 2018).

Recombinant strain of *Escherichia coli* with genes for nickel affinity transmembrane protein and metallothionein showed enhanced tolerance and hyperaccumulation of Ni^{2+} in heavy metal-contaminated mining sites (Kanaujia, 2017). Recombinant strain of *Burkholderia* sp. showed relatively higher potential towards the degradation of benzene, phenol, toluene, chloroform and naphthalene (Malik et al., 2021). Certain genetically engineered fungal species (*Aspergillus* sp., *Pleurotus ostreatus, Penicillium* sp., *Polyporus* sp. and *Ganoderma lucidum*) have been successfully implemented for bioremediation of mining pollutants such as toxic metal ions (Cu, Mn, Cd), anthracene, pyrene and chrysene respectively (Malik et al., 2021). Although, the advancements in genetic engineering has increased the microbial potential for contaminant tolerance and uptake, still the analysis at genetic level becomes a matter of concern that accelerated the discovery and advancements in multi-omics techniques for elucidating the mechanisms involved in bioremediation.

5.2 *Omics Technology*

Omics technology broadly refers to the development of efficient tools at molecular level that led to the discovery of novel pollution control and mitigation strategies. More than 99% of microbes found in disturbed habitats are uncultivable, and thus culture-independent omics techniques have accelerated the investigation of potent microbes and functional genes involved in degrading mining contaminants (Fig. 2). The integrated application of genomics, proteomics, transcriptomics and metabolomics aids in the in-situ bioremediation by providing information about correlating DNA sequences, mode of action towards different metabolites, potent m-RNA and proteins involved in bioremediation (Bala et al., 2022; Satapathy et al., 2024; Tripathy et al., 2024).

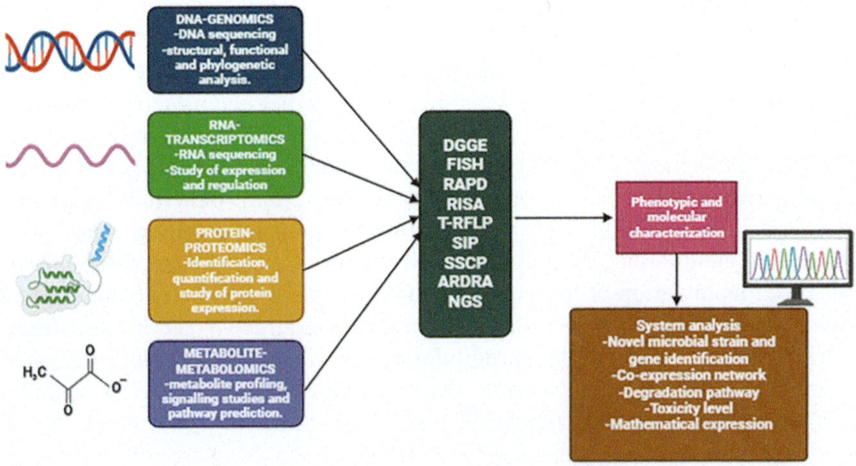

Fig. 2 Multi-omics approaches involved in bioremediation of mining pollutants

5.2.1 Metagenomics

Metagenomics is the promising technological advancement applied in environmental samples for directly analysing the genetic material followed by high-throughput sequencing. Metagenomics involves the study of DNA extracted from environmental sample followed by phylotyping for determining the microbial diversity and soil functionality (Rodríguez-Berbel et al., 2020). Metagenomic studies reveal variation in microbial community structure, nature of their relationships and physiological adaptations in disturbed ecosystems using sequence and function-based techniques (Romero et al., 2021). A typical metagenomic study workflow begins with DNA extraction from the environmental samples, followed by genomic library, sequencing and data analysis. Next-generation sequencing-based approach facilitates the identification of pollutant-degrading microbes, biocatalysts and specialized genes encoding for heavy metal resistance and degradation enzymes through source tracking and environmental monitoring. This method can be regarded as effective and ecofriendly technique that not only mitigates pollutants but also regulates the biogeochemical processes (Sharma, Sirohi, et al., 2021). Metagenomic based studies work in three principal ways such as SMASH, MEGAN and MGRAST (Soman et al., 2022).

Simple metagenomic analysis shell (SMASH) is a software-based tool designed for analysing the microbial community data generated from Sanger sequencing using SmashCell (Varsha et al., 2022), which facilitates comparison among multi-metagenome composition along with their phylogenetic and functional compositions proving the visual representation (Zulfiqar et al., 2023). Meta Genome Analyzer (MEGAN) is a commonly preferred tool used for the functional and taxonomical evaluation of metagenomics and metaproteomics data. The tool searches data on NCBI followed by functional analysis, assigning reads to KOGG, COG and SEED. The input data is provided in different formats and results are generated in readable forms, which is user friendly, efficient and highly compatible metagenome analysing bioinformatics tool (Devarapalli & Kumavath, 2015). Metagenome Subsystem Technology Rapid Annotation (MGRAST) provides knowledge about the microbial community dynamics on the basis of sequence analysis, quality control, clustering and protein prediction of microbial metagenomes. Integrated Microbial Genomes and Metagenomes (IMG/M) is a comparative data analysis tool that generates annotations and metabolism databases of microbial genomes (Soman et al., 2022). Overall, such metagenomic analysing bioinformatics tools allow the precise identification, analysis and the optimization of microbial consortium. Furthermore, the advancements in omics techniques at the genomic, proteomic and metabolic level have hasten the study exploring the role of microbes in bioremediation process.

5.2.2 Genomics

Whole genome sequencing of microbial strains highlighted degradation pathways of different aromatic, xenobiotic and halogenated compounds isolated and screened in mining sites (Mishra et al., 2021). Studies revealed that whole genome sequencing of *Pseudomonas putida* has predicted various pathways such as homogentisate pathway, phenylacetate pathway and ortho pathway involve in the detoxification of PAHs available in mining wastes (Paliwal et al., 2014). In addition, the gene clusters for catabolism of N-heterocyclic aromatic compounds have also been identified (Carmona et al., 2009). Some other microbes including *Pseudomonas alkaliphile* and *Sphingomonas paucimobilis* have been considered as the potent biphenyl and xenobiotic-degrading bacterial strains through whole genome sequencing studies (Petsas & Vagi, 2019).

5.2.3 Transcriptomics

Transcriptomic studies have been applied to explore genome-wide transcriptional activity, map operon structures, identify regulons, DNA-binding sites and conduct comparative genotyping across various microbes in mining site (Yunusa & Umar, 2022). Transcriptomics studies revealed that the engineered *Polaromonas* sp. showed greater degradation of cDCE (cis-dischloroethene) due to boosted expression of certain antioxidant proteins (Sharma et al., 2022). *Geobacter uraniireducens* found in uranium-contaminated soil contains cytochrome genes which are associated with degradation of iron and uranium (Yu & Cui, 2023).

5.2.4 Metaproteomics

Metaproteomics study is useful in the detection of changes in protein composition and physiological activity in response to different anthropogenic contaminants (Santhiyasri et al., 2023). Accordingly, the protein of interest is extracted using 2-D gel electrophoresis and subsequently identified using mass spectroscopy. Matrix-assisted laser desorption ionization time-of-flight (MALDI-TOF) and liquid chromatography connected to electrospray ionization source mass spectroscopy (LC-ESI-MS) is most preferred technique for protein identification amongst peptide fragments extracted from microbial strains from mining sites (Raman, 2024). Metaproteomic studies on *Mycobacterium vanbaalenii* predominantly found in contaminated mining sites showed elevated level of gene expression of putative monooxygenases, aldehyde dehydrogenase and catalase enhancing bioremediation of PAHs (Mishra et al., 2021). Nevertheless, *Pseudomonas putida* showed the activation of 80 novel proteins including thiolase, dioxygenases and hydrolases that were involved in benzoate degradation (Mohapatra et al., 2022).

5.2.5 Metabolomics

The study of primary and secondary metabolite produced and their interactions inside microbes in disturbed ecosystem is called metabolomics (Salam et al., 2023). Metabolomics involve impact assessment of different environmental stressors on the indigenous microbial communities. Metabolomics can be coupled with other omics technology such as genomics, proteomics and transcriptomics for the quantitative and qualitative investigation of metabolite profiles in presence of different environmental stimuli such as mining pollutants. It has been a promising tool for effective removal of heavy metals, hydrocarbons and dyes in mining sites (Shi et al., 2024). There are open-source tools useful to study metabolomics data such as MetaboAnalyst (Xia & Wishart, 2016), MetabooLights and OpenMS (Spjuth et al., 2020) and thus the multi-omics approach has gained more attention (Kumar & Shukla, 2024).

5.3 Bioinformatics Tools

Bioremediation of mining pollutants using bioinformatics involve computational tools and databases that provide information regarding the physico-chemical properties of pollutants, level of toxicity, enzymes and metabolites, pathway prediction and degradation process (Kumari & Kumar, 2021). Computational databases are classified into chemical and biodegradative databases (Sharma, Kumar, & Pandey, 2021). Chemical databases are used for identification and classification along with predicting the mutagenicity and carcinogenicity of mining pollutants. Chemical databases used in bioremediation include ACToR (Aggregated Computational Toxicology Resource), which comprise of 500,000 environmental chemical compounds and their toxicity-based databases (Kumari & Kumar, 2021). Another chemical database that comprises of around 370,000 chemicals is ChemIDplus (Polettini et al., 2008). PubChem is a repository composing of standardized data of diverse chemicals and their biological activities collected from various research studies. Besides, the biodegradative databases comprise of information on metabolic degradation pathway, xenobiotic-degrading microbes and active microbial enzymes engaged in the degradation pathway (Mishra et al., 2021). Nevertheless, UM-BBD (University of Minnesota Biocatalysis/Biodegradation Databases) is publicly accessible online database, contains data and information on microbial biocatalytic activities and associated biodegradation pathways (Gao et al., 2010). Additionally, it comprises of Biochemical Periodic Table (UM-BPT) and Pathway Prediction System (UM-PPS). Furthermore, the BioRadBase database contains information regarding the microbes that are beneficial in bioremediation of radioactive wastes (Cheema, 2023; Reena et al., 2012).

Predicting chemical toxicity by applying computational tools is an emerging trend used to elucidate pace and progress of bioremediation of mining wastes. QSAR (Quantitative structure-regulatory activity relationship) is an economic way to determine the toxicity level of harmful chemical pollutants based on their size, molecular weight and number of benzene rings based on mathematical algorithms (Dimitrov et al., 2016). Linear algorithms include linear regression, multiple regression and linear discrimination analysis, whereas non-linear model involves artificial neural network. Overall, the advancements in bioinformatics tools have emerged as transformative approach for addressing and mitigating the challenges posed by mining pollutant. By integrating computational tools, databases and predictive algorithms has facilitated deeper knowledge of the complex microbe-pollutant interactions and predict the microbial mediated degradation pathways.

5.4 Microarray Technique

With the advent of new genomic tools including metabolic engineering, whole genome sequencing, transcriptomics, protein engineering and gene transcriptional analysis have been extensively used to explore the pace and progress of microbial mediated bioremediation. The microarray technique has been designed to elucidate the microbial mediated metabolic pathways, regulatory cellular functions and gene expression involved in bioremediation of toxic metal pollutants effectively at the mining sites (Sharma, Sirohi, et al., 2021). The conventional gene expression profiling techniques including northern blotting, western blotting, southern blotting and reverse transcriptase polymerase chain reaction had suffered from different limitations to analyse single gene expression. However, the DNA microarrays have emerged as the high-throughput, holistic and effective tool used for analysing the total genes at once (Immanuel Suresh & Sri Janani, 2022). Microarray analysis involves the immobilization of DNA probes on solid matrix followed by the hybridization with complementary target DNA sequence. PhyloChip simultaneously detect both archaeal and bacterial species participated in the bioremediation of polluted sites. Moreover, it is a 16S rRNA based microarray technique that can identify the microbial diversity of around 10,993 subfamilies under 147 phyla. In addition, the GeoChip is a functional gene microarray technique that probes diversity of 152,000 genes from around 410 gene types (Techtmann & Hazen, 2016). Though the DNA microarray approach provides the comprehensive understanding of vast array of microbial genes simultaneously, it is limited to analyse the genes present on the chip and thereby sometimes microarray technique shows its limitations towards the discovery of new genes or pathway involved in bioremediation left in the sample.

5.5 Nanobioremediation

Nanotechnological advancements and the utilization of nanoparticles have offered more viable and sustainable approaches for environmental benefits including bioremediation (Bhatt et al., 2022). The method is highly effective, economic, ecofriendly and self-propelling technique, which is applied for the purification of ground water and surface water (Yogalakshmi et al., 2020) as well as for the decontamination of heavy metal and hydrocarbon polluted sediments and industrial effluents (Vázquez-Núñez et al., 2020). Nanoparticles (having highly compact size, large surface area, biocompatibility) serve as highly selective and effective sorbents, which can be applied for the bioremediation of heavy metal ions at low concentration (<100 mg/L) through biosorption, accumulation and degradation (Mallikarjunaiah et al., 2020). Numerous types of nanoparticles (including gold, platinum, zinc oxide, silicon oxide, manganese oxide) have been designed for detection and elimination of environmental contaminants (Dey, Bissoyi, Koteswara Reddy, & Das, 2024; Guerra et al., 2018; Mourdikoudis et al., 2018). Further, green nanoparticles synthesized by bacteria, algae, plants, fungi and actinomycetes have been widely used for the bioremediation applications (Bhatt et al., 2022).

Besides, biosynthesis of intracellular nanoparticles involves oxidation and reduction mechanism of metal ions within cell matrix. It is mediated by the carrier proteins, cofactors and reductive enzymes (Saravanan, Kumar, Vo, et al., 2021), which is a metabolism-dependent process involving the requirement of the microbe to be alive and biologically active. In contrast, the extracellular biosynthesis involves the integration of metal ions to the specialized functional groups found on the microbial cell membrane through the mechanism of biosorption (Garg et al., 2024). However, the process is dependent on pressure, medium temperature, culture medium used for their growth and proliferation, salinity and concentration of pollutants at the time of biosynthesis. Key steps involved in accumulation of heavy metals inside microbes includes metal uptake through cell wall and plasma membrane through physical adsorption followed by the assimilation inside cytoplasm based on chemisorption and detoxification mechanism. Nanoadsorbents such as activated carbon, zinc oxide, carbon nanotubes, ferric oxide, titanium oxides and graphene are used for the bioremediation of heavy metals (Se, Hg, Cd, Fe), inorganic and organic pollutants from harmful waste water (Malik et al., 2022). Metal or metal oxide-based nanoparticles have been widely applied for removal of harmful metal ions such as lead, arsenic, mercury, chromium and cadmium. They possess high surface area to volume ratio and specificity used for the selective adsorption of metal ions.

6 Advantages and Disadvantages of Bioremediation

Bioremediation is a great alternative to conventional remediation techniques that is applied on actual contamination site without causing major alterations (Agrawal et al., 2024). The method eliminates the necessity of transporting huge amounts of mine wastes off site, thereby reducing the risks associated during transportation process. It is a natural, cost friendly, ecofriendly and sustainable approach (Abatenh et al., 2017). Bioremediation helps in overall annihilation of pollutants as most of toxic pollutants such as radionuclides and rare-earth elements can be converted into innocuous products such as CO_2 and H_2O. This eliminates the tasks associated with the disposal and handling of mining pollutants. Due to their inherent role in environment, bioremediation is a simple, less labour intensive and economical approach (Tyagi & Kumar, 2021). Microbial and plant possess the inherent biological properties that enable their survive under harsh conditions to decontaminate pollutants (Ojuederie & Babalola, 2017). Phytoremediation is useful in enhancing soil quality, recovery of metals preventing leaching and metal erosion that provides large-scale volumetric treatment of huge quantity of mine wastes within short period of time.

Bioremediation also confronts various challenges. The process is best suitable for the degradation of biodegradable pollutants but the presence of radionuclides and chlorinated organic pollutants do not degrade completely (Kensa, 2011). The secondary contaminants or byproducts produced during detoxification are more toxic compared to primary pollutants. Some approaches such as nano-bioremediation are highly specific that allows the bioremediation of single pollutant at once. Factors that influence the success of bioremediation include the suitable growth condition, available nutrients, pollutants, metabolically active plants and microbial communities (Dey, Dash, & Das, 2024; Dey, Tripathy, et al., 2023). Full-scale field operations across different mine sites are not always possible (Tyagi & Kumar, 2021). Phytoremediation though effective is time consuming that depends upon numerous factors like the bioavailability of heavy metal tolerant plant species at mining sites, content and toxicity of pollutants, lack of specific selective enzymes required for the bioremediation of complex organic pollutants.

7 Conclusion and Future Perspectives

Owing to rapid rise in urbanization and increased human activities, there is a marked rise in mining activities that poses significant threats to healthcare and deteriorates ecological scenario. The intimidation of mining activities includes excess waste residues, fragmentation, vegetation loss, decline in biodiversity and unenthusiastic effect on society by generating wastelands. Bioremediation offers better substitute to decontaminate mining pollutants in a reliable, economical, ecofriendly, transformative and sustainable way compared to physico-chemical remediation strategies. Besides, microbial and plant-based bioremediation methods clutch promise to tackle mining pollutants. Integration of bioremediation with technological innovations has gained momentum that advance the cleanup of environmental pollution. With the advent of science techniques, the system biology, nanotechnology, multi-omics, gene editing, microarray profiling and bioinformatics tools have offered scalable solutions for effective bioremediation. Genetic engineering has unlocked dimensions for effective bioremediation with the discovery of potent microbial strains. Omics technology aids in analysing genes, metabolites and proteins along with the mechanism involved in bioremediation pathways for effective detoxification of pollutants. Although advanced bioremediation techniques, including omics-based approaches facilitate precise solutions for mitigation of specific contaminants, identification of novel genes and comparing gene and protein sequences that have the potential to eliminate specific contaminates need to be extensively studied in future bioremediation research. Moreover, the awareness to public at the national and global regulatory level would promote the utilization of natural sources as powerful bioremediation agents. Long-term application of such integrated approaches will lead to ecological stability and sustainability.

Acknowledgments I am grateful to the Department of Biotechnology and Bioinformatics, Sambalpur University, for providing the necessary platform for scientific studies.

Funding The authors, Miss. Amisha Mohanty and Miss. Ankita Agrawal, would like to acknowledge the Department of Science and Technology (DST), Govt. of India, New Delhi, for providing funding through DST/INSPIRE/03/2023/002053 and DST/INSPIRE FELLOWSHIP/2020/IF200208, and for providing necessary scientific knowledge.

Conflict of Interest None.

References

Abatenh, E., Gizaw, B., Tsegaye, Z., & Wassie, M. (2017). The role of microorganisms in bioremediation-A review. *Open Environmental Research Journal, 2*(1), 038–046.

Abioye, O. P., Ezugwu, B. U., Aransiola, S. A., & Ojeba, M. I. (2020). Phycoremediation of water contaminated with arsenic (as), cadmium (cd) and lead (Pb) from a mining site in Minna, Nigeria. *European Journal of Biological Research, 10*(1), 35–44.

Agarwal, P., Giri, B. S., & Rani, R. (2020). Unravelling the role of rhizospheric plant-microbe synergy in phytoremediation: A genomic perspective. *Current Genomics, 21*(5), 334–342.

Agrawal, A., Maharana, J. K., & Patel, A. K. (2024). Microbial inoculants and their potential application in bioremediation. In *Microbes based approaches for the management of hazardous contaminants* (pp. 321–344). Wiley.

Alao, J. O. (2023). Impacts of open dumpsite leachates on soil and groundwater quality. *Groundwater for Sustainable Development, 20*, 100877.

Ali, H., Khan, E., & Sajad, M. A. (2013). Phytoremediation of heavy metals—Concepts and applications. *Chemosphere, 91*(7), 869–881.

Anusha, P., & Natarajan, D. (2020). Bioremediation potency of multi metal tolerant native bacteria *Bacillus cereus* isolated from bauxite mines, Kolli hills, Tamil Nadu- A lab to land approach. *Biocatalysis and Agricultural Biotechnology, 25*, 101581.

Aranda, E., Godoy, P., Reina, R., Badia-Fabregat, M., Rosell, M., Marco-Urrea, E., & García-Romera, I. (2017). Isolation of Ascomycota fungi with capability to transform PAHs: Insights into the biodegradation mechanisms of *Penicillium oxalicum. International Biodeterioration and Biodegradation, 122*, 141–150.

Arulazhagan, P., Al-Shekri, K., Huda, Q., Godon, J. J., Basahi, J. M., & Jeyakumar, D. (2017). Biodegradation of polycyclic aromatic hydrocarbons by an acidophilic *Stenotrophomonas maltophilia* strain AJH1 isolated from a mineral mining site in Saudi Arabia. *Extremophiles, 21*, 163–174.

Ashrafi, Z., Heidari, P., & Mojerlou, S. (2022). Does the long-term contamination of lead (PbII) affect the bioremediation mechanisms of *microbacterium oxydans* strain CM3 and CM7. *Soil and Sediment Contamination, 31*(8), 959–973.

Ayilara, M. S., & Babalola, O. O. (2023). Bioremediation of environmental wastes: The role of microorganisms. *Frontiers in Agronomy, 5*, 1183691.

Azad, M. A. K., Amin, L., & Sidik, N. M. (2014). Genetically engineered organisms for bioremediation of pollutants in contaminated sites. *Chinese Science Bulletin, 59*, 703–714.

Azubuike, C. C., Chikere, C. B., & Okpokwasili, G. C. (2016). Bioremediation techniques–classification based on site of application: Principles, advantages, limitations and prospects. *World Journal of Microbiology and Biotechnology, 32*, 1–18.

Bala, K., Anshul, A., Verma, S., Dogra, A., Das, A. P., Kumar, R., & Nandini, A. (2024). Sustainable management for urban plastic waste generation. In *Sustainable management of urban plastic waste through circular economic approaches* (pp. 211–236). Routledge. https://doi.org/10.1201/9781003465447-13

Bala, S., Garg, D., Thirumalesh, B. V., Sharma, M., Sridhar, K., Inbaraj, B. S., & Tripathi, M. (2022). Recent strategies for bioremediation of emerging pollutants: A review for a green and sustainable environment. *Toxics, 10*(8), 484.

Balaji, S., Kalaivani, T., Rajasekaran, C., Shalini, M., Siva, R., Singh, R. K., & Akthar, M. A. (2014). *Arthrospira* (spirulina) species as bioadsorbents for lead, chromium, and cadmium–a comparative study. *CLEAN, 42*(12), 1790–1797.

Balzano, S., Sardo, A., Blasio, M., Chahine, T. B., Dell'Anno, F., Sansone, C., & Brunet, C. (2020). Microalgal metallothioneins and phytochelatins and their potential use in bioremediation. *Frontiers in Microbiology, 11*, 517.

Banala, U. K., Das, N. P. I., Padhi, R. K., & Toleti, S. R. (2021). Alkaliphilic bacteria retrieved from uranium mining effluent: Characterization, U sequestration and remediation potential. *Environmental Technology and Innovation, 24*, 101893.

Bastida, F., Jehmlich, N., Lima, K., Morris, B. E. L., Richnow, H. H., Hernández, T., & García, C. (2016). The ecological and physiological responses of the microbial community from a semiarid soil to hydrocarbon contamination and its bioremediation using compost amendment. *Journal of Proteomics, 135*, 162–169.

Basu, S., Rabara, R. C., Negi, S., & Shukla, P. (2018). Engineering PGPMOs through gene editing and systems biology: A solution for phytoremediation. *Trends in Biotechnology, 36*(5), 499–510.

Bhakat, K., Chakraborty, A., & Islam, E. (2019). Characterization of arsenic oxidation and uranium bioremediation potential of arsenic resistant bacteria isolated from uranium ore. *Environmental Science and Pollution Research, 26*, 12907–12919.

Bhatt, P., Pandey, S. C., Joshi, S., Chaudhary, P., Pathak, V. M., Huang, Y., & Chen, S. (2022). Nanobioremediation: A sustainable approach for the removal of toxic pollutants from the environment. *Journal of Hazardous Materials, 427*, 128033.

Brinza, L., Geraki, K., Cojocaru, C., Holdt, S. L., & Neamtu, M. (2020). Baltic fucus vesiculosus as potential bio-sorbent for Zn removal: Mechanism insight. *Chemosphere, 238*, 124652.

Candeias, C., Ávila, P., Coelho, P., & Teixeira, J. P. (2018). Mining activities: Health impacts. In *Reference module in earth systems and environmental sciences* (pp. 1–21). Elsevier.

Carmona, M., Zamarro, M. T., Blázquez, B., Durante-Rodríguez, G., Juárez, J. F., Valderrama, J. A., & Díaz, E. (2009). Anaerobic catabolism of aromatic compounds: A genetic and genomic view. *Microbiology and Molecular Biology Reviews, 73*(1), 71–133.

Chakraborty, R., Karmakar, S., & Ansar, W. (2022). Advances and applications of bioremediation: Network of omics, system biology, gene editing and nanotechnology. In *Environmental informatics: Challenges and solutions* (pp. 167–199). Springer.

Chasanah, U., Nuraini, Y., & Handayanto, E. (2018). The potential of mercury-resistant bacteria isolated from small-scale gold mine tailings for accumulation of mercury. *Journal of Ecological Engineering, 19*(2), 236–245.

Chaudhary, S., & Sindhu, S. S. (2024). Microbiome-mediated remediation of heavy metals: Impact on soil health, crop production, and ecosystem sustainability. In *Microbiome-assisted bioremediation* (pp. 257–312). Academic Press.

Chauhan, A., Saini, R., & Sharma, J. C. (2021). Plant growth promoting rhizobacteria and their biological properties for soil enrichment and growth promotion. *Journal of Plant Nutrition, 45*(2), 273–299.

Cheema, H. A. (2023). A review on bioremediation-an emerging technology for treatment of radionuclide waste. *Journal of Modern Agriculture and Biotechnology, 2*(1), 1–16.

Choudhary, S., & Sar, P. (2011). Uranium biomineralization by a metal resistant *Pseudomonas aeruginosa* strain isolated from contaminated mine waste. *Journal of Hazardous Materials, 186*(1), 336–343.

Ciszewski, D., & Grygar, T. M. (2016). A review of flood-related storage and remobilization of heavy metal pollutants in river systems. *Water, Air, and Soil Pollution, 227*, 1–19.

Cooke, C. A., Emmerton, C. A., & Drevnick, P. E. (2024). Legacy coal mining impacts downstream ecosystems for decades in the Canadian rockies. *Environmental Pollution, 344*, 123328.

da Silva, G. O. A., Southam, G., & Gagen, E. J. (2022). Accelerating soil aggregate formation: A review on microbial processes as the critical step in a post-mining rehabilitation context. *Soil Research, 61*(3), 209–223.

da Silva-Rêgo, L. L., de Almeida, L. A., & Gasparotto, J. (2022). Toxicological effects of mining hazard elements. *Energy Geoscience, 3*(3), 255–262.

Dada, E. O., Njoku, K. I., Osuntoki, A. A., & Akinola, M. O. (2015). A review of current techniques of physico-chemical and biological remediation of heavy metals polluted soil. *Ethiopian Journal of Environmental Studies and Management, 8*(5), 606–615.

Das, A., Dey, S., Bala, K., Gnanasekaran, R., & Das, A. P. (2024). Microbial bioremediation technology for sustainable treatment and management of synthetic microfiber waste. In *Sustainable microbial technology for synthetic and cellulosic microfiber bioremediation* (pp. 221–238). Springer. https://doi.org/10.1007/978-3-031-62600-5_12

Das, A., Dey, S., & Das, A. P. (2024). Ecotoxicological impacts of synthetic microfiber pollutants and development of sustainable mitigation strategies. *Environmental Chemistry and Ecotoxicology*. https://doi.org/10.1016/J.ENCECO.2024.12.003

Das, A., Dey, S., Gacem, A., Yadav, K. K., Nath, H., Ram, D. K., & Das, A. P. (2025). Revolutionizing solid hydrocarbon sourcing: Plastic's journey from waste to upcycled treasure. *Environmental Science and Engineering, 218*, 331–351. https://doi.org/10.1007/978-3-031-81615-4_14

Das, A. P., Behera, I. D., & Das, N. P. (2024). *Renewable energy generation and value addition from environmental microfiber pollution through advanced greener solution* (p. 278). Springer.

Das, A. P., & Ghosh, S. (2023). *Manganese mining microorganisms*. Elsevier.

Deepika Haritash, A. K. (2023). Cadmium uptake from soil by ornamental metallophytes: A meta-analytical approach. *Environmental Management, 71*(5), 1087–1097.

Dehkordi, M. M., Nodeh, Z. P., Dehkordi, K. S., Khorjestan, R. R., & Ghaffarzadeh, M. (2024). Soil, air, and water pollution from mining and industrial activities: Sources of pollution, environmental impacts, and prevention and control methods. *Results in Engineering, 23*, 102729.

Dell'Anno, F., Brunet, C., van Zyl, L. J., Trindade, M., Golyshin, P. N., Dell'Anno, A., & Sansone, C. (2020). Degradation of hydrocarbons and heavy metal reduction by marine bacteria in highly contaminated sediments. *Microorganisms, 8*(9), 1402.

Deng, J., Fu, D., Hu, W., Lu, X., Wu, Y., & Bryan, H. (2020). Physiological responses and accumulation ability of *Microcystis aeruginosa* to zinc and cadmium: Implications for bioremediation of heavy metal pollution. *Bioresource Technology, 303*, 122963.

Devarapalli, P., & Kumavath, R. N. (2015). Metagenomics-a technological drift in bioremediation. In *Advances in bioremediation of wastewater and polluted soil* (p. 9). IntechOpen.

Dey, S., Bissoyi, A., Koteswara Reddy, G., & Das, A. P. (2024). Synthetic microfibers: An urban emerging pollutant demanding for sustainable environmental solutions. In *Sustainable microbial technology for synthetic and cellulosic microfiber bioremediation* (pp. 107–121). Springer. https://doi.org/10.1007/978-3-031-62600-5_7

Dey, S., Das, A., Mallick, K., Sahu, A., & Das, A. P. (2023). *Environmental petroleum waste: Pollution, toxicity, sustainable remediation* (pp. 159–175). Springer. https://doi.org/10.1007/978-3-031-48220-5_7

Dey, S., Dash, A., & Das, A. P. (2024). Bioremediation as a sustainable and circular solution for mining waste management and bioenergy generation. In *Sustainable management of mining waste and tailings: A circular economy approach* (pp. 107–119). Routledge. https://doi.org/10.1201/9781003442455-6

Dey, S., Dash, N., Ray, S. K., Priyadarsini, S., Chhotaray, C., & Das, A. P. (2024). Occurrence and quantification of tire wear plastics pollutants from highway city dust of Bhubaneswar City of Odisha. In *Sustainable microbial technology for synthetic and cellulosic microfiber bioremediation* (pp. 77–89). Springer. https://doi.org/10.1007/978-3-031-62600-5_5

Dey, S., Tripathy, B., Kumar, M. S., & Das, A. P. (2023). Ecotoxicological consequences of manganese mining pollutants and their biological remediation. *Environmental Chemistry and Ecotoxicology*. https://doi.org/10.1016/J.ENCECO.2023.01.001

Dimitrov, S. D., Diderich, R., Sobanski, T., Pavlov, T. S., Chankov, G. V., Chapkanov, A. S., & Mekenyan, O. G. (2016). QSAR toolbox–workflow and major functionalities. *SAR and QSAR in Environmental Research, 27*(3), 203–219.

Dong, C., Shao, Q., Ran, Q., Li, X., & Han, Y. (2024). Interactions of rhizosphere microbiota–environmental factors–pharmacological active ingredients of *Eucommia ulmoides*. *Planta, 259*(3), 59.

Duc, H. D., Hung, N. V., & Oanh, N. T. (2021). Anaerobic degradation of endosulfans by a mixed culture of *pseudomonas* sp. and *staphylococcus* sp. *Applied Biochemistry and Microbiology, 57*(3), 327–334.

Dusengemungu, L., Kasali, G., Gwanama, C., & Ouma, K. O. (2020). Recent advances in biosorption of copper and cobalt by filamentous fungi. *Frontiers in Microbiology, 11*, 582016.

El Machi, A., El Berdai, Y., Mabroum, S., Safhi, A. E. M., Taha, Y., Benzaazoua, M., & Hakkou, R. (2024). Recycling of mine wastes in the concrete industry: A review. *Buildings, 14*(6), 1508.

El Rasafi, T., Nouri, M., & Haddioui, A. (2021). Metals in mine wastes: Environmental pollution and soil remediation approaches–a review. *Geological Engineering, 24*(3), 157–172.

Entwistle, J. A., Hursthouse, A. S., Marinho Reis, P. A., et al. (2019). Metalliferous mine dust: Human health impacts and the potential determinants of disease in mining communities. *Current Pollution Reports, 5*, 67–83.

Eskandari, S., Hoodaji, M., Tahmourespour, A., Abdollahi, A., Mohammadian-Baghi, T., Eslamian, S., & Ostad-Ali-Askari, K. (2017). Bioremediation of polycyclic aromatic hydrocarbons by *bacillus Licheniformis* ATHE9 and *Bacillus mojavensis* ATHE13 as newly strains isolated from oil-contaminated soil. *Journal of Geography, Environment and Earth Science International, 11*(2), 1–11.

Firoozi, A. A., Tshambane, M., Firoozi, A. A., & Sheikh, S. M. (2024). Strategic load management: Enhancing eco-efficiency in mining operations through automated technologies. *Results in Engineering, 24*, 102890.

Flores-Amaro, O. A., Ramos-Gómez, M. S., Guerrero-Barrera, A. L., Yamamoto-Flores, L., Romo-Rodríguez, P., Mitchell, K., & Avelar-González, F. J. (2024). Characterization and evaluation of the bioremediation potential of *Rhizopus microsporus* Os4 isolated from arsenic-contaminated soil. *Water, Air, and Soil Pollution, 235*(8), 529.

Fu, W., Xu, M., Sun, K., Hu, L., Cao, W., Dai, C., & Jia, Y. (2018). Biodegradation of phenanthrene by endophytic fungus *Phomopsis liquidambari in vitro* and *in vivo*. *Chemosphere, 203*, 160–169.

Gajić, G., Djurdjević, L., Kostić, O., Jarić, S., Mitrović, M., & Pavlović, P. (2018). Ecological potential of plants for phytoremediation and ecorestoration of fly ash deposits and mine wastes. *Frontiers in Environmental Science, 6*, 124.

Gao, J., Ellis, L. B., & Wackett, L. P. (2010). The University of Minnesota biocatalysis/biodegradation database: Improving public access. *Nucleic Acids Research, 38*(suppl_1), D488–D491.

Garg, V., Arora, A., & Prakash, A. (2024). A review on bioremediation using nanobiotechnology and microbial heavy metal resistance mechanisms. *Current Materials Science: Formerly: Recent Patents on Materials Science, 17*(4), 289–303.

Gasparotto, J., & Martinello, K. D. B. (2021). Coal as an energy source and its impacts on human health. *Energy Geoscience, 2*(2), 113–120.

Gaur, N., Narasimhulu, K., & PydiSetty, Y. (2018). Recent advances in the bio-remediation of persistent organic pollutants and its effect on environment. *Journal of Cleaner Production, 198*, 1602–1631.

Gautam, S., Patra, A. K., Sahu, S. P., & Hitch, M. (2018). Particulate matter pollution in opencast coal mining areas: Threat to human health and environment. *International Journal of Mining, Reclamation and Environment, 32*(2), 75–92.

Gayathiri, E., Prakash, P., Selvam, K., Awasthi, M. K., Gobinath, R., Karri, R. R., & Ravindran, B. (2022). Plant microbe-based remediation approaches in dye removal: A review. *Bioengineered, 13*(3), 7798–7828.

Ghosh, S., & Sarkar, B. (2023). Genetically modified bacteria for arsenic bioremediation. In *Genomics approach to bioremediation: Principles, tools, and emerging technologies* (pp. 467–483). Wiley.

Ghosh, S., Sharma, I., Nath, S., & Webster, T. J. (2021). Bioremediation—The natural solution. In *Microbial ecology of wastewater treatment plants* (pp. 11–40). Elsevier.

Gill, R., Naeem, M., Ansari, A. A., & Gill, S. S. (2023). Phytoremediation and management of environmental contaminants: An overview. *Phytoremediation: Management of Environmental Contaminants, 7*, 3–14.

Goutam, J., Sharma, J., Singh, R., & Sharma, D. (2021). Fungal-mediated bioremediation of heavy metal–polluted environment. *Microbial Rejuvenation of Polluted Environment, 2*, 51–76.

Guerra, F. D., Attia, M. F., Whitehead, D. C., & Alexis, F. (2018). Nanotechnology for environmental remediation: Materials and applications. *Molecules, 23*(7), 1760.

Hassan, S., Bhadwal, S. S., Khan, M., Nissa, K. U., Shah, R. A., Bhat, H. M., & Ganai, B. A. (2024). Revitalizing contaminated lands: A state-of-the-art review on the remediation of mine-tailings using phytoremediation and genomic approaches. *Chemosphere, 356*, 141889.

Hudson-Edwards, K. A., & Dold, B. (2015). Mine waste characterization, management and remediation. *Minerals, 5*(1), 82–85.

Husain, R., Vikram, N., Yadav, G., Kumar, D., Pandey, S., Patel, M., & Hussain, T. (2022). Microbial bioremediation of heavy metals by marine bacteria. In *Development in wastewater treatment research and processes* (pp. 177–203). Elsevier.

Ibrahim, A. A., Khalifa, U. A., Sani, A., Gado, A. M., Ismail, G., Ibrahim, M. A., & Adam, U. D. (2021). Bioremediation: A biological tool for environmental mitigation. *International Research Journal of Modernization in Engineering Technology and Science, 1*, 649–655.

Ibuot, A., Webster, R. E., Williams, L. E., & Pittman, J. K. (2020). Increased metal tolerance and bioaccumulation of zinc and cadmium in *Chlamydomonas reinhardtii* expressing a AtHMA4 C-terminal domain protein. *Biotechnology and Bioengineering, 117*(10), 2996–3005.

Immanuel Suresh, J., & Sri Janani, M. S. (2022). Molecular tools-a future perspective approach for monitoring landfill leachates and validating bioremediation process. In *Circular economy in municipal solid waste landfilling: Biomining & leachate treatment: Sustainable solid waste management: Waste to wealth* (pp. 95–113). Springer.

Irshad, S., Xie, Z., Mehmood, S., Nawaz, A., Ditta, A., & Mahmood, Q. (2021). Insights into conventional and recent technologies for arsenic bioremediation: A systematic review. *Environmental Science and Pollution Research, 28*, 18870–18892.

Ite, A. E., & Ibok, U. J. (2019). Role of plants and microbes in bioremediation of petroleum hydrocarbons contaminated soils. *International Journal of Environmental Bioremediation & Biodegradation, 7*(1), 1–19.

Jain, M. K., & Das, A. (2017). Impact of mine waste leachates on aquatic environment: A review. *Current Pollution Reports, 3*, 31–37.

Jakovljević, K., Ranđelović, D., & Mišljenović, T. (2021). Phytoremediation of mine waste disposal sites: Current state of knowledge and examples of good practice. In *Biotechnology for sustainable environment* (pp. 223–250). Elsevier.

Jia, Q. Y., Chen, C. M., Li, B. Z., & Wang, L. (2022). Construction and characterization of an engineered recombinant *Rhodopseudomonas* palustris to remove Cd^{2+}, Zn^{2+} and Cu^{2+}. *Journal of Environmental Chemical Engineering, 10*(3), 107698.

John, A., & Das, A. P. (2024). Government strategies for achieving circular economy for plastic waste management in Southeast Asia. In *Sustainable management of urban plastic waste through circular economic approaches* (pp. 72–82). Routledge. https://doi.org/10.1201/9781003465447-6

Kafle, A., Timilsina, A., Gautam, A., Adhikari, K., Bhattarai, A., & Aryal, N. (2022). Phytoremediation: Mechanisms, plant selection and enhancement by natural and synthetic agents. *Environmental Advances, 8*, 100203.

Kalaivanan, D., & Ganeshamurthy, A. N. (2016). Mechanisms of heavy metal toxicity in plants. In N. Rao, K. Shivashankara, & R. Laxman (Eds.), *Abiotic stress physiology of horticultural crops*. Springer.

Kanaujia, S. P. (2017). Understanding toxic metal–binding proteins and peptides. In *Handbook of metal-microbe interactions and bioremediation* (pp. 201–228). CRC Press.

Karatela, S., Caruana, S., & Paul, G. (2022). Prevalence of respiratory disease in the population of Queensland communities in proximity to coal mines and coal mining activities. *ternational Journal of Community Medicine and Public Health, 9*, 3014–3022.

Kaushik, S., Alatawi, A., Djiwanti, S. R., Pande, A., Skotti, E., & Soni, V. (2021). Potential of extremophiles for bioremediation. In *Microbial rejuvenation of polluted environment* (Vol. 1, pp. 293–328). Springer.

Kensa, V. M. (2011). Bioremediation-an overview. *Journal of Industrial Pollution Control, 27*(2), 161–168.

Khanna, N., Chatterji, T., Singh, S., & Chaturvedi, P. (2023, September). Microbial metabolism in bioremediation: A review. *AIP Conference Proceedings, 2800*, 020218.

Kidwai, M. K., Malik, A., Dhull, S. B., Rose, P. K., & Garg, V. K. (2022). Bioremediation potential of *Trichoderma* species for metal (loid)s. In *Bioremediation of toxic metal (loid)s* (pp. 137–152). CRC Press.

Kour, D., Khan, S. S., Kour, H., Kaur, T., Devi, R., Rai, P. K., & Yadav, A. N. (2022). Microbe-mediated bioremediation: Current research and future challenges. *Journal of Applied Biology & Biotechnology, 10*(2), 6–24.

Kumar, D., Malik, S., Rani, R., Kumar, R., & Duhan, J. S. (2023). Behavior, risk, and bioremediation potential of heavy metals/metalloids in the soil system. *Rendiconti Lincei. Scienze Fisiche e Naturals, 34*(3), 809–831.

Kumar, K., Shinde, A., Aeron, V., Verma, A., & Arif, N. S. (2023). Genetic engineering of plants for phytoremediation: Advances and challenges. *Journal of Plant Biochemistry and Biotechnology, 32*(1), 12–30.

Kumar, N., & Shukla, P. (2024). Microalgal multiomics-based approaches in bioremediation of hazardous contaminants. *Environmental Research, 147*, 118135.

Kumar, P., Gupta, S. B., & Soni, R. (2019). Bioremediation of cadmium by mixed indigenous isolates *Serratia liquefaciens* BSWC3 and *Klebsiella pneumoniae* RpSWC3 isolated from industrial and mining affected water samples. *Pollution, 5*(2), 351–360.

Kumari, P., & Kumar, Y. (2021). Bioinformatics and computational tools in bioremediation and biodegradation of environmental pollutants. In *Bioremediation for environmental sustainability* (pp. 421–444). Elsevier.

Leitão, A. L. (2009). Potential of *Penicillium* species in the bioremediation field. *International Journal of Environmental Research and Public Health, 6*(4), 1393–1417.

Li, Y., Zhao, X., Zhong, Z., Ju, F., Liu, H., Lin, N., & Zha, X. (2024). Interaction of microalgae, nitrogen and arsenic in a water environment co-polluted by nitrogen and arsenic: Current knowledge and future perspectives. *Aquatic Ecology, 59*, 323–337.

Liu, C. J., Deng, S. G., Hu, C. Y., Gao, P., Khan, E., Yu, C. P., & Ma, L. Q. (2023). Applications of bioremediation and phytoremediation in contaminated soils and waters: CREST publications during 2018–2022. *Critical Reviews in Environmental Science and Technology, 53*(6), 723–732.

Liu, S., Li, W., Qiao, W., Wang, Q., Hu, Y., & Wang, Z. (2019). Effect of natural conditions and mining activities on vegetation variations in arid and semiarid mining regions. *Ecological Indicators, 103*, 331–345.

Lopez, A. M. Q., Silva, A. L. D. S., Maranhão, F. C. D. A., & Ferreira, L. F. R. (2022). Plant growth promoting bacteria: Aspects in metal bioremediation and phytopathogen management.

In *Microbial biocontrol: Sustainable agriculture and phytopathogen management* (Vol. 1, pp. 51–78). Springer.

Ma, B., Song, W., Zhang, X., Chen, M., Li, J., Yang, X., & Zhang, L. (2023). Potential application of novel cadmium-tolerant bacteria in bioremediation of cd-contaminated soil. *Ecotoxicology and Environmental Safety, 255*, 114766.

Ma, Y., Prasad, M. N. V., Rajkumar, M., & Freitas, H. J. B. A. (2011). Plant growth promoting rhizobacteria and endophytes accelerate phytoremediation of metalliferous soils. *Biotechnology Advances, 29*(2), 248–258.

Mahar, A., Wang, P., Ali, A., Awasthi, M. K., Lahori, A. H., Wang, Q., & Zhang, Z. (2016). Challenges and opportunities in the phytoremediation of heavy metals contaminated soils: A review. *Ecotoxicology and Environmental Safety, 126*, 111–121.

Mala, J. G. S. M., Takeuchi, S., Sujatha, D., & Mani, U. (2020). Microbial chromate reductases: Novel and potent mediators in chromium bioremediation-a review. *Applied Microbiology: Theory & Technology, 1*, 32–44.

Malaviya, P., & Singh, A. (2016). Bioremediation of chromium solutions and chromium containing wastewaters. *Critical Reviews in Microbiology, 42*(4), 607–633.

Malik, G., Arora, R., Chaturvedi, R., & Paul, M. S. (2021). Implementation of genetic engineering and novel omics approaches to enhance bioremediation: A focused review. *Bulletin of Environmental Contamination and Toxicology, 108*, 1–8.

Malik, S., Kishore, S., Shah, M. P., & Kumar, S. A. (2022). A comprehensive review on nanobiotechnology for bioremediation of heavy metals from wastewater. *Journal of Basic Microbiology, 62*(3–4), 361–375.

Mallikarjunaiah, S., Pattabhiramaiah, M., & Metikurki, B. (2020). Application of nanotechnology in the bioremediation of heavy metals and wastewater management. In *Nanotechnology for food, agriculture, and environment* (pp. 297–321). Elsevier.

Mansoor, S., Ali, A., Kour, N., Bornhorst, J., AlHarbi, K., Rinklebe, J., & Chung, Y. S. (2023). Heavy metal induced oxidative stress mitigation and ROS scavenging in plants. *Plants, 12*(16), 3003.

Martinelli, N., Olivieri, O., & Girelli, D. (2013). Air particulate matter and cardiovascular disease: A narrative review. *European Journal of Internal Medicine, 24*(4), 295–302.

Mathiyazhagan, N., & Natarajan, D. (2011). Bioremediation on effluents from magnesite and bauxite mines using *Thiobacillus* sp. and *pseudomonas* sp. *Journal of Bioremediation & Biodegradation, 2*(115), 2.

Mesa, J., Mateos-Naranjo, E., Caviedes, M. A., Redondo-Gómez, S., Pajuelo, E., & Rodríguez-Llorente, I. D. (2015). Endophytic cultivable bacteria of the metal bioaccumulator *Spartina maritima* improve plant growth but not metal uptake in polluted marshes soils. *Frontiers in Microbiology, 6*, 1450.

Mishra, S., Bharagava, R. N., More, N., Yadav, A., Zainith, S., Mani, S., & Chowdhary, P. (2019). Heavy metal contamination: An alarming threat to environment and human health. In *Environmental biotechnology: For sustainable future* (pp. 103–125). Springer.

Mishra, S., Dash, D., & Das, A. P. (2024a). Isolation and molecular characterization of polyester-degrading bacterial strains from municipal wastewater and their application in microplastic degradation in contaminated wastewater. *Geomicrobiology Journal*. https://doi.org/10.1080/01490451.2024.2376101

Mishra, S., Dash, D., & Das, A. P. (2024b). Microbial degradation of polyester microfibers using indigenously isolated bacterial strain Exiguobacterium sp. *CLEAN, 52*(12), e202300343. https://doi.org/10.1002/CLEN.202300343

Mishra, S., Lin, Z., Pang, S., Zhang, W., Bhatt, P., & Chen, S. (2021). Recent advanced technologies for characterization of xenobiotic-degrading microorganisms and microbial communities. *Frontiers in Bioengineering and Biotechnology, 9*, 632059.

Mitra, A., Chatterjee, S., & Gupta, D. K. (2017). Potential role of microbes in bioremediation of arsenic. In *Arsenic contamination in the environment: The issues and solutions* (pp. 195–213). Springer.

Mohanty, J., Dey, S., Bala, K., & Das, A. P. (2024). Recent advances in microplastic pollution for its sustainable management. In *Sustainable management of urban plastic waste through circular economic approaches* (pp. 27–48). Routledge. https://doi.org/10.1201/9781003465447-3

Mohanty, M., Tripathy, B., Koteswara Reddy, G., & Das, A. P. (2024). Greener technologies for establishing restraint over microplastic pollution. In *Sustainable microbial technology for synthetic and cellulosic microfiber bioremediation* (pp. 239–258). Springer. https://doi.org/10.1007/978-3-031-62600-5_13

Mohapatra, B., Malhotra, H., Saha, B. K., Dhamale, T., & Phale, P. S. (2022). Microbial metabolism of aromatic pollutants: High-throughput OMICS and metabolic engineering for efficient bioremediation. In *Current developments in biotechnology and bioengineering* (pp. 151–199). Elsevier.

Mohapatra, B., & Phale, P. S. (2021). Microbial degradation of naphthalene and substituted naphthalenes: Metabolic diversity and genomic insight for bioremediation. *Frontiers in Bioengineering and Biotechnology, 9*, 602445.

Mohsin, M., Zhu, Q., Naseem, S., Sarfraz, M., & Ivascu, L. (2021). Mining industry impact on environmental sustainability, economic growth, social interaction, and public health: An application of semi-quantitative mathematical approach. *PRO, 9*(6), 972.

Mondal, S., & Palit, D. (2022). *Challenges in natural resource management for ecological sustainability* (pp. 29–59). Elsevier.

Mourdikoudis, S., Pallares, R. M., & Thanh, N. T. (2018). Characterization techniques for nanoparticles: Comparison and complementarity upon studying nanoparticle properties. *Nanoscale, 10*(27), 12871–12934.

Newsome, L., & Falagán, C. (2021). The microbiology of metal mine waste: Bioremediation applications and implications for planetary health. *GeoHealth, 5*(10), e2020GH000380.

Norgate, T., & Jahanshahi, S. (2010). Low grade ores–smelt, leach or concentrate. *Minerals Engineering, 23*(2), 65–73.

Nouri, M., El Rasafi, T., & Haddioui, A. (2017). Levels of metals in soils of AIT Ammar iron mine, Morocco: Human health risks. *Acta Chemica Iasi, 25*(2), 127–144.

Nwankwegu, A. S., Zhang, L., Xie, D., Onwosi, C. O., Muhammad, W. I., Odoh, C. K., & Idenyi, J. N. (2022). Bioaugmentation as a green technology for hydrocarbon pollution remediation. Problems and prospects. *The Journal of Environmental Management, 304*, 114313.

Ojuederie, O. B., & Babalola, O. O. (2017). Microbial and plant-assisted bioremediation of heavy metal polluted environments: A review. *International Journal of Environmental Research and Public Health, 14*(12), 1504.

Okoye, A., Chikere, C., & Okpokwasili, G. (2019). Fungal population dynamics associated with active-phase of hydrocarbon degradation in oil-polluted soil. *Journal of Advances in Microbiology, 19*, 1–12.

Oladipo, O. G., Awotoye, O. O., Olayinka, A., Ezeokoli, O. T., Maboeta, M. S., & Bezuidenhout, C. C. (2016). Heavy metal tolerance potential of *aspergillus* strains isolated from mining sites. *Bioremediation Journal, 20*(4), 287–297.

Oladipo, O. G., Ezeokoli, O. T., Maboeta, M. S., Bezuidenhout, J. J., Tiedt, L. R., Jordaan, A., & Bezuidenhout, C. C. (2018). Tolerance and growth kinetics of bacteria isolated from gold and gemstone mining sites in response to heavy metal concentrations. *Journal of Environmental Management, 212*, 357–366.

Olafisoye, O. B., Adefioye, T., & Osibote, O. A. (2013). Heavy metals contamination of water, soil, and plants around an electronic waste dumpsite. *Polish Journal of Environmental Studies, 22*(5), 1431–1439.

Onifade, M., Zvarivadza, T., Adebisi, J. A., Said, K. O., Dayo-Olupona, O., Lawal, A. I., & Khandelwal, M. (2024). Advancing toward sustainability: The emergence of green mining technologies and practices. *Green and Smart Mining Engineering, 1*(2), 157–174.

Paliwal, V., Raju, S. C., Modak, A., Phale, P. S., & Purohit, H. J. (2014). *Pseudomonas putida* CSV86: A candidate genome for genetic bioaugmentation. *PLoS One, 9*(1), e84000.

Paluchamy, B., Mishra, D. P., & Panigrahi, D. C. (2021). Airborne respirable dust in fully mechanised underground metalliferous mines–generation, health impacts and control measures for cleaner production. *Journal of Cleaner Production, 296*, 126524.

Parida, S. K., Satpathy, A., Dalai, A., Kullu, S., Hota, S., & Mishra, S. (2024). Novel methods and techniques for the remediation of mining waste residues. In *Sustainable management of mining waste and tailings* (pp. 1–29). CRC Press.

Patel, R., Mugunthan, J., Singh, P., Mukherjee, S., & Koka, R. (2022). Microbial bioremediation and biodegradation of radioactive waste contaminated sites. In *Microbes and microbial biotechnology for green remediation* (pp. 733–746). Elsevier.

Pattnaik, S., Mohapatra, S., Pati, S., Dash, D., Devadarshini, D., Tanaya, K., & Samantaray, D. (2022). Microbial bioremediation of Cr (VI)-contaminated soil for sustainable agriculture. In *Microbial biodegradation and bioremediation* (pp. 395–407). Elsevier.

Petsas, A. S., & Vagi, M. C. (2019). Trends in the bioremediation of pharmaceuticals and other organic contaminants using native or genetically modified microbial strains: A review. *Current Pharmaceutical Biotechnology, 20*(10), 787–824.

Polettini, A., Gottardo, R., Pascali, J. P., & Tagliaro, F. (2008). Implementation and performance evaluation of a database of chemical formulas for the screening of pharmaco/toxicologically relevant compounds in biological samples using electrospray ionization-time-of-flight mass spectrometry. *Analytical Chemistry, 80*(8), 3050–3057.

Priyadarshanee, M., & Das, S. (2021). Biosorption and removal of toxic heavy metals by metal tolerating bacteria for bioremediation of metal contamination: A comprehensive review. *Journal of Environmental Chemical Engineering, 9*(1), 104686.

Priyadarshini, S., Jagatee, S., & Das, A. P. (2024). Synthetic fabrics and microfiber pollution–an assessment of their global impact. *Environmental Science and Engineering, 2333*, 137–157. https://doi.org/10.1007/978-3-031-51792-1_8

Priyadarsini, S., & Das, A. P. (2025). Lithium bioleaching: A review on microbial-assisted sustainable technology for lithium bio-circularity. *Journal of Water Process Engineering, 69*, 106744. https://doi.org/10.1016/j.jwpe.2024.106744

Raklami, A., Meddich, A., Oufdou, K., & Baslam, M. (2022). Plants—Microorganisms-based bioremediation for heavy metal cleanup: Recent developments, phytoremediation techniques, regulation mechanisms, and molecular responses. *International Journal of Molecular Sciences, 23*(9), 5031.

Ramakrishnan, B., Megharaj, M., Venkateswarlu, K., Sethunathan, N., & Naidu, R. (2011). *Mixtures of environmental pollutants: Effects on microorganisms and their activities in soils* (pp. 63–120). Springer.

Raman, M. (2024). Bioinformatics and computational tools. In *Fish structural proteins and its derivatives: Functionality and applications* (pp. 107–134). Singapore.

Rana, S., Handa, S., Aggarwal, Y., Puri, S., & Chatterjee, M. (2023). Role of Candida in the bioremediation of pollutants: A review. *Letters in Applied Microbiology, 76*(9), ovad103.

Rathod, S. V., Saras, P., & Gondaliya, S. M. (2025). Environmental pollution: Threats and challenges for management. In *Eco-restoration of polluted environment* (pp. 1–34). CRC Press.

Ray, S. K., Dey, S., Das, A. P., & John, A. (2024). Microplastic waste management through greener technologies: A review. In *Sustainable management of urban plastic waste through circular economic approaches* (pp. 237–261). Routledge. https://doi.org/10.1201/9781003465447-14

Reddy Koteswara, G., Bhargavi, S., Yamuna, D., Anjani Devi, M., Jeevana Lakshmi Kumar, G., Manas, T., Adarsh, S., Bhavya Sai, V., Nath, H., Alreshidi, M. A., Togun, H., Ram, D. K., & Das, A. P. (2025). Recent advancements in the remediation of marine hydrocarbon and Plastic pollutants. *Environmental Science and Engineering, 218*, 127–142. https://doi.org/10.1007/978-3-031-81615-4_7

Reena, R., Majhi, M. C., Arya, A. K., Kumar, R., & Kumar, A. (2012). BioRadBase: A database for bioremediation of radioactive waste. *African Journal of Biotechnology, 11*(35), 8718–8721.

Rodríguez-Berbel, N., Ortega, R., Lucas-Borja, M. E., Solé-Benet, A., & Miralles, I. (2020). Long-term effects of two organic amendments on bacterial communities of calcareous mediterranean soils degraded by mining. *Journal of Environmental Management, 271*, 110920.

Rodriguez-Franco, C., & Page-Dumroese, D. S. (2021). Woody biochar potential for abandoned mine land restoration in the US: A review. *Biochar, 3*(1), 7–22.

Román-Ponce, B., Ramos-Garza, J., Arroyo-Herrera, I., Maldonado-Hernández, J., Bahena-Osorio, Y., Vásquez-Murrieta, M. S., & Wang, E. T. (2018). Mechanism of arsenic resistance in endophytic bacteria isolated from endemic plant of mine tailings and their arsenophore production. *Archives of Microbiology, 200*, 883–895.

Romero, M. F., Gallego, D., Lechuga-Jiménez, A., Martínez, J. F., Barajas, H. R., Hayano-Kanashiro, C., & Alcaraz, L. D. (2021). Metagenomics of mine tailing rhizospheric communities and its selection for plant establishment towards bioremediation. *Microbiological Research, 247*, 126732.

Saha, L., Tiwari, J., Bauddh, K., & Ma, Y. (2021). Recent developments in microbe–plant-based bioremediation for tackling heavy metal-polluted soils. *Frontiers in Microbiology, 12*, 731723.

Saharan, B. S., Chaudhary, T., Mandal, B. S., Kumar, D., Kumar, R., Sadh, P. K., & Duhan, J. S. (2023). Microbe-plant interactions targeting metal stress: New dimensions for bioremediation applications. *J Xenobiotics, 13*(2), 252–269.

Sahu, A., Dey, S., Mallick, K., Priyadarsini, S., Behera, I. D., & Das, A. P. (2024). A review on advanced technology for sustainable management of synthetic microplastic waste. In *Sustainable management of urban plastic waste through circular economic approaches* (pp. 262–279). Routledge. https://doi.org/10.1201/9781003465447-15

Sajjad, W., Zheng, G., Din, G., Ma, X., Rafiq, M., & Xu, W. (2019). Metals extraction from sulfide ores with microorganisms: The bioleaching technology and recent developments. *Transactions of the Indian Institute of Metals, 72*, 559–579.

Salam, U., Ullah, S., Tang, Z. H., Elateeq, A. A., Khan, Y., Khan, J., & Ali, S. (2023). Plant metabolomics: An overview of the role of primary and secondary metabolites against different environmental stress factors. *Life, 13*(3), 706.

Samuel, J., Paul, M. L., Ravishankar, H., Mathur, A., Saha, D. P., Natarajan, C., & Mukherjee, A. (2013). The differential stress response of adapted chromite mine isolates *Bacillus subtilis* and *Escherichia coli* and its impact on bioremediation potential. *Biodegradation, 24*, 829–842.

Santhiyasri, S., Romauld, S. I., Vivek, P., & Thiruvengadam, S. (2023). Proteomics monitoring of microbes in contaminated environments. In *Metagenomics to bioremediation* (pp. 723–735). Academic Press.

Saravanan, A., Kumar, P. S., Karishma, S., Vo, D. V. N., Jeevanantham, S., Yaashikaa, P. R., & George, C. S. (2021). A review on biosynthesis of metal nanoparticles and its environmental applications. *Chemosphere, 264*, 128580.

Saravanan, A., Kumar, P. S., Vo, D. V. N., Jeevanantham, S., Karishma, S., & Yaashikaa, P. R. (2021). A review on catalytic-enzyme degradation of toxic environmental pollutants: Microbial enzymes. *Journal of Hazardous Materials, 419*, 126451.

Sarma, U., Hoque, M. E., Thekkangil, A., Venkatarayappa, N., & Rajagopal, S. (2024). Microalgae in removing heavy metals from wastewater–an advanced green technology for urban wastewater treatment. *Journal of Hazardous Materials Advances, 15*, 100444.

Sarwan, J. (2022). Role of isolates of *bacillus* species for biodegradation of multiple contaminants. *Journal of Environmental Management, 1*(2), 292–298.

Satapathy, S., Kar, D., Priyadarsini, S., Das, A. P., & Kuanar, A. (2024). Application of microorganisms in emerging waste management strategies and remediation of synthetic and cellulosic microfibers. In *Sustainable microbial technology for synthetic and cellulosic microfiber bioremediation* (pp. 183–206). Springer. https://doi.org/10.1007/978-3-031-62600-5_10

Sharma, J., Goutam, J., Dhuriya, Y. K., & Sharma, D. (2021). Bioremediation of industrial pollutants. In *Microbial rejuvenation of polluted environment* (Vol. 2, pp. 1–31). Springer.

Sharma, K., Devi, P., Dey, S. R., & Kumar, P. (2024). Mercury phytovolatilization: An overview of the mechanism and mitigation. In *Role of green chemistry in ecosystem restoration to achieve environmental sustainability* (pp. 325–331). Elsevier.

Sharma, P., Bano, A., Singh, S. P., Sharma, S., Xia, C., Nadda, A. K., & Tong, Y. W. (2022). Engineered microbes as effective tools for the remediation of polyaromatic aromatic hydrocarbons and heavy metals. *Chemosphere, 306*, 135538.

Sharma, P., Kumar, S., & Pandey, A. (2021). Bioremediated techniques for remediation of metal pollutants using metagenomics approaches: A review. *Journal of Environmental Chemical Engineering, 9*(4), 105684.

Sharma, P., Sirohi, R., Tong, Y. W., Kim, S. H., & Pandey, A. (2021). Metal and metal (loids) removal efficiency using genetically engineered microbes: Applications and challenges. *Journal of Hazardous Materials, 416*, 125855.

Shi, C., Cheng, L., Yu, Y., Chen, S., Dai, Y., Yang, J., & Geng, N. (2024). Multi-omics integration analysis: Tools and applications in environmental toxicology. *Environmental Pollution, 360*, 124675.

Singh, A., Kumar, V., & Srivastava, J. N. (2013). Assessment of bioremediation of oil and phenol contents in refinery waste water via bacterial consortium. *Journal of Petroleum & Environmental Biotechnology, 4*(3), 1–4.

Singh, A. K., Bilal, M., Iqbal, H. M., & Raj, A. (2021). Trends in predictive biodegradation for sustainable mitigation of environmental pollutants: Recent progress and future outlook. *Science of the Total Environment, 770*, 144561.

Singh, N. S., Sharma, R., & Singh, D. K. (2019). Identification of enzyme (s) capable of degrading endosulfan and endosulfan sulfate using *in silico* techniques. *Enzyme and Microbial Technology, 124*, 32–40.

Singh, R., Sharma, A., Goswami, P., Pradhananga, D., Aryal, D., Pradhanang, S. M., & Kumar, R. (2023). Phytoremediation of organic contaminants: An eco-friendly approach-based application of aquatic macrophytes. In *Aquatic macrophytes: Ecology, functions and services* (pp. 175–205). Singapore.

Soman, R., Kavitha, M. H., & Shaji, H. (2022). Metagenomics: A genomic tool for monitoring microbial communities during bioremediation. In *Microbes and microbial biotechnology for green remediation* (pp. 813–821). Elsevier.

Spjuth, O., Capuccini, M., Carone, M., Larsson, A., Schaal, W., Novella, J. A., & Lampa, S. (2020). Approaches for containerized scientific workflows in cloud environments with applications in life science. *F1000Research, 10*, 513.

Srivastava, P. K., Vaish, A., Dwivedi, S., Chakrabarty, D., Singh, N., & Tripathi, R. D. (2011). Biological removal of arsenic pollution by soil fungi. *Science of the Total Environment, 409*(12), 2430–2442.

Talukder, P., Ray, R., Sarkar, M., Das, A., & Chakraborty, S. (2024). Adverse effects of mining pollutants on terrestrial and aquatic environment and its remediation. *Environmental Quality Management, 33*(4), 595–610.

Techtmann, S. M., & Hazen, T. C. (2016). Metagenomic applications in environmental monitoring and bioremediation. *Journal of Industrial Microbiology & Biotechnology, 43*(10), 1345–1354.

Tiwary, B. N., Das, R., & Paul, V. (2022). Bioremediation technology: A cumulative study on microbial bioremediation of heavy metals, aromatic hydrocarbons, acrylamide, and polyacrylamide. In *Industrial microbiology and biotechnology* (pp. 589–623). Springer.

Torimiro, N., Daramola, O. B., Oshibanjo, O. D., Otuyelu, F. O., Akinsanola, B. A., Yusuf, O. O., & Omole, R. K. (2021). Ecorestoration of heavy metals and toxic chemicals in polluted environment using microbe-mediated nanomaterials. *International Journal of Environmental Bioremediation & Biodegradation, 9*(1), 8–21.

Tripathy, B., Dey, S., Dubey, N. K., Lo, W.-C., & Das, A. P. (2024). *An omics-based approach for managing the risk of COVID-19-generated wastes* (pp. 51–66). Springer. https://doi.org/10.1007/978-3-031-50840-0_4

Tyagi, B., & Kumar, N. (2021). Bioremediation: Principles and applications in environmental management. In *Bioremediation for environmental sustainability* (pp. 3–28). Elsevier.

Upadhyay, N., Vishwakarma, K., Singh, J., Mishra, M., Kumar, V., Rani, R., & Sharma, S. (2017). Tolerance and reduction of chromium (VI) by *bacillus* sp. MNU16 isolated from contaminated coal mining soil. Front. *Plant Science, 8*, 778.

Upadhyay, S. K., Rani, N., Kumar, V., Mythili, R., & Jain, D. (2023). A review on simultaneous heavy metal removal and organo-contaminants degradation by potential microbes: Current findings and future outlook. *Microbiological Research, 273*, 127419.

Varsha, K., Kirthana, R., & Rajakumari, K. (2022). Metagenomics tools for assessment of microbial diversity in bioremediation: A novel statistical approach. In V. Kumar & I. S. Thakur (Eds.), *Omics insights in environmental bioremediation*. Springer.

Vázquez-Núñez, E., Molina-Guerrero, C. E., Peña-Castro, J. M., Fernández-Luqueño, F., & de la Rosa-Álvarez, M. G. (2020). Use of nanotechnology for the bioremediation of contaminants: A review. *PRO, 8*(7), 82.

Verâne, J., Dos Santos, N. C., da Silva, V. L., de Almeida, M., de Oliveira, O. M., & Moreira, Í. T. (2020). Phytoremediation of polycyclic aromatic hydrocarbons (PAHs) in mangrove sediments using *Rhizophora mangle. Marine Pollution Bulletin, 160*, 111687.

Wang, Y., Song, J., Zhao, W., He, X., Chen, J., & Xiao, M. (2011). In situ degradation of phenol and promotion of plant growth in contaminated environments by a single *Pseudomonas aeruginosa* strain. *Journal of Hazardous Materials, 192*(1), 354–360.

Wen, X., Du, C., Zeng, G., Huang, D., Zhang, J., Yin, L., & Wan, J. (2018). A novel biosorbent prepared by immobilized *bacillus licheniformis* for lead removal from wastewater. *Chemosphere, 200*, 173–179.

Widyati, E., Irianto, R. S., Dewi, T., Maftu'ah, E., Hartatik, S. E., & Parray, J. A. (2024). Emerging strategies for engineering microbial communities to augment bioremediation in Indonesia. In *Microbiome-assisted bioremediation* (pp. 401–427). Academic Press.

Worlanyo, A. S., & Jiangfeng, L. (2021). Evaluating the environmental and economic impact of mining for post-mined land restoration and land-use: A review. *Journal of Environmental Management, 279*, 111623.

Wu, T., Xu, J., Xie, W., Yao, Z., Yang, H., Sun, C., & Li, X. (2018). *Pseudomonas aeruginosa* L10: A hydrocarbon-degrading, biosurfactant-producing, and plant-growth-promoting endophytic bacterium isolated from a reed (*Phragmites australis*). *Frontiers in Microbiology, 9*, 1087.

Xia, J., & Wishart, D. S. (2016). Using MetaboAnalyst 3.0 for comprehensive metabolomics data analysis. *Current Protocols in Bioinformatics, 55*(1), 14–10.

Yadav, R., Arora, P., Kumar, S., & Chaudhury, A. (2010). Perspectives for genetic engineering of poplars for enhanced phytoremediation abilities. *Ecotoxicology, 19*, 1574–1588.

Yin, K., Wang, Q., Lv, M., & Chen, L. (2019). Microorganism remediation strategies towards heavy metals. *Chemical Engineering Journal, 360*, 1553–1563.

Yogalakshmi, K. N., Das, A., Rani, G., Jaswal, V., & Randhawa, J. S. (2020). Nano-bioremediation: A new age technology for the treatment of dyes in textile effluents. In *Bioremediation of industrial waste for environmental safety: Volume I: Industrial waste and its management* (pp. 313–347). Springer.

Yu, H., Zahidi, I., & Liang, D. (2023). Spatiotemporal variation of vegetation cover in mining areas of Dexing City. *China. Environ Res, 225*, 115634.

Yu, Q., & Cui, O. (2023). Enrichment and remediation of uranium by microorganisms: A review. *Open Journal of Environmental Biology, 8*(1), 20–38.

Yunusa, Y. R., & Umar, Z. D. (2022). Effective microbial bioremediation via the multi-omics approach: An overview of trends, problems and prospects. *UMYU Journal of Microbiology Research, 6*(1), 127–145.

Zehra, A., Dubey, M. K., Meena, M., Aamir, M., Patel, C. B., & Upadhyay, R. S. (2018). Role of *Penicillium* species in bioremediation processes. In *New and future developments in microbial biotechnology and bioengineering* (pp. 247–260). Elsevier.

Zeng, P., Guo, Z., Cao, X., Xiao, X., Liu, Y., & Shi, L. (2018). Phytostabilization potential of ornamental plants grown in soil contaminated with cadmium. *International Journal of Phytoremediation, 20*(4), 311–320.

Zhou, B., Zhang, T., & Wang, F. (2023). Microbial-based heavy metal bioremediation: Toxicity and eco-friendly approaches to heavy metal decontamination. *Applied Sciences, 13*(14), 8439.

Zine, H., Hakkou, R., Papazoglou, E. G., Elmansour, A., Abrar, F., & Benzaazoua, M. (2024). Revegetation and ecosystem reclamation of post-mined land: Toward sustainable mining. *International Journal of Environmental Science and Technology, 21*, 5.

Zulfiqar, U., Haider, F. U., Maqsood, M. F., Mohy-Ud-Din, W., Shabaan, M., Ahmad, M., & Shahzad, B. (2023). Recent advances in microbial-assisted remediation of cadmium-contaminated soil. *Plants, 12*(17), 3147.

Bioremediation as a Greener Alternative for the Sustainable Recycling of Mining Waste

Smruti Smaranika Nayak, Sudeshna Dey, Amel Gacem, Maha Awjan Alreshidi, Krishna Kumar Yadav, and Alok Prasad Das

1 Introduction

Mining is an activity that must be conducted to extract important minerals and metals serving a significant proportion of various industries ranging from construction to the technological sectors. In many cases, it has been associated with significant environmental challenges such as the generation of tremendous volumes of mining waste, habitat destruction, air, water, and soil pollution. This constant quest for effective, environmentally responsible waste management techniques provoked interest in environmental-friendly remediation technologies and resulted in the bioremediation of waste products from mines, which has vast potential (Singh et al. 2016). Bioremediation involves using living organisms, or the activities within these organisms, primarily microorganisms, to destroy, detoxify, or alter toxic

S. S. Nayak · A. P. Das (✉)
Department of Life Sciences, Rama Devi Women's University, Bhubaneswar, Odisha, India

S. Dey
Department of Life Sciences, Rama Devi Women's University, Bhubaneswar, Odisha, India

Centre for Biotechnology, Siksha O Anusandhan University, Bhubaneswar, Odisha, India

A. Gacem
Department of Physics, Faculty of Sciences, University 20 Août 1955, Skikda, Algeria

M. A. Alreshidi
Department of Chemistry, College of Science, University of Ha'il, Ha'il, Saudi Arabia

K. K. Yadav
Department of Environmental Science, Parul Institute of Applied Sciences, Parul University, Vadodara, Gujarat, India

Environmental and Atmospheric Sciences Research Group, Scientific Research Center, Al-Ayen University, Nasiriyah, Iraq

© The Author(s), under exclusive license to Springer Nature Switzerland AG 2026

A. P. Das (ed.), *Bioremediation Technology for Sustainable Mining Waste Management*, Sustainable Environmental Waste Management Strategies,
https://doi.org/10.1007/978-3-032-03468-7_11

wastes associated with mining in order to make them harmless or useful materials. Mining activities can cause significant volumes of waste (i.e., tailings and waste rock), which must be managed and dealt with to avoid environmental contamination. According to the 2019 research, more than 100 billion tons of solid waste debris are produced yearly through the 3500 major mining enterprises worldwide (Das et al., 2025; Das and Ghosh 2023a; Mohanty, Tripathy, et al. 2024; Priyadarshini et al., 2024; Reddy Koteswara et al., 2025; Sahu et al., 2024). In Sweden, the mining industry's contribution to waste production is over 80% annually. Mining produces enormous quantities of waste products as tailings, slag, overburden, and AMD. Tailings are crushed rock that remains after the removal of minerals, which often includes toxic heavy metals like arsenic, cadmium, and lead. AMD is feared especially when sulfide minerals oxidize, creating acidic conditions and results in the leaching of several metals into the nearby water bodies. These contaminants have some serious challenges attached to them in terms of ecosystems since they affect biodiversity, water quality as well as soil fertility (Lottermoser, 2010). Traditionally, waste management in the form of landfilling or chemical neutralization tends to be energy-intensive, costly, and unsustainable. Such traditional methods may lead to secondary pollution.

Bioremediation offers an environmental-friendly, renewable option that should well be able to meet these challenges by utilizing the naturally occurring biological processes for treatment of the hazardous waste (Das et al. 2023, 2021). Bioremediation exploits the biochemical capabilities of microorganisms, such as bacteria, fungi, and archaea, which degrade or transform contaminants. In doing so, microbes break up complex organic material, immobilize metals through biosorption or precipitation, and can recover valuable resources such as copper and gold. It is sustainable because it minimizes harmful chemicals and minimizes the carbon footprint of waste management operations (Gadd, 2010). Bioremediation aligns with the concepts of green chemistry and the circular economy: it advocates for the valorization of waste and the recovery of resources (Bala et al., 2024; Das, Dey, Bala, et al. 2024; Das, Behera, and Das 2024; Dey, Dash, & Das, 2024; John & Das, 2024; Mohanty, Dey, et al. 2024). For example, one of the bioremediation techniques, bioleaching, employs acidophilic microorganisms to extract metals from low-grade ores or tailings; therefore, there is a reduction of waste as well as the recovery of the resources. Such an application has already proved effective in big-scale copper and gold mining industries (Rawlings & Johnson, 2007). The new global drive toward sustainability is a shift in the approach of waste management in the mining industry. Bioremediation can be integrated into mining operations to significantly reduce the ecological footprint. For instance, in the case of having microbial consortia for the treatment of acid mine drainage, its adverse effects on the environment are reduced besides restored pH and concentrations of metals in water to stable ranges within the environments. Besides, there is phytoremediation. It is one form of bioremediation subgroup that incorporates the use of plants in absorbing and stabilizing contaminants. This would improve soil quality and inhibit further degradation of mining-affected lands (Mendez & Maier, 2008). A study indicated that mining has resulted in a loss of over 40,000 km^2 of land in China, and only the derelict land once occupied by the mine registers an increase of almost 330 km^2 (Das et al., 2020).

Advanced Science in Bioremediation Advanced bioremediation techniques such as the use of genetically engineered microbes create a new platform for mining wastes to be handled. Engineered microbes have become more capable, from metal detoxification to organic pollutant degradation to even synthesis of valuable compounds. Moreover, knowledge on microbial communities in mining environments is also generated through studies on microbial genomics and metagenomics, which now enables the designing of specific bioremediation strategies (Singh, Kumar, and Singh 2016). The types of mining waste include overburden, tailings, and other byproducts, with coal mining being a major contributor. Brazil alone produced 3.6 billion tons of solid mining debris between 2008 and 2019, accounting for 62% of the global mass of nonfuel minerals extracted in 2006. Mining copper produces about 110 tons of waste every ton of ore, while mining coal produces about 0.4 tons of waste per ton of coal. Another great challenge is that bioremediation processes need to be scaled up from laboratory applications to field uses. The future of bioremediation in the management of mining waste is in multilateral collaboration among scientists, engineers, policymakers, and the business community. Advances in biotechnology adequate enough with a suitable policy framework shall unlock mainstream penetration. Public–private partnerships may further stimulate innovation and investment in research in ways that ensure openness to accessibility and economically viable research areas for mining companies worldwide (Das et al., 2011; Das & Ghosh, 2023d, 2023f; Ghosh & Das, 2018; Mohanty et al., 2016). Climate change and tightened environmental concerns in addition to stakeholders' raising consciousness are strong imperatives for more "green technologies in mining processes." Bioremediation for instance over the environmental consideration even offers benefits to the organization from an economic aspect such as savings and recycling marketable byproduct (Rohwerder et al., 2003).

The effectiveness of the process depends on the type of contaminant, nature of environment, and adaptability of the microbial community. For example, extreme pH or temperature limits microbial activities in some mining sites; therefore, selection or engineering extremophiles for such activities becomes necessary. With environmental degradation and scarcity of resources being two-fold goals, bioremediation is a sustainable, environmentally friendly, and low-cost technique for the management of mining wastes. The novel technology exploited the use of biological activities to decrease the adverse impacts of mining practices while becoming part of the global plan toward sustainable development. As research and innovation grow with this sector, bioremediation can alter the outlook in the management of mining waste, thereby stimulating the greener culture of mining. While mining is an activity crucial to the extraction of useful raw materials, it produces considerable amounts of waste, and its impact is anything but insignificant. To formulate options for sustainable management, first, the identification of the types of mining wastes with their properties and thus their environmental impact is necessary.

1.1 Examples of Various Mining Wastes

Mining activities produce various types of wastes depending on the ore being mined, the processes involved in processing, and the site and geology of the mine. The wastes differ in their composition, toxicity, and environmental impacts. Here are some examples of various mining wastes, along with actual applications and their impacts.

One of the most critical types of mining wastes is tailings, which are residues that remain after the extraction of valuable metals or minerals from tailings ores. Tailings are usually impounded in tailings dams or ponds, which require a lot of space and handling. Tailings are, therefore, an important environmental issue mainly because of their large quantities and possibly toxic nature (Priyadarsini & Das, 2025b, 2026). The tailings usually contain finely ground materials. Mixing them with water creates a liquid paste. Whether the tailings have toxic substances such as heavy metals, like arsenic, cadmium, or mercury, and residues of reagents used in the processing process, like cyanide or sulfuric acid, depends on the type of ore being processed. The chemicals are then released into the soil and water bodies around the area, thereby contaminating the ecosystem (Das et al., 2015a, 2015b; Das & Ghosh, 2018; Ghosh et al., 2023).

Tailings carry high ecological risks. Tailings dam failures, as the Brumadinho tragedy in Brazil demonstrated in 2019, have been catastrophic in terms of adverse impacts resulting from bad waste management. While direct loss of life and ecosystem degradation, tailings can lead to secondary adverse effects on water and soil through heavy metals leaching and AMD (Lyu et al., 2019). The full extent of tailings pollution has been seen in the Ok Tedi gold and copper mine. Heavy metals like copper and zinc, which the tailings produced by this mine contained, were released directly into the Ok Tedi River. As a result, significant deforestation, riverbed sedimentation, and harmful effects on aquatic ecosystems have devastated biodiversity and local communities (Kirsch, 2014). Despite all the dangers of tailings, management often practices very traditional techniques including wet storage within tailings ponds. Dry stacking and backfilling mined out areas using the employment of tailings is a relatively recent development that promises a lot. The study on the remediation of tailings by microbes is fairly hopeful for the abatement of impacts resulting from tailings (Johnson & Hallberg, 2005).

Overburden is soil, rock, and vegetative materials that have to be removed to provide access to minerals buried below them. Although overburden is nontoxic per se, its removal drastically alters the original ecosystem and poses rehabilitation challenges to lands involved in removal activities. Overburden is primarily made up of material that cannot be mined and is left in the mining site. In some instances, overburden can contain sulfide minerals that oxidize upon exposure to air and water, causing AMD. Overburden displacement leads to deforestation, habitat destruction, and increased erosion, further exacerbating environmental concerns (Singh, Kumar, and Singh 2016). There will be minimal influence from the activities of overburden as appropriate removal and backfilling of the material is ensured. Sometimes, overburden can be recycled as a building material or even reutilized in restoring the lands. Techniques involving bioremediation like sowing of flora or the microbial consortium are expedient means for stabilization and reclamation of overburdened lands (Mendez & Maier, 2008). In 2014, the collapse of the tailings dam at Mount Polley mine left 25 million m^3 of tailing and contaminated water in the lakes and rivers near, which caused a huge ecological damage since it revealed different risks during the storage of tailings (Macklin et al. 2003). Overburden removal in the Jharia coalfields has committed crimes of major deforestation, soil erosion, and destruction of arable land. Rock and soil to be removed to expose the coal seams are deposited in large unstable heaps. Such heaps invite frequent landslide derailments. This dilutes the arable land for an impoverished population to farm (Saini & Raina, 2016).

Waste rock is all the material excavated in a mining operation that has not enough mineral value to be processed. Like overburden, waste rock is generally piled around large areas near the mine, posing monumental environmental challenges. Waste rock is heterogeneous, containing different contents of sulfide minerals; thus it poses a potential for AMD when not appropriately managed. The sulfide oxidation in waste rock results in the generation of sulfuric acid, which mobilizes heavy metals into the ground and surface waters (Rohwerder et al., 2003). Today, the waste rock has its controlled placement so as to minimize environmental exposure, and barriers that reduce AMD formation exist. Newer approaches for bioremediation include using sulfate-reducing bacteria, and possibly neutralization of acid production along with immobilization of heavy metals (Gadd, 2010). Bingham Canyon Mine is one of the biggest open-pit mines in the world. The enormous volumes of waste rock generated from this mine are dumped at the long dumps on either side of the mine. The slag from this mine has been observed to leach copper and other metals into surrounding water bodies, thus contributing to the contamination of the ground water (Parsons et al., 2001).

Acid mine drainage forms one of the worst forms of environmental problems affecting mining. It forms when mineral is exposed to air and water, which makes it to react with the formation of sulfuric acid to dissolve the metals and release them to the ecosystems. Forming AMD is a natural process that mining activities exacerbate. Once it has started, AMD can continue for decades or even centuries. Therefore, the long-term challenges for mining waste management relate to AMD. AMD is extremely acidic, often with a pH below 3, and commonly contains elevated levels of iron, aluminum, and other heavy metals. The ecological impacts of AMD are immense. Acidity causes destruction to aquatic life and biodiversity and affects the quality of water adversely by making it inappropriate for human intent or for irrigational purposes. In addition, contamination in soils further limits lands from being reclaimed (Nordstrom et al., 2015). Although lime neutralization is the primary conventional treatment used for AMD, it is equally expensive and generates secondary wastes. Other remediation processes, including constructed wetlands and sulfate-reducing bacteria, have been proposed to neutralize acidity and stimulate metal precipitation to treat AMD (Sheoran et al., 2010). The Sudbury region in Ontario is highly contaminated with nickel and copper smelting. Arsenic and lead are released into the slag from such a process. Impoundment failure of such a slag, results in contaminations through soil and water; however, it renders the contaminated environment open to reclamation (Moncur et al., 2006). The Berkeley Pit is one of the biggest acid mine drainage examples—a former copper mine. It fills with highly acidic water, having a pH of ~2.5, dissolved metals that include copper, zinc, and arsenic. These waters are a serious threat to both human health and the environment and need constant treatment to avoid penetration into adjacent water systems (Davis & Ashenberg, 1989).

These are the by-products of smelting ore to extract metal. The waste, or residue, is commonly referred to as slag. Most of the slags do not react. However, some contain remnant metals and toxic elements. Slags can leach these into the environment. It is usually heavy and glass-like in appearance. Its use can be conceived in construction, such as aggregates or cement additives (Parida et al., 2026). Slag that contains a high level of heavy metals or radionuclides requires proper disposal with proper monitoring. The exposures of leachates from the slag could have different effects on ecosystems and human health. Recycling of slags for industrial purposes is thus one of the feasible alternatives and might be applied after ensuring that the toxicity of leachate is very low and responsibly managed (Lottermoser, 2010). These include tailings, waste rock, and residues from the processing of ores containing radionuclides such as uranium-238 and thorium-232. It may give radioactive contamination that lasts tens of thousands of years, calling for long-term storage or containment. Exposures to mining wastes with radon increase risks for cancer and other adverse health impacts. Moreover, contamination of water and soil with radionuclides makes this worse, therefore there is the need for waste disposal and remediation (International Atomic Energy Agency 2010).

Cyanide is used abundantly in the extraction processes of gold and silver; its toxicity can strongly be an environmental and safety hazard. These cyanide-contaminated sludges are highly soluble and hence readily mobile, and thus a serious threat to aquatic environments. Cyanide is spontaneously biodegradable. However, this biodegradation might be relatively slow and not complete, leading to evidence of an operating process to regulate and monitor contamination. The Baia Mare cyanide spill in Romania 2000 illustrated the devastating consequence of poor waste management of cyanide. This polluted more than 2000 km of rivers and also caused severe ecological damage (Kováčová et al. 2002). Investigations into cyanide-free extractive processes as well as bioremediation technologies, for example, microbial degradation of cyanide can be considered the most viable remediation methods in order to prevent the devastating consequences of cyanide (Alkorta et al. 2003). In the year 2000, a collapse of tailings dam from a gold mine in Baia Mare resulted in releasing about 100,000 m^3 of cyanide-bearing water into the Danube River system. The unfortunate event was followed by deaths of aquatic flora and fauna, adulteration of sources of potable water supplies, and wrought grave ecological as well as societal impacts in more than one country (Kováčová et al. 2002).

Mining waste is highly diverse in both composition and environmental impact, from inert overburden to hazardous materials like AMD and cyanide-laden residues.

It is presently being managed effectively with a judicious mix of traditional engineering practice combined with innovative bioremediation techniques. Most critical is an understanding of the nature and risks pertaining to different types of waste, which the following chapter will illustrate. Bioremediation, with its great potential in recovery of resources and restoration of the environment, is among the solutions that can be crucial in these efforts (Das & Ghosh, 2023g; Ghosh et al., 2016, 2020; Ghosh & Das, 2017).

Jaduguda uranium mines in India have resulted in the generation of radioactive wastes as tailings and waste rock. Lack of proper containment of the wastes has resulted in contamination of the ground water and high health risks in the form of cancer and genetic defects for the inhabitants of the region. If mining has occurred in Goa, then this operation generates a lot of dust through blasting and transporting the ore. The dust particles contain iron and other pollutants, which have worsened pollution, impaired agriculture, and led to respiratory health issues for nearby residents.

The examples illustrated above are the types of mining wastes encountered, which present very different challenges. Each-from tailings to waste rock, radioactive wastes and cyanide-laden wastes-calls for special management strategies to combat their adverse impacts on the environment and society. Bioremediation has been much promising as a remedy to most of these problems, and it is very much an activity wherein the goals of environmental sustainability can be achieved without diminishing economic feasibility (Bal et al., 2018; Dey, Dash, Ray, et al., 2024; Ghosh et al., 2017; Mishra et al., 2022, 2023, 2024). Research and industry collaboration will prove to be very instrumental in coping with this increasingly serious problem of mining wastes. Table 1 represents mining waste types and their recycling applications.

Table 1 Mining waste types and their recycling applications

Mining waste type	Composition	Recycling potential	Application
Tailings	Fine-grained residues from ore processing, often containing metals and minerals	Recovery of residual metals, use in construction materials	Cement, bricks, road base
Slag	Byproduct of metal smelting, rich in silica, calcium, and iron	Aggregate for construction, metal recovery	Asphalt, concrete, soil amendments
Waste rock	Unwanted rock removed during mining, may contain trace metals	Crushed stone for road construction, aggregate in concrete	Roads, embankments, land reclamation
Acid mine drainage (AMD)	Sulfide minerals reacting with water and oxygen, producing acidic water with dissolved metals	Metal recovery, neutralization, water treatment	Metal extraction, water purification
Mine dust and fine particles	Fine airborne particles generated during mining operations	Recycling into industrial fillers, soil conditioners	Paints, plastics, cement

2 Bioremediation of Mining Wastes

Bioremediation is said to be a multi-disciplinary application and integrates biological approaches which are mainly constituted by a biological process comprising microbial activities for degradation, detoxification, or alteration of harmful elements into less or more useful but nontoxic and non-harmful forms. Bioremediation is based upon the natural ability of microorganisms where their metabolic potentials rely on indigenous mechanisms that microbes have natively designed and set for survival in poisoned environments. This chapter explores the fundamental principles and mechanisms of practical applications for the bioremediation approach to mining waste management. The electron acceptors for the anaerobic bacteria include those from nitrate or sulfate whenever the level of oxygen is scarce (Vidali, 2001). Bioremediation provides an environmentally friendly approach through which heavy metal contamination existing in mining waste is controlled in the presence of microorganisms or plants. The work of immobilizing or even extracting metals from contaminated soils or water occurs due to these organisms or flora. This process is a major one especially in dealing with mining wastes where old methods such as chemical treatment are expensive and resource-exploiting. Most strategies for mining waste management generally involve the use of the strategies of bioleaching and biosorption (Das, Dey, & Das, 2024; Das & Ghosh, 2022; Dey et al., 2023; Priyadarshini and Das 2024; Priyadarsini & Das, 2024, 2025a). Bioleaching utilizes specific bacteria and fungi, which will convert the metals to their soluble state, recoverable and then further processed or treated (Das, Dash, et al. 2023). Bioleaching microorganisms such as *Acidithiobacillus ferrooxidans* make the oxidation of metal ions, in which metals dissolve the residues left at the mine sites. On the other hand, the strategy in biosorption involves the use of non-living microbial biomass or plants absorbing and immobilizing the metal in wastewater. Both are useful in controlling the toxic heavy metals present in the mining effluents including cadmium, lead, arsenic, and copper, while a quantity of pollution is diminished along with the recovery of resources. There are broad ways of categorizing bioremediation techniques, including microbial remediation, phytoremediation, and mycoremediation. Different biological systems are used to remove or stabilize metal contaminants in mining waste.

2.1 *Environmental Conditions*

Alongside are pH, temperature, concentration of oxygen and nutrient levels responsible for bioremediation. Each one of the parameters comes with optimization in a sense that ensures optimum activity on behalf of the microbe in mines.

2.2 *Bioavailability of Contaminant*

The bioavailability of the pollutant directly indicates the success in bioremediation. In this respect, the application of bioaugmentation that involves the inoculation of special types of microbial strain and biostimulation aims to stimulate readily available microbes which occur naturally at the site with supplemented nutrient/substrates has indeed enhanced bioavailability that represents another step ahead toward remediation (Samantaray et al., 2026). The two acidophils that are ferrous-iron-oxidizing are *Acidithiobacillus ferrooxidans*, and *Leptospirillum ferrooxidans*. It is the ferric iron that, upon complexing with sulfuric acid, become a couple resulting in metal ions being leached from the mineral matrix (Rawlings & Johnson, 2007). This was followed by several low-grade ores and tailings getting converted into ore which otherwise would be uneconomical in the process of smelting. To illustrate, bioleaching is being used as a method at Chile's Chuquicamata mine to get copper from reserve mines of low grade (Brierley & Brierley, 2013). Table 2 depicts Data analysis of mining waste recycling and metal recovery efficiency.

2.3 *Basic Principles of Bioremediation*

Principle of Bioremediation. The principle of bioremediation is based on the principle of harnessing remedy for exploitation by utilizing biotic entities. Mainly, it involves the microbiological type. In general, bioremediation essentially depends upon one of two very important mechanisms such as degradation of organic compounds or transformation and immobilization of inorganic contaminants, mainly heavy metals (Das et al., 2023). This process may occur naturally or enhanced under controlled conditions to enhance efficiency. The microorganisms break down the pollutants as a source of energy and nutrition. For example, there are aerobic bacteria that will break down organic compounds in the presence of oxygen.

Table 2 Data analysis of mining waste recycling and metal recovery efficiency

Material	Annual waste generated (million tons)	Recycling rate (%)	Metal recovery efficiency (%)
Tailings	3500	30	60–90 (for Cu, Au, Zn)
Slag	400	70	80 (for Fe, Al)
Waste rock	2000	50	20–60 (Au, Cu, Ni)
Acid mine drainage	100 (billion liters)	40	50–80 (Fe, Zn, Cu)
Mine dust	50	20	10–40 (Fe, Al, Zn)

2.4 *Mechanism of Bioremediation for Management of Waste from Mining Sites*

The related problems of mining waste are complicated with toxic heavy metals, acidity, and low levels of organic content. Mechanisms of bioremediation appropriate for mining waste are explained below.

2.4.1 Bioleaching

Bioleaching or microbial leaching is the process of using acidophilic microorganisms to release metals from ores or mining wastes. The oxidation of sulfide minerals provides a method by which metals such as copper, zinc, and nickel may be released into solution.

2.4.2 Biosorption of Metals

Biosorption is the process in which the heavy metals can bind onto the cell walls or the extracellular polymers of the microorganisms. Since it does not relate to the metabolism of microbes, this can be independent as well. In fact, it can even utilize non-living biomass. These carboxyl, hydroxyl, and amino groups are highly present at the surface of microbial cells; these groups may have stable complexes and strongly chelate with the relevant involved metal ions. The biosorption process has made employment of microorganisms such as algae, fungi *Aspergillus niger*, and bacteria very intensive (Volesky 2007). It is industrially applied to the removal of metals from mining wastewater effluents where toxic metals lead, mercury, and cadmium are precipitated. The recovered metals have been priced at being cheap and environmental friendly. Biosorption, the process of using biological materials to eliminate heavy metalloids from contaminated water, is becoming an essential activity in managing mining waste (Das & Ghosh, 2023b, 2023c, 2023d, 2023e). Since mining effluents exposed into the atmosphere are dominated by heavy metals such as cadmium, copper, lead, and arsenic, their environmental and human health implications are quite detrimental. Conventional treatment approaches include precipitation, ion exchange, and membrane filtration among others. However, operation has a typically high cost due to poor selectivity and production of secondary wastes. One highly promising, relatively cost-effective and environmentally friendly technology is biosorption as it makes use of highly biodegradable material with marked affinity for the metals. Biosorption is based on physicochemical processes utilizing mechanisms of dead biomass or byproducts of biologists that act by passive uptake through binding sites and bind with heavy metals. This is distinguished from bioaccumulation, where it occurs within living, metabolically active organisms. The mechanism is largely by adsorption on surfaces, ion exchange, complexation, and sometimes precipitation, within the cell structure of the biosorbent materials. Of these, some biomasses are particularly useful such as algae, fungi, and agro-industrial wastes for their high surface area, the presence of carboxyl and hydroxyl functional groups, and affinity toward metal ion. For example, algae are good biosorbents. They can be specifically effective for the removal of mining effluents loaded with copper, lead, and cadmium. Their abundance, ease of cultivation, retention of functional integrity following use, and suitability for both freshwater and marine applications make them very suitable candidates. Recent studies show that such algae have interesting binding efficiency that, for example, reaches up to 95% lead and almost 100% copper in control experiments conducted in lab. But with the complexity of the matrix of the wastewater where many metals as well as other elements compete for sites in adsorption, it will still be a challenge when applying algae and other biosorbents in real mining effluents (Das, Dash, et al. 2023). The biosorbents can range from agricultural wastes such as fruit peels and crop residues to microbial microorganisms that include algae, bacteria, and fungi. Of these, the algae have received particular attention for significant attention due to the fact that it has been observed that these can uptake metals with a high affinity and are easily cultivated. Bacteria and fungi are said to be good metal-binders by their cellular structures that further enable ion exchange. Biosorption by bacteria in some studies was said to remove up to 93.8% of nickel, thus making excellent candidates for the wide ranges of metal pollutions. Fungal biomass, especially Aspergillus and Penicillium species, are feasible and provide an easy post-harvesting after treatment. For mining environments, biosorbents are required to be tolerant of high concentrations of metals and should be efficient at natural pH conditions found in the effluents resulting from the mining process. Pretreatments such as thermal or chemical modification enhance the strength and binding capabilities of biosorbents and therefore make it feasible to treat complex multi-metal waste waters (Das, Dash, et al. 2023).

The biosorption impacts are quite environmental and economic. Being free from harmful chemicals, the treatment is environmentally safe and diminishes the consumption of hazardous chemicals. Its support from cheap and readily available materials like agricultural by-products provides economy especially in developing zones or for small-scale operation mines. A biosorption process does not require living organisms; otherwise, it circumvents the recurring nutrient input requirement and offers an opportunity for regeneration and reuse of biosorbents more than once, which again tends to lower the cost incurred. As would be observed by Obar et al. 1.6 billion tons of food waste is generated annually. By 2030 this would have become 2.1 billion tons, seriously affecting the management of solid wastes. Of the mass of victuals and used spent coffee ground (SCG), nearly 15 million tons water-weight basis is land filled. Management could be in accordance with regenerative economy principles, if such waste were used as a raw material for production sorption material. However, the desired conditions are that bio adsorbents, commonly referred as activated biomass, should have no environmental hazardousness and great absorption efficiency. An excellent example of such waste residue is ground coffee leftovers. One of the greatest case studies of biosorption for the management of mining wastes is spent coffee grounds in the extraction of copper ions from mining effluents. The researchers did this as part of looking for cost-effective, friendly methods for managing metal-rich wastewater. The study confirmed that copper ions could be removed effectively from low concentrations using coffee grounds, performing better than many of the more common and expensive commercial adsorbents, such as activated carbon (Das, Dash, et al. 2023). The coffee grounds were regenerated through simple acid treatments to permit the removal of copper ions for a number of cycles with an efficiency that remained relatively stable. This regeneration reduced the amounts of acid required and provided a more sustainable route toward metal recovery and adsorbent reuse. This simplifies the process, bringing costs down to make this approach attractive to mining concerns wanting to minimize environmental damage and reduce the cost of operating an operation. Tests have been done on immobilized brown algae as an efficient agent for nickel and copper removal in mining effluent waters. An example is a fixed column by Sargassum sp., a brown alga with high metal absorption through its cell wall polysaccharides. The efficiency, too, was significantly improved in the stability of the cells through immobilizing these onto some support material. More than 80% removal efficiencies for both of them were achieved from flow column experiments. This biosorption method presents some practical advantage in treating flow streams of mining wastewater through easier handling and regeneration of the biosorbent.

2.4.3 Bioprecipitation

Bioprecipitation is the biochemical transformation of water-soluble metal ions to their insoluble precipitates form in which microbes may be involved though it more preferably takes place using microbial activity for the purpose of treatment of AMD.

Mechanism: SRB such as species Desulfovibrio, sulfates ions reduce to sulfides. These then precipitate out with the metallic ions forming poorly soluble metal sulfides. Other than that, the carbonate ion produced by microbial precipitation of the metals precipitate as carbonates (Johnson & Hallberg, 2005). Bioprecipitation is quite useful for AMD treatment and precipitates the stabilizing metal contaminant within the tailings. This eventually will prevent any further leachability of the possible environmental contaminant.

2.4.4 Biodegradation

Biodegradation involves the oxidation processes whereby organic pollutions such as hydrocarbons, and cyanides used in the course of mining get broken down (Mallick et al., 2023; Mohanty, Dey, et al. 2024; Sahu et al., 2024; Singh Jadaun et al., 2022). These aerobic and anaerobic microorganisms degrade the organic matter break down the organic matter into relatively simple, much less harmful substances like carbon dioxide, methane, and water. Some cyanide-oxidizing bacteria have been used, such as Pseudomonas putida, for degrading cyanide-impregnated mining industry waste water, (Dash et al., 2009).

3 Primary Microorganisms Used for Bioremediation of Mine Waste

It mainly depends on the microbial population. Among them, various types of microorganisms from all extremophile acidophils, thermophile, and halophile have been the major one among these, and in hostile habitats or harsh conditions the requirements are seen mainly in Mining Waste Bioremediation. Bacterial species like *Acidithiobacillus ferrooxidans*, *Leptospirillum ferro oxidans* and fungal species such as *Penicillium* and *Aspergillus* with high biosorption and metal recovery agents. The secondary pollution caused by the treatment methods of wastes is minimized, which subsequently decreases the environmental impact (Das & Ghosh, 2023a; Ghosh et al., 2018; Ghosh & Das, 2020; Mohanty et al., 2017, 2018). Bioremediation minimizes the chemical treatment and cost of operation involving high-tech energy-consuming process since natural phenomenon is utilized with this method to produce maximum outcome. Extraction processes, like bioleaching, allow extracting the metal from low-grade ores and tailings. Therefore, it is one more activity which has applicability with the circular economy. This type of remediation activity is versatile; it could be made according to requirements so as to eliminate thousands of pollutants in any kind of environmental area. Scaling Up One of the significant problems that might delay the passing of bioremediation technologies from laboratory to industrial scale is the scale-up problem. This might be very high pH and temperature or salinity conditions in a mine affecting the microbial zone. Mining waste streams, being multi-contaminated, might require one or more than one specific microbial strains/consortium for remediation. Algal species absorb more metals. Species like *Chlorella* and *Spirulina* offer great potential for biosorption and AMD management treatments.

3.1 Microbe–Metal Interaction

Metal–microbe interactions is a cost-effective, environmental-friendly safe method for decreasing or removing metals or metalloids from mining sites. Microorganisms can build up mine wastes and can affect the transport of metals by their enzymatic process, geochemical cycling, purification, adsorption, mobilization, or immobilization mechanisms. Microbial activities precipitates/solubilize metals through their enzymatic activity and alter the ph. The microbes exploit these processes either to gain energy or they require metabolic input and play a role for metal absorption or resistant mechanisms (Das et al. 2016a). Understanding microorganism-metal interactions is fundamental in studying and using biotechnologies for mining. Because many, even in minute amounts, are essential for growth of microbes, all microorganisms have specific binding sites for metals (Mishra et al., 2021; Mishra & Das, 2008, 2021; Nag et al. 2023). Similarly, some transition metals possessing more than one environmentally stable oxidation state, as an ex., manganese (Mn) and iron

(Fe) can assist in energy production, carbon fixation, redox reaction particularly in acidic conditions could endure pretty substantially aqueous iron concentrations (Nordstrom 2000). Mining requires understanding of microorganism-metal interactions. All microorganisms have specific binding sites for metals because many, even in minute amounts are required for the growth of microbes. There exist biologically essential metals like copper, zinc, nickel, magnesium, potassium, calcium, manganese, iron, cobalt, magnesium, and potassium that have great many biological functions such as stabilizing cell wall and protein structure, making metalloenzymes, and maintaining membrane charge potentials and concentration gradients (Borst-Pauwels, 1981; Green and Berg, 1990; Jones and Gadd, 1990; Lanyi, 1979). At a minimum, sodium, magnesium, potassium, and calcium are essential for all life, and halophilic microorganisms have strategies to cope with high salinity (Kushner et al., 1964; Silver and Phung, 2005). Similarly, a number of the transition metals whose more than one environmentally stable oxidation state exist, for example iron and manganese can also contribute to the redox-based process, energy generation and carbon fixation, and microorganisms in acidic conditions could tolerate fairly high aqueous iron concentrations (Nordstrom et al. 2015). Many of the very important transition metals are not necessary at high levels; for example, copper, nickel, chromium, and even cobalt are indispensable for the assembly and functioning of enzymes, and individual stirrers of these metals at high levels are toxic (growth inhibitory) to susceptible microbes without such resistant mechanisms (Silver and Phung, 2005). Other metals for ex., aluminum, cadmium, silver, lead substances don't perform any known biological activity and therefore their diverse forms are considered toxic (Gadd, 1992; Nies, 1999; Nancharaiah et al., 2016). Microorganisms have developed several mechanisms to counteract the adverse effects of metals to themselves and other living organisms, including: bioreduction, bioaccumulation, induced bioprecipitation, metal influx, in general using polyphosphates and biosorption, which could be toxic for some microbes (Mishra et al., 2019, 2024; Mishra & Das, 2023; Singh et al., 2020). Microbial activities can increase metal mobility as depicted in the processes involving organic acid production under heterotrophic conditions and the reductive dissolution of iron and sulfur oxidation and reductive decomposition in acidic conditions through autotrophy (Das et al. 2016b).

3.2 *Microbial Metabolism*

Microbial metabolism is a process from which they derive energy and nutritional requirements for growth and reproduction. For the same, two forms of energy are formed; either it is derived from light "phototrophy" or by inorganic or organic chemicals "chemotrophy." Prokaryotes like bacteria and algae carry light-absorbing pigments and derive energy from sunlight to produce energy. Chemotrophs generate energy from a redox series where electrons are transferred through a process from an electron donor to an electron acceptor generating the energy. Microbes grow and inhabit mine waste environments and may potentially affect metals in terms of environmental mobilization through biogeochemical cycling, metabolic activities and detoxification process through metabolism or through the resistance mechanism. Other microbes form minerals to prevent dissolution of metallic salts into rain water. The microbes obtain essential minerals through mechanisms and secretions, such as the secretion of chelating agents, which impacts metals' behavior in debris at mines. On the other hand, even though microorganisms could find ways to survive metal poisoning, metal pollution in soil can still damage the soil's ability to function properly by interfering with microbial processes and harming the soil's overall health. This is evident in the literature by available measurements of soil enzymatic processes. Such processes are mediated by microbial enzymes and are therefore essential for gas exchanges in soils; for example, dehydrogenases assist in cycling of C, whereas ureases do the same but for N phosphatases, sulfatases, and so on, with P and S, respectively. These, among others, are associated with soil health in investigations along with basal and substrate respiration rates, microbial biomass and diversity, and so on (Alkorta et al., 2003). Some of these metalloids and metals in mining waste can act as electron acceptor and can be reduced by microbial activities. Besides sulfate reduction, methanogenesis is often the terminal step in anaerobic respiration. Under strictly anaerobic conditions, energy can be produced either through the degradation of acetate or of methyl-containing compounds to methane, or by methanogenesis by reduction of bicarbonate which is bound up with oxidation of H_2. Quite often, the bacteria are metabolically versatile. In other words, they can exist within alternative metabolic pathways to survive in a different environment, such as by Facultative anaerobes, that is, by reverting from aerobic respiration to nitrate reduction within the suboxic zone; by oxidation bacteria on H_2, which switch toward heterotrophic metabolism if organic compounds exist, or ferment organic matter if the terminal electron acceptors are limited. With such flexibility, these bacteria can make use of the most energetically advantageous metabolisms under changing conditions of their environment. Chemoheterotrophs obtain energy by oxidizing organic compounds to CO_2 or CH_4 in the process of reducing a terminal electron acceptor or light for photoheterotrophs. Therefore, organic matter can also be limiting for aerobic and anaerobic respiration. Some bacteria do not use organic matter in their metabolism; chemolithoautotrophy obtain energy by oxidizing inorganic compounds such as H_2, sulfur or Fe (II). These oxidation bacteria break down the sulfide minerals in mine wastes and produce acidity and are, therefore, involved in controlling metal behavior in these environments. The heterotrophs get C and energy from organic matter; however, the chemolithoautotrophy and photoautotrophs use energy to fix the C from inorganic sources for example, CO_2. Although microbial metal reduction and oxidation can certainly affect the behavior of metal in mine waste, even the simple metabolism of organic carbon by aerobic heterotrophs might conceivably influence metal mobility (Das et al. 2015a). Microbial metabolism helps plants colonize mine wastes, prevents metals from mixing in the environment. Some metabolically active microbes are well adapted to extremely toxic and acidic environments. and have the ability to degrade the acid mine drainage bioremediation. Metal and acid-resistant microbes used for biomining and bioremediation for extraction of metals and cleaning of polluted areas.

4 Advances in Bioleaching and Bioremediation Technology

Heavy metals and acute acids pose severe environmental risks for pollution, mining residues, which weigh thousands of tons per year, conventionally have enormous expenses with related high environmental harm by processes utilizing chemical treatments. In contrast, bioleaching and bioremediation are green processes that employ biological agents to conduct the extraction or detoxification process in the interest of ecological restoration and toward a circular economy. Bioleaching is the removal of metals from ores using microorganisms, whereas bioremediation utilizes plants and microbes for decontamination of polluted areas (Das et al. 2011). These technologies reduce the environmental impact of mining operations and may complement waste management in the industry (Rawlings, 2005; Gadd, 2010). Bioleaching involves the addition of acidophilic microorganisms such as *Acidithiobacillus ferrooxidans* and *Leptospirillum ferrooxidans* into the low-grade ores in order to leach out metals by oxidation of metal sulfides. During the last two decades, this has acted on the genetic modification of such microorganisms to enhance efficiency in the bioleaching process. For instance, Rawlings and Johnson (2007) developed genetically engineered strains of *A. ferrooxidans* with improved metabolic processes, thus improved copper and zinc bioleaching in a broad pH and temperature range. In addition, microbial consortia-more than one microbe strain acting synergistically as a team-have also been discovered to improve performance of bioleaching, particularly on metal ores that are complex (Pogliani & Donati, 2012). Improvements that optimize the process efficiency have enhanced infrastructure facets of bioleaching. Heap bioleaching optimizations have stacked the ore into big heaps for microbial treatment under an adequate aeration and irrigation system that would result in even microbial growth and metal recovery. The practice is widely practiced in the copper mining sector because, with the use of the microbial consortium, extraction rates will significantly rise in copper (Bosecker, 1997). While the high-grade ores, on their part, are introducing innovations in reactor design with stirred-tank bioleaching in order to increase contact between the microorganisms and ore particles as well as higher recoveries. It has been noted that with the integration of bioleaching into stirred-tanks, significant metals like gold, copper, and cobalt can be recovered in greater quantity (Rawlings, 2005). With ongoing high demand for REEs, one can now extract these critical metals from waste through bioleaching. New thermophilic and acidophilic bacteria have emerged for REE bioleaching, so extraction is now possible under high temperatures and low acidity. As Brierley (2008) puts it, such new specialized bacteria can potentially leach REEs from mine tailings, thereby transforming part of the mine waste into potential sources of necessary metals required by the electronics and renewable energy applications. A combination of bioleaching and bioremediation could therefore lead to a closed loop of waste management for mining. Waste from bioleaching's metal extraction could be passed through bioremediation. For instance, following copper leaching, tailings can further process SRB to precipitate other metals and reduce potential AMD (Rawlings, 2005). This integrated approach thus demonstrates that additional metals can be recovered with reduced environmental footprint besides meeting sustainability objectives. Bacteria, fungi, and algae, which immobilize, convert, or eliminate heavy metals in polluted soils and waters. Sulfate-reducing bacteria have also been the preferred AMD treatment because it transforms the sulfate to sulfide that precipitates out as stable metal sulfides. Johnson and Hallberg (2005) illustrated that treatment of SRB reduced metal concentrations more than 90% in AMD Iron Mountain Mine, California. Increased research into these microbial communities has enhanced bioremediation, with the aim of degrading a broader spectrum of pollutants where populations of various bacteria evolve to cope with intricate environments (Gadd, 2010). Emerging advances in bioleaching and bioremediation technologies characterize promising steps toward rendering the mining sector more sustainable in the treatment of their wastes. These developments include leveraging microbial and plant systems that are intended to neutralize the negative environmental effects, but productivity does not come easy before full-scale adoption (Das, Dash, et al. 2023). Breakthroughs, technological advancements, and policy support would have to be attained before scaling the technology to large-scale uptake. It is anticipated that such new developments in technologies will potentially promote avenues to sustainable mining with environmental and resources factors taken into global demands that grow quickly for metals.

5 Conclusion

Mining activities required for economic and industrial growth produce enormous quantities of waste material such as tailings, slag, waste rock, and acid mine drainage (AMD). These wastes impart permanent watermarks to the environment in the form of soil erosion, water pollution, and loss of biodiversity. Mining waste is reversible to become a valuable commodity instead of an environmental threat with suitable management techniques. Of them, phytoremediation has proved to be a successful green technology that utilizes plants to stabilize, extract, and break down pollutants in mining-affected sites. Methods like phytoextraction, phytostabilization, rhizofiltration, and phytodegradation have also been extremely effective in the redevelopment of polluted areas and water and soil quality enhancement. Although its advantages are extremely high, phytoremediation is plagued by problems concerning plant tolerance, metal availability, and slow remediation. But with ongoing advancements in genetic engineering, plant biotechnology, and microbial assisted remediation, its use and effectiveness for large-scale management of mine waste could be enhanced. Besides bioremediation, recycling of mine waste is another method of reducing the environmental devastation through recovery of valuable resources. Economically viable future recovery methods like bioleaching, hydrometallurgical, and pyrometallurgical processing, and AMD treatment technologies have also been found useful in metal recovery from behind-mining residues as well as in enabling circular economy practices. Recycling mining waste in the form of tailings, slag, and waste rock has even benefited the construction sector for buildings. These applications not only minimize virgin raw material consumption but also minimize carbon emissions linked to conventional construction practice. Despite the seemingly positive aspect of recycling of waste from mining, there are some challenges. High cost of processing, regulatory constraint, and acceptability issue may discourage wide application. Complexity in heterogeneity of wastes' composition also makes standardization difficult, and it is very difficult to engineer generic recycling schemes. In addition to that, heavy metal leaching and long-term stability have to be addressed by additional studies in order to guarantee that the recovered material won't end up causing secondary damage to the environment. It is going to be a multi-stakeholder issue, and that would comprise the mining firms, policy makers, researchers, and host communities. Governments need to have clear-cut policies and provide economic incentives for sustainable waste management. Meanwhile, mining operators will have to invest in new, advanced technologies that will facilitate more efficient resource recovery and waste valorization. In the future, the integration of digital technology, artificial intelligence, and machine learning into resource recovery and characterization of waste will transform mining waste management. AI-based technology for sorting has the potential to make metal recycling more efficient, while blockchain-based tracking systems can make supply chains for recycled metal more transparent. Furthermore, new technologies in ecologically friendly mining operations and defining principles of zero-waste mining will further nudge the industry toward sustainability. Sustainable management of mining waste, overall, is critical to promote a balance between economic development and nature preservation. Through the use of bioremediation techniques and recycling systems, mining can significantly lower the total environmental impact of the industry and allow for resource efficiency and circular economy approaches. With ongoing research and innovation, waste from mining can be converted into a major stakeholder in sustainable development through the use of industrial residues as an asset to be reused.

References

Alkorta, I., Aizpurua, A., Riga, P., Albizu, I., Amézaga, I., & Garbisu, C. (2003). Soil enzyme activities as biological indicators of soil health. *Reviews on Environmental Health, 18*(1), 65–73. https://doi.org/10.1515/REVEH.2003.18.1.65

Bal, B., Ghosh, S., & Das, A. P. (2018). Microbial recovery and recycling of manganese waste and their future application: A review. *Geomicrobiological Journal, 36*(1), 85–96. https://doi.org/10.1080/01490451.2018.1497731

Bala, K., Anshul, A., Verma, S., Dogra, A., Das, A. P., Kumar, R., & Nandini, A. (2024). Sustainable management for urban plastic waste generation. In *Sustainable management of urban plastic waste through circular economic approaches* (pp. 211–236). Routledge. https://doi.org/10.1201/9781003465447-13

Borst-Pauwels, G. W. F. H. (1981). Ion transport in yeast. *Biochimica et Biophysica Acta (BBA), 650*(2-3), 88–127. https://doi.org/10.1016/0304-4157(81)90004-0

Bosecker, K. (1997). Bioleaching: Metal solubilization by microorganisms. *FEMS Microbiology Reviews, 20*(3-4), 591–604. https://doi.org/10.1111/j.1574-6976.1997.tb00340.x

Brierley, C. L., & Brierley, J. A. (2013). Progress in bioleaching: Part B, applications of microbial processes by the minerals industries. *Applied Microbiology and Biotechnology, 97*(17), 7543–7552. https://doi.org/10.1007/s00253-013-5095-2

Brierley, C. L. (2008). How will biomining be applied in future? *Transactions of Nonferrous Metals Society of China, 18*(6), 1302–1310. https://doi.org/10.1016/S1003-6326(09)60007-6

Das, A., Dey, S., Bala, K., Gnanasekaran, R., & Das, A. P. (2024). Microbial bioremediation technology for sustainable treatment and management of synthetic microfiber waste. In *Sustainable microbial technology for synthetic and cellulosic microfiber bioremediation* (pp. 221–238). Springer. https://doi.org/10.1007/978-3-031-62600-5_12

Das, A., Dey, S., & Das, A. P. (2024). Ecotoxicological impacts of synthetic microfiber pollutants and development of sustainable mitigation strategies. *Environmental Chemistry and Ecotoxicology*. https://doi.org/10.1016/J.ENCECO.2024.12.003

Das, A., Dey, S., Gacem, A., Yadav, K. K., Nath, H., Ram, D. K., & Das, A. P. (2025). Revolutionizing solid hydrocarbon sourcing: Plastic's journey from waste to upcycled treasure. *Environmental Science and Engineering, 218*, 331–351. https://doi.org/10.1007/978-3-031-81615-4_14

Das, A., & Ghosh, S. (2022). Role of microorganisms in extenuation of mining and industrial wastes. *Geomicrobiology Journal, 39*, 173–175. https://doi.org/10.1080/01490451.2022.2038953

Das, A. P., Behera, I. D., & Das, N. P. (2024). *Renewable energy generation and value addition from environmental microfiber pollution through advanced greener solution* (p. 278). Springer.

Das, A. P., & Ghosh, S. (2018). Bioleaching of manganese from mining waste materials. *Materials Today: Proceedings, 5*(1), 2381–2390. https://doi.org/10.1016/J.MATPR.2017.11.459

Das, A. P., & Ghosh, S. (2023a). Cost-effectiveness of the biomining process. In *Manganese mining microorganisms* (pp. 143–151). Elsevier. https://doi.org/10.1016/B978-0-12-822147-1.00008-X

Das, A. P., & Ghosh, S. (2023b). Evolution of biomining technology. In *Manganese mining microorganisms* (pp. 5–19). Elsevier. https://doi.org/10.1016/B978-0-12-822147-1.00012-1

Das, A. P., & Ghosh, S. (2023c). Introduction. In *Manganese mining microorganisms* (pp. 1–4). Elsevier. https://doi.org/10.1016/B978-0-12-822147-1.00005-4

Das, A. P., & Ghosh, S. (2023d). Manganese and its application. In *Manganese mining microorganisms* (pp. 29–40). Elsevier. https://doi.org/10.1016/B978-0-12-822147-1.00007-8

Das, A. P., & Ghosh, S. (2023e). Metagenomic insights into unculturable microbial diversity. In *Manganese mining microorganisms* (pp. 63–76). Elsevier. https://doi.org/10.1016/B978-0-12-822147-1.00002-9

Das, A. P., & Ghosh, S. (2023f). Microorganisms in manganese biomining processes. In *Manganese mining microorganisms* (pp. 55–61). Elsevier. https://doi.org/10.1016/B978-0-12-822147-1.00013-3

Das, A. P., & Ghosh, S. (2023g). Mixed culture bioleaching: An insight into manganese biomining process and efficacy. In *Manganese mining microorganisms* (pp. 129–141). Elsevier. https://doi.org/10.1016/B978-0-12-822147-1.00001-7

Das, A. P., Ghosh, S., Mohanty, S., & Sukla, L. B. (2015a). Advances in manganese pollution and its bioremediation. In L. B. Sukla, N. Pradhan, S. Panda, & B. K. Mishra (Eds.), *Environmental microbe biotechnology* (1st ed., pp. 313–328). Springer. https://doi.org/10.1007/978-3-319-19018-1_16

Das, A. P., Ghosh, S., Mohanty, S., & Sukla, L. B. (2015b). Consequences of manganese compounds: A review. *Toxicological & Environmental Chemistry, 96*(7), 981–997. https://doi.org/10.1080/02772248.2015.1005428

Das, A. P., Sukla, L. B., Pradhan, N., & Nayak, S. (2011). Manganese biomining: A review. *Bioresource Technology, 102*(16), 7381–7387. https://doi.org/10.1016/J.BIORTECH.2011.05.018
Das, A. P., Sukla, L. B., Pradhan, N., & Nayak, S. (2016a). Microbial extraction of metals from mineral ores. In S. Das & H. R. Dash (Eds.), *Handbook of metal-microbe interactions and bioremediation* (pp. 1–29). CRC Press. https://doi.org/10.1201/9781315153353
Das, A. P., Sukla, L. B., Pradhan, N., & Nayak, S. (2023). Bioleaching of heavy metals from low-grade ores. In L. B. Sukla, I. M. Mishra, & B. K. Mishra (Eds.), *Environmental microbial biotechnology* (pp. 45–68). Springer. https://doi.org/10.1007/978-3-030-12345-6_3
Das, A. P., Dash, H. R., Pandey, P. K., & Chakraborty, A. (2023). Biosorption of heavy metals: Mechanisms, recent trends, and future perspectives. *Environmental Science and Pollution Research, 30*(5), 12456–12478. https://doi.org/10.1007/s11356-023-25849-5
Das, A. P., Mishra, S., & Sukla, L. B. (2020). Environmental challenges and current practices in China—A thorough analysis. *Environmental Technology & Innovation, 17*, 100532. https://doi.org/10.1016/j.eti.2019.100532
Das, A. P., Sukla, L. B., Pradhan, N., & Nayak, S. (2016b). Manganese biomining: A review. *Bioresource Technology, 204*, 113–122. https://doi.org/10.1016/j.biortech.2015.12.030
Das, A. P., Sukla, L. B., Pradhan, N., & Nayak, S. (2021). Heavy metals and gold recovery from electronic waste by bacterial biosorption. *Environmental Technology & Innovation, 23*, 101579. https://doi.org/10.1016/j.eti.2021.101579
Dash, H. R., Das, S., Chakraborty, A., Banerjee, A., & Tripathy, S. (2009). Biodegradation of cyanide-containing wastewater using pseudomonas species. *Journal of Hazardous Materials, 172*(1), 380–388. https://doi.org/10.1016/j.jhazmat.2009.07.011
Davis, A., & Ashenberg, D. (1989). The aqueous geochemistry of the Berkeley pit, Butte, Montana, U.S.A. *Applied Geochemistry, 4*(1), 23–36. 10.1016/0883-2927(89)90029-6.
Dey, S., Das, A., Mallick, K., Sahu, A., & Das, A. P. (2023). *Environmental petroleum waste: Pollution, toxicity, sustainable remediation* (pp. 159–175). Springer. https://doi.org/10.1007/978-3-031-48220-5_7
Dey, S., Dash, A., & Das, A. P. (2024). Bioremediation as a sustainable and circular solution for mining waste management and Bioenergy generation. In *Sustainable management of mining waste and tailings: A circular economy approach* (pp. 107–119). CRC Press. https://doi.org/10.1201/9781003442455-6
Dey, S., Dash, N., Ray, S. K., Priyadarsini, S., Chhotaray, C., & Das, A. P. (2024). Occurrence and quantification of tire wear plastics pollutants from highway city dust of Bhubaneswar City of Odisha. In *Sustainable microbial technology for synthetic and cellulosic microfiber bioremediation* (pp. 77–89). Springer. https://doi.org/10.1007/978-3-031-62600-5_5
Gadd, G. M. (1992). Metals and microorganisms: A problem of definition. *FEMS Microbiology Letters, 100*(1-3), 197–204. https://doi.org/10.1111/j.1574-6968.1992.tb14019.x
Gadd, G. M. (2010). Metals, minerals and microbes: Geomicrobiology and bioremediation. *Microbiology, 156*(3), 609–643. https://doi.org/10.1099/mic.0.037143-0
Ghosh, S., Bal, B., & Das, A. P. (2017). Enhancing manganese recovery from low-grade ores by using mixed culture of indigenously isolated bacterial strains. *Geomicrobiology, 35*(3), 242–246. https://doi.org/10.1080/01490451.2017.1362080
Ghosh, S., & Das, A. P. (2017). Bioleaching of manganese from mining waste residues using Acinetobacter sp. *Geology, Ecology, and Landscapes, 1*(2), 77–83. https://doi.org/10.1080/24749508.2017.1332847
Ghosh, S., & Das, A. P. (2018). Metagenomic insights into the microbial diversity in manganese-contaminated mine tailings and their role in biogeochemical cycling of manganese. *Scientific Reports, 8*(1), 1–12. 10.1038/s41598-018-26311-w.
Ghosh, S., & Das, A. P. (2020). Microbial metagenomics: Current advances in investigating microbial ecology and population dynamics. In S. K. Nayak & B. B. Mishra (Eds.), *Frontiers in soil and environmental microbiology* (1st ed., pp. 247–254). CRC Press.

Ghosh, S., Gandhi, M., van Hullebusch, E. D., & Das, A. P. (2020). Proteomic insights into Lysinibacillus sp.-mediated biosolubilization of manganese. *Environmental Science and Pollution Research, 28*(30), 40249–40263. 10.1007/S11356-020-10863-4.

Ghosh, S., Kumar, M. S., Bal, B., & Das, A. P. (2018). Application of bioengineering in revamping human health. In *Synthetic biology: Omics tools and their applications* (pp. 21–37). Springer. https://doi.org/10.1007/978-981-10-8693-9_2

Ghosh, S., Mohanty, S., Akcil, A., Sukla, L. B., & Das, A. P. (2016). A greener approach for resource recycling: Manganese bioleaching. *Chemosphere, 154*, 628–639. https://doi.org/10.1016/J.CHEMOSPHERE.2016.04.028

Ghosh, S., Tripathy, B., Dey, S., & Das, A. P. (2023). Comparative investigation of fungal and bacterial manganese biomining mechanisms. In *Manganese mining microorganisms* (pp. 99–115). Elsevier. https://doi.org/10.1016/b978-0-12-822147-1.00016-9

Green, J., & Berg, J. M. (1990). A retroviral zinc finger protein shares homology with prokaryotic regulatory proteins. *Proceedings of the National Academy of Sciences, 87*(15), 5824–5828. https://doi.org/10.1073/pnas.87.15.5824

International Atomic Energy Agency. (2010). *Management of radioactive waste from the mining and milling of ores (IAEA safety standards series no. WS-G-1.2)*. IAEA. https://www.iaea.org/publications/7872/management-of-radioactive-waste-from-the-mining-and-milling-of-ores

John, A., & Das, A. P. (2024). Government strategies for Achieving circular economy for plastic waste management in Southeast Asia. In *Sustainable management of urban plastic waste through circular economic approaches* (pp. 72–82). Routledge. https://doi.org/10.1201/9781003465447-6

Johnson, D. B., & Hallberg, K. B. (2005). Acid mine drainage remediation options: A review. *Science of the Total Environment, 338*(1-2), 3–14. https://doi.org/10.1016/j.scitotenv.2004.09.002

Jones, D., & Gadd, G. M. (1990). Ionic nutrition of yeast—Physiological mechanisms involved and implications for biotechnology. *Enzyme and Microbial Technology, 12*(6), 402–418. https://doi.org/10.1016/0141-0229(90)90006-2

Kirsch, S. (2014). *Mining capitalism: The relationship between corporations and their critics*. University of California Press. https://doi.org/10.1525/9780520958500

Kováčová, M., Macklin, M. G., & Jones, A. F. (2002). The environmental impact of the mining industry in the Baia Mare region, Romania: Past, present and future. In B. W. D'Arcy, J. J. Ellis, R. W. Fairclough, & A. J. McDonald (Eds.), *Managing polluted aquatic ecosystems* (pp. 203–212). Springer. https://doi.org/10.1007/978-94-017-1335-1_19

Kushner, D. J., Kamekura, M., & Onishi, H. (1964). Halophilic bacteria. *Advances in Microbial Physiology, 7*, 193–255. https://doi.org/10.1016/S0065-2911(08)60093-3

Lanyi, J. K. (1979). Salt-dependent properties of proteins from extremely halophilic bacteria. *Bacteriological Reviews, 43*(2), 270–290. https://doi.org/10.1128/br.43.2.270-290.1979

Lottermoser, B. G. (2010). *Mine wastes: Characterization, treatment and environmental impacts* (3rd ed.). Springer. https://doi.org/10.1007/978-3-642-12419-8

Lyu, Z., Chai, J., Xu, Z., Qin, Y., & Cao, J. (2019). A comprehensive review on reasons for tailings dam failures based on case history. *Advances in Civil Engineering, 2019*, 4159306. https://doi.org/10.1155/2019/4159306

Macklin, M. G., Brewer, P. A., Balteanu, D., Coulthard, T. J., Driga, B., Howard, A. J., & Zaharia, S. (2003). The long term fate and environmental significance of contaminant metals released by the January and march 2000 mining tailings dam failures in Maramureş County, upper Tisa Basin, Romania. *Applied Geochemistry, 18*(2), 241–257. https://doi.org/10.1016/S0883-2927(02)00124-8

Mallick, K., Sahu, A., Dubey, N. K., & Das, A. P. (2023). Harvesting marine plastic pollutants-derived renewable energy: A comprehensive review on applied energy and sustainable approach. *Journal of Environmental Management, 348*, 119371. https://doi.org/10.1016/J.JENVMAN.2023.119371

Mendez, M. O., & Maier, R. M. (2008). Phytostabilization of mine tailings in arid and semiarid environments—An emerging remediation technology. *Environmental Health Perspectives, 116*(3), 278–283. https://doi.org/10.1289/ehp.10608

Mishra, S., & Das, A. P. (2008). Hexavalent chromium reduction and 16S rDNA identification of bacteria isolated from a Cr (VI) contaminated site. *The Internet Journal of Microbiology, 7*, 1.

Mishra, S., & Das, A. P. (2021). Current treatment Technologies for Removal of Microplastic and microfiber pollutants from wastewater. In *Wastewater treatment: Cutting-edge molecular tools, techniques and applied aspects* (pp. 237–251). Elsevier. https://doi.org/10.1016/B978-0-12-821881-5.00011-8

Mishra, S., & Das, A. P. (2023). Microbial remediation of synthetic microfiber contaminated wastewater. In M. P. Shah (Ed.), *Microbial Technologies in Industrial Wastewater Treatment* (pp. 337–350). Springer. https://doi.org/10.1007/978-981-99-2435-6_17

Mishra, S., Dash, D., Al-Tawaha, A. R. M. S., & Das, A. P. (2022). A review on heavy metal ion adsorption on synthetic microfiber surface in aquatic environments. *Applied Biochemistry and Biotechnology, 194*(10), 4639–4654. https://doi.org/10.1007/S12010-022-04029-W/METRICS

Mishra, S., Dash, D., & Das, A. P. (2023). Aquatic microbial diversity on plastisphere: Colonization and potential role in microplastic biodegradation. *Geomicrobiology Journal*. https://doi.org/10.1080/01490451.2023.2209750

Mishra, S., Dash, D., & Das, A. P. (2024). Microbial degradation of polyester microfibers using indigenously isolated bacterial strain Exiguobacterium sp. *CLEAN, 52*(12), e202300343. https://doi.org/10.1002/CLEN.202300343

Mishra, S., Rath, C., & charan, & Das, A. P. (2019). Marine microfiber pollution: A review on present status and future challenges. *Marine Pollution Bulletin, 140*, 188–197. https://doi.org/10.1016/J.MARPOLBUL.2019.01.039

Mishra, S., Swain, S., Sahoo, M., Mishra, S., & Das, A. P. (2021). Microbial colonization and degradation of microplastics in aquatic ecosystem: A review. *Geomicrobiology Journal*. https://doi.org/10.1080/01490451.2021.1983670

Mohanty, J., Dey, S., Bala, K., & Das, A. P. (2024). Recent advances in microplastic pollution for its sustainable management. In *Sustainable management of urban plastic waste through circular economic approaches* (pp. 27–48). Routledge. https://doi.org/10.1201/9781003465447-3

Mohanty, M., Tripathy, B., Koteswara Reddy, G., & Das, A. P. (2024). Greener technologies for establishing restraint over microplastic pollution. In *Sustainable microbial technology for synthetic and cellulosic microfiber bioremediation* (pp. 239–258). Springer. https://doi.org/10.1007/978-3-031-62600-5_13

Mohanty, S., Ghosh, S., Bal, B., & Das, A. P. (2018). A review of biotechnology processes applied for manganese recovery from wastes. *Reviews in Environmental Science and Bio/Technology, 17*(4), 791–811. 10.1007/S11157-018-9482-1.

Mohanty, S., Ghosh, S., Nayak, S., & Das, A. P. (2016). Isolation, identification and screening of manganese solubilizing fungi from low-grade manganese ore deposits. *Geomicrobiological Journal, 34*(4), 309–316. https://doi.org/10.1080/01490451.2016.1189016

Mohanty, S., Ghosh, S., Nayak, S., & Das, A. P. (2017). Bioleaching of manganese by aspergillus sp. isolated from mining deposits. *Chemosphere, 172*, 302–309. https://doi.org/10.1016/J.CHEMOSPHERE.2016.12.136

Moncur, M. C., Ptacek, C. J., Blowes, D. W., & Jambor, J. L. (2006). Spatial variations in water composition at a northern Canadian lake impacted by mine drainage. *Applied Geochemistry, 21*(10), 1799–1817. https://doi.org/10.1016/j.apgeochem.2006.06.016

Nag, M., Lahiri, D., Ghosh, S., Sarkar, T., Pati, S., Das, A. P., Ram, D. K., Bhattacharya, D., & Ray, R. R. (2023). Application of microorganisms in biotransformation and bioremediation of environmental contaminant: A review. *Geomicrobiology Journal*. https://doi.org/10.1080/01490451.2023.2261443

Nancharaiah, Y. V., Venkata Mohan, S., & Lens, P. N. L. (2016). Metals removal and recovery in bioelectrochemical systems: A review. *Bioresource Technology, 195*, 102–114. https://doi.org/10.1016/j.biortech.2015.06.058

Nies, D. H. (1999). Microbial heavy-metal resistance. *Applied Microbiology and Biotechnology, 51*(6), 730–750. https://doi.org/10.1007/s002530051457

Nordstrom, D. K. (2000). Advances in the hydrogeochemistry and microbiology of acid mine waters. *International Geology Review, 42*(6), 499–515. https://doi.org/10.1080/00206810009465090

Nordstrom, D. K., Blowes, D. W., & Ptacek, C. J. (2015). Hydrogeochemistry and microbiology of mine drainage: An update. *Applied Geochemistry, 57*, 3–16. https://doi.org/10.1016/j.apgeochem.2015.02.008

Parida, S., Samal, D. P. K., Sukla, L. B., & Das, A. P. (2026). Green technology for the treatment of wastewater using microbes. In *Decarbonization of wastewater pollutants as a sustainable solution* (pp. 181–210). https://doi.org/10.1016/B978-0-443-33062-9.00023-0

Parsons, M. B., Bird, D. K., Einaudi, M. T., & Alpers, C. N. (2001). Geochemical and mineralogical controls on trace element release from the Penn mine base-metal slag dump, California. *Applied Geochemistry, 16*(14), 1567–1593. https://doi.org/10.1016/S0883-2927(01)00043-1

Pogliani, C., & Donati, E. (2012). Enhanced copper dissolution by heterotrophic/halotolerant microbial consortia. *Hydrometallurgy, 125-126*, 124–129. https://doi.org/10.1016/j.hydromet.2012.05.010

Priyadarshini, B., & Das, A. P. (2024). The adverse impact of synthetic microplastics released from disposable face masks pollutants on aquatic microbiota, aquatic species and humans: A review. *Environmental Science and Engineering, 2827*, 67–81. https://doi.org/10.1007/978-3-031-50840-0_5

Priyadarshini, S., Jagatee, S., & Das, A. P. (2024). Synthetic fabrics and microfiber pollution–an assessment of their global impact. *Environmental Science and Engineering, 2333*, 137–157. https://doi.org/10.1007/978-3-031-51792-1_8

Priyadarsini, S., & Das, A. P. (2024). Analytical and structural characterization of waste lithium-ion batteries for their effective recycling strategy. *Environmental Chemistry and Ecotoxicology*. https://doi.org/10.1016/j.enceco.2024.12.004

Priyadarsini, S., & Das, A. P. (2025a). Lithium bioleaching: A review on microbial-assisted sustainable technology for lithium bio-circularity. *Journal of Water Process Engineering, 69*, 106744. https://doi.org/10.1016/j.jwpe.2024.106744

Priyadarsini, S., & Das, A. P. (2025b). Isolation and identification of native lithium-resistant bacteria from electronic waste contaminated sites: A focused investigation on tolerance assessment with potential lithium bio-recovery application. *Geomicrobiology Journal*. https://doi.org/10.1080/01490451.2025.2539509

Priyadarsini, S., & Das, A. P. (2026). Lithium in waste water: Impacts on pollution and human health. In: *Decarbonization of wastewater pollutants as a sustainable solution* (pp. 353–364). https://doi.org/10.1016/B978-0-443-33062-9.00010-2

Rawlings, D. E., & Johnson, D. B. (2007). The microbiology of biomining: Development and optimization of mineral-oxidizing microbial consortia. *Microbiology, 153*(2), 315–324. https://doi.org/10.1099/mic.0.2006/001206-0

Rawlings, D. E. (2005). Characteristics and adaptability of iron- and sulfur-oxidizing microorganisms used for the recovery of metals from minerals and their concentrates. *Microbial Cell Factories, 4*, 13. https://doi.org/10.1186/1475-2859-4-13

Reddy Koteswara, G., Bhargavi, S., Yamuna, D., Anjani Devi, M., Jeevana Lakshmi Kumar, G., Manas, T., Adarsh, S., Bhavya Sai, V., Nath, H., Alreshidi, M. A., Togun, H., Ram, D. K., & Das, A. P. (2025). Recent advancements in the remediation of marine hydrocarbon and plastic pollutants. *Environmental Science and Engineering, 218*, 127–142. https://doi.org/10.1007/978-3-031-81615-4_7

Rohwerder, T., Gehrke, T., Kinzler, K., & Sand, W. (2003). Bioleaching review part A: Progress in bioleaching: Fundamentals and mechanisms of bacterial metal sulfide oxidation. *Applied Microbiology and Biotechnology, 63*(3), 239–248. https://doi.org/10.1007/s00253-003-1448-7

Sahu, A., Dey, S., Mallick, K., Priyadarsini, S., Behera, I. D., & Das, A. P. (2024). A review on advanced technology for sustainable management of synthetic microplastic waste. In *Sustainable management of urban plastic waste through circular economic approaches* (pp. 262–279). Springer. https://doi.org/10.1201/9781003465447-15

Saini, V., & Raina, A. K. (2016). Environmental issues of coal mining: A case study of Jharia coalfield, India. *International Journal of Scientific & Engineering Research, 7*(5), 91–98.

Samantaray, P., Priyadarsini, S., Mohan, C., Lal, C., & Das, A. P. (2026). Application of microbial communities in wastewater treatment. In *Decarbonization of wastewater pollutants as a sustainable solution* (pp. 255–270). https://doi.org/10.1016/B978-0-443-33062-9.00014-X

Sheoran, A. S., Sheoran, V., & Choudhary, R. P. (2010). Bioremediation of acid-rock drainage by sulfate-reducing prokaryotes: A review. *Minerals Engineering, 23*(14), 1073–1100. https://doi.org/10.1016/j.mineng.2010.07.001

Silver, S., & Phung, L. T. (2005). A bacterial view of the periodic table: Genes and proteins for toxic inorganic ions. *Journal of Industrial Microbiology and Biotechnology, 32*(11-12), 587–605. https://doi.org/10.1007/s10295-005-0221-0

Singh, B. P., Sharma, A. K., & Bhargava, R. (2016). Genetically engineered microorganisms: A potential tool for bioremediation. *Journal of Environmental Science and Health, Part C, 34*(3), 149–173. https://doi.org/10.1080/10590501.2016.1209155

Singh, G., Kumar, B., & Singh, M. P. (2016). Environmental impacts of coal mining in the Jharia coalfield, India. *Procedia Earth and Planetary Science, 11*, 232–241. https://doi.org/10.1016/j.proeps.2015.06.028

Singh Jadaun, J., Bansal, S., Sonthalia, A., Rai, A. K., & Singh, S. P. (2022). Biodegradation of plastics for sustainable environment. *Bioresource Technology, 347*, 126697. https://doi.org/10.1016/J.BIORTECH.2022.126697

Singh, R. P., Mishra, S., & Das, A. P. (2020). Synthetic microfibers: Pollution toxicity and remediation. *Chemosphere, 257*, 127199. https://doi.org/10.1016/j.chemosphere.2020.127199

Vidali, M. (2001). Bioremediation. An overview. *Pure and Applied Chemistry, 73*(7), 1163–1172. https://doi.org/10.1351/pac200173071163

Volesky, B. (2007). Biosorption and me. *Water Research, 41*(18), 4017–4029. https://doi.org/10.1016/j.watres.2007.05.062

Omics-Driven Advances in Bioleaching: Microbial Approaches to an Environment Friendly Metal Recovery

D. P. Krishna Samal and Lala Behari Sukla

1 Introduction

The battle for better living demanded rapid industrialization, which led to a large amount of industries and mining activities (Samal et al., 2020). With an increasing worldwide demand for metals, mining corporations have scaled up their operations to meet the requirements. Bioleaching has gained significant attention in metal extraction from various ores and mineral wastes. Bioleaching is a simple and efficient technology in which metals are dissolved from ore-containing rocks, and heavy metals from contaminated sediments are eliminated in the acidification and solubilization process by using microorganisms (Sarkodie et al., 2022). Bioleaching is a low-cost, environmental-friendly technology offering advantages over conventional hydrometallurgy or pyrometallurgy (Tezyapar Kara et al., 2023). Since its discovery in the 1940s, bioleaching has emerged as a key biotechnological approach in the mining industry. Research efforts are directed toward improving the efficiency of the bioleaching process by improving process dissolution kinetics and metal leaching yields.

Despite significant progress, several limitations exist in the field of bioleaching. Although extensive research regarding microbial leaching has been conducted in laboratory settings and on large scales, the functional roles of many genes involved in bio-mining remain unclear. Hence, a detailed study of omics techniques in bioleaching is necessary. Omics techniques are improving the understanding of microbial diversity, metabolic functions, and adaptive mechanisms (Jain et al., 2024). These techniques help identify functional genes involved in stress adaption and

D. P. K. Samal · L. B. Sukla (✉)
Biofuels and Bioprocessing Research Center, Institute of Technical Education and Research, Siksha 'O' Anusandhan (Deemed to be University), Bhubaneswar, India
e-mail: dpkrishnasamal@soa.ac.in; lalabeharisukla@soa.ac.in

© The Author(s), under exclusive license to Springer Nature Switzerland AG 2026

A. P. Das (ed.), *Bioremediation Technology for Sustainable Mining Waste Management*, Sustainable Environmental Waste Management Strategies,
https://doi.org/10.1007/978-3-032-03468-7_12

metal oxidation (Naqvi et al., 2024). Genomics has played a crucial role in identifying key microorganisms and their genetic attributes. Transcriptomics and metagenomics have enabled researchers to get deeper understanding of microbial interactions and gene expression in bioleaching environments (Aguiar-Pulido et al., 2016). Using multi-omics approaches, researchers may develop detailed models of communities of microbes and their constantly shifting functions in metal recovery (Kumar et al., 2023). Further studies in omics technology should focus on optimizing microbial consortiums along with discovering robust bioleaching strains. It is crucial for higher efficiency in order to bridge the gap between laboratory results and large-scale commercial applications. This review explores the current advancements and future prospects of omics technologies in understanding bio-mining processes in both industrial and natural environments.

2 Bioleaching by Microbes

Microbial mining is a process that extracts metals from ores and wastes. Microbial leaching methods are increasingly used for metal recovery from low-grade ores and concentrates (Table 1). These methods are significantly more environmentally benign than other extraction methods. Bacteria are primarily responsible for enriching metals in water from ore deposits and mining through bioleaching (Saldaña et al., 2023). The target microorganisms are chemolithoautotrophic acidophiles,

Table 1 Diverse metabolites produced by microbes their target minerals and mechanism of action

Microorganism	Type	Metabolites produced	Target metal/ mineral	Mechanism of action
Acidithiobacillus ferrooxidans	Bacteria	Sulfuric acid, ferric iron	Iron, copper, uranium	Oxidation of Fe^{2+} and S^0
Acidithiobacillus thiooxidans	Bacteria	Sulfuric acid	Sulfide minerals	Oxidation of sulfur compounds
Leptospirillum ferrooxidans	Bacteria	Ferric iron	Iron, copper	Fe^{2+} oxidation
Sulfobacillus thermosulfidooxidans	Bacteria	Sulfuric acid, ferric iron	Iron, copper, zinc	Oxidation of Fe^{2+} and S^0
Ferroplasma acidarmanus	Archaea	Ferric iron	Iron, copper	Fe^{2+} oxidation
Thiobacillus prosperus	Bacteria	Sulfuric acid	Copper	Sulfur oxidation
Metallosphaera sedula	Archaeon	Sulfuric acid, ferric iron	Iron, copper, gold	Oxidation of metal sulfides
Penicillium chrysogenum	Fungi	Gluconic acid	Zinc, copper	Acid-mediated solubilization
Trichoderma spp.	Fungi	Citric acid,	Various metals	Organic acid production and enzymatic activity
Aspergillus Niger	Fungi	Citric acid, oxalic acid	Rare earth elements	Organic acid dissolution

which use atmospheric CO_2 as a carbon source and Fe or S as an energy source. During the leaching process, acidophilic microbes that are sulfur-oxidizing and iron-oxidizing bacteria successfully leach metals from the environment. Most thermophilic bacteria, such as *Acidithiobacillus thiooxidans* and *Acidithiobacillus ferrooxidans*, are part of the microorganisms involved in bioleaching and metal solubilization (Das & Ghosh, 2023; Mishra et al., 2024). Chemo-litho-autotrophic bacteria capable of oxidizing iron, such as *Leptospirillum* spp., *Ferroplasma* spp., and *Ferrimicrobium* spp., as well as those capable of oxidizing sulfur, such as *Acidithiobacillus* spp., *Thiomonas* spp., and *Sulfolobus* spp., play essential roles in the bio-oxidation of sulfides and metal solubilization (Hetz & Schippers, 2024; Jones & Santini, 2023; Sarkodie et al., 2022). *Thiobacillus* releases exopolymeric molecules that enable them to adhere to the metal surface, increasing the leaching rate (Tonietti et al., 2024). These microorganisms also release organic and inorganic acids, which help to dissolve metals (Sethi et al., 2025). Regularly employed genera aid in the bio-oxidation of sulfides and release the necessary metals into the liquid phase. These include chemo-litho-autotrophic bacteria that can oxidize sulfur, such as *Acidithio bacillus* spp. *Thiomonas* spp., and *Sulfolobus* spp., or bacteria that can oxidize iron, such as *Leptospirillum* spp., *Ferroplasma* spp., and *T. cuprinus* is a facultative chemolithoautotrophic bacterium that oxidizes metal sulfides but does not oxidize ferrous iron (Table 1).

Fungi have also been investigated for bioleaching because of their capacity to leach metals. Species such as *Aspergillus* spp., *Penicillium* spp., and *Ganoderma* spp. have been proven involved in metal extraction (Ahmed et al., 2024). Filamentous fungi are heterotrophs capable of solubilizing metals through the secretion of organic acids (Sarkodie et al., 2022). Metals solubilization with fungi involves acidolysis, redoxolysis, bioaccumulation, chelate, and complex formation mechanisms (Bala et al., 2024; Das, Dey, Bala, et al., 2024; Ghosh, Tripathy, et al., 2023). Fungal leaching of metals has been documented, with *Aspergillus* and *Penicillium* genera fungi being among the most effective for biological leaching (Nkuna & Matambo, 2024). The capacity of microorganisms to tolerate metals has been examined, with *Aspergillus foetidus* being a multi-metal tolerant fungus.

3 Current Advances in Omics Technology in Bioleaching

Genome sequencing has significantly advanced our understanding of bioleaching microbes (Fig. 1). Bioinformatic analyses reveal their genetic and metabolic potential, highlighting roles in iron and sulfur oxidation, nitrogen metabolism, quorum sensing, and heavy metal resistance. These insights enhance our understanding of their adaptive mechanisms in diverse environments. The first genome of a bioleaching microbe, *Acidithiobacillus ferrooxidans* ATCC 23270, was published in 2000, leading to predictions about its metabolic pathways (Muñoz-Villagrán et al., 2022). Over the years, additional genomes have been sequenced, expanding knowledge about other bioleaching microbes such as *A. caldus*, *A. thiooxidans*, and *Sulfobacillus*

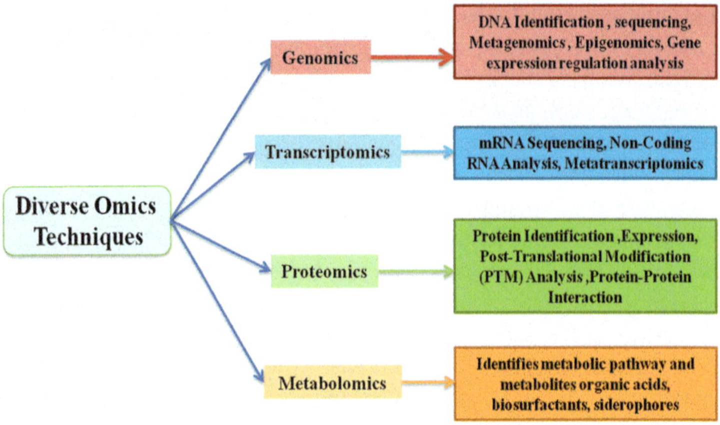

Fig. 1 Overview of diverse omics techniques and their applications

spp. Microbes' origin and ecological succession in bioleaching heaps remain key research areas. While some heaps are inoculated with microbial cultures, many are colonized naturally by acidophiles present in ore deposits (Das, Behera, & Das, 2024; John & Das, 2024; Mohanty, Dey, et al., 2024; Sahu et al., 2024). Often, chemolithoautotrophs use iron and sulfur as energy sources. The acidic environment develops as sulfide minerals react with air and water, potentially awakening dormant microbial communities. As bioleaching heaps heat up due to microbial activity, thermophiles dominate. However, their source remains unclear. Some may exist in a dormant state within the ore and become active at higher temperatures. The transition of microbial communities as the temperature rises, particularly in iron and sulfur oxidation, is still poorly understood. Future metagenomics and transcriptomics research could clarify these microbial interactions and improve bioleaching efficiency (Das & Ghosh, 2022; Das & Ghosh, 2023; Ghosh & Das, 2020; Ghosh, Sinha, et al., 2023). Both bioleaching heaps and AMDs support chemolithoautotrophs due to nutrient limitations, promoting primary biological production. The microbial community composition in bioleaching heaps is influenced by fluctuating metal and proton concentrations, leading to variations from AMD microbial populations.

3.1 Genomics

Diverse microbial communities help in bioleaching (Fig. 1). To understand how these microbes works in metal leaching environment is essential for enhancing the leaching process. Genomic is an advanced approach of microbes involved in bioleaching and their molecular characterization. Due to advancements in genomics and molecular techniques, there is now more research on extreme mineral leaching conditions. Microbial population dynamics can be monitored using technologies including DGGE, FISH, and qPCR (Martinez et al., 2015). Genomic research has revealed complex RISC oxidation pathways in bacteria and archaea (Wang et al., 2020). Novel metabolic features have been identified in microbes like *A. ferrooxidans*, *L. ferriphilum*, *A. caldus*, and *S. thermosulfidooxidans*. These findings will boost the effectiveness of industrial bioleaching (Arias et al., 2024; Christel et al., 2018; Valdés et al., 2008).

It was once thought that *Acidithiobacillus ferrooxidans* was the most significant bacterium responsible for metal sulfide bioleaching. However, more recent study has identified other microbes which may have economic application. Researchers discovered *Acidiplasma cupricumulans* in heap solids and mining solutions. The Agnes gold mine in South Africa was habitat to the most common *Sulfobacillus* species (Martinez et al., 2015). These findings have improved the comprehension of acidophilic biodiversity. Research on microbial diversity has brought about significant findings. The most frequently encountered microorganisms in recent lab-scale bioleaching investigations of low-grade copper sulfide ore have been identified to be Acidithiobacillus and Sulfobacillus. *Leptospirillum* species were found in the tailings from the copper mine La Andina in Chile. Although *Anoxybacillus* is known to be an alkaliphilic genus, it was also found in acidic environment (Diaby et al., 2007). Investigation indicates that it might accommodate colonies of acid-leaching microorganisms in specific circumstances. Future developments in genomics will give researchers more control over the bioleaching process. The next-generation sequencing is going to enhance our understanding of microbial metabolic pathways (Wensel et al., 2022). Due to technological models, researchers now have a clear understanding of how iron and sulfur compounds oxidize.

3.2 Transcriptomics

Transcriptomics studies RNA molecules transcribed by an organism using high-throughput technologies such as microarrays and RNA sequencing (Lowe et al., 2017). This field provides insights into gene expression dynamics, particularly in bioleaching microorganisms, under varying environmental conditions and stressors (Fig. 1). DNA microarray technology has usually been employed to know the transcriptomic responses in bioleaching (Zhou et al., 2018). It plays a crucial role in determining regulatory responses when these microorganisms are exposed to different stressors like extreme pH or organic extractant. Study shows when *Acidithiobacillus ferrooxidans ATCC 23270* comes in contact with the organic extractant Lix984n, it resulted in the upregulation of genes (Zhou et al., 2012). These genes can be involved in the pentose phosphate pathway and synthesis of fatty acids. Moreover, genes that encoding electron transport proteins such as petI, petII, cyo, and cyd increases in *A. ferrooxidans* under high stress conditions (Li & Wen, 2021). In extreme pH *A. ferrooxidans* uphold cytoplasmic homeostasis through hydrogen absorption and nitrogen fixation. RNA-seq has more accuracy and sensitivity than DNA microarrays. This technique is significantly contributed to understand the diverse mechanism in response to various metal ions. Microbes can adapt to various energy sources, different pH levels and heavy metal stress. Studies reveal different adaptive mechanism in *Acidithiobacillus. A. ferriphilus SCUT-1* under different energy sources exhibit genes involved in ferrous ion oxidation pathways. *A. thiooxidans ATCC 19377* has a strong metabolic reaction to sulfur-based energy sources in a similar manner (Ibáñez et al., 2023). *A. caldus* and *A. thiooxidans* under acidic condition exhibit gene regulation for flagellar assembly, sulfur and carbon metabolism, and fatty acid production. RNA-Seq also demonstrated how microbes responds to magnesium stress. In *Acidithiobacillus ferrooxidans*, the type IV fimbriae and carbon fixing genes are down regulated (Yadavalli et al., 2020). Furthermore, transcriptome studies of genetically modified strains have also provide information on microbial survival in extremely acidic conditions. For instance, *A. caldus* MTH-04 fur knockout strain give valuable insight into microbe survival in extreme acidic conditions (Shi et al., 2023).

3.3 Metatranscriptomics

Metatranscriptomics is an advanced method which employs sequencing of entire RNA or mRNA (Aguiar-Pulido et al., 2016). It contributes to the knowledge of gene expression in microbial communities. This technique allows researchers to investigate microbial structure, function, and responses to environmental changes. For example, transcriptome analysis of *Leptospirillum* spp. biofilms revealed metabolic differences between biofilm-attached and planktonic cells (Moreno-Paz et al., 2010). Biofilm-attached cells prefer mixed acid fermentation, whereas planktonic cells use a more active TCA cycle. There are significant differences in the microbial biofilms in natural environment and those in lab bioreactors. This demonstrates how environmental influences affect the variety of biofilms (Das, Dey, & Das, 2024; Dey, Das, et al., 2023; Dey, Dash, & Das, 2024; Lahiri et al., 2021). Additionally, metatranscriptomic analysis sheds light on the functions that microbes play in biomining. In early bio-mining phases, studies on a six-strain microbial consortium revealed strong gene expression for catalytic activities in *A. caldus* and cell proliferation in *Sulfobacillus thermosulfidooxidans* (Christel et al., 2018). Genes associated with localization in *A. caldus*, transport functions in *S. thermosulfidooxidans*, and stress response and signal transduction in *Leptospirillum ferriphilum* were elevated in later stages (Li & Wen, 2021). Adaptive measures against metal toxicity and metabolite buildup are shown in these modifications.

3.4 Proteomics

The quantitative and qualitative study of proteins in microbes under stress conditions is known as proteomics (Al-Amrani et al., 2021). It is commonly used to investigate the proteins involved in stress adaptation in microbes (Fig. 1). Environmental changes such as pH, nutrient availability, temperature, and high metal stress conditions significantly influence microbial responses (Jagatee et al., 2024; Priyadarshini & Das, 2024; Satapathy et al., 2024). These stressors can modify in protein expressions and structures. Hence a comparative proteomics study is essential to understand these modifications and analyze the adaptive mechanism of microbes under stress conditions. The proteomic investigations on *A. ferrooxidans* have revealed diverse proteins involved in stress adaptation (Bellenberg et al., 2019). Likewise *A. caldus* and *L. ferriphilum* exhibit specific protein expression patterns in response to stress such as pH, NaCl, and arsenic exposure (Panyushkina et al., 2020).

3.5 Metaproteomics

Metaproteomics is an analytical approach of study of entire protein composition of microbial communities in the ecosystem (Shrestha et al., 2021). Unlike proteomics which focuses on individual microbes, it is a complete analysis of collective proteins involved in different microbial diversity and its interactions with the environments (Fig. 1). It can uncover the role of collective proteins involved in biogeochemical cycling, degradation, and industrial bioprocess recycling (Dey, Bissoyi, et al., 2024; Dey, Tripathy, et al., 2023; Mohanty, Tripathy, et al., 2024; Ray et al., 2024). For complex ecosystem like mining soil, drainage water, industrial effluents, acid mine drainage, this technique is very much useful. It provides insights into how proteins work into different environments. By knowing collective protein expression profiles of diverse microbial ecosystem metaproteomics provides helpful interpretations into their functional dynamics (Xiong et al., 2015). Advanced methods like isotope leveling enhance the protein identification and quantification.

3.6 Metabolomics

Proteomics and metaproteomics have provided insights into microbial survival strategies in metal-rich environments. But the emerging field of metabolomics offers new possibilities for process optimization through biomarker identification.

Metabolomics is a detailed understanding of various metabolic alterations in microbes. Metabolomics contribute to the knowledge regarding bio-mining microbial metabolism particularly in harsh conditions (Tang, 2011). It also offers valuable insights in the formation of biofilms in different microbes. The bioleaching process happens when sulfur and iron is oxidized by microbes. Environmental stress factors such as high acidic conditions, osmotic pressure, and heavy metals affects their metabolic activities (Priyadarsini & Das, 2025). Metabolomics enables the researchers to evaluate physiological responses and microbial adaptations at the time of bioleaching (Li & Wen, 2021). These are crucial for application of microbes in industries as well as diverse environmental conditions. Some researchers also investigated adaptation of microbes to acidic metal-rich environment utilizing advanced metabolomic methods (Mosier et al., 2013). Some of the samples were taken from acid mine drainage sites and analyzed using fluorescence in situ hybridization (FISH) and liquid chromatography-mass spectrometry (LC-MS). A total of 56 metabolites have been identified with more than over 90% remain uncharacterized. Some significant metabolites that are discovered are phosphatidylethanolamine lipids, taurine, and hydroxyectoine that contributes to stress adaptability (Mosier et al., 2013). Most of the acidophilic microbes have phosphatidylethanolamine metabolites. Those help them to tolerate high acidic environment. Hydroxyectoine is a metabolite found in microbes for adapting osmotic stress (Czech et al., 2018; Mosier et al., 2013). Taurine is identified as osmoprotectant, mostly found in *sulfobacilli* (Martinez et al., 2015; Mosier et al., 2013). The presence of osmoprotectant metabolites during biofilm development indicates bacteria encounters harsher circumstances. Many of these stress responsive metabolites have potential as indicator for enhancing industrial bio-mining operations and environmental monitoring (Das, Dey, Bala, et al., 2024).

4 Future Scope and Challenges

Advancing in omics-driven bioleaching in future can contribute to sustainable and eco-friendly metal recovery solutions. Future studies should focus on integrating these techniques to develop predictive models for bioleaching efficiency. Additionally, refining omics-based approaches for process control and optimization will enhance the industrial viability of bioleaching (Pattanaik et al., 2021). Despite many advantages, individual omics approaches have limitations. Transcriptomics and metatranscriptomics have been used less frequently in bioleaching than other omics methods. Metatranscriptomics may not be adequate for monitoring the metabolic pathways and expression of genes of numerous species. This is due to technological constraints such as short sequencing reads, difficulties reconstructing full-length transcripts, and sequencing RNA modifications. Advanced techniques like spatial transcriptomics (CITE-seq) and long-read transcriptomics could potentially help to address all of these challenges. These advancements might help researchers better understand gene expression, regulatory networks, and metabolic flexibility in complicated microbial environmental systems. Overcoming the challenges of effectively linking DNA sequencing, RNA sequencing, and mass spectrometry data will be key to advancing multi-omics research.

5 Conclusion

The bioleaching process has been significantly improved using omics technology. Omics technology plays a crucial role in exploring the mechanism that enables bioleaching microbes to survive in extreme environments. It assists in identifying each microbe's functional elements and stress-resistance mechanisms. Advancing omics technologies paves the pathways for development of more efficient bioleaching. Modern molecular technologies such as metagenomics, transcriptomics, metatranscriptomics, proteomics, metaproteomics, metabolomics, and genomics are integrated to uncover novel compounds, investigate interactions between genes, proteins, metabolic pathways and contribute to evolutionary relationships. It highlights the complex relationships between microbial communities and how environmental conditions affect how they operate. Additionally these methods also emphasize the crucial functions that low-abundance species play in the microbial ecology and aid in their identification. Furthermore, researching the dynamics of microbial communities offers insights into cooperative relationships to maximize the effectiveness of bioleaching. By bridging the gaps integration of multi-omics technology will contribute to more sustainable and efficient bioleaching process.

Acknowledgements The authors are grateful to Prof. (Dr.) Manojranjan Nayak, President of Siksha 'O' Anusandhan University, for providing infrastructure and encouragement throughout the research process.

Declaration of Competing Interest The authors declare that they have no known competing financial interests or personal relationships that could have appeared to influence the work reported in this paper.

Credit Authorship Contribution Statement D. P. Krishna Samal: Conceptualization, Writing—original draft, review. Lala Behari Sukla: Supervision, editing.

References

Aguiar-Pulido, V., Huang, W., Suarez-Ulloa, V., Cickovski, T., Mathee, K., & Narasimhan, G. (2016). Metagenomics, metatranscriptomics, and metabolomics approaches for microbiome analysis. *Evolutionary Bioinformatics, 12*(Suppl 1), 5–16. https://doi.org/10.4137/EBO.S36436

Ahmed, B. M., Mohammed, A. A., Kawady, N. A., et al. (2024). Cost-effective bioleaching of valuable metals from low-grade ore using *aspergillus nidulans*. *International Journal of Environmental Science and Technology, 21*, 5469–5482. https://doi.org/10.1007/s13762-023-05355-0

Al-Amrani, S., Al-Jabri, Z., Al-Zaabi, A., Alshekaili, J., & Al-Khabori, M. (2021). Proteomics: Concepts and applications in human medicine. *World Journal of Biological Chemistry, 12*(5), 57–69. https://doi.org/10.4331/wjbc.v12.i5.57

Arias, D., Zepeda, V., Nancucheo, I., Saldaña, M., & Galleguillos, P. A. (2024). Osmotic response in Leptospirillum ferriphilum isolated from an industrial copper bioleaching environment to sulfate. *Frontiers in Microbiology, 15*, 1369244. https://doi.org/10.3389/fmicb.2024.1369244

Bala, K., Anshul Verma, S., Dogra, A., Das, A. P., Kumar, R., & Nandini. (2024). Sustainable management for urban plastic waste generation. In *Sustainable management of urban plastic waste through circular economic approaches* (pp. 211–236). Routledge. https://doi.org/10.1201/9781003465447-13

Bellenberg, S., Huynh, D., Poetsch, A., Sand, W., & Vera, M. (2019). Proteomics reveal enhanced oxidative stress responses and metabolic adaptation in Acidithiobacillus ferrooxidans biofilm cells on pyrite. *Frontiers in Microbiology, 10*, 592. https://doi.org/10.3389/fmicb.2019.00592

Christel, S., Herold, M., Bellenberg, S., Buetti-Dinh, A., El Hajjami, M., Pivkin, I. V., Sand, W., Wilmes, P., Poetsch, A., Vera, M., & Dopson, M. (2018). Weak iron oxidation by Sulfobacillus thermosulfidooxidans maintains a favorable redox potential for chalcopyrite bioleaching. *Frontiers in Microbiology, 9*, 3059. https://doi.org/10.3389/fmicb.2018.03059

Czech, L., Hermann, L., Stöveken, N., Richter, A. A., Höppner, A., Smits, S. H. J., Heider, J., & Bremer, E. (2018). Role of the extremolytes ectoine and hydroxyectoine as stress protectants and nutrients: Genetics, phylogenomics, biochemistry, and structural analysis. *Genes, 9*(4), 177. https://doi.org/10.3390/genes9040177

Das, A., Dey, S., Bala, K., Gnanasekaran, R., & Das, A. P. (2024). Microbial bioremediation technology for sustainable treatment and management of synthetic microfiber waste. In *Sustainable microbial technology for synthetic and cellulosic microfiber bioremediation* (pp. 221–238). Springer. https://doi.org/10.1007/978-3-031-62600-5_12

Das, A., Dey, S., & Das, A. P. (2024). Ecotoxicological impacts of synthetic microfiber pollutants and development of sustainable mitigation strategies. *Environmental Chemistry and Ecotoxicology*. https://doi.org/10.1016/J.ENCECO.2024.12.003

Das, A., & Ghosh, S. (2022). Role of microorganisms in extenuation of mining and industrial wastes. *Geomicrobiology Journal, 39*, 173–175. https://doi.org/10.1080/01490451.2022.2038953

Das, A. P., Behera, I. D., & Das, N. P. (2024). *Renewable energy generation and value addition from environmental microfiber pollution through advanced greener solution* (p. 278). Springer.

Das, A. P., & Ghosh, S. (2023). Evolution of biomining technology. In *Manganese mining microorganisms* (pp. 5–19). Elsevier. https://doi.org/10.1016/B978-0-12-822147-1.00012-1

Dey, S., Bissoyi, A., Koteswara Reddy, G., & Das, A. P. (2024). Synthetic microfibers: An urban emerging pollutant demanding for sustainable environmental solutions. In *Sustainable microbial technology for synthetic and cellulosic microfiber bioremediation* (pp. 107–121). Springer. https://doi.org/10.1007/978-3-031-62600-5_7

Dey, S., Das, A., Mallick, K., Sahu, A., & Das, A. P. (2023). *Environmental petroleum waste: Pollution, toxicity, sustainable remediation* (pp. 159–175). Springer. https://doi.org/10.1007/978-3-031-48220-5_7

Dey, S., Dash, A., & Das, A. P. (2024). Bioremediation as a sustainable and circular solution for mining waste management and bioenergy generation. In *Sustainable management of mining waste and tailings: a circular economy approach* (pp. 107–119). Springer. https://doi.org/10.1201/9781003442455-6

Dey, S., Tripathy, B., Kumar, M. S., & Das, A. P. (2023). Ecotoxicological consequences of manganese mining pollutants and their biological remediation. *Environmental Chemistry and Ecotoxicology*. https://doi.org/10.1016/J.ENCECO.2023.01.001

Diaby, N., Dold, B., Pfeifer, H.-R., Holliger, C., Johnson, D. B., & Hallberg, K. B. (2007). Microbial communities in a porphyry copper tailings impoundment and their impact on the geochemical dynamics of the mine waste. *Environmental Microbiology, 9*, 298–307. https://doi.org/10.1111/j.1462-2920.2006.01138.x

Ghosh, S., & Das, A. P. (2020). Microbial metagenomics: Current advances in investigating microbial ecology and population dynamics. In S. K. Nayak & B. B. Mishra (Eds.), *Frontiers in soil and environmental microbiology* (1st ed., pp. 247–254). CRC Press.

Ghosh, S., Sinha, J. K., Ghosh, S., Vashisth, K., Han, S., & Bhaskar, R. (2023). Microplastics as an emerging threat to the global environment and human health. *Sustainability, 15*(14), 10821. https://doi.org/10.3390/SU151410821

Ghosh, S., Tripathy, B., Dey, S., & Das, A. P. (2023). Comparative investigation of fungal and bacterial manganese biomining mechanisms. In *Manganese mining microorganisms* (pp. 99–115). Elsevier. https://doi.org/10.1016/b978-0-12-822147-1.00016-9

Hetz, S. A., & Schippers, A. (2024). Do ferrous iron-oxidizing acidophiles (*Leptospirillum* spp.) disturb aerobic bioleaching of laterite ores by sulfur-oxidizing acidophiles (*Acidithiobacillus* spp.)? *Frontiers in Microbiology, 15*, 1359019. https://doi.org/10.3389/fmicb.2024.1359019

Ibáñez, A., Garrido-Chamorro, S., Coque, J. J. R., & Barreiro, C. (2023). From genes to bioleaching: Unraveling sulfur metabolism in Acidithiobacillus genus. *Genes, 14*(9), 1772. https://doi.org/10.3390/genes14091772

Jagatee, S., Priyadarshini, S., Rath, C. C., & Das, A. P. (2024). Synthetic microfiber: An enduring environmental problem linked to sustainable development. *Environmental Science and Engineering, 2333*, 93–112. https://doi.org/10.1007/978-3-031-51792-1_6

Jain, A., Sarsaiya, S., Singh, R., Gong, Q., Wu, Q., & Shi, J. (2024). Omics approaches in understanding the benefits of plant-microbe interactions. *Frontiers in Microbiology, 15*, 1391059. https://doi.org/10.3389/fmicb.2024.1391059

John, A., & Das, A. P. (2024). Government strategies for achieving circular economy for plastic waste management in Southeast Asia. In *Sustainable management of urban plastic waste through circular economic approaches* (pp. 72–82). Routledge. https://doi.org/10.1201/9781003465447-6

Jones, S., & Santini, J. M. (2023). Mechanisms of bioleaching: Iron and sulfur oxidation by acidophilic microorganisms. *Essays in Biochemistry, 67*(4), 685–699. https://doi.org/10.1042/EBC20220257

Kumar, U., Raj, S., Sreenikethanam, A., Maddheshiya, R., Kumari, S., Han, S., Kapoor, K. K., Bhaskar, R., Bajhaiya, A. K., & Gahlot, D. K. (2023). Multi-omics approaches in plant–microbe interactions hold enormous promise for sustainable agriculture. *Agronomy, 13*(7), 1804. https://doi.org/10.3390/agronomy13071804

Lahiri, D., Nag, M., Dey, A., Sarkar, T., Joshi, S., Pandit, S., Das, A. P., Pati, S., Pattanaik, S., Tilak, V. K., & Ray, R. R. (2021). Biofilm mediated degradation of petroleum products. *Geomicrobiology Journal*. https://doi.org/10.1080/01490451.2021.1968979

Li, M., & Wen, J. (2021). Recent progress in the application of omics technologies in the study of bio-mining microorganisms from extreme environments. *Microbial Cell Factories, 20*(1), 178. https://doi.org/10.1186/s12934-021-01671-7

Lowe, R., Shirley, N., Bleackley, M., Dolan, S., & Shafee, T. (2017). Transcriptomics technologies. *PLoS Computational Biology, 13*(5), e1005457. https://doi.org/10.1371/journal.pcbi.1005457

Martinez, P., Vera, M., & Bobadilla-Fazzini, R. A. (2015). Omics on bioleaching: Current and future impacts. *Applied Microbiology and Biotechnology, 99*, 8337–8350. https://doi.org/10.1007/s00253-015-6903-8

Mishra, S., Dash, D., & Das, A. P. (2024). Microbial degradation of polyester microfibers using indigenously isolated bacterial strain Exiguobacterium sp. *CLEAN, 52*(12), e202300343. https://doi.org/10.1002/CLEN.202300343

Mohanty, J., Dey, S., Bala, K., & Das, A. P. (2024). Recent advances in microplastic pollution for its sustainable management. In *Sustainable management of urban plastic waste through circular economic approaches* (pp. 27–48). Routledge. https://doi.org/10.1201/9781003465447-3

Mohanty, M., Tripathy, B., Koteswara Reddy, G., & Das, A. P. (2024). Greener technologies for establishing restraint over microplastic pollution. In *Sustainable microbial technology for synthetic and cellulosic microfiber bioremediation* (pp. 239–258). Springer. https://doi.org/10.1007/978-3-031-62600-5_13

Moreno-Paz, M., Gómez, M. J., Arcas, A., & Parro, V. (2010). Environmental transcriptome analysis reveals physiological differences between biofilm and planktonic modes of life of the iron oxidizing bacteria Leptospirillum spp. in their natural microbial community. *BMC Genomics, 11*, 404. https://doi.org/10.1186/1471-2164-11-404

Mosier, A. C., Justice, N. B., Bowen, B. P., Baran, R., Thomas, B. C., Northen, T. R., & Banfield, J. F. (2013). Metabolites associated with adaptation of microorganisms to an acidophilic, metal-rich environment identified by stable-isotope-enabled metabolomics. *mBio, 4*(2), e00484. https://doi.org/10.1128/mBio.00484-12

Muñoz-Villagrán, C., Grossolli-Gálvez, J., Acevedo-Arbunic, J., Valenzuela, X., Ferrer, A., Díez, B., & Levicán, G. (2022). Characterization and genomic analysis of two novel psychrotolerant Acidithiobacillus ferrooxidans strains from polar and subpolar environments. *Frontiers in Microbiology, 13*, 960324. https://doi.org/10.3389/fmicb.2022.960324

Naqvi, R. Z., Mahmood, M. A., Mansoor, S., Amin, I., & Asif, M. (2024). Omics-driven exploration and mining of key functional genes for the improvement of food and fiber crops. *Frontiers in Plant Science, 14*, 1273859. https://doi.org/10.3389/fpls.2023.1273859

Nkuna, R., & Matambo, T. S. (2024). Determining the metabolic processes of metal-tolerant fungi isolated from mine tailings for bioleaching. *Minerals, 14*(3), 235. https://doi.org/10.3390/min14030235

Panyushkina, A., Matyushkina, D., & Pobeguts, O. (2020). Understanding stress response to high-arsenic gold-bearing sulfide concentrate in extremely metal-resistant acidophile Sulfobacillus thermotolerans. *Microorganisms, 8*(7), 1076. https://doi.org/10.3390/microorganisms8071076

Pattanaik, A., Samal, D. P. K., Sukla, L., & Pradhan, D. (2021). Advancements and use of omic technologies in the field of bioleaching: A review. *Biointerface Research in Applied Chemistry, 11*, 10185–10204. https://doi.org/10.33263/BRIAC113.1018510204

Priyadarshini, B., & Das, A. P. (2024). The adverse impact of synthetic microplastics released from disposable face masks pollutants on aquatic microbiota, aquatic species and humans: A review. *Environmental Science and Engineering, 2827*, 67–81. https://doi.org/10.1007/978-3-031-50840-0_5

Priyadarsini, S., & Das, A. P. (2025). Lithium bioleaching: A review on microbial-assisted sustainable technology for lithium bio-circularity. *Journal of Water Process Engineering, 69*, 106744. https://doi.org/10.1016/j.jwpe.2024.106744

Ray, S. K., Dey, S., Das, A. P., & John, A. (2024). Microplastic waste Management through greener technologies: A review. In *Sustainable management of urban plastic waste through circular economic approaches* (pp. 237–261). Springer. https://doi.org/10.1201/9781003465447-14

Sahu, A., Dey, S., Mallick, K., Priyadarsini, S., Behera, I. D., & Das, A. P. (2024). A review on advanced technology for sustainable management of synthetic microplastic waste. In *Sustainable management of urban plastic waste through circular economic approaches* (pp. 262–279). Springer. https://doi.org/10.1201/9781003465447-15

Saldaña, M., Jeldres, M., Galleguillos Madrid, F. M., Gallegos, S., Salazar, I., Robles, P., & Toro, N. (2023). Bioleaching modeling—A review. *Materials, 16*(10), 3812. https://doi.org/10.3390/ma16103812

Samal, D. P. K., Sukla, L. B., Pattanaik, A., & Pradhan, D. (2020). Role of microalgae in treatment of acid mine drainage and recovery of valuable metals. *Materials Today: Proceedings, 30*(Part 2), 346–350. https://doi.org/10.1016/j.matpr.2020.02.165

Sarkodie, E. K., Jiang, L., Li, K., Yang, J., Guo, Z., Shi, J., Deng, Y., Liu, H., Jiang, H., Liang, Y., Yin, H., & Liu, X. (2022). A review on the bioleaching of toxic metal(loid)s from contaminated soil: Insight into the mechanism of action and the role of influencing factors. *Frontiers in Microbiology, 13*, 1049277. https://doi.org/10.3389/fmicb.2022.1049277

Satapathy, S., Kar, D., Priyadarsini, S., Das, A. P., & Kuanar, A. (2024). Application of microorganisms in emerging waste management strategies and remediation of synthetic and cellulosic microfibers. In *Sustainable microbial technology for synthetic and cellulosic microfiber bioremediation* (pp. 183–206). Springer. https://doi.org/10.1007/978-3-031-62600-5_10

Sethi, S., Jonwal, H., & Parihar, R. (2025). Metal solubilization through bioleaching: Microbial action. In M. P. Shah (Ed.), *Extraction of precious metals from industrial waste using microbial technology. Environmental science and engineering*. Springer. https://doi.org/10.1007/978-3-031-83623-7_2

Shi, Y., Wu, W., Yang, Y., Liu, X., Lin, J., Liu, X., Lin, J., & Pang, X. (2023). Gene knockout of glutathione reductase results in increased sensitivity to heavy metals in Acidithiobacillus caldus. *Frontiers in Microbiology, 14*, 1250330. https://doi.org/10.3389/fmicb.2023.1250330

Shrestha, H. K., Appidi, M. R., Villalobos Solis, M. I., et al. (2021). Metaproteomics reveals insights into microbial structure, interactions, and dynamic regulation in defined communities as they respond to environmental disturbance. *BMC Microbiology, 21*, 308. https://doi.org/10.1186/s12866-021-02370-4

Tang, J. (2011). Microbial metabolomics. *Current Genomics, 12*(6), 391–403. https://doi.org/10.2174/138920211797248619

Tezyapar Kara, I., Kremser, K., Wagland, S. T., et al. (2023). Bioleaching metal-bearing wastes and by-products for resource recovery: A review. *Environmental Chemistry Letters, 21*, 3329–3350. https://doi.org/10.1007/s10311-023-01611-4

Tonietti, L., Esposito, M., Cascone, M., Barosa, B., Fiscale, S., Muscari Tomajoli, M. T., Sbaffi, T., Santomartino, R., Covone, G., Cordone, A., Rotundi, A., & Giovannelli, D. (2024). Unveiling the bioleaching versatility of *Acidithiobacillus ferrooxidans*. *Microorganisms, 12*(12), 2407. https://doi.org/10.3390/microorganisms12122407

Valdés, J., Pedroso, I., & Quatrini, R. (2008). Acidithiobacillus ferrooxidans metabolism: From genome sequence to industrial applications. *BMC Genomics, 9*, 597. https://doi.org/10.1186/1471-2164-9-597

Wang, P., Li, L. Z., Qin, Y. L., Liang, Z. L., Li, X. T., Yin, H. Q., Liu, L. J., Liu, S. J., & Jiang, C. Y. (2020). Comparative genomic analysis reveals the metabolism and evolution of the thermophilic archaeal genus Metallosphaera. *Frontiers in Microbiology, 11*, 1192. https://doi.org/10.3389/fmicb.2020.01192

Wensel, C. R., Pluznick, J. L., Salzberg, S. L., & Sears, C. L. (2022). Next-generation sequencing: Insights to advance clinical investigations of the microbiome. *Journal of Clinical Investigation, 132*(7), e154944. https://doi.org/10.1172/JCI154944

Xiong, W., Abraham, P. E., Li, Z., Pan, C., & Hettich, R. L. (2015). Microbial metaproteomics for characterizing the range of metabolic functions and activities of human gut microbiota. *Proteomics, 15*(20), 3424–3438. https://doi.org/10.1002/pmic.201400571

Yadavalli, S. S., Goh, T., Carey, J. N., Malengo, G., Vellappan, S., Nickels, B. E., Sourjik, V., Goulian, M., Yuan, J., & Silhavy, T. J. (2020). Functional determinants of a small protein controlling a broadly conserved bacterial sensor kinase. *Journal of Bacteriology, 202*, e00305.

Zhou, S., Gan, M., Zhu, J., Liu, X., & Qiu, G. (2018). Assessment of bioleaching microbial community structure and function based on next-generation sequencing technologies. *Minerals, 8*(12), 596. https://doi.org/10.3390/min8120596

Zhou, Z., Fang, Y., Li, Q., Yin, H., Qin, W., Liang, Y., Li, Q., Li, N., Liu, X., Qiu, G., & Liu, X. (2012). Global transcriptional analysis of stress-response strategies in Acidithiobacillus ferrooxidans ATCC 23270 exposed to organic extractant–Lix984n. *World Journal of Microbiology and Biotechnology, 28*(3), 1045–1055. https://doi.org/10.1007/s11274-011-0903-3

Turning Waste into Wealth: Challenges, Opportunities, and Strategies in Mining Waste Management

Suchismita Sahu, M. Sai Subham Priya, Amel Gacem, Sailaja Priyadarsini, Maha Awjan Alreshidi, Krishna Kumar Yadav, and Alok Prasad Das

1 Introduction

1.1 Overview of Mining Waste Generation

Make it tailling is any unwanted material that is result from the extraction, processing and benefication of any mineral ore (Das et al., 2023; Das & Ghosh, 2018, 2022; Ghosh & Das, 2017). Mining waste can be solid, semi-solid, or liquid, contained vapours. It is one of the largest streams of waste production contains good amount of hazardous harmful substances (Dey, Tripathy, et al., 2023; Mohanty et al., 2017; Pradhan et al., 2023). Mining waste can be of many types such as waste rock, tailings, overburden, slag, gaseous waste, mine water, etc. Mining is a major industry for 7% of global GDP.

Many abandoned mines are present in various parts like Canada, Australia, South Africa, China having 10,000, 50,000, 6000, 9500, respectively. It is predicted that the abandoned mines in China will reach up to 15,000 by 2050. The main sources

S. Sahu · M. S. S. Priya · S. Priyadarsini · A. P. Das (✉)
Department of Life Sciences, Ramadevi Women's University, Bhubaneswar, Odisha, India

A. Gacem
Department of Physics, Faculty of Sciences, University 20 Août 1955, Skikda, Algeria

M. A. Alreshidi
Department of Chemistry, College of Science, University of Ha'il, Ha'il, Saudi Arabia

K. K. Yadav
Department of Environmental Science, Parul Institute of Applied Sciences, Parul University, Vadodara, Gujarat, India

Environmental and Atmospheric Sciences Research Group, Scientific Research Center, Al-Ayen University, Nasiriyah, Thi-Qar, Iraq

© The Author(s), under exclusive license to Springer Nature Switzerland AG 2026

A. P. Das (ed.), *Bioremediation Technology for Sustainable Mining Waste Management*, Sustainable Environmental Waste Management Strategies,
https://doi.org/10.1007/978-3-032-03468-7_13

of mine waste or mine tailings are from metals, gypsum, metallurgical, steel slags, etc. The countries with major mining outputs such as China, India, USA, Australia, South Africa, Russia etc., have over 1,30,000 of coal mines present and these coal mines generate about 1000 to 2500 million tons (Mt) of mine tailings yearly (Bal et al., 2018; Das et al., 2015; Mohanty et al., 2018).

After extensive activities by the mining industries, leakage from tailing areas often occurs and eventually enters the ecosystem, primarily contaminating nearby water bodies (Dey, Dash, & Das, 2024; Ghosh, Mohanty, Nayak, et al., 2016; Sanket et al., 2016). These non-coal mine tailings/wastes comprise of 33% tailings, 17% dump, and 50.25% surface and underground wastes (1987–2000) as per the records. Quarry waste in the circular economy comprises of the concept of reusing, recycling, and recovering of the product considered as waste (Szczepańska & Twardowska, 2004).

The highest quantity of 632 million tons ferrous ore tailings is generated by Australia, followed by Canada, USA, and Africa which is less than one million tons. In Brazil, about 95% of mining tailings are preordained to 676 containment dams. Most of the tailings that are generated cannot get recycled is about 20% of the total waste or tailings produced (Lèbre et al., 2017).

Data regarding mining waste research are as follows: India obtains eighth position as one of the most generative nations in mining waste research and with 170 number of total articles per year, 0.128 number of total articles per 1.2 million inhabitants, 1492 annual number of total articles, and finally 8.8 citations by article. The USA hold the highest position with 613 annual number of total articles, 1897 articles per one million inhabitants; 13,646 annual number of articles for all articles and finally 22.3 citation by article. The most frequently used words in mining waste research are mine tailings, mining, sustainable waste management, heavy metals, soils, soil pollution, metals such as copper, industrial waste, iron, soil pollutants, etc. Waste generated/produced from mining mostly affects (pollutes) our three environmentally important sources, which are air, water, and soil (Das, Ray et al., 2024; Dey, Dash, Ray, et al., 2024; Ghosh, Mohanty, Akcil, et al., 2016). Most of the articles on mining waste research are published in the year 2017 which is in the year 2017 which is 108 annual number of articles. In terms of effective mining waste management, the United States tops the list, whereas India is seventh position.

1.2 Importance of Sustainable Mining Practices

Mining waste management means the proper disposal of the unwanted substances. Proper waste management means it make sure the long-term stability and, it also minimizes the water pollutants and soil pollutants from acid and base drainage and straining of heavy metals (Das & Ghosh, 2023b, 2023c; Ghosh et al., 2023).

Tailings reduction can be done by returning of the waste to the void by developing selective mining methods or by designing new mining methods to minimize waste output. Reduce, reprocess, stockpile, downcycle, and disposing waste (Das & Ghosh, 2023d, 2023e, 2023g).

The waste can be disposed in tailing dams, exhausted in mine pits, in piles, dry stacking (in case of cold climates), in paste underground back filling, STDs (submarine waste disposal), etc. Waste can be rather used such as copper, iron ore, platinum group metals; acid mine drainage (AMD); phosphonyls; slag; electric oven powers, and lime stone power, etc. can be used as "construction materials". Acid mine drainage (AMD) causing tailings can be used as geopolymers. Sand based tailings, coal burning and copper slag can be used as landfills and source of rare earth elements.

In case of metal wastes, the main recycling/reuse process can be flotation, solvent extraction, bioleaching (biolixiviation), amalgamation, etc. (Priyadarsini & Das, 2024, 2025). Main recycling processes for gypsum waste are solvent extraction and acid leaching. And for metallurgical wastes are pyrometallurgical processes, hydrometallurgical processes, and electrometallurgical processes. Geo-polymerization technique is being researched recently to convert wastes or dump or waste into usable raw materials (Aznar-Sánchez et al., 2019).

Sustainable green technology is one of the techniques whose efforts on current waste management system should be given for getting cost-effective and high-yielding methods of 100% utilization of wastes (Das & Ghosh, 2023a, 2023f). Previously, the tailings produced by steel making process were dumped in an open space. Then shortage of dumping space led to stopping the traditional practices. The most industries were producing toxic chemical waste which is dangerous to environment (Das et al., 2024).

The green technology was also known as Geo-polymer technology is a green technology mainly adapted by steel and mining industries for the sustainable management Sustainable management of mining wastes (Singh et al., 2020).

1.3 Concept of Turning Waste into Wealth

The concept of circular economy plays a very important role in turning waste to wealth. This concept shows a very important urge of promoting sustainable development and thus reducing waste by positively impacting environment, people's heath as well as our countries' economy (Bala et al., 2024; Dey, Das, et al., 2023; Singh Jadaun et al., 2022). It acts as an alternative to the old/traditional method of resource extraction, utilization, and waste disposal which is not sustainable (Dey, Bissoyi, Koteswara Reddy, & Das, 2024; Mohanty, Dey, et al., 2024; Paul et al., 2023). Here comes the circular economy which is a strategy or a key plan to allay global waste. There are many policies of circular economy at various governance levels working on perceptive of ecological parameters. Till the year 2018, the estimated waste generated is about 2.01 billion tons. It is unlike the traditional methods of mineral extraction. Unlike unsustainable practices, the circular economy offers a sustainable approach in which waste and mine tailings are recycled and reused. Although significant efforts are being made to promote this increasingly popular concept, many people still struggle to understand its policies and applications. The circular economy is being implemented across industries, landfills, and even waterways (Cui et al., 2020).

Relationship between economic activity and environmental pollution is being studied way Environmental Kuznets Curve (EKC) framework.

Pollution at first goes exponential with income and eventually decreases as per the increase in income as the waste is finally being reused and recycled for economic uses. But the drawback is that EKC is rejected in countries like Denmark, Spain, United Kingdom, and so on as well as in European countries.

Then comes the concept of logistic operations, some Asian countries showed that logistic operations specifically the clearance process helps in generating income and as well as sustainable mining and manufacturing. The use of green gas came to use at that time which proved to be a cost-effective method (Magazzino et al., 2021).

Also, the ANNs computational method is used to reduce and reprocesses waste. This is made up of processing units. It has a multilayer network. As per this ANNs method Denmark is so close of achieving a circular economy. Urbanization has a negligible effect for GHG emissions to waste generation, while Denmark has an urban waste management system that is quite efficient in conversion of waste into energy.

2 Types and Characteristic of Mining Waste

2.1 Tailings and Sludge

Mining tailings are the residues/leftovers/remaining after any mining operations. There are mainly two kinds of mining: placer mining in which water and gravity is used for extraction of metal from ores. Another kind is hard rock mining: in this type of mining, chemical reactions go on in a series for concentration of sought after the metal extraction. The process of grounding particles into small particles ranging from size to size this process is known as "comminution". After this process, a slurry is produced from which through a mill mining waste/mining tailings are formed. These tailings include heavy metals, some toxic substance, sulphides as well as radioactive elements.

Few examples of tailings are; during the extraction of sulphide minerals the mine tailings large amount of pyrites are produced, which are harmless in underground but if not treated with proper care and precaution, it leads to the production and acid mining drainage.

Coal residues which contain fly ashes, iron leaching, manganese and aluminium residues when mixed with water leads to acid mine drainage.

Red mud or bauxite residue is an industrial waste produced during "Bayer's process". It is also produced during industrial production of aluminium.

Gypsum waste also as mining tailing produced during phosphate rock mining (Tayebi-Khorami et al., 2019).

Gold mine tailings is a type of heavy metal waste. According to some researchers, these heavy metal wastes are highly mobile in soil environment. These heavy metal wastes when produced from goldmines and not disposed or treated subsequently bioaccumulates which in turn causes health hazards to humans, other live stocks, as well as to the ecosystem.

The metals are said to be carcinogenic and causes many health problems such as nausea, vomiting, decreased sperm count, skin damage, kidney damage, etc. Gold mining is heavily carried out in African regions for more than 120 years. From these gold mine tailings radionucleotides are found which is a factor of concern as these will emit radiations harmful for the skin and health of the humans as well as for the animals. Researchers are now determining heavy metals in the soil samples of their areas as well as in the ecosystem. Main focus is about the carcinogenic effects (Ngole-Jeme & Fantke, 2017).

Arsenic mine tailings are the amount of arsenic present in mining tailings. These arsenic mining tailings causes acid mine drainage. Arsenic as well as lead solubility is found in gold mine tailings. Arsenic in mining tailings is found in redox states. Determination of these arsenic and lead content is generally done by spectrophotometry and diffractometry(Roussel et al., 2000).

Sludge is mainly mud which is a mining waste of iron ore mining or any other heavy metal extraction mining. It is a thick semiliquid slurry. It has components like sand, clay, gravel, and water. It is also called as slickens. Some miners want to reprocess this sludge to extract residual metals like gold or any other heavy metal remaining. In times sludge caused many damages to the live stocks thus came to the study by the researchers and scientists. They are working for the proper disposal and treatment of these mining tailings and sludge.

2.2 *Overburden and Waste Rock*

Waste rock dumped from a mining site in a heavy amount is known as "overburden". It is mainly the mining waste of some industries like mostly coal and iron industries. This overburden waste can be used as raw materials for construction works. For example, the coal mine waste rock is considered as overburden, but it is seen that it's a great alternative for the sea sand/natural sand used in construction works (Mishra et al., 2022).

These overburden dump wastes and waste rocks are also used in filling underground mines. This is required as filling the underground mines helps achieving stability to the land. This in traditional methods used fly ash and river sand, now using mine tailings, waste rock and overburden waste foundry sand. The industries which are lacking sand such as coal mining industries these waste rocks and overburden wastes can be considered as a great replacement (Gupta & Paul, 2015).

This sustainable management of these overburden waste rocks is very necessary because it some time contains toxic elements, such as high amount of sulphur, which potentially leads to acidic leachates. These when left on the soil or in the environment causes soil degradation and soil erosion as well as environmental pollution. These waste materials, which are named as overburden, are generated from mostly open mining than in underground distillation processes. In open cast mining processes, waste rocks and other waste materials are produced in a larger amount (Islam et al., 2021).

The chemical composition of overburden waste rock is mainly metal oxides whose compositions are as follows: Na_2O is about 1.4% followed by MgO which is of 6.3%, Al_2O_3 is of 4.6%, SiO_2 which is of 42.8%, P_2O_5 is of 0.1%, CuO is of 0.14%, and many other metal oxides. These concentrations are identified using X-ray fluorescence (XRF) investigations.

For the making of the "M_{40}" grade concrete water binder in 0.4, 0.5 ratio and overburden waste rock which is used as a replacement of sand is added in the ratio of 0%, 20%, then 40%, then 60%, 80% and finally 100%. IS1119 will determine the working of the concrete(Mishra et al., 2022). These concretes are filled in concrete moulds and its compressive strengths are measured. Then finally the replacement percentage is calculated. The compressive strength V/s replacement percentage graph is made for testing. The back sand elimination method is used in this process. Apart from these sustainable waste management methods, there are many challenges and development concerns are related to overburden waste rock management.

2.3 *Hazardous and Non-Hazardous Mining by-Products*

There is a high amount of hazardous waste found in UK, Sweden, Germany, and Greece thus they have many disposal centres in underground mines. They have a project going on named low risk disposal technology (low risk DT) for the management of these wastes. Abandoned mines are the main sites of these hazardous waste. These have harmful effects on plants, animals, soil, as well as water. In this, water is again and again most frequently being polluted by abandoned mines.

The waste contains sediments, acidity, hydrocarbons, and brine leaching which harms underground water quality and aquatic habitats. It also contains incinerated ash, low level radioactive waste, contaminated soil. Many well-developed countries such as Austria have 62% waste in hazardous waste generation, while Denmark have 57% decrease in waste production. Ukraine hazardous waste production decreased by 38% from year 1996 and 2000. In Russia waste quantity increased by 32%. Hazardous waste list contains counties like UK, Germany, Europe, etc. Construction waste is 0.7652 kg per capita; manufacturing waste is about 0.3857 kg, municipality's waste is about 0.6872 kg, harmful waste is about 0.8960 kg per capita of waste generation (Popov & Pusch, 2006).

There are about 200–300 million non-harmful factory's tailings in the USA. 120 million tons and 200 billion KWh of non-hazardous waste is produced yearly in between the year 1990 and 2016. In these non-hazardous wastes, solid waste is about 0.23 billion tons/year. This is about 3–5 times of the waste. Minerals like lime (CaO), silica (SiO_2), and salt (NaCl) are the minerals catagorised as non-hazardous waste.

2.4 Environmental and Health Impacts of Mining Waste

The environmental impacts of mining include the deterioration of water quality and quantity, degradation of air and soil quality, along with significant social consequences. Tailings, waste rock, laboratory tailings, and chemical which are the solid wastes, blasting compounds, slags, leached ores, ore stockpiles cause water pollution. Due to the extraction metal from the mine, a huge amout of mine tailings are generated, which cause many environmental and health impacts. These waste rocks are being disposed in the pit due to which the surface has been contaminated up to a large area of about tense of squire miles. Again, these water rocks are treated using chemicals, acids, etc. And no barriers are made for the control of these acids, contaminated leachates due to which soil has been contaminated. The tailings generated from mineral processing activities have been contaminated the oceans in the areas of lax oversight, and are disposed in the oceans, which potentially contains toxic concentration of various metals and non-metals as well as organic compounds.

Along with these environmental impacts, tailings have a huge impact on human health and other organisms. It causes respiratory diseases like asthma, lung infections, skin rashes, and toxic reactions due to airborne contaminants (Das, Dey, Bala, et al., 2024; Das, Dey, & Das, 2024; Mallick et al., 2024; Ray et al., 2024). The health of humans as well as the organisms will be affected due to the consumption of contaminated food grown on contaminated soil. Aquatic animals will also die or get diseases due to the treatment of toxic waste into the oceans and other water resources. Toxic metals like lead, cobalt, mercury are hazardous to flora and fauna. Emissions of these air borne contaminants from industries of Europe and the USA causes acid rain. Figure 1 shows the Sources, impact, and implementation of various sustainable waste management.

3 Challenges in Mining Waste Management

Due to inadequate infrastructure in many cities, town, villages lack the facilities to treat and dispose of waste. This can lead to waste accumulation in open dumps. Improving waste management at a global level requires addressing several challenges of waste collection and disposal infrastructure, lack of community engagement, climate change, lagging technological advancements, shifting preferences of consumers, and some other financial constraints.

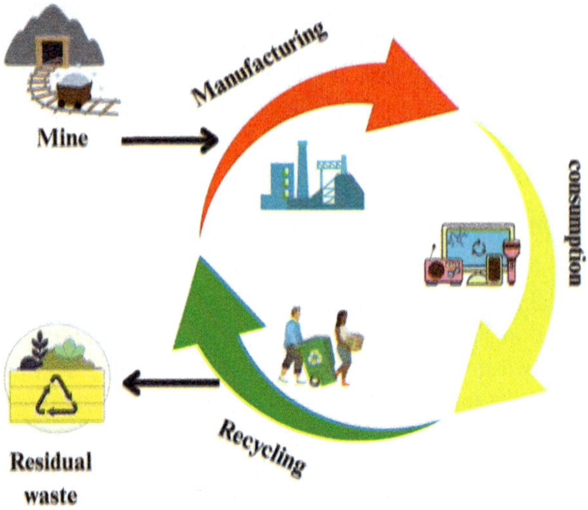

Fig. 1 Describe the circular economy model for mining waste management

3.1 Technical Challenges with Waste Handling

Developing nations often face challenges such as inadequate funding for effective tailings management systems. Overburden or waste rock is a large volume of initial soil and rock must be excavated to accumulate mineral deposits. Mining operations use a significant amount of water from local suppliers. So, water management is also a challenge. A large volume of initial soil and rock must be excavated to accumulate mineral deposits. Acid drainage is also a challenge. Chemical and microbial properties such as the chemical nature of mine tailing can differ depending upon the type of metal being filtered. Companies that hold regulatory compliance can lessen their impact on the environment, improve their operational efficiency, and fulfil the requirements of regulatory agencies. It faces significant challenges related to sustainability, demand uncertainty, technological disruption, work force skill and operational costs. Eight important sustainable challenges are carbon footprint, water conservation, site rehabilitation, community relations, biodiversity preservation, materials efficiency, supply chain sustainability, regulatory compliance (Moharana et al., 2024; Pradhan et al., 2024; Sahu et al., 2024).

Over 379 million city people live in 7937 town and cities and produce 63 million tons of solid waste per year. Improper disposal of hazardous waste can lead to long lasting environmental damage. Five ways to make mining more sustainable: lower impact mining techniques, reusing mining waste, ecofriendly equipment, rehabilitating mining sites, shutting down illegal mining. Main challenges are as follows: climate changes, drought and water scarcity, biodiversity loss, resource depletion, pollution, and deforestation.

Environmental and regulatory constraints are water pollution: mining can cause H_2O population by increasing the impurity of H_2O, which can lower the pH levels of rivers. This can lead to dissolution of metal oxide. Air pollution: improper disposal of mining waste can emit toxic pollutants into air. Acid mine drainage: mine wastes can release large number of heavy metals and acids, which can contaminate the environment. Erosion: mining can cause erosion, which can lead to sink house and loss of biodiversity heavy metal contamination and leaching mining waste can contain high concentration of heavy metals, which can contaminate nearby soils and water bodies. Regulatory compliance: mining companies must navigate a dynamic regulatory landscape to adhere to environmental standards. Tailing: tailing are the biggest waste streams in mining and they can create long term environmental liabilities. The recycling and reuse of mining waste are often constrained by economic limitations such as cost-effectiveness, waste volume, legislation, and environmental impacts, with each type of waste requiring its own suitable method of recycling.

3.2 *Environmental and Regulatory Constraints*

Processing waste: in the phase, an ore mill is constructed in the distillation site to output first economic products. At this stage, an ore mill is constructed at the distillation site to produce the first economically valuable products. However, the process generates various types of waste, including aqueous solutions, wastewater, and slurry, which are unsafe for direct discharge into water bodies as they pose serious hazards to marine life (Das et al., 2025; Mohanty, Mohanty, et al., 2024). Their habitat is going to be affected by this. Mining waste can affect the natural habitat green problem of both surface soil and water. It can make the soil water (ground and surface) infected with chemicals. It can lead to the loss of biodiversity.

3.2.1 By drilling & excavation the land resource

High foot print mining requires a lot of water, which leads to environmental pollution. Tailing: residual material from the mineral enrichment process, which is a fine slurry with a major H_2O content. Majority of products are failed to be recycled. High water uses in washing process can lead to reduced facilities for locals to get fresh H_2O cause locals to suffer from water streams.

The USA has one of the highest mineral productions after China. Water used for mining only makes about 1.3% of the total nations water use. There are many documents that highlight environmental hazards caused by mining operations, which are often attributed to the leakage of mining waste. Many tailing materials are stored in large tailing dams to prevent environmental hazards, as they may be radioactive, acidic, or toxic.

Table. 1 The type of waste and its environmental concerns

Types of waste	Description	Environmental concerns
Overburden	Soil and rock removed to access mineral deposits	Soil erosion, landslides, and habitat destruction
Slag	Waste rock from smelting and refining processes	Heavy metal contamination and air pollution
Heap leach residues	Toxic chemicals from the cheap leaching process	Ground water contamination, soil pollution
Tailing	Finely ground rock and minerals left after processing	Water pollution, sedimentation, acid mine drainage

As a result of strict international regulations, pollutions caused by mining has been dramatically reduced. In artisanal mining, large-scale clearing of vegetation and soil leads to significant CO_2 emissions and the release of greenhouse gases. Additionally, the high energy consumption associated with mining activities further contributes to greenhouse gas (GHG) emissions. Table 1 describes the environmental concerns of different types of wastes.

3.3 *Economic and Regulatory Constraints*

Recycling of mining waste often faces economic challenges, as developing and implementing new technologies for recycling and resource recovery can be costly. Market dynamics: its market dynamics and regulatory framework may not always support circular economy initiatives. Volumes: some waste, like that generated during the extraction cycle, is large in volume and has little chance of being reused economically. Economic interest: some waste like rock waste, may not have a metal concentration of economic interest. Other challenges include cultural and organizational change, ensuring sustainability of process and preventing contaminants from migration from waste repositories. Recycling can be energy inefficient promote reliance on unsustainable materials and lead to pollution, high processing costs and lower quality recycled products. Only 10% of plastic is recycled. Reuse does not involve any harmful impact on the environment. Although recycling limits the drastic impact on the environment. It does cause harm in some case.

A total of ten benefits of recycling are protecting ecosystem and wildlife, conserve natural resources, saving energy, reduce the need to harvest, reduce carbon emission. Educate locals about the importance of saving the environment. “3R” rules are reduce, reuse, and recycle. Each person adds to waste management process cost; The cost of recycling mine waste may not be cost effective. Quality: the mine waste can vary depending on the type of ore and mining process. Contaminants: mine waste may contain contaminants that can affect the durability of the durability of the final product. Standardization: there may be a lack of standardized guideline for using mineral waste. Technical challenges: mine waste may have different physical and chemical properties than traditional matter. Market acceptance: consumers and the construction industry may be hesitant to acceptance. Logistic: the remote locations of mines can make it difficult to transport waste. Environmental impacts: the environmental impacts of reusing mine waste can vary depending on the source rock.

Some ways to address these challenges investing in research and development can be costly but investing in research and development can help overcome these barriers. Advocacy and collaboration with policy maker and industry stake holders can help create conductive environment that help overcome these barriers.

Education training and leadership commitment are essential to drive a shift in mindset and culture within mining organization. Recycling too often downcycling, recycle underused as new raw material, material degradation, product design not setup of costly process, pollution, human behaviour, recycling compromised quality of materials, cost financial and environmentally friendly recycling.

3.4 Social Acceptance and Community Concerns

Mining has a number of social and environmental impact on local communities. Displacement: mining often requires relocating people which can disrupt their way of life, culture, and social structures. Health risks: communities near mining sites can be exposed to hazardous substances, such as airborne pollutants and chemicals which can lead to serious health issue. Environmental degradation: mining can contaminate local streams and wet lands and increase the concentration of heavy metal surrounding soil or water resource. Loss of livelihood: mining can lead to loss of traditional livelihood.

Land use change: mining can result in land use change increase in deforestation and erosion. Noise and dust: mining can increase in noise and dust level. Loss of biodiversity; mining can lead to a loss of biodiversity. Reduced soil microbial: mining can reduce soil microbial population and diversity (Parida et al., 2026, Samantaray et al., 2026).

Mining makes significant contributions to the gross domestic product of countries where it's practiced and it also plays a major role in consumer driven societies. The co-existence of mining generations and indigenous populations necessities a cordial relationship between minor and surrounding communities. Corporate social responsibility a/c to (Jenkins, 2004) aims to balance the demand of community.

4 Opportunities in Mining Waste Utilization

There are mining waste management procedures which can be used to turn waste into wealth. Mining waste can be used as an aggregate in concrete. Minerals and useful mining materials can be extracted from mine tailings by reprocessing it, can also be used as secondary sources of minerals and metals in future use. For example, coal gangue which is made up of organic and inorganic compounds can be used as a replacement of clay for cement production.

4.1 Recovery of Valuable Minerals and Metals

An increased/high consumption of minerals can be noticed in these years. Mine tailings undergo hydrometallurgical processes which results in the recovery of strategically important critical minerals (SICMs) which have ecological benefits. Radioactive metals such as magnesium (Mg), potassium(K), lithium (Li) will have a high demand as predicted by 2050. These earth metals have applications in widespread high-tech industries such as telecommunications, agriculture renewable energy, aerospace, medical, and many more. Tailings contain ore deposits from which only the desired or host metals—such as zinc (Zn), copper (Cu), and iron (Fe)—have been extracted. Technologies like hydrometallurgical processes, solvent extraction, acidic leaching, and liquid–liquid extraction are commonly employed for metal recovery. Ammonium thiosulfate–copper leaching, in particular, is used to extract valuable metals such as gold (Au), silver (Ag), and copper (Cu). Approximately 80% of gold, 75% of silver, and 75% of copper are estimated to remain within these mine tailings.

There is another process known as green recovery of metal minerals. There are three main steps in this process leaching, purification, and finally sedimentation. This process is used for separating metals from ores that are polymetallic. For improved metal recovery, both polymetallurgy process and hydrometallurgy process are used in combination. Individually, these are not efficient to separate valuable metals. For example, leaching process is not efficient for vanadium extraction.

Ultrasound technology can be used for the extraction of rare metals, such as lithium (Li), cobalt (Co), and nickel (Ni). Manganese (Mn), used in the manufacturing of lithium-ion batteries, can be recovered via organic acid leaching processes. Calcification and carbonization methods are used for dissolving low-grade bauxite aluminium. MOF nanofiller membranes are used for separation of lithium (Li) and magnesium (Mg) from salt lakes. Techno-economic value of producing valuable minerals and materials from my tailings. Around 477 Chilean tailings are found which results in CRMs production. CRMs production is calculated using discounted cash flow (DCF) method (Priyadarsini & Das, 2025, 2026)

4.2 Application in Construction and Infrastructure

In recent years, advancements in technology have enabled the utilization of mine residues or tailings in the construction industry. Many construction materials—such as cement, bricks, and mortars—are now being produced using mine residues, particularly in small-scale mining communities like artisanal and small-scale mining (ASM), which represents a simplified form of ore extraction. Materials that were traditionally regarded as mine tailings or waste products are now being recycled and repurposed for various construction applications. This shift has brought mine tailings recycling (MTR) into the spotlight, gaining significant attention and popularity. For instance, recycled tailings are now used in the production of construction materials such as cinder blocks.

Environmental and health risk assessments consist of four main steps; in case of MTR, first the orientation and concentration of heavy metals in mine tailings would be recognized, then exposure of assessment addresses occurs. Differentiation of mine tailings into most importantly two categories cancer causing and non-cancer causing. Finally risk characterization occurs.

Mine tailings have similar characteristics as compared to crusher sand used in engineering construction. Thus tailings are used to replace sand in concrete mixtures. Another very important application is geo polymerization, A class of materials synthesized by alkaline activation of aluminosilicate source at ambient or higher temperature. Some geopolymers are meta-kaolinite. Geopolymers are the tri-dimensional aluminosilicates that are formed at low temperature and in a short period of time.

Geopolymers can naturally occur mixing naturally occurring materials with NAOH and water. It is the analogue of zeolites synthesis since the chemistry involved is composition and structure. The raw materials used in this process includes leaching, diffusion, condensation process and finally hardening of the raw material occurs for the formation of geopolymers. It's patented many times by some scientists.

Portland cement comprises of clinker, and other minor components like lime stone and gypsum are about 0–10% of the mixture and gypsum is about 1–3% of the mixture. Mine tailing sands are used in conventional concrete, both the crushed mine residues and the sands are extracted from the "River Pindos". The disposal of these mining tailings sand is very necessary.

The chemical components present in the cement type GU are Portland cement, calcium sulphate, calcium carbonate, calcium oxide, magnesium oxide, silica crystals, etc. The physical state of the cement is dust. The pH of the cement is about 12–13, which is alkaline in nature. The solubility of the cement in water is about 0.1–1.0%. Life cycle assessment (LCA) is a method used to measure the environmental impact of a product system. The tailings have a major impact on water bodies, human localities, river flow, and overall environment.

4.3 Potential in Renewable Energy Technologies

As per the predictions, the global energy demand will rise by the year 2050. The energy demand will rise from 500 to 1000 Ej as per the business perspective. The renewable reseources will provide the energy for day to day needs when the mines resources will gradually end. This will surely occur due to the exploitation done by the human beings. The climate change also plays a very important role in affecting the renewable the renewable energy. The greenhouse gases like methane, CFC, etc. will certainly rise for a certain period of time.

IEA is also a type of energy technology. In some old industries renewable energies or bioenergy are used. United Nations Industrial Development Organisation (UNIDO) is working for the employment of the renewable bioenergy. By 2035, it is predicted that the bioenergy/renewable energy will grow exponentially by 20%. The growth and development of renewable energy sources used are so far good.

Turkey has a good amount of fossil fuel, but due to significantly increasing energy prices, the energy shortage will certainly see in Turkey. But now it has one of the largest domestic energy sources. Here the main renewable source is biomass and animal waste.

In China, due to increasing population, the energy demand is as well rapidly increasing. It is one of the major coals producing countries which bought challenges to decreasing greenhouse gases (GHGs). Denmark also came into knowledge for having the most sustainable strategies.

4.4 Circular Economy Approaches in Mining

The concept of circular economy is well known to the research scholars and lab practitioners by means of sustainable waste management. Circular economy has both supporters and critiques present. There are mainly two types of economics dismissed till now; one is linear economy and other one is circular economy. In case of circular economy every bit of the base/raw material are being used. The traditional “3R” framework, now became the “9R” framework. The “9R” are refuse, rethink, reduce, reuse, repair, refurbish, remanufacture, repurpose, recycle, and recover from R0 to R9, respectively (Kirchherr et al., 2017). Any economy trying to reach at the goal for becoming circular economy have some targets to acheive, such as efficiency targets of recycling and reuse of water, energy and materials. There are many approaches and initiatives performed by the government to achieve these targets. They also have the target of achieving circular economy at macro as well as in micro-levels. Many strategies are made in order to achieve circular economy. This includes using the materials in a proper way, extended life span of products and the manufacturing of smart products(Morseletto, 2020).

There is a major use of circular economy in mining; there are about 3500 active mine tailing dams and waste rock dumping stations present all over the world.

Circular economy aspects are as follows: improving mining tailings reuse and using of modern innovative technologies. Allowing the subsequent reuse and recycling of mining tailings by-products by making easy-to-repair and made-to-reuse products. Balancing two important properties of human need supply and demand the idea of circular economy is fulfilled. The concept of re-thinking mining waste states that except recycling and circularity it should be supplied with proper infrastructure, legislation, and economical support. This will make a huge difference in utilization of every bit of mining tailings. Many companies can be involved in making of products out of these mining wastes.

5 Strategies for Effective Mining Waste Management

5.1 Technological Innovations in Waste Treatment

Smart waste bins: these waste bins can detect when they are full and alert users to dispose of their trash. This can help people manage their disposable habits and optimizes trash Pickup schedules.

AI recycling roberts: these roberts use sensors and machine learning algorithms to identify recyclable materials and sort through waste.

Digesters: these machines use heat and pressure to convert biogas from waste into energy. E-waste kiosks: these kiosks are used to dispose of electronic waste, which can contain harmful substance that can leak into environment if disposed of improperly.

Prismatic waste pipes: these pipes transport waste from building to collection points instead of using traditional garbage truck.

Solar powered trash compactors; these compactors can hold up to five times more than traditional trash bins, which can reduce greenhouse emissions from garbage trucks. Smart waste management: systems are considered use connected sensors, data analytics and automation to optimize waste collection, route planning, and resource allocation. Recycling technology: it made it possible to recycle a wide range of materials including plastics, paper glass, and metals.

5.2 Policies and Regulatory Frameworks

Every country has formulated several regulatory and legal mechanisms to enhance environmentally sustainable mining. Environmental impact assessment, forest clearance provision, coastal zone maintenance, waste air pollution management act.

Environmental protection act in 1986 gives the power to regulate all types of waste. National mineral policy (NMP) 2019, this policy focuses on sustainably mining. Indian bureau of mining (IBM) promotes conservation, scientific development of mineral resources. Mining policy framework make good practices for environmental, social, and economic governance of mining sector. Waste management law governs possibility method and location disposal of particular waste. Some other law included such as.

clean air act (CAA), clean water act (CWA), National environmental policy act (NEPA), resource conservation and recovery act (RCRA), toxic substances control act (TSCA).

5.3 Industry Collaboration and Best Practices

The 5Rs of waste management are refuse, reduce, reuse, repurpose, and recycle. Industry collaboration and sustainable practices are important for mining waste management. Source reduction; mine closure a detailed mine closure plan can be developed to protect remaining resources. Waste reduction and material use: sustainable mining practices minimize waste and produces material from waste like dry stacking can be used for tailing.

By the help of the technology, it can reduce the danger of collapsing of tailing dam and water contamination. In the year 2000 World lithium-ion battery production is recorded to be the highest with 500 million metric tons. And in the year 2010 it is 4.6 billion units. Large amount of LIB is used/handled causing production of hazardous electronic residues/electronic waste. About 200–500 tons of consumer battery wastes are found per year during 2002–2006. Technical solutions/e-waste management is implemented by hydrometallurgical or pyrometallurgical processes/techniques.

6 Economic and Environmental Impact Assessment

6.1 Cost Benefit Analysis of Waste Utilization

For the waste utilization, environment protection mechanisms are being implemented as well as the expected annual penalty cost is accepted. This expected cost is of rare earth element (REEs), strategic metals (SMs) or other precious metals (PM) recovery methods. The result from these analysis helps acquire an industrial mode. Metal recovery plays a crucial role in this. Cost benefit analysis is done for any business decision as well as the economic value of the product before being implemented. It is based on the signal of economic rate of returned (ERR) and long-term perspective (LTP).

CBA in term of circular economy is most important in mining waste management activities and acidic character for the reduction of final waste disposal site. Data regarding the implementation of cost benefit analysis are as follow. To identify the environment rehabilitation cost, metal ions concentration is first identified. To determine the reusability of these metal ions of tailing waste recoveries, its economical assessment is performed. The annual volume of waste, the capacity for environmental rehabilitation, and the establishment of industrial installation units are also evaluated (Campbell et al., 2023).

In countries such as Estonia, a population tax related to landfill impacts was implemented on June 10, 1998. Similarly, in Hungary, environmental charges are regulated under the Environmental Protection Act, with funds allocated accordingly. Additionally, industries that utilize or occupy environmental components are required to pay mining royalties as stipulated under Section 20 of the Mining Act. Many more countries like Latvia, Bulgaria, Lithuania, Slovenia are also included in the list. Increase in capitalized cost attributed to the mines also increases the predicted/estimated environmental costs (Worlanyo & Jiangfeng, 2021).

6.2 *Environmental Restoration through Mining Waste Reuse*

40,000 km^2 of land in China has covered by these abandoned mines. The soil present in these abandoned mine areas lack nitrogen (N), phosphorus (p), Potassium (K) and rich in toxic chemicals, poor physical structure as well as high in pH value.

In Africa, large-scale mining started in colonial period. In Kangwa, according to 2008 data 350 tons of waste are generated, in Zambia (Africa), according to the estimate about 791 million tons of tailing are presence (Festin et al., 2018).

Restoring mean reviving anything to it' s previous/former state most importantly to a perfect condition. Some natural methods to rehabilitate/revive the areas are to differentiate the horizons of the soil. It is a lengthy process as there are many physiochemical reactions taking place.

The underground water conservation quantity is 11,832.036 m^2/a. Water and soil conservation quantity is about 14,946.4 m^2/a. The overall indirect economic is evaluated to be 1.64105 yuan/a (Bradshaw, 1997).

6.3 *Long-Term Sustainability Impacts*

For the long-term assessment of mining some crucial points are required to be thoroughly undergone for throughout the life the mine. The by-products of mining tailing are basically of two types: economic and non-economic. Also, there are two categories of disposal of these tailing those are direct and indirect. Direct disposal is again categorized into two types: submarine tailings disposal (STD) and riverine tailings disposal (RTD). Indirect methods are making by product from these mining tailings some of them have economic value which can be sold by recycling and reusing them(Zou, 2019).

The co-efficient associated to the long-term mining waste trade amount is 0.15% and as of the short-term mining waste trade amount is of 0.07%. Some long-term strategies are such as sustainable area of land use, green development, reforestation strategies, investing in clean technologies. The very known mine environment neutral drainage (MEAD) which was implemented by Canadian government is a great involvement in sustainable mine tailing management research. Still in many other areas, it is required to be improved in case of the environmental performance.

Another great example is NSERC-Polytechnique UQAT chair on environment and mine management. In integrated solid, liquid waste management during operation (Adiansyah et al., 2015).

7 Challenges and Future Perspectives

7.1 Addressing Gaps in Technology and Infrastructure

Some ways to address gaps in technology and infrastructure in mining Industry are as follows: it can be improved by improving digital infrastructure, in remote mining areas improving internet connectivity, power supply and communication networks. Public–private partnership can help with this investment. New technology such as automation, digitalization, electrification can improve mining efficiency and safety. For example, autonomous vehicles, automated drilling and funnel baring system, drones, Smart sensors, mechanical cutting, sensor-based sorting, etc. Sustainable practices such as lower impact mining techniques, reusing mining waste, eco-friendly equipment, rehabilitating mining sites can be implemented. A.

Advancements in exploration technologies—such as geospatial analysis, drone mapping, surveying, and artificial intelligence (AI)—have transformed the mining sector. Innovative mining technologies now enable companies to automate processes related to location awareness, machine learning, and robotics, enhancing efficiency and precision in mineral exploration and extraction. By automatic process like are site monitoring, delivery and transport, and drilling and ventilation systems minor don't have jeopardize their health and safety by venturing into mines and materials.

There are four types of main mining methods: underground, open surface, places, in situ mining, and many more.

By optimizing transports logistic, mining companies can streamline their supply chains, reduce delay and improve overall operation efficiency.

Method for surface mining—Surface mining contribute majority of coal produce and the challenges of planning, design, operation management of large-scale opencast projects involves adaption of new technologies software and management.

Land acquisition—Mass production using surface miner.

Dump planning—Organization of dump material, angle of internal friction, cohesion of foundation material, and angle of internal friction in foundation material.

7.2 *Opportunities for Innovation in Waste Management*

Smart tailings management is important as per the environmental protection agency roughly 76% of the waste steam in the United States in recyclable, but only about 33% of recyclable material actually get recycled. AI is increasingly being integrated into recycling and waste management systems. Technologies such as smart waste brine management, waste level sensors, garbage truck weighing mechanisms, pneumatic waste transport pipes, and solar-powered trash compactors are enhancing efficiency, monitoring, and sustainability in modern waste handling operations.

Methodology for underground mining—Underground mining is going to play a major role in meeting the future production, requirements considering the growing limitation of surface mining technology, namely environment management and land availability apart from other factors. Some of key issue needing immediate attention are discussed below.

Mining tailings can be used as nanomaterials that can be used in electronics, medicine, environmental remediation, and more. Cascade recycling is a process in which batteries can be reused for less demanding application before being recycled. Mine waste such as coal mine waste rock can be used in road embarkments. The mining industry may increase recycling efforts to increase production in line with circular economy efforts.

Enhancing recycling efforts in the mining sector presents an opportunity to create value through sustainable practices. Strengthening supply chain resilience in the mining industry through AI is quite acceptable.

References

Adiansyah, J. S., Rosano, M., Vink, S., & Keir, G. (2015). A framework for a sustainable approach to mine tailings management: Disposal strategies. *Journal of Cleaner Production, 108*, 1050–1062. https://doi.org/10.1016/J.JCLEPRO.2015.07.139

Aznar-Sánchez, J. A., Velasco-Muñoz, J. F., Belmonte-Ureña, L. J., & Manzano-Agugliaro, F. (2019). Innovation and technology for sustainable mining activity: A worldwide research assessment. *Journal of Cleaner Production, 221*, 38–54. https://doi.org/10.1016/J.JCLEPRO.2019.02.243

Bal, B., Ghosh, S., & Das, A. P. (2018). Microbial recovery and recycling of manganese waste and their future application: A review. *Geomicrobiological Journal, 36*(1), 85–96. https://doi.org/10.1080/01490451.2018.1497731

Bala, K., Anshul, Verma, S., Dogra, A., Das, A. P., Kumar, R., & Nandini. (2024). Sustainable management for urban plastic waste generation. In *Sustainable management of urban plastic waste through circular economic approaches* (pp. 211–236). Routledge. https://doi.org/10.1201/9781003465447-13

Bradshaw, A. (1997). Restoration of mined lands—Using natural processes. *Ecological Engineering, 8*(4), 255–269. https://doi.org/10.1016/S0925-8574(97)00022-0

Campbell, M. D., Martinez-Frias, J., Machairas, E., & Varouchakis, E. A. (2023). Cost–benefit analysis and risk assessment for mining activities in terms of circular economy and their environmental impact. *Geosciences, 13*(10), 318. https://doi.org/10.3390/GEOSCIENCES13100318

Cui, C. Q., Wang, B., Zhao, Y. X., & Xue, L. M. (2020). Waste mine to emerging wealth: Innovative solutions for abandoned underground coal mine reutilization on a waste management level. *Journal of Cleaner Production, 252*, 119748. https://doi.org/10.1016/J.JCLEPRO.2019.119748

Das, A., Dey, S., Bala, K., Gnanasekaran, R., & Das, A. P. (2024). Microbial bioremediation technology for sustainable treatment and management of synthetic microfiber waste. In *Sustainable microbial technology for synthetic and cellulosic microfiber bioremediation* (pp. 221–238). Springer. https://doi.org/10.1007/978-3-031-62600-5_12

Das, A., Dey, S., & Das, A. P. (2024). Ecotoxicological impacts of synthetic microfiber pollutants and development of sustainable mitigation strategies. *Environmental Chemistry and Ecotoxicology*. https://doi.org/10.1016/J.ENCECO.2024.12.003

Das, A., Dey, S., Gacem, A., Yadav, K. K., Nath, H., Ram, D. K., & Das, A. P. (2025). Revolutionizing solid hydrocarbon sourcing: Plastic's journey from waste to upcycled treasure. *Environmental Science and Engineering, 218*, 331–351. https://doi.org/10.1007/978-3-031-81615-4_14

Das, A., & Ghosh, S. (2022). Role of microorganisms in extenuation of mining and industrial wastes. *Geomicrobiology Journal, 39*, 173–175. https://doi.org/10.1080/01490451.2022.2038953

Das, A., Ray, S. K., Mohanty, M., Mohanty, J., Dey, S., & Das, A. P. (2024). Ecotoxicity of microplastic wastes and their sustainable management: A review. *Environmental Chemistry and Ecotoxicology, 6*, 144–152. https://doi.org/10.1016/J.ENCECO.2024.05.003

Das, A. P., & Ghosh, S. (2018). Bioleaching of manganese from mining waste materials. *Materials Today: Proceedings, 5*(1), 2381–2390. https://doi.org/10.1016/J.MATPR.2017.11.459

Das, A. P., & Ghosh, S. (2023a). Cost-effectiveness of the biomining process. In *Manganese mining microorganisms* (pp. 143–151). Elsevier. https://doi.org/10.1016/B978-0-12-822147-1.00008-X

Das, A. P., & Ghosh, S. (2023b). Evolution of biomining technology. In *Manganese mining microorganisms* (pp. 5–19). Elsevier. https://doi.org/10.1016/B978-0-12-822147-1.00012-1

Das, A. P., & Ghosh, S. (2023c). Manganese and its application. In *Manganese mining microorganisms* (pp. 29–40). Elsevier. https://doi.org/10.1016/B978-0-12-822147-1.00007-8

Das, A. P., & Ghosh, S. (2023d). Mechanism of manganese bioleaching. In *Manganese mining microorganisms* (pp. 87–98). Elsevier. https://doi.org/10.1016/B978-0-12-822147-1.00014-5

Das, A. P., & Ghosh, S. (2023e). Metagenomic insights into unculturable microbial diversity. In *Manganese mining microorganisms* (pp. 63–76). Elsevier. https://doi.org/10.1016/B978-0-12-822147-1.00002-9

Das, A. P., & Ghosh, S. (2023f). Mixed culture bioleaching: An insight into manganese biomining process and efficacy. In *Manganese mining microorganisms* (pp. 129–141). Elsevier. https://doi.org/10.1016/B978-0-12-822147-1.00001-7

Das, A. P., & Ghosh, S. (2023g). Role of metabolites and metabolomics for understanding of manganese bioleaching. In *Manganese mining microorganisms* (pp. 117–127). Elsevier. https://doi.org/10.1016/B978-0-12-822147-1.00011-X

Das, A. P., Ghosh, S., Mohanty, S., & Sukla, L. B. (2015). Advances in manganese pollution and its Bioremediation. In L. B. Sukla, N. Pradhan, S. Panda, & B. K. Mishra (Eds.), *Environmental microbe biotechnology* (1st ed., pp. 313–328). Springer. https://doi.org/10.1007/978-3-319-19018-1_16

Das, A. P., Mallavarapu, M., & Ghosh, S. (2023). Ecotoxicity of mining pollutants on the environment and their remediation. *Environmental Chemistry and Ecotoxicology, 5*, 165–167. https://doi.org/10.1016/J.ENCECO.2023.08.002

Dey, S., Bissoyi, A., Koteswara Reddy, G., & Das, A. P. (2024). Synthetic microfibers: An urban emerging pollutant demanding for sustainable environmental solutions. In *Sustainable microbial technology for synthetic and cellulosic microfiber bioremediation* (pp. 107–121). Springer. https://doi.org/10.1007/978-3-031-62600-5_7

Dey, S., Das, A., Mallick, K., Sahu, A., & Das, A. P. (2023). *Environmental petroleum waste: Pollution, toxicity, sustainable remediation* (pp. 159–175). Springer. https://doi.org/10.1007/978-3-031-48220-5_7

Dey, S., Dash, A., & Das, A. P. (2024). Bioremediation as a sustainable and circular solution for mining waste management and bioenergy generation. In *Sustainable management of mining waste and tailings: A circular economy approach* (pp. 107–119). CRC Press. https://doi.org/10.1201/9781003442455-6

Dey, S., Dash, N., Ray, S. K., Priyadarsini, S., Chhotaray, C., & Das, A. P. (2024). Occurrence and quantification of tire wear plastics pollutants from highway city dust of Bhubaneswar City of Odisha. In *Sustainable microbial technology for synthetic and cellulosic microfiber bioremediation* (pp. 77–89). Springer. https://doi.org/10.1007/978-3-031-62600-5_5

Dey, S., Tripathy, B., Kumar, M. S., & Das, A. P. (2023). Ecotoxicological consequences of manganese mining pollutants and their biological remediation. *Environmental Chemistry and Ecotoxicology*. https://doi.org/10.1016/J.ENCECO.2023.01.001

Festin, E. S., Tigabu, M., Chileshe, M. N., Syampungani, S., & Odén, P. C. (2018). Progresses in restoration of post-mining landscape in Africa. *Journal of Forestry Research, 30*(2), 381–396. https://doi.org/10.1007/S11676-018-0621-X

Ghosh, S., & Das, A. P. (2017). Bioleaching of manganese from mining waste residues using Acinetobacter sp. *Geology, Ecology, and Landscapes, 1*(2), 77–83. https://doi.org/10.1080/24749508.2017.1332847

Ghosh, S., Mohanty, S., Akcil, A., Sukla, L. B., & Das, A. P. (2016). A greener approach for resource recycling: Manganese bioleaching. *Chemosphere, 154*, 628–639. https://doi.org/10.1016/J.CHEMOSPHERE.2016.04.028

Ghosh, S., Mohanty, S., Nayak, S., Sukla, L. B., & Das, A. P. (2016). Molecular identification of indigenous manganese solubilising bacterial biodiversity from manganese mining deposits. *Journal of Basic Microbiology, 56*(3), 254–262. https://doi.org/10.1002/JOBM.201500477

Ghosh, S., Tripathy, B., Dey, S., & Das, A. P. (2023). Comparative investigation of fungal and bacterial manganese biomining mechanisms. In *Manganese mining microorganisms* (pp. 99–115). Elsevier. https://doi.org/10.1016/b978-0-12-822147-1.00016-9

Gupta, A. K., & Paul, B. (2015). A review on utilisation of coal mine overburden dump waste as underground mine filling material: A sustainable approach of mining. *International Journal of Mining and Mineral Engineering, 6*(2), 172–186. https://doi.org/10.1504/IJMME.2015.070380

Jenkins, H. (2004). Corporate social responsibility and the mining industry: conflicts and constructs. Corporate Social Responsibility and Environmental Management, 11(1), 23–34. https://doi.org/10.1002/CSR.50

Islam, N., Rabha, S., Subramanyam, K. S. V., & Saikia, B. K. (2021). Geochemistry and mineralogy of coal mine overburden (waste): A study towards their environmental implications. *Chemosphere, 274*, 129736. https://doi.org/10.1016/J.CHEMOSPHERE.2021.129736

Kirchherr, J., Reike, D., & Hekkert, M. (2017). Conceptualizing the circular economy: An analysis of 114 definitions. *Resources, Conservation and Recycling, 127*, 221–232. https://doi.org/10.1016/J.RESCONREC.2017.09.005

Lèbre, É., Corder, G. D., & Golev, A. (2017). Sustainable practices in the management of mining waste: A focus on the mineral resource. *Minerals Engineering, 107*, 34–42. https://doi.org/10.1016/J.MINENG.2016.12.004

Magazzino, C., Mele, M., Schneider, N., & Sarkodie, S. A. (2021). Waste generation, wealth and GHG emissions from the waste sector: Is Denmark on the path towards circular economy? *Science of the Total Environment, 755*, 142510. https://doi.org/10.1016/J.SCITOTENV.2020.142510

Mallick, K., Sahu, S., Sahu, A., Dey, S., & Das, A. P. (2024). A critical review of marine microfiber pollution routes, toxicity, and its sustainable remediation. *Environmental Science and Engineering, 2333*, 189–211. https://doi.org/10.1007/978-3-031-51792-1_10

Mishra, A., Das, S. K., & Reddy, K. R. (2022). Processing coalmine overburden waste rock as replacement to natural sand: Environmental sustainability assessment. *Sustainability, 14*(22), 14853. https://doi.org/10.3390/SU142214853/S1

Mohanty, J., Dey, S., Bala, K., & Das, A. P. (2024). Recent advances in microplastic pollution for its sustainable management. In *Sustainable management of urban plastic waste through circular economic approaches* (pp. 27–48). Routledge. https://doi.org/10.1201/9781003465447-3

Mohanty, M., Mohanty, J., Dey, S., Dutta, K., Shah, M. P., & Das, A. P. (2024). The face mask: A tale from protection to pollution and demanding sustainable solution. *Emerging Contaminants, 10*(2), 100298. https://doi.org/10.1016/J.EMCON.2023.100298

Mohanty, S., Ghosh, S., Bal, B., & Das, A. P. (2018). A review of biotechnology processes applied for manganese recovery from wastes. *Reviews in Environmental Science and Bio/Technology, 17*(4), 791–811. https://doi.org/10.1007/S11157-018-9482-1

Mohanty, S., Ghosh, S., Nayak, S., & Das, A. P. (2017). Bioleaching of manganese by aspergillus sp. isolated from mining deposits. *Chemosphere, 172*, 302–309. https://doi.org/10.1016/J.CHEMOSPHERE.2016.12.136

Moharana, S., Dey, S., Priyadarsini, S., Kumar, M. S., & Das, A. P. (2024). Identification and characterization of microplastic pollutants from the marine sediments of Paradeep coast of bay of Bengal, India for their sustainable management. *Environmental Science and Engineering, 2333*, 57–73. https://doi.org/10.1007/978-3-031-51792-1_4

Morseletto, P. (2020). Targets for a circular economy. *Resources, Conservation and Recycling, 153*, 104553. https://doi.org/10.1016/J.RESCONREC.2019.104553

Ngole-Jeme, V. M., & Fantke, P. (2017). Ecological and human health risks associated with abandoned gold mine tailings contaminated soil. *PLoS One, 12*(2), e0172517. https://doi.org/10.1371/JOURNAL.PONE.0172517

Parida, S., Samal, D. P. K., Sukla, L. B., & Das, A. P. (2026). Green technology for the treatment of wastewater using microbes. Decarbonization of Wastewater Pollutants as a Sustainable Solution, 181–210. https://doi.org/10.1016/B978-0-443-33062-9.00023-0

Paul, A., Dey, S., Ram, D. K., & Das, A. P. (2023). Hexavalent chromium pollution and its sustainable management through Bioremediation. *Geomicrobiology Journal*. https://doi.org/10.1080/01490451.2023.2218377

Popov, V., & Pusch, R. (2006). *Disposal of hazardous waste in underground mines* (pp. 1–288). WIT Press.

Pradhan, G., Dey, S., Priyadarsini, S., Kumar, M. S., & Das, A. P. (2024). Characterization and quantification of microplastics pollutants in sediment samples from Daya River of Odisha state in India for their appropriate management. *Environmental Science and Engineering, 2333*, 75–92. https://doi.org/10.1007/978-3-031-51792-1_5

Pradhan, G., Tripathy, B., Ram, D. K., Digal, A. K., & Das, A. P. (2023). Bauxite mining waste pollution and its sustainable management through bioremediation. *Geomicrobiology Journal*. https://doi.org/10.1080/01490451.2023.2235353

Priyadarsini, S., & Das, A. P. (2024). Analytical and structural characterization of waste lithium-ion batteries for their effective recycling strategy. *Environmental Chemistry and Ecotoxicology*. https://doi.org/10.1016/j.enceco.2024.12.004

Priyadarsini, S., & Das, A. P. (2025). Lithium bioleaching: A review on microbial-assisted sustainable technology for lithium bio-circularity. *Journal of Water Process Engineering, 69*, 106744. https://doi.org/10.1016/j.jwpe.2024.106744

Priyadarsini, S., & Das, A. P. (2026).Lithium in waste water: Impacts on pollution and human health. Decarbonization of Wastewater Pollutants as a Sustainable Solution, 353–364. https://doi.org/10.1016/B978-0-443-33062-9.00010-2

Ray, S. K., Dey, S., Das, A. P., & John, A. (2024). Microplastic waste management through greener technologies: A review. In *Sustainable management of urban plastic waste through circular economic approaches* (pp. 237–261). Routledge. https://doi.org/10.1201/9781003465447-14

Roussel, C., Néel, C., & Bril, H. (2000). Minerals controlling arsenic and lead solubility in an abandoned gold mine tailings. *Science of the Total Environment, 263*(1–3), 209–219. https://doi.org/10.1016/S0048-9697(00)00707-5

Samantaray, P., Priyadarsini, S., Mohan, C., Lal, C., & Das, A. P. (2026). Application of microbial communities in wastewater treatment. Decarbonization of Wastewater Pollutants as a Sustainable Solution, 255–270. https://doi.org/10.1016/B978-0-443-33062-9.00014-X

Sahu, A., Dey, S., Mallick, K., Priyadarsini, S., Behera, I. D., & Das, A. P. (2024). A review on advanced technology for sustainable management of synthetic microplastic waste. In *Sustainable management of urban plastic waste through circular economic approaches* (pp. 262–279). Springer. https://doi.org/10.1201/9781003465447-15

Sanket, A. S., Ghosh, S., Sahoo, R., Nayak, S., & Das, A. P. (2016). Molecular identification of acidophilic manganese (Mn)-solubilizing bacteria from mining effluents and their application in mineral beneficiation. *Geomicrobiological Journal, 34*(1), 71–80. https://doi.org/10.1080/01490451.2016.1141340

Singh Jadaun, J., Bansal, S., Sonthalia, A., Rai, A. K., & Singh, S. P. (2022). Biodegradation of plastics for sustainable environment. *Bioresource Technology, 347*, 126697. https://doi.org/10.1016/J.BIORTECH.2022.126697

Singh, S., Sukla, L. B., & Goyal, S. K. (2020). Mine waste & circular economy. *Materials Today: Proceedings, 30*, 332–339. https://doi.org/10.1016/J.MATPR.2020.01.616

Szczepańska, J., & Twardowska, I. (2004). Mining waste. *Waste Management Series, 4*(C), 319–385. https://doi.org/10.1016/S0713-2743(04)80015-1

Tayebi-Khorami, M., Edraki, M., Corder, G., & Golev, A. (2019). Re-thinking mining waste through an integrative approach led by circular economy aspirations. *Minerals, 9*(5), 286. https://doi.org/10.3390/MIN9050286

Worlanyo, A. S., & Jiangfeng, L. (2021). Evaluating the environmental and economic impact of mining for post-mined land restoration and land-use: A review. *Journal of Environmental Management, 279*, 111623. https://doi.org/10.1016/J.JENVMAN.2020.111623

Zou, H. (2019). Study on soil ecological environment restoration strategy of abandoned mining area. *Arabian Journal of Geosciences, 12*(23), 1–7. https://doi.org/10.1007/S12517-019-4873-3/METRICS

Hazardous Impact of Mining Pollutants on Biodiversity in Terrestrial and Aquatic Environments: Prevention and Bioremediation

Vivek Raj, Santosh Kumar Karn, Moni Kumari, Rakesh Kumar, Aman Prakash, Arun Kumar, Arun Lal Srivastav, and Vandana

1 Introduction

Pollution and environmental degradation are two of the biggest problems facing the modern world, and they are mostly brought on by the growing buildup of dangerous materials.

These pollutants present a significant danger to ecosystems, leading to their destruction when they come into contact with toxic particles (Das & Ghosh, 2023; Mohanty, Dey, et al., 2024; Priyadarshini et al., 2024; Sahu et al., 2024). The infiltration of these contaminants into soil and water bodies disrupts ecological balance,

V. Raj (✉)
Department of Biotechnology, Patna Women's College (Autonomous), Patna University, Patna, Bihar, India

S. K. Karn · M. Kumari
Department of Biochemistry and Biotechnology, Sardar Bhagwan Singh University, Dehradun, Uttarakhand, India

R. Kumar
School of Pharmaceutical Sciences, Lovely Professional University, Phagwara, Punjab, India

A. Prakash
Department of Marine Biology, Leon H. Charney School of Marine Sciences, University of Haifa, Haifa, Israel

A. Kumar
Mahavir Cancer Sansthan and Research Centre, Patna, Bihar, India

A. L. Srivastav
Chitkara University School of Engineering and Technology, Chitkara University, Barotiwala, Himachal Pradesh, India

Vandana
Institute of Management Studies, Indus University, Ahmedabad, Gujarat, India

© The Author(s), under exclusive license to Springer Nature Switzerland AG 2026

A. P. Das (ed.), *Bioremediation Technology for Sustainable Mining Waste Management*, Sustainable Environmental Waste Management Strategies,
https://doi.org/10.1007/978-3-032-03468-7_14

affecting plants, animals, and habitats that depend on these resources. This disturbance results in harmful effects such as the accumulation of toxins in the food chain (biomagnification), biodiversity loss, and even widespread mortality, ultimately destabilizing entire ecosystems (Das, Dey, & Das, 2024; Dey, Bissoyi, et al., 2024; Sahu et al., 2024). This chapter examines the environmental effects of mining, an activity that greatly damages the areas where it is executed. Mining involves the extraction of minerals and geological materials through deep excavation, often resulting in severe environmental degradation. The use of heavy machinery during mining disrupts the soil structure, causing significant erosion and making the land unsuitable for plant growth. Additionally, mining byproducts and waste materials seep into the soil and water, posing serious risks to nearby ecosystems (Worlanyo & Jiangfeng, 2021). Moreover, soil loosening decreases its stability, increasing the likelihood of landslides, which can have disastrous consequences (Briffa et al., 2020). To reduce these negative impacts, it is crucial to implement preventive measures and develop innovative mining techniques to limit environmental harm. This chapter highlights the ecological consequences of mining and explores potential solutions to decrease the presence of toxic substances at mining sites, addressing the environmental challenges posed by this industry.

2 Primary Mining Pollutant Sources

The process of mining involves the removal of both organic and inorganic materials from the earth's surface. This process entails excavating at multiple sites and employing mechanical equipment like drills, explosives, and other machinery to extract these materials. Mining is essential for extracting valuable ores, it also presents significant environmental risks. Among the main issues is the adverse effect on ecosystems, as mining operations generate hazardous byproducts that can harm living organisms. For instance, water pollution, commonly known as "acid mine drainage" occurs when acidic waste materials from mining sites enter water bodies, threatening aquatic life (Ghosh et al., 2023; Mohanty, Tripathy, et al., 2024; Ray et al., 2024). Plants and animals exposed to these acidic pollutants may experience harmful effects. Additionally, mining can increase the concentration of toxic metals, leading to water contamination and poisoning of aquatic species. Hazardous metals such as arsenic and mercury, often released from mining sites, pose severe health risks to both humans and wildlife (Kumar et al., 2023; Yadav et al., 2022). Abandoned coal mines pose significant environmental dangers, including acid mine drainage, heavy metal contamination, groundwater pollution, and soil degradation, which threaten ecosystems and human health (Alhamed & Wohnlich, 2014). The effects are caused by the oxidation of iron disulfide minerals, which release iron, sulfate, and hydrogen. The presence of carbonate-rich materials in landfill waste has resulted in the release of calcium, magnesium, and bicarbonate ions, thereby increasing the alkalinity of the affected water. Considerable regional differences in water contamination were observed, highlighting the need for necessary measures to reduce their impact (Fig. 1). Several other major sources of mining pollutants are described below.

2.1 *Suspended Particulate Matter*

Assessing air quality is a vital aspect when examining mining activities. This process involves measuring the levels of suspended particulate matter and other emissions released from mining sites, which leads to air pollution. Airborne particulate matter is comprised of tiny solid and liquid particles, mainly emanating from excavations and the functioning of mining machinery. These activities generate dust and toxic substances, leading to an increase in harmful airborne particles. When inhaled by humans and animals, these pollutants can lead to respiratory disorders and cardiovascular diseases (Dwernychuk et al., 2002). Therefore, implementing strict regulations and control strategies is essential to reduce the emission of these pollutants into the atmosphere.

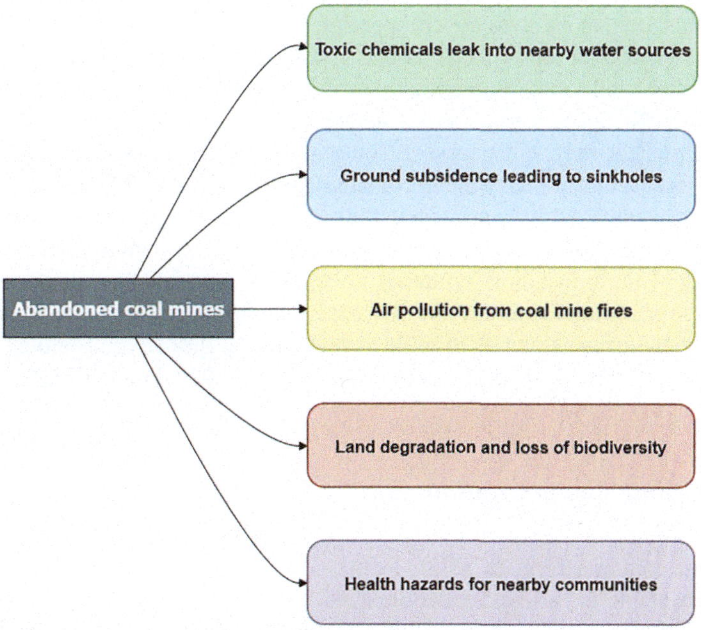

Fig. 1 Environmental hazards posed by abandoned coal mines

2.2 Heavy Metals Pollution

Heavy metal contamination is a significant cause of environmental pollution and degradation, posing severe threats to ecosystems and living organisms. These inorganic pollutants mainly originate from industrial waste and mining operations, where ore extraction plays a key role. Inadequate waste disposal and incorrect disposal of metal byproducts cause the build-up of toxins, interfering with the natural biological processes of plants and animals (Dri et al., 2018; Koda & Podlasek, 2023). Additionally, human exposure to heavy metals, including arsenic, mercury, and lead can have toxic health consequences (Briffa et al., 2020). Heavy metals are found in the Earth's crust, but they are released into the environment primarily as a result of the activities associated with mining and metal extraction industries. Mining activities significantly increase their concentration, allowing them to leach into surrounding water bodies and ecosystems (Dey, Das, et al., 2023; Dey, Tripathy, et al., 2023; Paul et al., 2023). Both natural processes like rock weathering and erosion as well as human activity like mining allow these toxins to reach the ecosystem. Since heavy metals do not degrade over time, they remain in the soil for extended periods, interacting with other soil components and sometimes becoming even more toxic. A notable example is methylmercury, a highly toxic compound that forms when inorganic mercury undergoes bacterial transformation in water, sediments, and soil (Azaroff et al., 2020). Isolated mining locations and places that have previously been subjected to metal-based insecticides frequently have high quantities of heavy metals. In such locations, only metal-tolerant plant species can survive due to the toxic conditions. One approach to reducing contamination in these areas is capping, a method that involves covering polluted sites with an impermeable layer and adding fresh soil on top. This technique prevents heavy metals from leaching into groundwater while also protecting vegetation. In the past, pesticides containing metals such as copper, arsenic, lead, and chromium were widely used, and traces of these substances may still persist in some regions (Alonso et al., 2020). Another major consequence of heavy metal contamination is water pollution, particularly in rivers and aquatic systems near industrial and mining zones. Mining waste seeps into these water bodies, where heavy metals often accumulate at the bottom. The solubility of these metals is primarily affected by the acidity or alkalinity of the water. As metal concentrations rise, the water becomes more acidic, reducing metal solubility and causing metals to settle as precipitates, further contributing to water pollution (Das, Dey, Bala et al., 2024; Mallick et al., 2024; Pradhan et al., 2024).

2.3 Sulfate Contamination Due to Mining

The most common form of sulfur in the crust of the Earth is sulfate, which is extremely stable in its (+6) oxidation state. It enters groundwater through various natural and human-induced processes, including mineral dissolution, atmospheric deposition, and mining activities. While sulfate is not highly toxic, excessive intake through food and water can lead to health issues such as dehydration, diarrhea, and digestive discomfort in both humans and animals. Plants and microorganisms absorb sulfates, allowing them to enter the food chain and potentially cause adverse effects. Even though babies are more vulnerable to exposure to sulfates, no definitive evidence directly links sulfate concentrations in drinking water to serious health risks. However, individuals consuming water with high sulfate levels may initially experience mild laxative effects (Wang et al., 2022).

Sulfate is an essential component of the sulfur cycle, which involves both atmospheric and terrestrial processes. The cycle begins with the weathering of sulfate-rich rocks and minerals, known as evaporites, which release stored sulfides into the environment. When exposed to air, sulfate ions (SO_4^{2-}) form and are taken up by plants and microorganisms, contributing to the production of organosulfur compounds. These compounds move through the food chain as organisms consume sulfur-containing materials. When plants and animals decompose, sulfate is reintroduced into the environment. In wetlands and coastal regions, decaying organic matter releases hydrogen sulfide (H_2S), which later oxidizes into sulfate in aquatic environments. Another major natural source of sulfur is volcanic activity, where gases such as hydrogen sulfide and sulfur dioxide (SO_2) are released into the environment. Over time, these gases react with water to produce sulfate ions, which may eventually lead to the formation of evaporite deposits. Mining activities are a significant contributor to sulfate pollution, leading to air, water, and soil contamination. Mineral extraction and processing chemicals can have a detrimental effect on the health of mine employees and the surrounding community, raising the possibility of exposure-related diseases. The accumulation of sulfate in the environment highlights the significance of proper management and monitoring techniques to reduce pollution and safeguard ecosystems.

2.4 Cyanide

Cyanide is widely used in gold extraction from ore. In its pure form, it is a colorless substance with a distinct almond-like odor, though this scent may disappear when mixed with other chemicals. Cyanide can exist as a gas, liquid, or powder, and even small doses can be highly toxic if ingested. Exposure to low levels of cyanide over an extended period may lead to health concerns including the development of goiter, a swelling of the throat often linked to malnutrition. During gold mining, waste storage ponds can sometimes rupture, leading to the release of cyanide into nearby water bodies (Dey, Dash, & Das, 2024; Priyadarsini & Das, 2025). While mining companies argue that cyanide becomes inactive in water, this only occurs under conditions with sufficient oxygen and sunlight (Acheampong et al., 2010). However, even in such cases, cyanide may still retain hazardous chemical properties. If cyanide seeps into underground water sources or if environmental factors such as rain or cloud cover reduce oxygen and sunlight exposure, it remains highly toxic for long periods. Aquatic life is seriously threatened by this pollution, which also renders the water unfit for human consumption. Due to its significant environmental and health risks, some countries have implemented bans on the use of cyanide in mining operations.

2.5 *Sulfuric Acid*

Copper mining relies heavily on sulfuric acid, a highly toxic substance that further exacerbates acid mine drainage by reacting with water and heavy metals throughout the mining process. Sulfuric acid smells strongly of eggs and, when it comes into close contact with the skin, can inflict blindness, serious burns, or even death. Furthermore, mining activities generate a lot of wastewater that contains hazardous chemicals and waste particles. When these byproducts are discharged into water bodies, they pose significant threats to aquatic ecosystems.

Industrial emissions from mining also affect aquatic environments by releasing smoke and soot into the atmosphere. When sulfur dioxide from mining reacts with water vapor, it forms sulfuric acid, which falls as acid rain. This acidic precipitation and runoff lower the pH of water bodies, making it difficult for fish to absorb essential nutrients, salts, and oxygen. Reduced pH levels also disrupt calcium absorption, which is vital for aquatic organisms. As a result, fish eggs may fail to develop properly due to calcium deficiency, making them brittle or weak. Additionally, calcium shortages lead to fragile fish bones and spines, as well as weakened exo-skeletons in species like crayfish. Acid rain further depletes the soil of essential minerals such as aluminum, negatively affecting fish and crustacean reproduction (Abril et al., 2008). Moreover, low pH levels significantly disrupt aquatic food chains. Many insect species, including mayflies and stoneflies, cannot survive in waters with a pH below 6, leading to imbalances in the ecosystem (Conant et al., 2008). The combined effects of acid mine drainage, industrial pollution, and acid rain emphasize the need for stringent environmental regulations in copper mining to minimize its harmful impact on aquatic life and ecosystems.

3 Mechanisms of Contamination of the Terrestrial Environment

Mining has been practiced for centuries, with trace elements entering the environment not only through the extraction process but also through managing resources such tailings, waste rock, and smelting operations (Mwesigye et al., 2016; Odumo et al., 2014). To minimize environmental and health risks, responsible mining companies follow safety and sustainability guidelines set by organizations like the International Council of Mining and Metals. But mining continues to be a major cause of soil contamination (Abhilash & Singh, 2008). Ore extraction often involves the release of trace elements that are naturally present within rocks. Even if these elements are present in low concentrations that are not economically viable for extraction, mining waste can still pose environmental and health hazards (Bala et al., 2024; John & Das, 2024; Mohanty, Tripathy, et al., 2024). Certain extraction methods use chemicals such as cyanide, producing large volumes of waste that contribute to pollution.

The residues left after ore processing, known as tailings, contain trace elements that can mix with rainwater or nearby water sources, forming liquid slurries. These tailings, which pose serious threats to terrestrial environments, are typically stored in on-site basins with dams. The solid waste is allowed to dry out, with the discarded water either reused or discharged. Soil contamination occurs when tailings dams collapse or when wind erosion spreads fine mineral particles unless proper protective measures, such as vegetation cover, are in place (Karaca et al., 2018). Studies on global tailings dam failures indicate that many incidents occur due to poor maintenance. Disasters such as those in Aznalcóllar, Spain (1998) (Simón et al., 1999), Baia Mare, Romania (2000), and Brazil (2015) (Hatje et al., 2017; Queiroz et al., 2018) released toxic waste containing high levels of cadmium (Cd), arsenic (As), lead (Pb), copper (Cu), and zinc (Zn), contaminating nearby soils and water sources (Abraham et al., 2017). Furthermore, mining waste deposits, particularly those with sulfur-based minerals like pyrite, contribute to soil pollution through wind erosion and leaching. When exposed to air and water, these minerals produce acid mine drainage (AMD), which dissolves metals, creating highly acidic wastewater containing iron, sulfates, and trace metals (Alemayehu et al., 2020). If not properly managed, this wastewater can seep into soil and water sources, causing long-term pollution (Ribeiro et al., 2013).

In addition to extraction and waste disposal, pollution can result from ore transportation and smelting. Spills during transport can contaminate soil, while smelting emissions release trace elements that settle in nearby and distant areas (He et al., 2019). Some of the most severe soil contamination cases have been linked to the extraction of coal, gold, uranium, tungsten, tin, and polymetallic sulfides (Duggan & Dhawan, 2007). Several regions, including the Tibagi River in Brazil, Ledo in India, Ptolemais-Amynteon in Greece, and Barapukuria in Bangladesh, have reported high levels of trace elements in soil near coal mines (Sahoo et al., 2016).

Mining of gold is particularly associated with hazardous pollutants such as mercury (Hg), arsenic (As), cyanide (CN), lead (Pb), zinc (Zn), and nickel (Ni) (Fashola et al., 2016; Odumo et al., 2014). Mercury, commonly used in gold extraction, is often discharged into the environment through mining waste. It can evaporate or contaminate soil and water, posing significant health risks. Around 4000 tons of mercury were discharged into the environment by artisanal gold mining in the Amazon basin between the late 1980s and the early 1990s (de Lacerda, 2003). Research on global mercury emissions revealed that human activities have significantly raised mercury levels in organic soils by approximately 150,000 tons and in mineral soils by roughly 800,000 tons. Notably, small-scale gold mining accounts for up to 15% of this mercury pollution, with a notable presence in South America,

East Asia, and Southeast Asia. About 2700 tons of mercury are released into the soil annually only from large-scale gold mining. Uranium mining also produces radioactive waste. Worldwide production of uranium mill tailings in 2006 reached approximately 938 million m^3 (Abdelouas, 2006). These tailings contain high levels of radionuclides, including thorium-232 (^{232}Th), potassium-40 (^{40}K), uranium-235 (^{235}U), and uranium-238 (^{238}U), along with their decay byproducts (Smičiklas & Šljivić-Ivanović, 2016). Even after mining operations cease, these sites remain sources of pollution. Without proper long-term management, rock waste dumps and tailings dams are vulnerable to wind dispersal, water erosion, and weathering, further contaminating surrounding soils (Iavazzo et al., 2012).

4 Mechanisms of Contamination of the Aquatic Environment

Naturally occurring heavy metals infiltrate aquatic ecosystems via geological formations, which contain them in forms including silicates, carbonates, oxides, hydroxides, and sulfides. Besides natural sources, human activities like fertilizer and pesticide use, sewage discharge, and mining waste also contribute to heavy metal pollution. Included among these contaminants are fine metal-bearing particles measuring less than 30 μm, as well as precipitation deposits (Alamdar et al., 2014). Waste from mining procedures, particularly those containing arsenic (As), zinc (Zn), and iron (Fe), can significantly affect aquatic organisms and human health. In water bodies, heavy metals exist in different states dissolved, complex or particulate, and distributed throughout compartments such as water, suspended solids, sediments, and aquatic organisms. Metal ions and complexes are released during weathering, erosion, and runoff, which is a key method that metals and metalloids from mining waste enter water systems (Das et al., 2011; Das & Mishra, 2008; Mishra et al., 2019, 2021; Mishra & Das, 2021). Once in water reservoirs, these metals accumulate in sediments without breaking down, affecting their solubility, toxicity, and bioavailability based on their chemical form (Barrera Olivarez et al., 2021).

Sediments act as both storage and transport mediums for inorganic and organic pollutants in aquatic systems. They effectively adsorb metals and metalloids, limiting their movement in water. Heavy metals are mainly present in fine sediment particles, such as iron humus, ranging in size from 0.001 to 0.1 μm. These particles contain different chemical species, including aluminum (Al) and manganese (Mn) oxyhydroxides, aluminosilicates, and moderately soluble salts like calcium carbonate ($CaCO_3$).

The presence of hazardous metals and metalloids in water influences the movement of trace elements from the vadose zone (the unsaturated soil layer) into aquifers, rivers, and tributaries. These metals also affect crucial chemical processes, such as redox reactions, sorption, complexation, hydrolysis, and chelation. Among these, adsorption is the key mechanism regulating the movement of trace metals in water environments. Negatively charged clays and humic substances attract and bind positively charged metal ions (cations), while oxyhydroxides with variable charge on surface can also adsorb cations. Adsorption facilitates the movement of metals at the interface between the liquid and solid phases. To further affect metal mobility in water bodies, negatively charged ions (anions) can also be adsorbed (Adriano, 2001).

5 Processes Involved in Heavy Metal Transfer in Natural Waters

Metals and metalloids exhibit significant movement in aquatic ecosystems, influencing the regulation of pH levels, mineral formation and breakdown, adsorption–desorption reactions, hydrolysis, and oxidation–reduction processes. Co-adsorption has a significant impact on metal mobility. Graphene oxide, a highly soluble substance frequently detected in water bodies, comes into contact with inorganic impurities such as hematite and goethite. At a pH range of 3–5, around 92% of graphene oxide undergoes co-adsorption with these minerals. The extent of metal adsorption and desorption is determined by factors including the length of time for which the adsorbent and adsorbate are in contact with each other, as well as the aging of oxides resulting from weathering processes. While both adsorption and precipitation can occur simultaneously, depending on the kind of metals used and the environmental (Mondal et al., 2018).

When a solid precipitate forms, incorporating elements from both dissolved minerals and the surrounding water, it is known as a co-precipitate. The mobility of metals and metalloids in water varies depending on their chemical forms. Oxyanions such as sulfate (SO_4^{2-}), arsenate (AsO_4^{3-}), molybdate (MoO_4^{2-}), and chromate (CrO_4^{2-}) exhibit moderate mobility. In contrast, metal cations like zinc (Zn), copper (Cu), cadmium (Cd), and aluminum (Al) are highly mobile in acidic conditions, while alkaline or polar conditions further enhance their movement, increasing the risk of their spread in aquatic environments (Alloway, 1995).

6 Adsorption of Cations

Cation exchange adsorption or physisorption, also referred to as nonspecific adsorption, is an electrostatic process that occurs on the surfaces of various minerals, including olivine, pyroxene, mica, feldspar, and several types of clay, specifically kaolinite, montmorillonite, illite, vermiculite, and chlorite. In either a 1:1 or 2:1 ratio, stacked microcrystalline layers of octahedral aluminum sheets and tetrahedral silica sheets make up clays. Cation exchange mainly occurs in the outer sphere clusters, it uses weak covalent bonds to bind positively charged ions to the negatively charged clay surfaces. This process is both rapid and reversible, as smaller ions from the water phase are replaced by larger ions on the clay surface. Diffusion and electrostatic forces primarily control this exchange mechanism. A common example of this process is the substitution of silicon (Si^{4+}) with magnesium (Mg^{2+}) or aluminum (Al^{3+}) with silicon (Si^{4+}). The effectiveness of clay in exchanging cations is measured by its cation exchange capacity (CEC), which indicates the number of cations that can be reversibly adsorbed per unit weight of the material (Montes-Avila et al., 2019).

7 Particular Adsorption

Specific adsorption, also called surface complexation or chemisorption, mainly occurs with aluminum (Al), iron (Fe), and manganese (Mn) oxides. Strong covalent connections are formed between cations or anions and oxygen atoms or hydroxyl (OH) groups on the surface of these oxides during this process. The charge of Fe and Mn oxides is influenced by the adsorption or desorption of hydrogen (H^+) ions at the mineral-solution interface. Due to their amphoteric nature, these oxides can display both positive and negative charges. Their surface charge is neutralized at a particular pH level, known as the point of zero charge (pzc), which varies for different oxides. For Fe oxides, the pzc typically ranges between pH 7.0 and 8.5, resulting in a predominantly positive charge at these levels. In contrast, Mn oxides have a lower pzc, between pH 1.5 and 4.6, giving them a net negative surface charge. The adsorption of ions depends on pH-sensitive interactions between functional metal-oxygen (MO) groups and electrostatic forces on the oxide surface, which vary according to the charge properties of the oxide (Lindsay et al., 2015).

7.1 Hydrolysis

At 25 °C, pH of pure water is 7.0, which is neutral, where the concentrations of hydronium ions [H_3O^+] and hydroxide ions [OH^-] are both 1.0×10^{-7}M. When sodium chloride (NaCl) dissolves in water, it undergoes minimal hydrolysis, keeping the pH at 7.0. However, dissolving sodium acetate ($C_2H_3NaO_2$) increases the pH above 7.0, indicating a higher [OH^-] concentration than [H_3O^+]. In contrast, NH_4Cl, or ammonium chloride, decreases pH below 7.0. meaning [H_3O^+] exceeds [OH^-]. With the exception of beryllium (Be^{2+}), which does not significantly hydrolyze, the majority of salts include alkali or alkaline earth metal ions. Therefore, a hydrolysis process happens when a material dissolves in water and one of its ions lowers the pH below 7.0. In tailing deposits rich in sulfides, such as pyrite and pyrrhotite, iron (Fe^{2+}) dissolves through oxidation, reacting with water and undergoing hydrolysis, which lowers the pH and increases acidity. Similarly, when Fe^{2+} oxidizes to Fe^{3+} at a pH above 5, a hydrolysis reaction releases H^+ ions and forms iron hydroxide, leading to oxyhydroxide precipitation. However, when mining waste enters aquatic environments, an increase in acidity does not always result in a lower pH, as carbonate minerals in the surroundings can neutralize the acidity, reducing its impact (Fergusson, 1990).

7.2 *Oxidation–Reduction*

Sulfide minerals in mining tailings have some resistance to weathering. However, physicochemical weathering processes can alter water composition, leading to leaching or infiltration into mining waste (Alekseenko et al., 2018). When oxygen-rich surroundings are introduced to sulfide minerals, they oxidize due to their thermodynamic instability under such conditions. The most prevalent sulfide minerals, pyrite and pyrrhotite, are among the first to oxidize. This process creates ferrous iron (Fe^{2+}) and acidic conditions by reacting one mole of pyrite and one mole of pyrrhotite, which function as electron acceptors, with oxygen and water. Other sulfide minerals, such as sphalerite, chalcopyrite, galena, and arsenopyrite, also contribute to acid drainage formation. The oxidation of metal sulfides generates significant heat, with pyrite oxidation alone releasing 1440 KJ per mole, making it one of the most energy-intensive oxidation reactions. In highly permeable topsoil, this exothermic reaction accelerates sulfide oxidation and enhances oxygen transport through convection. However, when mining waste becomes water-saturated, convection slows, and diffusion becomes the primary means of oxygen transfer. Moreover, wind can introduce oxygen from the reservoir's sides, enabling it to move upward or penetrate deeper through advection, convection, and diffusion. In regions with high sulfide oxidation but low permeability, oxygen concentrations remain low, reducing oxidation rates. If a tailings dam has poor aeration and insufficient oxygen supply, sulfide oxidation can become extreme, causing the water pH to drop below 1. This results in high concentrations of sulfate and dissolved metals, including iron (Fe), zinc (Zn), and lead (Pb). In such highly acidic conditions, ferric iron (Fe^{3+}) plays a crucial role in further oxidizing metal sulfides, intensifying the acidification of the surrounding environment (Romero et al., 2007).

8 Role and Process of Bioremediation of Contamination

Industrialization has resulted in significant heavy metal pollution, posing serious risks to public health because of the toxic effects of these metals. Conventional methods for removing heavy metals are often expensive and generate hazardous byproducts that further damage the environment. Consequently, eco-friendly technologies have gained increasing attention for both heavy metal extraction and recovery. Various biological organisms, including algae, bacteria, yeasts, and fungi, have demonstrated high efficiency, affordability, and widespread availability in metal remediation. Among these approaches, bacterial bioaccumulation is considered the safest method for treating non-biodegradable contaminants such as heavy metals. Additionally, bioremediation and phytoremediation, either individually or in combination, provide sustainable and environmental-friendly solutions for removing heavy metal pollutants from ecosystems (Al-Awadhi et al., 1996; Raj et al., 2025).

8.1 Phytoremediation (Plant/Plant-Based Products)

Soil reclamation techniques focus on providing environmentally friendly solutions for restoring contaminated land. Agro-remediation, botano-remediation, or green remediation are other names for phytoremediation, which is an environmentally responsible and sustainable substitute for traditional soil remediation techniques. First introduced in 1983, While testing is still ongoing, phytoremediation has demonstrated great promise in treating soils with low to moderate levels of heavy metal pollution. Additionally, it may be used with conventional cleanup methods to increase the effectiveness of pollution removal (Shen et al., 2022). Fast-growing plants are employed in the process of phytoremediation to remove, stabilize, or eliminate harmful pollutants from soil and water. This approach is economical, aesthetically pleasing, and ecologically conscious. The plant species' capacity to absorb and retain significant levels of heavy metals in their shoot tissues determines how effective it is. Phytoremediation consists of five key subcategories: phytostabilization, phytostimulation, phytotransformation, hemofiltration, and phytoextraction.

8.2 Phytostabilization

Phytostabilization, also called phytoimmobilization or photo restoration, is an eco-friendly approach for managing metal-contaminated soils. This method focuses on reducing the mobility and bioavailability of heavy metals, preventing their leaching into groundwater or accumulation in the food chain. Unlike soil amendments, plants play a limited role in phytostabilization. Instead, this process primarily relies on the physical and chemical stabilization of metal pollutants by adsorbing them onto plant roots and immobilizing them using soil additives. Effective soil additives for heavy metal immobilization include organic matter, biosolids, phosphate fertilizers, and clay minerals. The major plant functions in phytostabilization are to reduce migration of contaminant, control erosion of soil, minimize direct pollutant interactions, and regulate water infiltration. However, because metal contaminants are not completely eliminated, they persist in the soil, phytostabilization acts as a containment strategy rather than a complete remediation solution.

8.3 Phytostimulation

The breakdown of organic contaminants in the rhizosphere as a result of enhanced microbial activity is called rhizodegradation or phytostimulation. The rhizosphere is the narrow soil zone surrounding plant roots, where root interactions influence microbial processes and enhance contaminant degradation (Ratajczak et al., 2023). Several key factors contribute to this increased microbial activity:

(a) Plant roots contribute to oxygen regulation in the rhizosphere, supporting the aerobic degradation of pollutants.
(b) Root exudates, rich in carbohydrates and amino acids, provide vital nutrients that promote the development and activity of microorganisms.
(c) The accumulation of root biomass increases organic carbon availability, further stimulating microbial processes.
(d) Mycorrhizal fungi assist in breaking down complex organic compounds that bacteria alone cannot degrade.
(e) Plants create a suitable environment that supports the growth of beneficial microbial communities, improving contaminant degradation.

8.4 Phytotransformation

Plants use their metabolic processes or the creation of enzymes to break down organic contaminants without the aid of microbes. This process is called phytotransformation or phytodegradation. Due to their ability to detoxify contaminants, plants are often referred to as the "green liver" of the biosphere. This method is particularly effective in degrading organic compounds. Phytovolatilization is a similar process in which pollutants are absorbed by plants, converted into volatile chemicals, and then released into the atmosphere by transpiration. This process is particularly useful in organic pollutants and specific heavy metals, such as mercury (Hg) and selenium (Se), which can be volatilized and removed from the soil environment.

8.5 Phytofiltration

Hemofiltration is a plant-based water purification technique used to filter low levels of pollutants from surface water, groundwater, and wastewater (Singh et al., 2024). In this method, plants are first exposed to contaminated water to acclimate to the pollutants before being deployed for site remediation. Once their roots reach saturation, they are harvested for further processing or disposal. Hemofiltration consists of three main approaches: rhizofiltration (using roots), phytofiltration (using whole plants), and caulofiltration (using plant shoots). Contaminants are removed through absorption, adsorption, or precipitation, preventing their spread into groundwater.

Furthermore, root exudates can alter the pH of the rhizosphere and affect metal precipitation, which improves pollution removal. In rhizofiltration, fast-growing terrestrial and aquatic plants are especially good in removing heavy metals like nickel (Ni), copper (Cu), lead (Pb), zinc (Zn), chromium (Cr), and cadmium (Cd). Terrestrial plants with extensive, fibrous root systems are preferred due to their superior ability to capture and retain contaminants.

8.6 *Phytoextraction*

Phytoextraction is a plant-based method for eliminating heavy metals from water and soil using fast-growing plant species. This process follows two main approaches: continuous (natural) phytoextraction and chemically induced phytoextraction. In continuous phytoextraction, plant roots absorb heavy metals and transport them to above-ground tissues. Once a sufficient concentration of metals accumulates, the plants are harvested and processed. The harvested biomass can be burned to recover metals, converted into biogas, or used in brick manufacturing. If not repurposed, it is disposed of in designated waste sites. This method, also referred to as phytomining or biomining, effectively reduces contamination of plant roots and shoots by heavy metals without appreciably changing the composition of the soil. Phytoextraction is an environmentally sustainable soil remediation technique that also offers economic potential through metal recovery (Medina-Díaz et al., 2024).

8.7 *Metal May Be Restored by Using Harvestable Plant Components*

Continuous cycle of planting and harvesting can gradually decrease pollutant concentrations in the soil. This innovative green technology is a cost-effective alternative to traditional remediation methods. For phytoextraction to be effective, plants must demonstrate rapid growth, large root systems, significant biomass output, and a potent capacity to absorb and store heavy metals. Continuous phytoextraction relies on natural hyperaccumulators, which are commonly found in metal-rich environments. The process involves several key stages:

- Some metal contaminants adhere to the outer surface of plant roots.
- Bioavailable metals enter root cells through the membrane.
- Certain absorbed metals are stored in vacuoles.
- Mobile metals travel through the xylem.
- Metals are transported to stems and leaves.

Phytoremediation is the most cost-efficient method for treating heavy metal-contaminated soils, reducing costs by 60%–80% compared to traditional physicochemical techniques. After phytoextraction, harvested plant biomass can be converted into sustainable energy sources. However, challenges such as slow remediation rates, limited metal bioavailability, and the slow growth of metal-accumulating plants remain. Since phytoremediation is still in its early research phase, outcomes may vary between greenhouse experiments and real-world field applications (Costa et al., 2019). Environmental factors including pollutant distribution, soil pH, microbial activity, nutrient availability, moisture levels, and temperature—significantly impact the efficiency of the process in field conditions (Sarwar et al., 2017). Before large-scale implementation, field testing is necessary to ensure effectiveness. The successful commercialization of this technology depends on identifying plant species that can accumulate high levels of metals while producing significant biomass (Farinella et al., 2007).

8.8 Bioremediation

Reducing metal bioavailability and immobilizing pollutants are the primary objectives of microbial remediation for heavy metal contamination. Microorganisms can change the physical and chemical characteristics of heavy metals to make them less harmful, even if they are unable to break them down (Douay et al., 2007; Raj et al., 2025). Depending on the kind of heavy metal and the particular microbes involved, this method may or may not work. A number of processes are involved in microbial remediation, such as precipitation, oxidation-reduction reactions, intracellular accumulation, and extracellular complexation. Microbial leaching is an additional technique used to remediate heavy metal-contaminated soils. Biosorption, in which heavy metals bind to anionic functional groups on the surface of microbial cells, is one of the most significant microbial remediation methods. This leads to metal immobilization through interactions with extracellular materials, including Van der Waals forces, covalent bonding, and electrostatic attractions. Additionally, microorganisms generate siderophores, which are chelating substances produced by bacteria, plants, and fungi. These compounds bind to heavy metals, reducing their bioavailability and limiting their environmental impact. Additionally, biosurfactants produced by microbes facilitate heavy metal removal by altering metal solubility and mobility. Heavy metals dissolve more quickly through complexation, which results in the creation of metal salts that are less soluble. Both enzymatic and non-enzymatic methods can be used by microorganisms to eliminate heavy metals (Mejias Carpio et al., 2018).

8.9 Bacterial Biomass in the Elimination of Harmful Heavy Metals

An efficient biosorbent for eliminating heavy metals from aqueous solutions is bacterial biomass, even at extremely low concentrations. Both living and non-living bacterial cells have been successfully used to adsorb toxic metals. The presence of anionic structures on the negatively charged bacterial cell surface makes it easier for the bacteria to interact with metal cations. Various bacterial species can bioaccumulate hazardous metal ions, like uranium (U), lead (Pb), and copper (Cu) from contaminated water sources, whether in free-floating or immobilized forms. The chemical makeup of bacterial cell walls is essential for biosorption, especially when it comes to the extracellular absorption of metals by dormant or non-living cells. To enhance and optimize the biosorption process, it is essential to determine which functional groups on the surfaces of microbial biomass are in charge of metal absorption. Different functional groups contribute to metal adsorption, influencing the overall efficiency of microbial remediation (Priyadarshanee & Das, 2021).

8.10 Extracellular Polymeric Substances (EPS) and Bacterial Biofilm in the Removal of Metal

A biofilm forms when microbial cells adhere to a surface and become enclosed in an extracellular polymeric matrix. Compared to free-floating bacteria, biofilm-associated microbes can withstand extreme environmental conditions because they are more hardy. Biofilm technology is becoming more and more popular in wastewater treatment because of its high metal sorption capability and affordability. Because they provide functional groups that act as binding sites for pollutants, particularly heavy metals, extracellular polymeric substances (EPS) are essential for the development of biofilms. The biopolymer within microbial communities effectively captures and adsorbs metal ions from polluted water. During remediation, these metallic ions become trapped within the bacterial biofilm, aiding in their removal from the environment. Research has shown that microbial biofilms are highly effective in adsorbing heavy metals. EPS is made up of low-molecular-weight non-polymeric materials as well as high-molecular-weight organic components such proteins, polysaccharides, nucleic acids, and phospholipids. According to studies, interactions between positively charged metal cations and negatively charged functional groups in the EPS layer allow EPS to bind harmful metal ions. This process occurs without requiring microbial metabolism or energy input. EPS is essential for biofilm formation and stability, enabling the immobilization of heavy metal ions within the biofilm. This enhances metal toxicity control and facilitates their removal (Priyadarshanee & Das, 2021).

9 Conclusion

Mining, while beneficial for resource extraction, can also harm land and water ecosystems by releasing toxic byproducts. These pollutants accumulate in high concentrations, posing health risks to all life forms by disrupting the food chain and causing biological damage. To minimize environmental harm, various remediation techniques must be implemented, along with ground teams to prevent geological disasters like landslides and mine collapses. This study highlights the negative impacts of mining pollution and explores both traditional and modern remediation technologies. It emphasizes the vital role of bioremediation as an effective method to counteract mining-induced pollution, enhancing our understanding of its molecular and biochemical mechanisms and broadening its potential applications.

References

Abdelouas, A. (2006). Uranium mill tailings: Geochemistry, mineralogy, and environmental impact. *Elements, 2*(6), 335–341.

Abhilash, P. C., & Singh, N. (2008). Distribution of hexachlorocyclohexane isomers in soil samples from a small scale industrial area of Lucknow, North India, associated with lindane production. *Chemosphere, 73*(6), 1011–1015.

Abraham, J., Dowling, K., & Florentine, S. (2017). Risk of post-fire metal mobilization into surface water resources: A review. *Science of the Total Environment, 599*, 1740–1755.

Abril, J. M., García-Tenorio, R., Enamorado, S. M., Hurtado, M. D., Andreu, L., & Delgado, A. (2008). The cumulative effect of three decades of phosphogypsum amendments in reclaimed marsh soils from SW Spain: 226Ra, 238U and Cd contents in soils and tomato fruit. *Science of the Total Environment, 403*(1-3), 80–88.

Acheampong, M. A., Meulepas, R. J., & Lens, P. N. (2010). Removal of heavy metals and cyanide from gold mine wastewater. *Journal of Chemical Technology & Biotechnology, 85*(5), 590–613.

Adriano, D. C. (2001). *Trace elements in terrestrial environments: Biogeochemistry, bioavailability, and risks of metals* (Vol. 860). Springer.

Alamdar, A., Syed, J. H., Malik, R. N., Katsoyiannis, A., Liu, J., Li, J., Zhang, G., & Jones, K. C. (2014). Organochlorine pesticides in surface soils from obsolete pesticide dumping ground in Hyderabad City, Pakistan: Contamination levels and their potential for air–soil exchange. *Science of the Total Environment, 470*, 733–741.

Al-Awadhi, N., Al-Daher, R., Al-Nawawy, A., & Salba, M. T. (1996). Bioremediation of oil-contaminated soil in Kuwait. I. Landfarming to remediate oil-contaminated soil. *Soil and Sediment Contamination, 5*(3), 243–260.

Alekseenko, V. A., Bech, J., Alekseenko, A. V., Shvydkaya, N. V., & Roca, N. (2018). Environmental impact of disposal of coal mining wastes on soils and plants in Rostov oblast, Russia. *Journal of Geochemical Exploration, 184*, 261–270.

Alemayehu, Y. A., Asfaw, S. L., & Tirfie, T. A. (2020). Management options for coffee processing wastewater. A review. *Journal of Material Cycles and Waste Management, 22*(2), 454–469.

Alhamed, M., & Wohnlich, S. (2014). Environmental impact of the abandoned coal mines on the surface water and the groundwater quality in the south of Bochum, Germany. *Environmental Earth Sciences, 72*, 3251–3267.

Alloway, B. J. (1995). *Heavy metals in soils* (2nd ed.). Chapman and Hall.

Alonso, D. L., Pérez, R., Okio, C. K., & Castillo, E. (2020). Assessment of mining activity on arsenic contamination in surface water and sediments in southwestern area of Santurbán paramo, Colombia. *Journal of Environmental Management, 264*, 110478.

Azaroff, A., Urriza, M. G., Gassie, C., Monperrus, M., & Guyoneaud, R. (2020). Marine mercury-methylating microbial communities from coastal to Capbreton canyon sediments (North Atlantic Ocean). *Environmental Pollution, 262*, 114333.

Bala, K., Anshul, A., Verma, S., Dogra, A., Das, A. P., Kumar, R., & Nandini, A. (2024). Sustainable management for urban plastic waste generation. In *Sustainable management of urban plastic waste through circular economic approaches* (pp. 211–236). Routledge. https://doi.org/10.1201/9781003465447-13

Barrera Olivarez, M., Murillo Tovar, M. A., Vergara Sánchez, J., Betancourt, M. L. G., Martín Romero, F., Ramírez Arteaga, A. M., Moeller Chávez, G. E., & Saldarriaga Noreña, H. A. (2021). Mobility of heavy metals in aquatic environments impacted by ancient mining-waste. In *Water quality-Factors and impacts*. IntechOpen.

Briffa, J., Sinagra, E., & Blundell, R. (2020). Heavy metal pollution in the environment and their toxicological effects on humans. *Heliyon, 6*(9), e04691.

Conant, R. T., Drijber, R. A., Haddix, M. L., Parton, W. J., Paul, E. A., Plante, A. F., Six, J., & Steinweg, J. M. (2008). Sensitivity of organic matter decomposition to warming varies with its quality. *Global Change Biology, 14*(4), 868–877.

Costa, J. M., Berckmoes, E., Beerling, E. A. M., Nicol, S., Jose, J., Garcia, J., & Cáceres, R. (2019). *Water use in greenhouse horticulture: Efficiency and circularity*. EIP-AGRI.

Das, A., Dey, S., & Das, A. P. (2024). Ecotoxicological impacts of synthetic microfiber pollutants and development of sustainable mitigation strategies. *Environmental Chemistry and Ecotoxicology*. https://doi.org/10.1016/J.ENCECO.2024.12.003

Das, A. P., & Ghosh, S. (2023). *Manganese mining microorganisms*. Elsevier.

Das, A. P., & Mishra, S. (2008). Hexavalent chromium (VI): Health hazards and environmental pollutant. *Journal of Environment & Development, 2*(3), 386–392.

Das, A. P., Sukla, L. B., Pradhan, N., & Nayak, S. (2011). Manganese biomining: A review. *Bioresource Technology, 102*(16), 7381–7387. https://doi.org/10.1016/J.BIORTECH.2011.05.018

De Lacerda, L. (2003). Updating global hg emissions from small-scale gold mining and assessing its environmental impacts. *Environmental Geology, 43*, 308–314.

Dey, S., Bissoyi, A., Koteswara Reddy, G., & Das, A. P. (2024). Synthetic microfibers: An urban emerging pollutant demanding for sustainable environmental solutions. In *Sustainable microbial technology for synthetic and cellulosic microfiber bioremediation* (pp. 107–121). Springer. https://doi.org/10.1007/978-3-031-62600-5_7

Dey, S., Das, A., Mallick, K., Sahu, A., & Das, A. P. (2023). *Environmental petroleum waste: Pollution, toxicity, sustainable remediation* (pp. 159–175). Springer. https://doi.org/10.1007/978-3-031-48220-5_7

Dey, S., Dash, A., & Das, A. P. (2024). Bioremediation as a sustainable and circular solution for mining waste management and bioenergy generation. In *Sustainable management of mining waste and tailings: A circular economy approach* (pp. 107–119). Routledge. https://doi.org/10.1201/9781003442455-6

Dey, S., Tripathy, B., Kumar, M. S., & Das, A. P. (2023). Ecotoxicological consequences of manganese mining pollutants and their biological remediation. *Environmental Chemistry and Ecotoxicology*. https://doi.org/10.1016/J.ENCECO.2023.01.001

Douay, F., Roussel, H., Fourrier, H., Heyman, C., & Chateau, G. (2007). Investigation of heavy metal concentrations on urban soils, dust and vegetables nearby a former smelter site in Mortagne du Nord, northern France. *Journal of Soils and Sediments, 7*(3), 143–146.

Dri, M., Canfora, P., Antonopoulos, I. S., & Gaudillat, P. (2018). *Best environmental management practice for the waste management sector*. JRC science for policy report.

Duggan, J., & Dhawan, A. (2007). Speciation and vertical distribution of lead and lead shot in soil at a recreational firing range. *Soil and Sediment Contamination: An International Journal, 16*(4), 351–369.

Dwernychuk, L. W., Cau, H. D., Hatfield, C. T., Boivin, T. G., Hung, T. M., Dung, P. T., & Thai, N. D. (2002). Dioxin reservoirs in southern Viet Nam—A legacy of agent Orange. *Chemosphere, 47*(2), 117–137.

Farinella, N. V., Matos, G. D., & Arruda, M. A. Z. (2007). Grape bagasse as a potential biosorbent of metals in effluent treatments. *Bioresource Technology, 98*(10), 1940–1946.

Fashola, M. O., Ngole-Jeme, V. M., & Babalola, O. O. (2016). Heavy metal pollution from gold mines: Environmental effects and bacterial strategies for resistance. *International Journal of Environmental Research and Public Health, 13*(11), 1047.

Fergusson, J. E. (1990). *The heavy elements: Chemistry, environmental impact and health effects*. Wiley.

Ghosh, S., Tripathy, B., Dey, S., & Das, A. P. (2023). Comparative investigation of fungal and bacterial manganese biomining mechanisms. In *Manganese mining microorganisms* (pp. 99–115). Elsevier. https://doi.org/10.1016/b978-0-12-822147-1.00016-9

Hatje, V., Pedreira, R. M., de Rezende, C. E., Schettini, C. A. F., de Souza, G. C., Marin, D. C., & Hackspacher, P. C. (2017). The environmental impacts of one of the largest tailing dam failures worldwide. *Scientific Reports, 7*(1), 10706.

He, B., Zhao, X., Li, P., Liang, J., Fan, Q., Ma, X., Zheng, G., & Qiu, J. (2019). Lead isotopic fingerprinting as a tracer to identify the pollution sources of heavy metals in the southeastern zone of Baiyin, China. *Science of the Total Environment, 660*, 348–357.

Iavazzo, P., Adamo, P., Boni, M., Hillier, S., & Zampella, M. (2012). Mineralogy and chemical forms of lead and zinc in abandoned mine wastes and soils: An example from Morocco. *Journal of Geochemical Exploration, 113*, 56–67.

John, A., & Das, A. P. (2024). Government strategies for achieving circular economy for plastic waste management in Southeast Asia. In *Sustainable management of urban plastic waste through circular economic approaches* (pp. 72–82). Routledge. https://doi.org/10.1201/9781003465447-6

Karaca, O., Cameselle, C., & Reddy, K. R. (2018). Mine tailing disposal sites: Contamination problems, remedial options and phytocaps for sustainable remediation. *Reviews in Environmental Science and Bio/Technology, 17*, 205–228.

Koda, E., & Podlasek, A. (2023). Sustainable use of construction and demolition wastes in a circular economy perspective. In *Sustainable and circular management of resources and waste towards a green deal* (pp. 137–147). Elsevier.

Kumar, A., Ali, M., Raj, V., Kumari, A., Rachamalla, M., Niyogi, S., Kumar, D., Sharma, A., Saxena, A., Panjawani, G., & Jain, P. (2023). Arsenic causing gallbladder cancer disease in Bihar. *Scientific Reports, 13*(1), 4259.

Lindsay, M. B., Moncur, M. C., Bain, J. G., Jambor, J. L., Ptacek, C. J., & Blowes, D. W. (2015). Geochemical and mineralogical aspects of sulfide mine tailings. *Applied Geochemistry, 57*, 157–177.

Mallick, K., Sahu, S., Sahu, A., Dey, S., & Das, A. P. (2024). A critical review of marine microfiber pollution routes, toxicity, and its sustainable remediation. *Environmental Science and Engineering, 2333*, 189–211. https://doi.org/10.1007/978-3-031-51792-1_10

Medina-Díaz, H. L., López-Bellido, F. J., Alonso-Azcárate, J., Fernández-Morales, F. J., & Rodríguez, L. (2024). A new hyperaccumulator plant (Spergularia rubra) for the decontamination of mine tailings through electrokinetic-assisted phytoextraction. *Science of the Total Environment, 912*, 169543.

Mejias Carpio, I. E., Ansari, A., & Rodrigues, D. F. (2018). Relationship of biodiversity with heavy metal tolerance and sorption capacity: A meta-analysis approach. *Environmental Science & Technology, 52*(1), 184–194.

Mishra, S., & Das, A. P. (2021). Current treatment technologies for removal of microplastic and microfiber pollutants from wastewater. In *Wastewater treatment: Cutting-edge molecular tools, techniques and applied aspects* (pp. 237–251). Elsevier. https://doi.org/10.1016/B978-0-12-821881-5.00011-8

Mishra, S., Rath, C., & Das, A. P. (2019). Marine microfiber pollution: A review on present status and future challenges. *Marine Pollution Bulletin, 140*, 188–197. https://doi.org/10.1016/J.MARPOLBUL.2019.01.039

Mishra, S., Swain, S., Sahoo, M., Mishra, S., & Das, A. P. (2021). Microbial colonization and degradation of microplastics in aquatic ecosystem: A review. *Geomicrobiology Journal*. https://doi.org/10.1080/01490451.2021.1983670

Mohanty, J., Dey, S., Bala, K., & Das, A. P. (2024). Recent advances in microplastic pollution for its sustainable management. In *Sustainable management of urban plastic waste through circular economic approaches* (pp. 27–48). Routledge. https://doi.org/10.1201/9781003465447-3

Mohanty, M., Tripathy, B., Koteswara Reddy, G., & Das, A. P. (2024). Greener technologies for establishing restraint over microplastic pollution. In *Sustainable microbial technology for synthetic and cellulosic microfiber bioremediation* (pp. 239–258). Springer. https://doi.org/10.1007/978-3-031-62600-5_13

Mondal, P., de Alcântara Mendes, R., Jonathan, M. P., Biswas, J. K., Murugan, K., & Sarkar, S. K. (2018). Seasonal assessment of trace element contamination in intertidal sediments of the meso-macrotidal Hooghly (Ganges) river estuary with a note on mercury speciation. *Marine Pollution Bulletin, 127*, 117–130.

Montes-Avila, I., Espinosa-Serrano, E., Castro-Larragoitia, J., Lázaro, I., & Cardona, A. (2019). Chemical mobility of inorganic elements in stream sediments of a semiarid zone impacted by ancient mine residues. *Applied Geochemistry, 100*, 8–21.

Mwesigye, A. R., Young, S. D., Bailey, E. H., & Tumwebaze, S. B. (2016). Population exposure to trace elements in the Kilembe copper mine area, Western Uganda: A pilot study. *Science of the Total Environment, 573*, 366–375.

Odumo, B. O., Carbonell, G., Angeyo, H. K., Patel, J. P., Torrijos, M., & Rodríguez Martín, J. A. (2014). Impact of gold mining associated with mercury contamination in soil, biota sediments and tailings in Kenya. *Environmental Science and Pollution Research, 21*, 12426–12435.

Paul, A., Dey, S., Ram, D. K., & Das, A. P. (2023). Hexavalent chromium pollution and its sustainable management through bioremediation. *Geomicrobiology Journal*. https://doi.org/10.1080/01490451.2023.2218377

Pradhan, G., Dey, S., Priyadarsini, S., Kumar, M. S., & Das, A. P. (2024). Characterization and quantification of microplastics pollutants in sediment samples from Daya River of Odisha state in India for their appropriate management. *Environmental Science and Engineering, 2333*, 75–92. https://doi.org/10.1007/978-3-031-51792-1_5

Priyadarshanee, M., & Das, S. (2021). Biosorption and removal of toxic heavy metals by metal tolerating bacteria for bioremediation of metal contamination: A comprehensive review. *Journal of Environmental Chemical Engineering, 9*(1), 104686.

Priyadarshini, S., Jagatee, S., & Das, A. P. (2024). Synthetic fabrics and microfiber pollution–an assessment of their global impact. *Environmental Science and Engineering, 2333*, 137–157. https://doi.org/10.1007/978-3-031-51792-1_8

Priyadarsini, S., & Das, A. P. (2025). Lithium bioleaching: A review on microbial-assisted sustainable technology for lithium bio-circularity. *Journal of Water Process Engineering, 69*, 106744. https://doi.org/10.1016/j.jwpe.2024.106744

Queiroz, H. M., Nóbrega, G. N., Ferreira, T. O., Almeida, L. S., Romero, T. B., Santaella, S. T., Bernardino, A. F., & Otero, X. L. (2018). The Samarco mine tailing disaster: A possible time-bomb for heavy metals contamination? *Science of the Total Environment, 637*, 498–506.

Raj, V., Kumar, A., Srivastav, A. L., Kumar, R., & Karn, S. K. (2025). Bioremediation: An emerging and sustainable biotechnological method for restoring polluted water ecosystems. In *Biotechnologies for wastewater treatment and resource recovery* (pp. 123–133). Elsevier.

Ratajczak, K., Sulewska, H., Panasiewicz, K., Faligowska, A., & Szymańska, G. (2023). Phytostimulator application after cold stress for better maize (Zea mays L.) plant recovery. *Agriculture, 13*(3), 569.

Ray, S. K., Dey, S., Das, A. P., & John, A. (2024). Microplastic waste management through greener technologies: A review. In *Sustainable management of urban plastic waste through circular economic approaches* (pp. 237–261). Springer. https://doi.org/10.1201/9781003465447-14

Ribeiro, J., Taffarel, S. R., Sampaio, C. H., Flores, D., & Silva, L. F. (2013). Mineral speciation and fate of some hazardous contaminants in coal waste pile from anthracite mining in Portugal. *International Journal of Coal Geology, 109*, 15–23.

Romero, F. M., Armienta, M. A., & González-Hernández, G. (2007). Solid-phase control on the mobility of potentially toxic elements in an abandoned lead/zinc mine tailings impoundment, Taxco, Mexico. *Applied Geochemistry, 22*(1), 109–127.

Sahoo, P. K., Equeenuddin, S. M., & Powell, M. A. (2016). Trace elements in soils around coal mines: Current scenario, impact and available techniques for management. *Current Pollution Reports, 2*, 1–14.

Sahu, A., Dey, S., Mallick, K., Priyadarsini, S., Behera, I. D., & Das, A. P. (2024). A review on advanced technology for sustainable management of synthetic microplastic waste. In *Sustainable management of urban plastic waste through circular economic approaches* (pp. 262–279). Springer. https://doi.org/10.1201/9781003465447-15

Sarwar, N., Imran, M., Shaheen, M. R., Ishaque, W., Kamran, M. A., Matloob, A., Rehim, A., & Hussain, S. (2017). Phytoremediation strategies for soils contaminated with heavy metals: Modifications and future perspectives. *Chemosphere, 171*, 710–721.

Shen, X., Dai, M., Yang, J., Sun, L., Tan, X., Peng, C., Ali, I., & Naz, I. (2022). A critical review on the phytoremediation of heavy metals from environment: Performance and challenges. *Chemosphere, 291*, 132979.

Simón, M., Ortiz, I., Garcıa, I., Fernández, E., Fernández, J., Dorronsoro, C., & Aguilar, J. (1999). Pollution of soils by the toxic spill of a pyrite mine (Aznalcóllar, Spain). *Science of the Total Environment, 242*(1–3), 105–115.

Singh, S., Kaushik, A., Bendi, A., Chetal, A., Ramakrishna, D. S., & Praveen, P. L. (2024). Constructed wetlands as bioeconomic solutions: Rhizofiltration with macrophytes for heavy metal removal. *Emergent Materials, 8*, 75–83.

Smičiklas, I., & Šljivić-Ivanović, M. (2016). Pollutants mobility with implication to remediation strategies. In *Soil contamination: Current consequences and further solutions* (p. 253). IntechOpen.

Wang, J., Dai, J., Chen, G., & Jiang, F. (2022). Role of sulfur biogeochemical cycle in mercury methylation in estuarine sediments: A review. *Journal of Hazardous Materials, 423*, 126964.

Worlanyo, A. S., & Jiangfeng, L. (2021). Evaluating the environmental and economic impact of mining for post-mined land restoration and land-use: A review. *Journal of Environmental Management, 279*, 111623.

Yadav, A. K., Shirin, S., Hopke, P. K., Pal, D., & Jamal, A. (2022). Process to reduce particulate matter in ambient air using bubbles of sodium palmitate. *Chemical Engineering & Technology, 45*(8), 1497–1500.

Overburden Management in Iron Ore Mines: A Key Challenge for Achieving Sustainability at the Iron Ore Belt of Odisha, India

Himansu Sekhar Patra, Aditya Kishore Dash, Amarendra Harichandan, and Manoj Kumar Meher

1 Introduction

Mineral extraction plays an important role for the economic development in developing nations at global scenario. After agriculture, mining is the second biggest industry in worldwide and has been quite essential for the development of human kind (Lodha et al., 2009). Mining has been instrumental in accelerating economic development, advancement of civilization, growth of infrastructure and technological progress. Worldwide, mining supplies raw materials required for different industries ranging from construction to electronics, thus fueling economic growth of a country. In India, mining activity provides employment and livelihood to millions thus deeply connected to the nation's economic structure and immensely contributing to the GDP. Though metals and minerals are seen as the backbone of the economic development in a country, its exploration, excavation, and processing directly affects surrounding natural resources such as land, water, air, and biodiversity. Mining is a process of extraction of raw materials from earth's crust starting from inspection, exploration and continuing through production. Besides the other sources, mining activities like blasting, extraction, transportation of mineral, ore processing causes air pollution. Monitoring of ambient air, determination of air

H. S. Patra (✉)
Environment and Sustainability, MSP Steel & Power Ltd, Haldiaguna, Odisha, India

A. K. Dash
Chemistry Department, ITER, Siksha 'O' Anusandhan University, Bhubaneswar, Odisha, India

A. Harichandan
Department of Basic Science, HiTech Institute of Technology, Khordha, Odisha, India

M. K. Meher
Department of Geography, Maa Manikeshwari University, Bhawanipatna, Odisha, India

© The Author(s), under exclusive license to Springer Nature Switzerland AG 2026

A. P. Das (ed.), *Bioremediation Technology for Sustainable Mining Waste Management*, Sustainable Environmental Waste Management Strategies,
https://doi.org/10.1007/978-3-032-03468-7_15

quality index and study of air dispersion models plays a vital role for future prediction of the situation (Dash et al., 2019, 2023; Dash & Dash, 2017, 2018a, 2018b; Kumar et al., 2018; Parida et al., 2015; Sahoo et al., 2017; Sahu et al., 2015). However, mining activity, particularly in opencast mining has number of externalities on surrounding environment thus degrading surrounding ecological value of landscape. Various parameters such as drainage, air quality, soil, water, biodiversity, noise levels, human health are impacted by opencast mining activities. Extraction of different valuable minerals produces waste rocks or overburden as wastes whose management is of prime importance. Overburden dumps/stacks stability is catching global attention for quite some time. After the extraction of ore reserves, the landscape altered by mining must be scientifically reclaimed to abate the environmental effects of opencast mining and restore the land through rehabilitation (Kuter, 2013). The waste produced by mineral extraction can be in the form of solid waste, tailings, or slurry, with tailings, waste rock, and tail ends being the most common types of waste. In some cases, vegetation and overburden are also considered waste (Lottermoser, 2010; Romero et al., 2007). Government agencies and regulatory bodies like State Pollution Control Board, MoEF and Department of Mines and Geology have shown concern regarding the way overburden dumps are being disposed/managed in most of the mines. Lack of proper management of over burden has resulted in river pollution due to surface runoff in rainy season, subsidence of dump and loss of natural vegetation. So appropriate scientific overburden dump management is crucial for minimizing the environmental impacts of various mining activities. Implementing scientific processes in overburden dump management ensures the safe disposal, stabilization, and reclamation of these dumps, reducing the risk of landslides and pollution. Timely and effective management restores the land for future use, enhancing its ecological value and preventing long-term environmental damage (Das et al., 2025; Reddy Koteswara et al., 2025; Sahu et al., 2024). Given the increasing scale of mining operations, the urgency of adopting such methods is essential for sustainable development and environmental conservation.

2 Mining Scenario in India and Odisha

Iron ore processing involves various steps such as crushing, screening, washing, beneficiation, and agglomeration. While crushing and screening are followed for sizing of the ore, the gangue materials are removed through beneficiation process. Before dispatch to market, Iron ore lumps and fines are processed through screening as well as beneficiation process. The fines are transformed into sinters, which are used in steel plants, pellets are produced from concentrates or fines and are used both for internal consumption in steel industries and for export (Bala et al., 2024;

Das, Dey, Bala, et al., 2024; Mohanty, Dey, et al., 2024). During 2021–2022, India had a total of 1319 mines reporting mineral production, excluding the minor, fuel as well as atomic minerals. With 262 mines, Madhya Pradesh tops the list followed by Gujarat—147, Karnataka—132, Odisha—128, and Chhattisgarh—114. Eleven states of India accounts for 93% of the country's total mining operation. Private sector played a crucial role in India's mineral production, contributing 55.37%, (Rs. 73,502 crore) including metallic and non-metallic minerals during 2021–2022. The private industry was solely responsible for mining or recovering garnet, leads concentrate, lime shell, siliceous earth, wollastonite, zinc concentrate, lead and zinc ore, marl, and molding sand. Additionally, a significant portion of limestone, vermiculite, chromite, iron ore, manganese ore, kyanite, and bauxite production came from the private sector. The public sector contributed 44.63% (Rs. 59,246 crore) including metallic and non-metallic minerals during 2021–2022. The public sector exclusively mined or recovered copper ore and concentrate, diamonds, graded fluorite, rock salt, silver, and sulfur. It also accounted for substantial quantities of gold ore (99.08%), tin concentrate (94.39%), phosphorite (92.87%), selenite (90.75%), graphite (63.24%), and magnesia (54.03%) during the same period (Indian Mineral Year Book 2022). Table 1 shows the no of reported mines in India during the year 2019–20 and 2020–21.

The state of Odisha is blessed with a rich bounty of mineral such as chromite, iron, bauxite, and manganese, limestone, graphite, heavy minerals, coal, gemstone, nickel, cobalt, dimensional, and decorative stone. The states rich mineral deposits is attribute to its geological settings In Orissa, Precambrian metamorphic rocks of Archaean and Proterozoic age, contain the most of the minerals, cover approximately 72.5% of the area. Odisha leads the estimated value of mineral production in India and had the share of 47.20% output in the national level (Indian Mineral Year Book 2022). Odisha was a key player in supplying iron ore to the various Indian steel industry, accounting for over 55% of the country's iron ore production during 2021–2022. The primary iron ore deposits are located along the Bonai-Keonjhar Horse Shoe-shaped synclinorium, which spans an area of approximately 60 × 25 km (Acharya, 2008). Iron ore bands are found in layered Banded Iron Formation (BIF) bands, along with volcano-sedimentary rock sequences known as the Iron Ore Super Group. Typically, four types of ore are found in Odisha such as: hard massive, laminated, lateritic, and blue dust.

Table 1 Number of reporting mines in India

Mineral sector	Year (2019–2020)	Year (2020–2021)	Year (2021–2022) (P)
Metallic mineral	602	608	545
Non-metallic mineral	783	767	774
All types of minerals	1385	1375	1319

[a] This doesn't include atomic minerals fuel mineral and also minor minerals (www.ibm.nic.in)

3 Overview of Mining Wastes: A Global Concern

Mining wastes are produced during extraction/processing and beneficiation of iron ore. The first step is the mineral extraction process, where ore are taken out of the earth. It is achieved through the process of blasting, in which large waste volume (soil, rubble, overburden, and other materials) are generated, compared to waste generated from underground mines, opencast mines is more pollutant-intensive as open-pit mining produces 8–10 times more waste than underground mining (Anon, 2006). These wastes are found piled up within and near to the mining lease area, as it is useless to the mineral sector. Though overburden and other wastes are mostly inert, some of these harmful hazardous heavy metals like Hg, As, Pb, Zn, and Cd, are found in the ore which can contaminate the nearby ecosystem if they seep out of the garbage piles. Enormous volumes of waste are produced in the process of ore extraction (Das & Ghosh, 2022, 2023; Das & Singh, 2011; Ghosh, Sinha, et al., 2023; Paul et al., 2023). As the better grade mineral deposits are reduced, this sector is producing an increasing amount of trash. In the USA, copper ore contains just 0.51% usable metal matched to 2.5% at the turn of the twentieth century (Gardner & Sampat, 1998). Although there is no estimate of the amount of garbage produced by industry on a worldwide scale, everyone believes that, it is unfathomably large. For instance, 110 tons of scrap ore and 200 tons of overburden are generated during the manufacture of 1 ton of copper (Anon, 2006). As per the Indian Bureau of Mines, the average limestone mines stripping ratio in India is 1:05; whereas for iron ore mines, it hovers around 1:2. This indicates that twice as much trash is generated for every ton of iron ore extracted. According to Khatua et al. (2006), Indian bauxite ores are generally harder and have a greater stripping ratio in comparison to their Australian counterparts. Table 2 provides the waste generation during the processing of different metal ores.

Table 2 Waste generation from metal ores processing (Anon, 2005)

Metal types	% of metal in ore	Mined ore (million tons)	Waste (%)	Waste: ratio
Fe	33.00	25,503	60	5.2:1
Cu	0.6	11,026	99	450:1
Au	0.0004	7235	99.99	950,000:1
Pb-Zn	3.7–5.0	1077	97.5	32:1
Al	8–20	856	70	3:1
Ag	0.03	N.A.	99.97	N.A.
W	0.5	N.A.	99.5	N.A.
Mo	0.2	N.A.	99.8	N.A.

4 Classification of Mining Wastes

Mining, Minerals, and Sustainable Development project report states that there are about 3500 sites worldwide that is actively producing mining wastes. The valuable mineral or resource excavated from the mines is outnumbered by the generation of waste. At Global scenario, mineral and metal mining per year produces around 100 billion tons of trash. Materials left after the required minerals or metals extracted from the ore at the time of mining operations are considered to as mining waste.

Several types of mining wastes are generated at the time of mining activities like:

- Waste rock (overburden): Waste rock, also known as overburden, in iron mines refers to the material that lies above or around the ore body and is removed to access the iron ore. This material does not contain sufficient concentrations of iron to be processed economically and includes soil, sand, gravel, and other non-ore rock formations. Due to no commercial value and environmental implications, managing waste rock is a critical aspect of mining operations. Its improper disposal can lead to vegetation loss, land degradation, dust pollution, and potential contamination of nearby water sources. Therefore, for sustainable mining practices, careful handling, storage, and rehabilitation of waste rock dumps are quite essential (Das & Mishra, 2009; Mishra & Das, 2008).
- Tailings: Tailings are generated as the finely ground waste materials left over after the extraction of iron from its ore during the beneficiation process. Iron mines tailings contain silica, residual iron particles, and other impurities which were separated from the valuable iron content during beneficiation process. Tailings can pose significant environmental and safety risks if not properly managed due to mis-management of tailing pond and dams. Water contamination due to dam failure, seepage, and dust generation are major concerns associated with tailings. To address the environmental impact from tailings, modern mining operations focus on safe tailings disposal, water recycling, and the rehabilitation of tailings storage areas (Bal et al., 2018; Ghosh et al., 2016).
- Mine water: Due to seepage of groundwater, surface water infiltration, large amount of water is accumulated inside the mine pit which is called as mine water. Mine water can contain high traces of suspended and dissolved solids and minerals, sometimes traces of heavy metals or other contaminants like oil and grease. Mine water can lead to environmental issues like surface and ground water pollution, soil degradation, and negative impacts on nearby aquatic ecosystems if not managed properly. Effective mine water management involves collection, treatment, and safe discharge or reuse of the water, ensuring mining activities do not harm surrounding water bodies and comply with environmental standards. Unscientific disposal and management of mine wastes can cause contamination

of water environment as it contains various toxic substances. It can have both short and long term effects on the ecosystem and on human life. Treatment of contaminants from water/waste water through physico-chemical methods are cost intensive, hence the scientist have search for low cost biological treatment methods (Amreen et al., 2024; Dash, 2012; Dash & Mishra, 1996a, 1996b, 1999a, 1999b; Jena et al., 2022; Mahapatra et al., 2019; Parida et al., 2023; Pradhan et al., 2013, 2017; Pradhan, Dash, et al., 2015; Shreya et al., 2022). Figure 1 shows the various stages of mining and the generation of waste through different steps.

Mining operations generate waste materials such as overburden and mine development rock. While Overburden is the by-product of opencast mining, mine development rock comes from the extraction process in underground mines. Typically, overburden and waste rubble is piled in large heaps on the ground without proper precautions to prevent environmental impact. While backfilling the mined areas

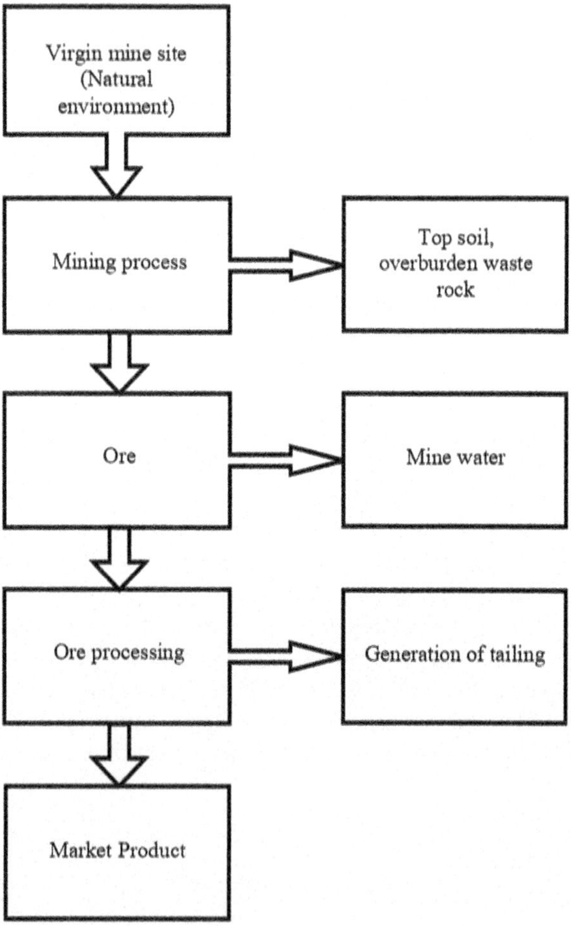

Fig. 1 Generation of waste during mining process

with waste rock is the best solution, it is hardly implemented as mining companies continue to create new mine faces without reclaiming the old ones. In India, many mines manage overburden in unscientific way resulting in generation of huge debris which risks to both local habitants and workers. Overburden leads to affect the productive agricultural land through soil degradation. It can also negatively affect surrounding lands through leachate, altered drainage patterns, and changes to groundwater levels. Various mining activities causes soil pollution and land degradation. However, fly ash generated from coal burning which contain the essential nutrients can support the growth of plant (Dash et al., 2015; Pradhan, Sahu, & Dash, 2015). Soil fertility study (Nayak et al., 2018) and evaluation of enzyme activities (Panigrahi et al., 2024) plays a vital role in determining the soil quality of a contaminated area.

5 Environmental Impact of Mine Overburden

Waste piles (overburden), which consist of waste rock, overburden substrata ore, and heavy metals are some of the externalities associated with active mining operations. Mining overburden has impact on the surrounding environment. It alters the surrounding land use by affecting the natural vegetation. The waste piles produced are unstable in nature with a steep slope. State like Goa, Odisha, and Chattisgarh experiences high intensity rainfall, which leads to the generation of surface runoff during rainy season thus polluting the nearby water bodies. Removal of overburden leads to significant soil degradation as removal of top nutrient-rich layer results in decreased soil fertility and health (Das, Dey, & Das, 2024; Dey et al., 2024; Ghosh, Tripathy, et al., 2023; Mohanty, Tripathy, et al., 2024). The neighboring settlements are worst affected due to the dispersal of waste during transportation. Silt transported from the waste piles are accumulated inlayer in the adjacent agricultural fields as most of the mines and the waste piles are generally located near to the residences and agricultural fields, thus making it infertile. The removal of ore contains the material that does not permit vegetation to grow well on the waste piles. When overburden is displaced, it can release sediments and pollutants into nearby streams, rivers, and groundwater thus affecting local water bodies. The surface runoff can release heavy metals, sulfides, and other toxic substances leading to a decrease in water quality affecting the life of aquatic animals. Overburden removal and deposit disrupts local ecosystem, leading to habitat loss and fragmentation. Furthermore, the alteration of land use changes the dynamics of local flora and fauna, potentially leading to a decline in biodiversity. In some cases, invasive species may thrive in disturbed areas, further threatening native species and ecosystem.

The iron and manganese mines of Odisha are located in dense forest regions which serve as river catchment areas and habitats of tribal communities. Mining operations generate vast amounts of unstable waste piles, making them prone to erosion, especially during the monsoon season. As a result, silt accumulation in agricultural fields progressively degrades soil quality, making cultivation difficult.

For each tonne of iron ore extracted, approximately three tonnes of waste (overburden) are produced, leading to the accumulation of huge volume of overburden, low-grade ore, and tailings in every passing year. Improper storage and management of these waste materials have severe environmental consequences, including deforestation and landscape alteration (Patra & Sethy, 2014a). Mining also significantly Impacts Rivers, as overburden is frequently dumped into valleys, clogging streams and altering their natural flow. Deforestation further accelerates runoff, increasing the risk of severe flooding. A large portion of India's mineral reserves lies near river sources, making them vulnerable to contamination. The solid waste generated primarily consists of overburden the material above the ore body and the mine development rock, which must be removed to quantify the ore. The stripping ratio, which measures overburden removal per unit of ore extracted, typically ranges from 2:1 to 2.5:1 in Odisha's iron ore mines (BBY Limited Resources, POSCO).

Deforestation due to mining, combined with settlement expansion, has accelerated forest degradation, particularly in Keonjhar district. The accumulation of waste rock dumps has severely impacted natural plant communities, making the environment unsuitable for vegetation growth. Opencast mining inflicts extensive damage on landscapes and biodiversity, significantly reducing forest cover. Changes in land use patterns, such as agricultural land conversion and wasteland formation, highlight the urgent need for intervention (Patra & Sethy, 2014b). Soil pollution, a major consequence of mining, alters the physico-chemical properties of the soil, negatively impacting flora and fauna (Dash et al., 2016). The higher presence of herbaceous species in mined areas compared to non-mined regions suggests that certain plants can better adapt to degraded environments (Dash & Patra, 2016). Given the similar climatic and edaphic conditions in mined and non-mined areas, the differences in species composition can largely be attributed to mining activities (Patra & Sethy, 2013, 2014a).

6 Legal Aspects for Overburden Dump Management in India

For addressing the potential environmental and safety risks of mines over burden dumps, various regulations have been introduced by government agencies such as the State Pollution Control Board, the MoEF, and the Department of Mines and Geology for mining sustainable activities. The Mines and Minerals (Development and Regulation) (MMDR) Act, 1957, along with the Mineral Conservation & Development Rules (MCDR), 1988, mandates the lease holders to adopt various types of strict environmental safeguards and pollution control measures during operations. Proper management of topsoil and overburden for ensuring sustainable mining practices and minimizing ecological impact has specified in Rules 32 and 33 of the MCDR. Rule 33 specifically governs the storage and management of overburden, waste rock, rejects, fines, tailings, slimes, and other mining and beneficiation waste. It mandates proper storing of overburden dumps at separate and secured areas for prevention of environmental degradation, land and water pollution, and mineral loss. It mandates the lease holder to properly secure the overburden dump to prevent the escape of harmful materials, with afforestation or vegetative cover. The MMDR Act, 1957 was amended as the MMDR Amendment Act, 2021, that was effective on March 28, 2021. This amendment introduced significant provisions to improve dump management and promote sustainable mining practices. Additionally, the MCDR, 2017 enforces responsible waste disposal by mandating that mining operations adopt sustainable waste management strategies to minimize environmental impacts. Mining leaseholders are mandated with taking various precautionary measures to control air, water and noise pollution, prevention of the discharge of toxic substances to surroundings, and controlling surface subsidence. To further enhance environmental accountability, Indian Bureau of Mines has started a star rating system to assess the environmental and social responsibility of mines.

7 Sustainable Management of Mining Overburden

The waste, sub-grade run-of-mine, and fine dumps produced in iron ore mining are having high content of rock fragment, less moisture retention, high bulk density, low nutrient, and pH value, along with higher metal concentrations. An effective revegetation program involving suitable plant species in aligns with site conditions and a characteristic is essential for enhancing the long-term mechanical and ecological stability of the mining over burden. Effective over burden dump management is required for ensuring environmental sustainability, land stability, and compliance with statutory regulatory norms (Dey, Das, et al., 2023; Dey, Tripathy, et al., 2023).

Overburden must be carefully handled in order to prevent issues like erosion, water pollution, sedimentation, slope failure, and dust pollution. Management of overburden states with the identification of proper site for dumping considering various factors such as drainage, topography, and proximity of the site to mining area. Then dumps are constructed in benches with suitable slopes for stability, compaction, developing drainage channels and retaining walls to enhance the stability. Progressive re-habitation through topsoil spreading and vegetation cover is crucial for controlling soil erosion and restoring ecological balance. A well planned and scientific overburden management contributes to mining safety and achieving sustainability in long term.

Effective management of overburden dump requires several critical steps to ensure stability, environmental safety, and sustainable land use. The overall dump management beings with dump design followed by dump construction and monitoring, dump rehabilitation, followed by dump stabilization through vegetation. The precious Topsoil, approximately 20–30 cm in depth, should be removed separately and stored in a separate heap for future use. The topsoil must be planted to maintain its fertility and biomass. The OB and waste rock dumps should be located away from natural water bodies such as springs and rivers, in impervious and non-mineralized areas, to reduce water pollution. These dumps should be adequately dressed, benched, and sloped at a low angle with terracing. Barricades with bamboo, retaining walls, stone barriers at the base, and gully plugging techniques should be used to prevent soil erosion, particularly during the monsoon. Additionally, vegetation should be established on the top and slopes of the dumps. To check fine particles from washing off the OB/IB/waste rock dumps during rains, deep garland-type trenches should be constructed around the waste dumps annually before the monsoon season. The collected water in the garland drains should be allowed to pass through check dams or settling tanks for allowing silt to settle before being discharged into the surrounding environment. Proper de-silting arrangements must also be made at the outlets of garland drains before the monsoon begins.

8 Dump Design

Soil pollution caused through wash-off with rains and air pollution from blown the dust through wind action are commonly observed at dumping site thus presenting an ugly and repulsive look. In order to minimize the land, air, and water pollution caused due to external dumps, the design of the waste dumps should allow for progressive rehabilitation thus minimizing the disturbed area at any given time and facilitating the completion of final rehabilitation as soon as possible. Reuse of some of dumping material for alternative uses, such as in landfill or road construction is another best way of its management. Securing additional land beyond the mining lease area is also required to accommodate a planned external dump.

9 Basic Objectives for Waste Dump to be Followed in the Planning Phase

The design of waste rock dumps should consider factors such as the available land area, the general topography, and existing vegetation in the area. The height, area, as well as shape of the dump must be planned accordingly to ensure stability and long-term resistance to erosion. Completed dump surfaces should be stable enough to withstand erosion over time. If any portion of the waste dump can be used for construction purposes or for road material, it should be carefully separated and stacked to prevent mixing with unusable materials, allowing for easy transport to its intended location when needed. Additionally, topsoil (approximately 20 cm) should be removed from the dump area in advance, stored in low-height dumps, and covered with vegetation to maintain its fertility and biomass. Stockpiled subsoil and topsoil needs to be spread over completed dump surfaces, where possible, and re-vegetated with appropriate plant species. The waste dump design should ensure that the slopes of the completed dumps should not make more than 20° from the horizontal. Proper drainage systems, including garland drains and individual terrace drains, should be constructed to manage heavy rainfall events. These drainage systems should guide rainwater to the main drainage channel, preventing the wash-off of dump materials.

Design of Dumps Dumps that cannot be used for the backfilling old pits or other purposes must be afforested to promote greenery and ensure land stability. The outer slopes of these dumps should be very gentle, ideally not exceeding 28°, as steeper slopes make biological rehabilitation more challenging. The relationship between the slop angle of the dump slope and the efficiency of revegetation, as illustrated in the figure below, should be considered as general guidance when designing these slopes. Figure 2 shows an idea about the different erosion risk area with respect the angles.

Dump Construction and Monitoring In hilly areas, valleys act as natural drains, and in mines located in such terrain, where no other areas are available for waste disposal, deep valleys can be utilized for dumping waste material. In this case, the base of the dump, or the bottom of the valley, should first be filled with large boulders. Above these, progressively smaller boulders should be placed, with finer materials dumped on top. This arrangement allows rainwater to flow through the voids between the larger boulders. The size of the boulders depends on factors such as the valley size, local precipitation, and the water catchment area for the valley. Some fines may be carried away with the rainwater, necessitating the construction of check dams downstream to allow the fines to settle, while clear water flows out via hume pipes. Check dams must be periodically cleared to prevent the fines from being carried further downstream once they accumulate to their full capacity. On top of this filled material, soil is not spread, and vegetation can be carried out to enhance the area's aesthetic appeal. This method not only facilitates the successful disposal of waste but also transforms barren valley rocks into lush green areas. This tech-

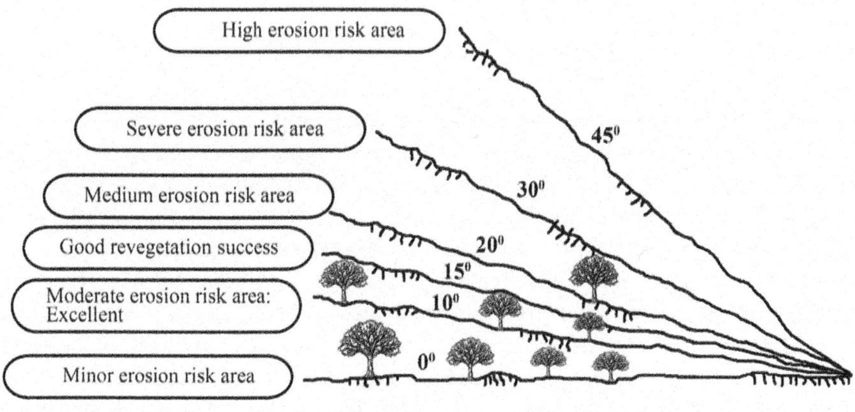

Fig. 2 Erosion risk area with respect the angles

nique is successfully implemented in a wollastonite mine at Sirohi district, Rajasthan.

Dump Rehabilitation Dump rehabilitation relies on establishing vegetative cover as the most effective long-term strategy for stabilization and erosion control. The process begins by replacing the topsoil to a depth similar to its original removal (200–300 mm). If topsoil is limited, it can be placed in strips to improve surface regeneration areas. To enhance vegetation establishment, rehabilitation efforts should focus on increasing rainfall infiltration through a method known as "water harvesting." With keeping a fundamental approach to maintain the surface area of dump tough, various techniques have been developed to enhance rehabilitation efforts. Deep rigging along the contour after spreading the top soil is one of the techniques to be adopted. The rough surface helps in water percolation thus providing spaces for seeds to germinate thus reducing the wind erosion. Pre-stripped vegetation further aids in this process thus creating a surface which promote water harvesting and promoting revegetation process. Ripping the surface and faces of dumps is quite essential to break the composition and helps in water penetration.

10 Dump Stabilization

To prevent soil erosion, landslides and environmental degradations, stabilization of overburden dump is quite essential. This dump stabilization process involves various techniques such as slope grading, compaction, terracing followed by vegetation growth on it. Geotextiles, retaining walls, and drainage systems are often used to control water flow and prevent excessive moisture buildup, which can weaken the dump structure. A stabilized dump not only ensures safety but also contribute to sustainable land management and environmental conservation.

11 Dump Stabilization through Revegetation

For controlling soil erosion and stabilizing dump slopes, revegetation is a well-established method (Akers & Muter, 1974), thus helping to restoration of the ecological balance (Jorgensen, 1994). Vegetation growth on dump slopes brings positive impact on both hydro-geological and mechanical processes (Cherubini & Geiasi, 1997). Plant roots play an important role in managing rainwater interception, evaporation, reducing the pore pressure thus enhancing the dump stability (Hussain, 1995). Vegetation growth also reinforces the loose dump material through intense root structure thus increasing the strength of the material. Stability of dump is closely linked with factors such as root density, depth, and strength. Indigenous plants such as *Shorea robusta*, *Dalbergia sissoo*, *Pongamia pinnata*, and *Azadirachta indica* control soil erosion by exhibiting strong binding capacities thus helping and improving the dump stability. Apart from this, these species help enhance the fertility of soil through nitrogen fixing capabilities. They possess root architectures that enhance drought resistance (Tewari, 1994). Overall vegetation not only brings stability to dump but it also establishes the ecological health of the area.

It is proofed that, forests play an important role in soil protection, and deforestation can lead to erosion and other forms of soil movement (Gray, 1973). It has been observed (Sternberg, 1975) that, in soils from the same rock type, the maximum response of angle of slopes with vegetation was approximately 60°, while in the absence of vegetation, it dropped to about 36°. Vegetation plays a vital role in protecting soil surfaces from erosion and facilitating the fine particles accumulation (Conesa et al., 2007). By developing extensive root systems, plants can reverse the degradation process and stabilize soils. For successful revegetation, plants must be chosen based on their ability toward survive and regenerate under harsh conditions, considering dump material nature, the exposure on dump surface, and their capacity of soil structure to stabilize (Madejon et al., 2006). The typical practices for revegetation involve selecting drought resistant, fast growing species or fodder plants that can thrive in nutrient deficient soils. In areas where acidity is a limiting factor, plants must be tolerant toward metal contaminants in the area (Caravaca et al., 2002). The use of exotic species for reclamation requires suitable consideration, as they may become invasive and problematic in other environments. Therefore, species selected for revegetation should be evaluated to prevent them from becoming harmful weeds in local ecosystems. It is generally recommended to use indigenous species, as they are better adapted to the local climate and ecosystem.

Grasses are commonly used as an early vegetation establishment on mine lands. They provide both +ve and −ve effects: while they help stabilize soils, they can also compete with woody vegetation. Grasses are particularly tolerant of drought, less soil nutrient, and environmental stresses. Their fibrous root systems help in slow erosion, and over time, they contribute to soil formation, stabilize the soil, and conserve moisture. Initial cover of grasses should allow for the development of different, self-sustaining plant diversity (Shu et al., 2002). Plants can further improve soil conditions through various processes, such as increasing soil organic matter, biological N_2 fixation, nutrient uptake, and enhancing infiltration and storage of water. Trees also reduce nutrient loss because of erosion and leaching, improve soil physical properties, decrease soil acidity, and enhance soil biological activity. The accumulation of plant litter and root exudates supports nutrient cycling, contributing to the formation of new, self-sustaining topsoil (Padmavathiamma & Li, 2007). To maintain healthy vegetation growth, regular nitrogen fertilizer applications is essential as it is often a limiting nutrient in mine spoils. Introduction of N_2fixing plants, such as legumes is an effective alternative of it as it can significantly enhance soil fertility. These species improved nutrient cycling through the production of nutrient rich litter, fine roots turnover, and the formation of nodules. Nitrogen rich litter releases nutrients after decomposition which becomes available to neighboring plants thus supporting the self-sustaining ecosystem development (Zhang et al., 2001). Research by Singh et al. (2002) indicates that local leguminous species are more effective in improving soil fertility than non-leguminous native species.

12 Plant Species Selection: A Vital Component for Ecological Restoration Success

Choosing suitable plant species plays an important role in the success of ecological restoration of dump and it depends on several factors, such as the climatic conditions, characteristics of the soil, and the land use pattern. Indigenous species like Shorea *robusta*, *Pongamia pinnata*, *Syzygium cumini*, *Dalbergia sissoo*, *Croton roxburghii*, *Phyllanthus emblica*, and *Azadirachta indica* are more tolerant to adverse soil moisture, nutrients content, pH, and stress of soil in comparison to grasses and legumes. These plants are particularly useful for creating windbreaks, visual shields, and for stabilizing steep, rocky terrains or coarse waste. Micro- and macro-climate, insect and disease resistance, competition, growth patterns, and the ability of the species propagation are some of key factors for successful development of vegetation on waste dumps, especially during hot and dry summers. Watering of the plant is direly needed in beginning period to ensure plant survival in summer till the vegetation matures after which the indigenous plants adapt to the dry summer conditions.

Species like *Pongamia pinnata*, *Dalbergia sissoo*, *Azadirachta indica* have interactions that promote soil fertility, and are comparatively fast growers, well-suited to environmental factors in more arid zones, such as intense heat and sunlight. They possess root adaptations that enhance drought tolerance (Lowry et al., 1994). Growth of indigenous plants depends not only on the soil quality but also on the organic matter content of the dump material. Leaf debris has a vital role in enhancing the organic matter of the dump material, which is essential for the successful application of revegetation. Table 3 provides the details of plant list suitable for revegetation near mine overburden with their name and utilization potential.

Table 3 Details of plant list suitable for revegetation near mine overburden

S. No.	Botanical name	Family	Common name	English name	Utilization
1.	*Ceriscoides turgid*	Rubiaceae	Kurud	Mountain gardenia	Medicinal value
2.	*Acacia catechu*	Fabaceae	Khaira	Black catechu	Medicinal value, as a dye and tanning agent
3.	*Haldinia cordifolia*	Rubiaceae	Kumm	Yellow teak	Timber production
4.	*Aegle marmelas*	Rutaceae	Bael	Indian bael	Food preparation, traditional medicine, therapeutic application
5.	*Albizia lebbeck*	Febaceae	Siriso	Lebbeck tree	Medicinal, fuel wood, timber, aegro forestry
6.	*Albizia procera*	Febaceae	Garkhair	White siris	Ornamental, soil improver, fuel wood
7.	*Anogeissus latifolia*		Dhou	Axle wood tree	Antimicrobial, antioxidant, antidiabetic, wound healing,

Table 3 (continued)

S. No.	Botanical name	Family	Common name	English name	Utilization
8.	*Anthocephalus chinensis*	Rubiaceae	Kadamba	Burflower tree	Match factory, particleboard industry
9.	*Artocarpus heterophyllus*	Moraceae	Panasa	Jackfruit	Food, fodder, timber, medicine
10.	*Azadirachta indica*	Meliaceae	Neem	Neem tree	Traditional medicine, insecticidal and pesticidial application
11.	*Bauhinia variegata*	Fabaceae	Kanchana	Mountain ebony	Ornamental, edible part, medicinal value
12.	*Bombox ceiba*	Bombacaceae	Simuli	Silk cotton tree	Edible part, ecological importance, traditional medicine
13.	*Bambusa tulda*	Poaceae	Bamboo	Indian timber bamboo	Construction, pulp paper, furniture
14.	*Bridelia retusa*	Phyllanthaceae	Kasi	Spiny kino tree	Medicinal, wound hilling, fodder
15.	*Buchanania lanzan*	Anacardiaceae	Char	Charoli nut	Digestive, wound treatment
16.	*Butea monosperma*	Fabaceae	Palas	Flame of the forest	Medicinal use, pigment
17.	*Careya arborea*	Lecythidaceae	Kumbhi	Patina oak	Medicinal value, decorative
18.	*Cassia fistula*	Fabaceae	Sumari	Indian laburnum	Medicinal, timber value
19.	*Chloroxylon swietiana*	Rutaceae	Bheru	Ceylon satinwood	Furniture, construction, traditional medicine
20.	*Cleistanthus collinus*	Phyllanthaceae	Karada	Toxic gooseberry	Cattle and fish poisoning, used as suicidal
21.	*Crateva magna*	Capparidacear	Baruna	Tree leaved caper	Medicinal value
22.	*Dalbergia sissoo*	Fabaceae	Sissoo	Shisham	Traditional medicine, timber, fuel
23.	*Ficus racemosa*	Moracear	Dimiri	Indian fig tree	Medicinal value
24.	*Dendrocalamus strictus*	Poaceae	Banso	Solid bamboo	Furniture, paper, construction
25.	*Diospyros melanoxylon*	Ebenaceae	Kendu	East Indian ebony	Medicinal importance, ecological value
26.	*Phyllanthus emblica*	Phyllanthaceae	Anola	Indian gooseberry	Food, medicine
27.	*Ficus religiosa*	Moraceae	Papal	Peepal tree	Medicinal value
28.	*Gmelina arborea*	Lamiaceae	Gambari	Kashmir tree	Construction, furniture, pulp
29.	*Madhuca indica*	Sapotaceae	Mahula	Indian butter tree	Medicinal, food, fuel, industry
30.	*Lannea coromandelica*	Anacardiaceae	Mai	Indian ash tree	Medicinal, gum, sprouts

13 Factors Affecting the Selection of Species for Revegetation

The selection of species for revegetation should be based on several key parameters, including: rapid growth, resistance to diseases and pests, the ability to compete with low desirable species, flexibility to local soil conditions, and compatibility with the local climate (Morgan, 2005). When trees are chosen for revegetation, Young (1997) outlines additional qualities for species aimed at improving or restoring soil fertility. These include: the ability to grow on poor soil nutrient condition, a high rate of N_2 fixation, a balanced nutrient in foliage litter, less lignin and polyphenol content in litter, rapid or moderate litter decay, a well-developed tap root, high leafy biomass, and a dense fine root network capable of mycorrhizal associations.

Species selection must consider local soil as well as climatic conditions, but the goal should be to create a uniform vegetation pattern to avoid concentrated runoff and localized soil erosion. Achieving this ideal can be difficult, especially where soils having low water retention ability, toxic, or where environmental factors like cold or drought hinder plant growth. The selection of vegetation with beneficial effects is therefore critical (Morgan, 2005). For example, in the reclamation of mine spoil mounds on the Jos Plateau in Nigeria, Eucalyptus species were chosen for their utility as a source of fuel and timber. However, this choice led to a significant decline in soil pH, negatively affecting the growth of other plants (Morgan, 2005).

Eroded ecosystems revegetation should be carried out with species selected based on their capacity to survive, regenerate, or reproduce under the harsh conditions of the dump material and the exposed dump surface, as well as their capacity for soil structure stabilization (Madejon et al., 2006). In general, revegetation efforts focus on selecting resistant to drought, fast growth rate species or fodder plants that thrive in nutrient poor soils. Chosen plants should be easy to establish, fast growing, dense canopies and roots. In some areas, soil acidity may limit vegetation growth. Additionally, plants should be tolerant to metal pollution in such areas (Caravaca et al., 2002).

14 Dump Stabilization Through Vertiver and Lemon Grass

Vetiver grass, an indigenous plant of India is being widely used for slope protection, reinforcement of embankments since the early 1900s. With progress of time, this knowledge spread to West Indies, South Africa, (National Research Council, 1993), Brazil (Grimshaw, 1994), and Fiji (Truoung & Gawander, 1996) as Indians migrated overseas, and vetiver usage resumed in these area. During 1931, vetiver was grown in Serdang, near Kuala Lumpur, Malaysia, where it was known for its effectiveness in stabilizing steep banks. Studies have identified key factors of vertiver grass contributing to slope stability. Vetiver Grass Technology (VGT), developed for water and soil conservation on farmlands, has gained recognition for its unique, physiological, morphological and ecological characteristics thus protecting the environment. The Vetiver grass has large, cleared structured, deeper root system, ability of reaching depths of 1.2–1.5 m. VGT process typically involves planting a live, stiff vetiver grass barrier, less than 1 m wide, across the slope. Dump stabilization through vertiver and lemon grass plantation is effective on slopes ranging from less than 1% to over 100% as the deep root system of them increases the shear strength of soil, improves the steep slopes stability.

Lemongrass, a tropical grass originally from India is an aromatic species has the potential option for slope stabilization and soil erosion control. Lemongrass is a perennial grass having many stiff, leafy stems that arise from short rhizomatous roots and the leaves can develop up to 2 m in height. Lemongrass has an economic lifespan of approximately 5 years. Though it can grow on a wide range of soils, but the best yields are found in well drained sandy to loamy soils. Lemongrass is often planted on poor or erosion prone soils by soil conservation department. Its deep root system allows it to survive even in the absence of water until the next rainy season. Root system of lemongrass grows in a uniaxial direction, thus help stabilizing the upper and deep layers of dump slopes due to deep penetration. Grasses, such as lemongrass, quickly bind the surface, while shrubs provide considerable strength to the slope's stability in the early stages.

15 Dump Stabilization Through Hydroseeding and Hydromulching

Hydroseeding and hydromulching are increasingly being used to stabilize the barren area, unstable surfaces and steep slopes. The technique involves applying a mixture of water, seeds, fertilizer, and mulch in a fine spray over the surface. Hydroseeding has proven effective for creating vegetative cover on rocky or stony ground, and can also be used to transplant trees in such areas. Any organic material that can be shredded, such as bark waste mixed with fines, can be used in hydroseeding to prevent clogging. Hydroseeding is particularly beneficial for establishing grass cover, though some tree and shrub species. For higher slopes, prototype hydroseeding devices have been developed to tackle barren, rocky surfaces. A variety of hydraulic mulching materials, including both chemical and biopolymer based as well as natural fiber based options, have been tested worldwide for their effectiveness in these applications.

16 Chemical Stabilizers

In order to enhance soil stability, erosion control, and boost fertility in overburden management within mining environments, chemical stabilizers such as polymeric binders, calcium-based materials such as lime, and magnesium salts are used. This application not only improves soil cohesion but also help in reducing the dust generation. Chemical stabilizers create favorable conditions for plant establishment by increasing the organic matter content, balancing soil pH, and improving overall soil structure. The chemical stabilizers effectively controls dust emission, apart from promoting vegetation on dump. Despite their advantages such as cost effective erosion control and accelerated vegetation recovery the use of chemical stabilizers must be carefully managed as it can alter the soil pH, reduced nutrient availability, and the need for periodic reapplication. Nevertheless, when chosen and applied appropriately, chemical stabilizers are an important tool for sustainable rehabilitation of overburden, thus contributing to long-term ecological stability.

Geotextiles Coir-mat offers the requisite mechanical strength for stabilization of soil in overburden and prevention of erosion. This netting not only disperses runoff from heavy rainfall but also reduces the energy of flowing of rain water thus minimizing the soil displacement. The coir mat aids in vegetation growth by retain of moisture and preventing topsoil from drying out. The Coir mats and netting which are prepared from coconut husk fibers control sediment, stream bank, and wetland shore protection thus creating a perfect foundation for plant establishment. Coir geotextiles after few years of use break down into humus thus eliminating the need for removal or post-installation maintenance.

17 Mining Overburden and Circular Economy

Environmental impact of mining activities, particularly the management of overburden, poses significant challenges to the surrounding environment. As the world is increasingly shifting toward a circular economy, a model aimed at minimizing waste and making judicious uses of resources, the effective management of mining overburden becomes crucial. In the traditional linear economy model "take-make-dispose" is practiced, where natural resources are extracted, used, and then discarded as waste, circular economy circular economy promotes sustainable practices that allow for the continual use of materials and resources. Through integration of the principles of a circular economy into mining operations, the management of overburden can take on a transformative role, shifting from a waste product to a valuable resource. In circular economy model, closed loop systems are practiced which reduces the need for the new resource extraction and processing. Within the mining sector, following the 3R principle can make a significant contribution.

The key points in achieving circular economy in mining sector can be achieved through;

- Mining and mineral processing recovery rate Improvement.
- Reduction and minimizing of tailings, waste water, and gas emissions.
- Suitable and feasible options development for poor grade ores.
- Better planning for future use and reuse, the life of any resource, material, and product is extended.

Recycling and repurposing of overburden materials is one of the promising approaches to advancing the circular economy in mining sector. Composition of mining wastes is evaluated to identify its potential applications. Materials such as clay, sand, and rock components of overburden can be reused in construction purposes, development of landscaping and rehabilitation of mined areas. Companies can reduce the disposal costs through development of marketing of these by-products thus promoting resource sustainability. Innovative technologies also play a key role in supporting this circular approach. Recent developments in sorting and processing techniques have made it possible to recover valuable minerals from overburden that were previously considered uneconomical or technically challenging to extract. These technologies not only improve resource efficiency but also help minimize the environmental footprint of mining by reducing the dependency on new extraction activities (John & Das, 2024; Priyadarsini & Das, 2025; Ray et al., 2024). These strategies contribute to more sustainable and economically viable mining practices.

Involving local community in management of mining overburden is quite crucial part of circular economy as they have valuable insights into sustainable practices and potential uses for overburden materials. Mining companies can develop strategies through collaboration with local communities which will flourish local economic development, enhance the social acceptance and ensure sharing of benefits of mining among local community. Governments can incentivize companies to adopt circular approaches to overburden management through taxation benefits or grants, encouraging innovation and investment in sustainable practices. Management of over burden presents an opportunity before the mining companies to achieve and contribute to the goals of the circular economy through reusing the materials, financial investing in technology, engaging local affected communities. Through this, the mining sector can convert the overburden management from liability in to an asset. This shift not only supports environmental sustainability but also fosters a more resilient and responsible economy, paving the way for future generations to benefit from the earth's resources. Reusing of overburden in road making, infrastructure projects and backfilling of the exhausted mines are some of the key activities to be undertaken by mining companies for achieving environmental sustainability in mining sector.

18 Conclusion

Effective scientific management of overburden in iron ore mining is essential, more particularly in regions where mineral extractions plays a vital role in the local economy and have detrimental impacts on the surrounding environment. Adopting appropriate overburden management practices has become crucial in order to address the environmental degradation, improve operational efficiency, and limit long-term consequences on the nearby communities. Improper or unregulated disposal of overburden is affecting the surrounding agriculture and posing risks to community health through deforestation, soil erosion, biodiversity loss, soil and water resource contamination.

Apart from these mining activities often result in the displacement of indigenous communities, intensifying social tensions and putting added pressure on local infrastructure. The traditional slope stabilization methods using native plant species have now evolved into more advanced and effective techniques. Revegetation has emerged as the most sustainable approach in ecologically sensitive and visually prominent areas—thus offering both environmental restoration and long-term stability. In order to stabilize dump slopes in iron ore mines, various ecological restoration methods such as coir matting, plantation of native species of plant, lemon and vertiver grass plantation, miamaki plantation are being practiced. While grasses and shrubs provide surface stabilization, deep rooted trees like *Dalbergia sissoo* offer more substantial long term stability due to their ability to reinforce dump structures. Native trees such as *Shorea robusta*, *Pongamia pinnata*, and *Azadirachta indica* and Bamboo are quite effective in prevention of soil erosion and improvement in dump stability. It is essential to conduct a preliminary thorough site analysis, taking into account factors such as climate, soil type, and annual rainfall for getting success in revegetation of dumps. For ensuring both ecological sustainability and structural safety, engaging horticultural experts and botanists are quite crucial to choose the appropriate plant species as per the site conditions. Vegetation plays a crucial role in reducing the impact of rainfall, surface runoff during monsoon thus minimizing the erosion risks apart from covering the area with green foliage. A multifaceted approach such as adoption of innovative technologies, such as waste reprocessing and the use of overburden in reclamation projects, to mitigate environmental impacts is need of the hour to manage the voluminous overburden generated from iron mines in Odisha. Strengthening regulatory frameworks, enhancing monitoring systems, and fostering community engagement in decision-making processes are also critical for ensuring the sustainability of mining operations. A proper balance between industrial growth and environment stewardship is the need of time for achieving sustainable growth in mining sector.

References

Acharya, S. (2008). *Genesis of banded iron formations, associated iron and manganese ore deposits of Orissa—Jharkhand region, India and its role as exploration guide; seminar on iron ores—Genesis and exploration techniques organized by SGAT, Bhubaneswar* (pp. 1–31).

Akers, J. D., & Muter, B. R. (1974). Gob pile stabilization and reclamation. In *Proceeding of the forth mineral waste utilization symposium* (pp. 229–239). IIT Research Institute.

Amreen, S., Dash, S. K., Naik, B., Dash, A. K., Verma, A. K., & Pradhan, A. (2024). Marine macro algae-derived activated carbon an efficient bio-sorbent for malachite green removal from wastewater. *Environmental Quality Management*. https://doi.org/10.1002/tqem.22213

Anon. (2005). *Minerals yearbook*. United States Bureau of Mines (USBM).

Anon. (2006). *Dirty metal, mining communities and environment* (p. 4). Earthworks.

Bal, B., Ghosh, S., & Das, A. P. (2018). Microbial recovery and recycling of manganese waste and their future application: A review. *Geomicrobiological Journal, 36*(1), 85–96. https://doi.org/10.1080/01490451.2018.1497731

Bala, K., Anshul, A., Verma, S., Dogra, A., Das, A. P., Kumar, R., & Nandini, A. (2024). Sustainable management for urban plastic waste generation. In *Sustainable management of urban plastic waste through circular economic approaches* (pp. 211–236). Routledge. https://doi.org/10.1201/9781003465447-13

Caravaca, F., Hernandez, M. T., Garcia, C., & Roldan, A. (2002). Improvement of rhizosphere aggregates stability of afforested semi-arid, plant species subjected to mycorrhizal inoculation and compost addition. *Geoderma, 108*, 133–144.

Cherubini, C., & Geiasi, C. (1997). The influence of vegetation on slope stability. In P. G. Marinos, G. C. Koukis, G. C. Tsiambaos, & G. C. Stournaras (Eds.), *Engineering geology and environment* (Vol. I, pp. 67–71). Balkema.

Conesa, H. M., Schulin, R., & Nowack, B. (2007). A laboratory study on revegetation and metal uptake in native plant species from neutral mine tailings. *Water, Air, & Soil Pollution, 183*(1-4), 201–212.

Das, A., Dey, S., Bala, K., Gnanasekaran, R., & Das, A. P. (2024). Microbial bioremediation technology for sustainable treatment and management of synthetic microfiber waste. In *Sustainable microbial technology for synthetic and cellulosic microfiber bioremediation* (pp. 221–238). Springer. https://doi.org/10.1007/978-3-031-62600-5_12

Das, A., Dey, S., & Das, A. P. (2024). Ecotoxicological impacts of synthetic microfiber pollutants and development of sustainable mitigation strategies. *Environmental Chemistry and Ecotoxicology*. https://doi.org/10.1016/J.ENCECO.2024.12.003

Das, A., Dey, S., Gacem, A., Yadav, K. K., Nath, H., Ram, D. K., & Das, A. P. (2025). Revolutionizing solid hydrocarbon sourcing: Plastic's journey from waste to upcycled treasure. *Environmental Science and Engineering, 218*, 331–351. https://doi.org/10.1007/978-3-031-81615-4_14

Das, A., & Ghosh, S. (2022). Role of microorganisms in extenuation of mining and industrial wastes. *Geomicrobiology Journal, 39*, 173–175. https://doi.org/10.1080/01490451.2022.2038953

Das, A. P., & Ghosh, S. (2023). Introduction. In *Manganese mining microorganisms* (pp. 1–4). Elsevier. https://doi.org/10.1016/B978-0-12-822147-1.00005-4

Das, A. P., & Mishra, S. (2009). Hexavalent chromium [Cr (VI)]: Yellow water pollution and its remediation. *ENVIS Newsletter on Wetland Ecosystems, 5*(2), 1–8.

Das, A. P., & Singh, S. (2011). Occupational health assessment of chromite toxicity among Indian miners. *Indian Journal of Occupational and Environmental Medicine, 15*(1), 6. https://doi.org/10.4103/0019-5278.82998

Dash, A. K. (2012). Impact of domestic wastewater on seed germination and physiological parameters of wheat and rice. *International Journal of Applied Science, 12*(2), 280–286.

Dash, A. K., & Dash, S. K. (2017). Atmospheric pollution load assessment through air quality index: A case study. *Indian Journal of Environmental Protection, 37*(9), 736–741.

Dash, A. K., & Mishra, P. C. (1996a). Changes in pigment and protein content of *Westiellopsis prolifica*, a blue-green alga, grown in paper mill waste water. *Microbios, 85*, 257–266.

Dash, A. K., & Mishra, P. C. (1996b). Changes in biomass, pigment and protein content of *Westiellopsis prolifica*, a blue-green alga, grown in nutrient manipulated paper mill wastewater. *Cytobios, 88*, 11–16.

Dash, A. K., & Mishra, P. C. (1999a). Growth response of the blue-green alga, *Westiellopsis prolifica* in sewage enriched paper mill waste water. *Revista Internacional de Contaminación Ambiental, 15*(2), 79–83.

Dash, A. K., & Mishra, P. C. (1999b). Role of the blue—Green alga, *Westiellopsis prolifica* in reducing pollution load from paper mill wastewater. *Indian Journal of Environmental Protection, 19*(1), 1–15.

Dash, A. K., & Patra, H. S. (2016). Floristic composition and biological Spectrum study near the mining cluster at Keonjhar District, Odisha, India. *International Journal of Engineering Sciences & Research Technology, 5*(2), 441–447.

Dash, A. K., Pradhan, A., Das, S., & Mohanty, S. S. (2015). Fly ash as a potential source of soil amendment in agriculture and a component of integrated plant nutrient supply system. *Journal of Industrial Pollution Control, 31*(2), 251–259.

Dash, A. K., Pradhan, A., & Patra, H. (2016). Soil physico-chemical characteristics near the mining areas of Keonjhar District, Odisha, India. *International Journal of Current Research and Review, 8*(4), 30306–30309.

Dash, S. K., & Dash, A. K. (2018a). Air pollution tolerance index to assess the pollution tolerance level of plant species in industrial areas. *Asian Journal of Chemistry, 30*(1), 219–222. https://doi.org/10.14233/ajchem.2018.20991

Dash, S. K., & Dash, A. K. (2018b). Seasonal variation of air quality index and assessment. *Global Journal of Environmental Science and Management, 4*(4), 483–492. https://doi.org/10.22034/gjesm.2018.04.008

Dash, S. K., Pradhan, A., Panda, B. P., & Dash, A. K. (2023). Ambient air quality monitoring and assessment of an industrial area as a tool for sustainable environmental management. *Materials Today: Proceedings*. https://doi.org/10.1016/j.matpr.2023.10.024

Dash, S. K., Sahu, H. B., & Dash, A. K. (2019). Atmospheric dispersion modeling using AERMOD to predict the impact of PM_{10} near Bileipada, Odisha. *Indian Journal of Environmental Protection, 39*(4), 299–306.

Dey, S., Das, A., Mallick, K., Sahu, A., & Das, A. P. (2023). *Environmental petroleum waste: Pollution, toxicity, sustainable remediation* (pp. 159–175). Springer. https://doi.org/10.1007/978-3-031-48220-5_7

Dey, S., Dash, A., & Das, A. P. (2024). Bioremediation as a sustainable and circular solution for mining waste management and Bioenergy generation. In *Sustainable management of mining waste and tailings: A circular economy approach* (pp. 107–119). Routledge. https://doi.org/10.1201/9781003442455-6

Dey, S., Tripathy, B., Kumar, M. S., & Das, A. P. (2023). Ecotoxicological consequences of manganese mining pollutants and their biological remediation. *Environmental Chemistry and Ecotoxicology*. https://doi.org/10.1016/J.ENCECO.2023.01.001

Gardner, G., & Sampat, P. (1998). *Mind over matter: Recasting the role of materials in our lives* (p. 18). Worldwatch Institute.

Ghosh, S., Mohanty, S., Akcil, A., Sukla, L. B., & Das, A. P. (2016). A greener approach for resource recycling: Manganese bioleaching. *Chemosphere, 154*, 628–639. https://doi.org/10.1016/J.CHEMOSPHERE.2016.04.028

Ghosh, S., Sinha, J. K., Ghosh, S., Vashisth, K., Han, S., & Bhaskar, R. (2023). Microplastics as an emerging threat to the global environment and human health. *Sustainability, 15*(14), 10821. https://doi.org/10.3390/SU151410821

Ghosh, S., Tripathy, B., Dey, S., & Das, A. P. (2023). Comparative investigation of fungal and bacterial manganese biomining mechanisms. In *Manganese mining microorganisms* (pp. 99–115). Elsevier. https://doi.org/10.1016/b978-0-12-822147-1.00016-9

Gray, D. H. (1973). *Effects of forest clear cutting on the stability of natural slopes: Results of field studies*. National Science Foundation, University of Michigan.

Grimshaw, R. G. (1994). Vetiver grass—Its use for slope and structure stabilization under tropical and semitropical conditions. In *Vegetation and slopes* (pp. 26–35). Institution of Civil Engineers.

Hussain, A. (1995). Fill compaction-erosion study in reclaimed areas. *The Indian Mining & Engineering Journal, 34*(6), 19–21.

Jena, P. S., Pradhan, A., Nanda, S. P., Dash, A. K., & Naik, B. (2022). Biosorption of heavy metals from wastewater using *Saccharomyces cerevisiae* as a biosorbent: A mini review. *Materials Today: Proceedings, 67*(8), 1140–1146. https://doi.org/10.1016/j.matpr.2022.07.306

John, A., & Das, A. P. (2024). Government strategies for achieving circular economy for Plastic waste Management in Southeast Asia. In *Sustainable management of urban plastic waste through circular economic approaches* (pp. 72–82). Springer. https://doi.org/10.1201/9781003465447-6

Jorgensen, S. E. (1994). Models as instruments for combination of ecological theory and environmental practice. *Ecological Modelling, 75-76*, 5–20. https://doi.org/10.1016/0304-3800(94)90003-5

Khatua, A., et al. (2006). Ecological debt: A case study from Orissa, India. In *Ecological debt: The people of the South are the creditors* (p. 145).

Kumar, D. S., Bhushan, S. H., & Kishore, D. A. (2018). Atmospheric dispersion model to predict the impact of gaseous pollutant in an industrial and mining cluster. *Global Journal of Environmental Science and Management, 4*(3), 351–358. https://doi.org/10.22034/gjesm.2018.03.008

Kuter, N. (2013). *Reclamation of degraded landscapes due to opencast mining advances in landscape architecture*. IntechOpen. https://doi.org/10.5772/55796

Lodha, R. M., Purohit, J. K., & Yadav, H. S. (2009). *Environment and mining. A peep in to deep*. Vedams Books.

Lottermoser, B. G. (2010). *Mine wastes: Characterization, treatment and environmental impacts* (3rd ed., pp. 204–240). Springer.

Lowry, J. B., Prinsen, H., & Burrows, D. M. (1994). Albizia lebbeck: A promising forage tree for semiarid regions. In R. C. Gutteridge & H. M. Shelton (Eds.), *Forage tree legumes in tropical agriculture* (pp. 75–83). CAB International.

Madejon, E., de Mora, A. P., Felipe, E., Burgos, P., & Cabrera, F. (2006). Soil amendments reduce trace element solubility in a contaminated soil and allow re-growth of natural vegetation. *Environmental Pollution, 139*, 40–52.

Mahapatra, B., Dhal, N. K., Dash, A. K., Panda, B. P., Panigrahi, K. C. S., & Pradhan, A. (2019). Perspective of mitigating atmospheric heavy metal pollution: Using mosses as biomonitoring and indicator organism. *Environmental Science and Pollution Research, 26*(29), 29620–29638. https://doi.org/10.1007/s11356-019-06270-z

Mishra, S., & Das, A. P. (2008). Hexavalent chromium reduction and 16S rDNA identification of bacteria isolated from a Cr (VI) contaminated site. *The Internet Journal of Microbiology, 7*, 1.

Mohanty, J., Dey, S., Bala, K., & Das, A. P. (2024). Recent advances in microplastic pollution for its sustainable management. In *Sustainable management of urban plastic waste through circular economic approaches* (pp. 27–48). Springer. https://doi.org/10.1201/9781003465447-3

Mohanty, M., Tripathy, B., Koteswara Reddy, G., & Das, A. P. (2024). Greener technologies for establishing restraint over microplastic pollution. In *Sustainable microbial technology for synthetic and cellulosic microfiber bioremediation* (pp. 239–258). Springer. https://doi.org/10.1007/978-3-031-62600-5_13

Morgan, R. P. C. (2005). *Soil erosion and conservation* (3rd ed., pp. 169–170). Blackwell Publishing.

National Research Council, Global Affairs, Board on Science, Technology for International Development, & Panel on Vetiver. (1993). Vetiver grass: a thin green line against erosion. National Academies Press.

Nayak, K. P., Mishra, A., Pradhan, A., Dash, A. K., & Saren, S. (2018). Soil fertility study as a tool for sustainable agriculture. *NPI International Research Journal of Industrial and Environmental Biotechnology, 1*(1), 57–74.

Padmavathiamma, P. K., & Li, L. Y. (2007). Phytoremediation technology: Hyper accumulation metals in plants. *Water, Air, and Soil Pollution, 184*(1-4), 105–126.

Panigrahi, B. R., Pradhan, S. C., Dash, A. K., & Mishra, P. C. (2024). Evaluation of protease activity in the soil of municipal solid waste dumping sites in Balasore district of Odisha, India. *Pollution Research, 43*(3–4), 267–269.

Parida, B., Dash, A. K., & Verma, A. (2023). Technologies for treatment of landfill leachate: A brief review. *Recent Innovations in Chemical Engineering, 16*(4), 260–284. https://doi.org/10.2174/0124055204276847231013040437

Parida, S., Das, M., & Dash, A. K. (2015). Ambient air study near the industrial complex 'nisha' at Angul, Odisha, India. In *Advances in environmental sciences and engineering* (pp. 514–537). Daya Publishing House.

Patra, H. S., & Sethy, K. M. (2013). Assessment of impact of iron ore mining impact by using remote sensing: A case study from Keonjhar district of Odisha, India. *International Journal of Environmental Engineering and Management, 4*(1), 17–24.

Patra, H. S., & Sethy, K. M. (2014a). Assessment of impact of opencast mine on surrounding forest: A case study from Keonjhar district of Odisha, India. *Journal of Environmental Research and Development, 9*(1), 249–254.

Patra, H. S., & Sethy, K. M. (2014b). Assessing the impact of open cast mining on floristic diversity: A case study from Gandhamardan hill of Keonjhar, Odisha. *International Journal of Environment and Natural Sciences, 3*, 8–15.

Paul, A., Dey, S., Ram, D. K., & Das, A. P. (2023). Hexavalent chromium pollution and its sustainable management through bioremediation. *Geomicrobiology Journal*. https://doi.org/10.1080/01490451.2023.2218377

Pradhan, A., Dash, A. K., Mohanty, S. S., & Das, S. (2015). Potential use of fly ash in floriculture: A case study on the photosynthetic pigments content and vegetative growth of *Tagetes erecta* (MARIGOLD). *Ecology, Environment and Conservation, 21*(Suppl), 369–376.

Pradhan, A., Kumari, S., Dash, S., Biswal, D. P., Dash, A. K., & Panigrahi, K. C. S. (2017). Heavy metal absorption efficiency of two species of mosses (*Physcomitrella patens* and *Funaria hygrometrica*) studied in mercury treated culture under laboratory condition. *IOP Conference Series: Materials Science and Engineering, 225*, 012225. https://doi.org/10.1088/1757-899X/225/1/012225

Pradhan, A., Sahu, S. K., & Dash, A. K. (2013). Changes in pigment content (chlorophyll and carotenoid), enzyme activities (catalase and peroxidase), biomass and yield of rice plant (*Oriza sativa*. L) Following irrigation of rice mill wastewater under pot culture conditions. *International Journal of Scientific Engineering and Research, 4*(6), 2706–2717.

Pradhan, A., Sahu, S. K., & Dash, A. K. (2015). Changes in Physico-chemical characteristics of soil following application of Rice mill wastewater under field irrigation conditions. In *Advances in environmental sciences and engineering* (pp. 480–492). Daya Publishing House.

Priyadarsini, S., & Das, A. P. (2025). Lithium bioleaching: A review on microbial-assisted sustainable technology for lithium bio-circularity. *Journal of Water Process Engineering, 69*, 106744. https://doi.org/10.1016/j.jwpe.2024.106744

Ray, S. K., Dey, S., Das, A. P., & John, A. (2024). Microplastic waste management through greener technologies: A review. In *Sustainable management of urban plastic waste through circular economic approaches* (pp. 237–261). Routledge. https://doi.org/10.1201/9781003465447-14

Reddy Koteswara, G., Bhargavi, S., Yamuna, D., Anjani Devi, M., Jeevana Lakshmi Kumar, G., Manas, T., Adarsh, S., Bhavya Sai, V., Nath, H., Alreshidi, M. A., Togun, H., Ram, D. K., & Das, A. P. (2025). Recent advancements in the remediation of marine hydrocarbon and Plastic pollutants. *Environmental Science and Engineering, 218*, 127–142. https://doi.org/10.1007/978-3-031-81615-4_7

Romero, F. M., Armienta, M. A., & Gonzales-Hernandez, G. (2007). Solid-phase control on the mobility of potentially toxic elements in an abandoned lead/zinc tailings impoundment, Taxco, Mexico. *Applied Geochemistry, 22*, 109–127.

Sahoo, D., Dash, A. K., & Sahu, S. K. (2017). Ambient air quality monitoring and health impact study of air pollution near Joda of Keonjhar, Odisha, India. *International Journal of Engineering Sciences & Research Technology, 6*(1), 429–434. https://doi.org/10.5281/zenodo.260069

Sahu, A., Dey, S., Mallick, K., Priyadarsini, S., Behera, I. D., & Das, A. P. (2024). A review on advanced technology for sustainable management of synthetic microplastic waste. In *Sustainable management of urban plastic waste through circular economic approaches* (pp. 262–279). Rouledge. https://doi.org/10.1201/9781003465447-15

Sahu, S. K., Mishra, A., & Dash, A. K. (2015). Ambient air quality around OCL India limited at Rajgangpur, Odisha, India. In *Advances in environmental sciences and engineering* (pp. 459–479). Daya Publishing House.

Shreya, Verma, A. K., Gilron, J., Oren, Y., Ronen, Z., Dash, A. K., Bhunia, P., & Dash, R. R. (2022). Effect of biological treatment on perfluoroalkyl and polyfluoroalkyl (PFASs) substances degradation. In *Current developments in biotechnology*. Elsevier. https://doi.org/10.1016/B978-0-323-99906-9.00004-8

Shu, W. S., Xia, H. P., Zhang, Z. Q., & Wong, M. H. (2002). Use of vetiver and other three grasses for revegetation of Pb/Zn mine tailings: Field experiment. *International Journal of Phytoremediation, 4*(1), 47–57.

Singh, A. N., Raghubanshi, A. S., & Singh, J. S. (2002). Plantations as a tool for mine spoil restoration. *Current Science, 82*(12), 1436–1441.

Sternberg, H. O. (1975). Effects of deforestation on slopes. *Journal of the Geotechnical Engineering Division, 101*, 147–165.

Tewari, D. N. (1994). *A monograph on Dalbergia sissoo Roxb*. International Book Distributors.

Truoung, P., & Gawander, J. S. (1996). Back from the future: Do's and don'ts after 50 years of vetiver utilization in Fiji. In *Proceedings of the first international conference on vetiver* (pp. 12–17). Office of the Royal Development Projects Board.

Young, A. (1997). *Agroforestry for soil management, effects of trees on soils* (Vol. 23). CAB International.

Zhang, Z. Q., Shu, W. S., Lan, C. Y., & Wong, M. H. (2001). Soil seed bank as an input of seed sources in vegetation of lead/zinc mine tailings. *Restoration Ecology, 9*(4), 1–8.

Recent Advances in Microbial Upcycling of Mining Waste Into Value-Added Products

Ankita Priyadarshini, Annapurna Maharana, Amel Gacem, Sudeshna Dey, Alok Prasad Das, Maha Awjan Alreshidi, and Krishna Kumar Yadav

1 Introduction

Any substance that is left behind after extracting minerals or metals from the ore is referred to as mining waste. The forms include waste rock, slag, overburden, and tailings. Depending on the geology, extraction methods, and kind of mining operation, mining waste will vary in quantity and form. The need for metals has grown due to the rapid advancement of society and industry. However, the rapid industrial expansion and the extraction of enormous mine resources add to the environmental contamination of tailings, which is growing at an alarming rate (Das et al., 2025; Reddy Koteswara et al., 2025). In the long run, tailings' effects on the ecosystem will either directly or indirectly harm human health. A chemical burden of

A. Priyadarshini · A. Maharana · A. P. Das (✉)
Department of Life Sciences, Rama Devi Women's University, Bhubaneswar, Odisha, India

A. Gacem
Department of Physics, Faculty of Sciences, University 20 Août 1955, Skikda, Algeria

S. Dey
Department of Life Sciences, Rama Devi Women's University, Bhubaneswar, Odisha, India

Centre for Biotechnology, Siksha 'O' Anusandhan, (Deemed to be University), Bhubaneswar, Odisha, India

M. A. Alreshidi
Department of Chemistry, College of Science, University of Ha'il, Ha'il, Saudi Arabia

K. K. Yadav
Department of Environmental Science, Parul Institute of Applied Sciences, Parul University, Vadodara, Gujarat, India

Environmental and Atmospheric Sciences Research Group, Scientific Research Center, Al-Ayen University, Nasiriyah, Iraq

© The Author(s), under exclusive license to Springer Nature Switzerland AG 2026
A. P. Das (ed.), *Bioremediation Technology for Sustainable Mining Waste Management*, Sustainable Environmental Waste Management Strategies, https://doi.org/10.1007/978-3-032-03468-7_16

contaminants has been left in the soil and sediment by human activity for more than 2000 years (Biswas et al., 2018; Dey, Dash, & Das, 2024). Due to the air dispersion of air-dried tailings, the leaching of potentially toxic chemical species, erosion and uptake by the aquatic system, and bioaccumulation, tailings contain hazardous amounts of toxic substances that may pose serious health and environmental issues (Gao et al., 2021). Exploration, mine development, metal extraction, mineral beneficiation, smelting, refining, reclamation, and remediation are all part of the complex process of mining. Large volumes of waste are produced throughout the metal value extraction process. These wastes are often divided into three categories: liquid wastes, such as waste water and effluents; solid wastes, such as waste rock, dust, sludge, and slags; and gaseous emissions. (Kalisz et al., 2022; Shome et al., 2021). Every stage generates waste, which varies according to mining practices, ore kinds, geological configurations, and processing procedures. Every day, large-scale mines generate approximately one million tons of ore and rubbish, whereas small-scale mines produce thousands of tonnes (Das & Ghosh, 2022, 2023b, 2023e, 2023g). For example, one tonne of coal mine produces roughly 0.4 tonnes of tailings and waste, but 1 ton of copper mining produces about 110 tons of waste and 200 tons of overburden. The yearly production of mill tailings from various ore concentration methods, including lead-zinc, copper, and iron, exceeds 500 million tonnes. One of the biggest environmental problems in South Africa is mining and metallurgical waste (Das & Ghosh, 2023a; Das, Mallavarapu, & Ghosh, 2023; Priyadarshini, Jagatee, & Das, 2024). If these mining and metal extraction activities are not properly regulated, the anthropogenic repercussions. Therefore, increasing the recycling and reusing of the various waste types is one surefire way to address the environmental and health issues caused by these waste streams (Dey, Dash, & Das, 2024; Dey, Tripathy, et al., 2023; Paul et al., 2024). By reusing and recycling waste materials, the mining and metallurgical industries can work together to create a zero-waste circular economy, notwithstanding the negative environmental effects of these waste kinds. Each of the waste streams can actually be thought of as a secondary source of valuable metals and other minerals. Metal mining, the smelting of lead and zinc, and the disposal of hazardous waste are examples of industrial activities that pollute the soil with a toxic mixture of heavy metals, leaving an environment with a harmful tag. This toxic legacy includes not just lead and zinc but also arsenic, cadmium, and mercury, all of which are unique hazards to soil health and ecological stability (Das, Dey, & Das, 2024; Das, Mallavarapu, & Ghosh, 2023; Das, Ray, Mohanty, et al., 2024; Dey, Tripathy, et al., 2023; Paul et al., 2024). The impact of the metal assault extends beyond superficial surface-level harm, since it penetrates deeply into the soil and disrupts vital ecological functions (Agboola et al., 2020; Matinde et al., 2018).

Mining tailings contamination with heavy metals has a significant negative influence on the environment and human health. Molybdenum was found to be the main contaminant in the Luhun Reservoir in Henan Province, China, with a pollution factor of 2.05. Groundwater in the Free State Province of South Africa has been tainted by gold mine tailings near Welkom and Virginia (Dey, Dash, Ray, et al., 2024; Satapathy et al., 2024). According to analysis, 100% of samples had faecal coliform bacteria, 50% had *E. coli* levels over safe drinking limits, 40% had lead levels above safe drinking limits, and 63% had elevated iron levels (Belle et al., 2021; Song et al., 2021). One of the main sources of drinking water is this contaminated groundwater, which endangers locals' health and damages the ecosystem. Even after years, research on a copper mine in Vicosa do Ceara/CE, Brazil, revealed substantial ecological harm and copper remains (Lei et al., 2024; Perlatti et al., 2021). Large amounts of bauxite residue are produced during the manufacturing of aluminium from bauxite ore, which presents serious problems for the environment and public health. These leftovers, which are frequently called "red mud," contribute to the deterioration of soil quality, which has a negative impact on crop yields and agricultural output. Because it causes long-term ecological imbalances, heavy metal, iron oxide, and aluminium hydroxide compound contamination of soil and water resources is a serious concern (Bala, Anshul, et al., 2024; Bala, Dogra, et al., 2024; Das, Dey, Bala, et al., 2024; Mohanty, Dey, et al., 2024). Moreover, human populations face significant health hazards due to the build up of dangerous materials in the food chain (Pradhan et al., 2024). The undesired byproducts that remain after the extraction of economically valuable ores are known as mine wastes, and they can be solid, liquid, or gaseous. Mining has been going on for thousands of years, and almost every nation has a history of mining waste that contains hazardous substances such different metals and metalloids. An estimated 20–25 Gt of solid mine waste are produced globally each year. Mine wastes, or waste rock; minerals from mined ore that are not recovered effectively; tailings; and chemicals and their residues from in-plant processes that are stored or were previously frequently discharged into rivers are among them (Abbadi & Mucsi, 2024).

Metal contamination will decrease the efficiency of soil functioning and the cycling of critical elements because bacteria will have to expend energy on cell maintenance and repair. However, when the contaminated metals are necessary components that are subject to homeostasis or have a near biochemical analogue, microbial populations survive and even thrive in the presence of metal contamination (Das & Ghosh, 2023a, 2023c, 2023f). In order to remediate various metals linked to mine wastes, it is advantageous to stimulate microbially reducing environments, such as in artificial wetlands. The effectiveness of this method has been shown in the lab and on a pilot field scale under low pH, circumneutral pH, and high pH circumstances. Interrupted microbial communities impair soil fertility and nutrient cycling, and reduced plant growth jeopardizes biodiversity and agricultural productivity (Moghimi Dehkordi et al., 2024). A sustainable solution is provided by microbial upcycling, which uses microorganisms to transform mining waste into useful products including biofuels, bioplastics, and refined metals via bioleaching and biomineralization (Das et al., 2015; Das, Mohanty, et al., 2023; Dey, Bissoyi, Koteswara Reddy, & Das, 2024; Dey, Dash, Ray, et al., 2024; Mohanty, Tripathy, et al., 2024). By combining biotechnology and environmental science, this novel strategy promises more environmental-friendly resource recovery and mining methods. The impact of metal mining wastes on the health of the earth may be lessened by the ability of microbial activity to mitigate these effects. In later phases of mineral extraction, mine waste is typically regarded as useless, but growing societal demand is predicted to make it useful. Decades of extensive mineral extraction have harmed the ecosystem by producing large amounts of mining waste, contaminated water supplies, and air. Millions of tonnes of solid byproducts are produced annually by mining businesses around the world from a variety of operations, including as blasting, drilling, and transportation (Das & Ghosh, 2018; Newsome & Falagán, 2021). As a result, using the best waste management techniques is essential to effectively recycling the enormous volumes of mining waste generated each year by various mining operations worldwide. This research critically assesses solid mining wastes and recommends the best waste management techniques, taking into account financial and regulatory limitations, to reduce hazards and optimize resource potential using a thorough end-to-end mining process approach (Jawadand & Randive, 2021). The extraction of rich minerals and resources required for human survival makes the mining industry an essential component of the global economy. However, the process generates a great deal of trash, which exacerbates sustainability and environmental challenges. The valuable components and minerals that are hidden underneath this garbage are only waiting to be released and turned into useful stuff. In addition to reducing the negative effects on the environment, this also makes use of the mining waste's economic potential. An increasing number of people are interested in investigating innovative solutions for mining waste (Godbole et al., 2024).

2 Microbial Upcycling: A Sustainable Solution

Microbial upcycling is an innovative approach worldwide using microorganisms like bacteria, fungi, and algae, which are capable of transforming waste materials into valuable products through their metabolism by recycling organic and inorganic compounds, industrial or mining byproducts, thereby reducing pollution, minimizing greenhouse gas emissions. This lowers the possible environmental impact of improper waste disposal and management, thus leading to a circular economy by reducing the dependence on raw materials (Aggarwal et al., 2023; Tsapekos et al., 2021). Various bioengineering processes, such as synthetic biology, fermentation technology, increase the microbial remediation of mining waste and are a better alternative to conventional practices, and promote economic viability through management and resource recovery of mining wastes (Dey, Das, et al., 2023; Jagatee et al., 2024; Mohanty, Mohanty, et al., 2024; Pradhan et al., 2024; Sahu, Dey, et al., 2024; Singh Jadaun et al., 2022). This innovative approach minimizes the global issue of waste management and transformation of waste into valuable products, thus promoting economic as well as environmental sustainability (Vo et al., 2024).

Sustainable mining practices provide an efficient production opportunity to deal with the negative impact on ecosystems and economic issues through the utilization of the natural metabolism of microbes that detoxify and stabilize hazardous substances present within mining waste. Moreover, microbial upcycling facilitates the recovery of valuable metals such as gold, copper, and nickel through biological processes like bioleaching and biomining, upgrading waste into economically valuable material (Das & Ghosh, 2022; Mohanty et al., 2017). This not only leads to a reduction in waste volume from mining but also has a basis in circular economy theory, reducing the need for fresh resource extraction. As opposed to chemical extraction or smelting, microbial processes are more cost-effective and usually require less energy and inputs of chemicals, while cutting greenhouse gas emissions by millions of tons, thus making mining a lesser carbon activity (Mallick et al., 2023, 2024; Mohanty, Mohanty, et al., 2024; Priyadarsini & Das, 2025; Sahu, Mallick, & Das, 2024). Microbial upcycling is very versatile and scalable, meaning it can be applied on different types of mining wastes and under different environmental conditions. This adaptability renders it a pragmatic solution for any kind of diverse mining activities around the world. Secondly, microbial techniques also result in the reduction of expansive land use such as vast tailings ponds or large landfills due to less volume of waste that needs to be stored (Nair et al., 2024; Yu, Zahidi, Chow, et al., 2024). Apart from all these technical and environmental benefits, microbial upcycling creates a better public image of mining companies because it provides evidence of their commitment towards sustainable practices and environmental stewardship. It will maintain an equilibrium in economic benefits and environmental conservation in the mining industry, ushering in more responsible utilization of resources in mining operations. This novel approach meets the urgent challenge of eliminating waste accumulation but also satisfies global sustainability targets, promoting cleaner and more efficient mining processes (Newsome & Falagán, 2021; Samarasekere, 2024).

3 Mechanisms of Microbial Biotransformation

Microbial biotransformation is one of the emerging techniques with much promise to address environmental pollution issues, wherein a large spectrum of contaminants could be degraded due to the versatility of microbial metabolism. In the current era, microbes and their catabolic pathways are being employed in bioremediation technologies to clean pollutants, such as polycyclic aromatic hydrocarbons, pharmaceuticals, radionuclides, hydrocarbons, and polychlorinated biphenyls from the environment (Das, Behera, & Das, 2024; John & Das, 2024; Priyadarshini & Das, 2024; Priyadarshini, Santhosh Kumar, & Das, 2024). The natural detoxifying features, fast growth rate, metabolic diversity of an organism significantly allow for the degradation of structurally diverse pollutants noticeable in microbes. Adaptation to dynamic environmental conditions through synergistic interactions between microbial consortia, horizontal gene transfer enables greater microbial resistance which is further complemented by biofilm formation and thereby offers efficiency in their biodegradation capabilities also develop eco-friendly and sustainable strategies for environmental cleanup (Singh Jadaun et al., 2022). Microbial biotransformation offers more advantages over traditional chemical methods through microbial enzymes that catalyse the conversion of organic compounds and thereby detoxify hazardous compounds and recover valuable metals in different processes like oxidation, reduction, hydrolysis, isomerization and also provides economic stability (Pradhan et al., 2024; Smitha et al., 2017). Microorganisms like sulphate-reducing bacteria (SRB), *Geobacter*, *Arthrobacter*, *Bacillus*, and *Shewanella*, *Aspergillus*, *Penicillium* reduces metal ions thus improving metal bioavailability for recovery thereby preventing environmental contamination, aids in soil rehabilitation, reduces toxicity and enhance nutrient cycling (Das & Mishra, 2010; Paul et al., 2024; Ruiz-Fresneda et al., 2023; Thakare et al., 2021).

4 Key Microorganisms in Mining Waste Valorisation

Microorganisms play promising role in stabilizing hazardous waste, convert toxic components present into less toxic forms and upcycling of mining waste into economically valuable materials through biotechnological processes than the traditional methods, offering a sustainable alternative for resource recovery and waste management thereby mitigating environmental impacts (Rathna & Nakkeeran, 2020). *Acidithiobacillus* spp., *Leptospirillum ferrooxidans*, fungus like Aspergillus and Penicillium are known to extract important metals like copper, gold, and nickel from sulphide minerals, helps in remediate acid mine drainage and improve metal recovery through detoxification (Pattanaik et al., 2021; Rathna & Nakkeeran, 2020). Cyanobacteria and green algae form biogenic minerals such as calcium carbonate through processes such as bio-precipitating and biomineralizing mining waste. *Sulfolobus* and *Thermoplasma* thrive in the harsh conditions thereby contributing to efficient waste treatment, removal of some heavy metals, decrease their bioavailability to a measurable degree and convert many harmful substances into less toxic forms (Mishra et al., 2022; Mishra & Das, 2022; Mishra, Dash, & Das, 2023; Mishra, Ghosh, et al., 2023). These microbial strategies help in production of secondary raw materials from mining waste, adhering to circular economy principles and promoting sustainable mining practices.(Levett et al., 2021; Nayak, 2023).

Emerging technologies such as bioleaching enhanced by nanoparticles and biosensor-mediated approaches represent a leap forward in this field. These advancements have expanded the potential of bioleaching for efficient and targeted extraction of metals (Das & Ghosh, 2018, 2023i; Ghosh et al., 2016; Mohanty, Tripathy, et al., 2024). The ongoing exploration of scaling these processes for commercial applications is accelerating their adoption and integration into industrial practices. The synergy between microbial metabolic processes and innovative nanotechnology tools is expected to revolutionize the field, enhancing efficiency and sustainability (Das & Ghosh, 2022). Beyond metal recovery, microbes are used to produce biofuels, bioplastics, and industrial materials from organic waste. For example, bacteria that generate methane degrade organic waste into biogas, a renewable energy source, while fungi and bacteria can produce biopolymers for packaging and construction. The integration of microbial technologies supports the circular economy by transforming mining waste into economic opportunities. Advances in genetic engineering and metabolic pathway optimization have enhanced the efficiency and

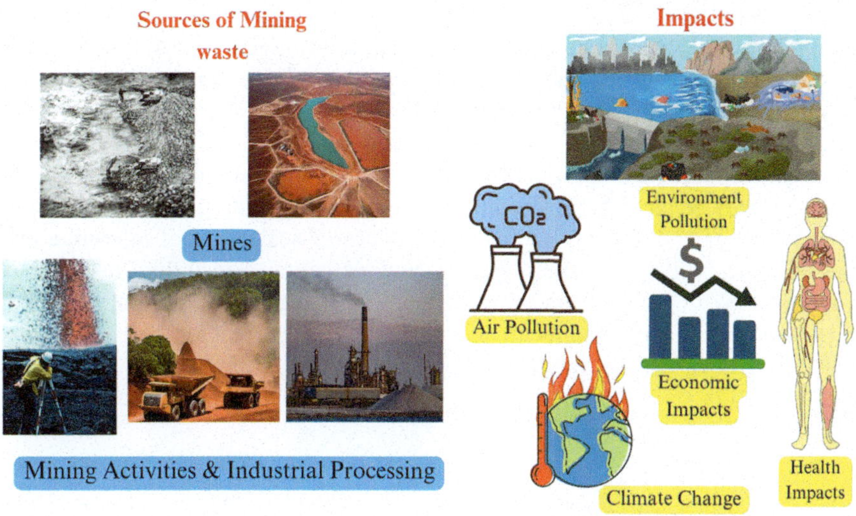

Fig. 1 Sources and impacts of mining waste

specificity of these microbes (Bal et al., 2018; Ghosh, Tripathy, et al., 2023). Acidophilic bacteria, sulphate-reducing bacteria, and other specialized microbes thrive in harsh mining waste conditions, contributing to metal recovery and waste detoxification (Ghosh & Das, 2020; Mohanty et al., 2018). Emerging research also explores microbial biofilms for sequestering rare earth elements from mining residues, offering a sustainable solution for their extraction. a sustainable solution for their extraction (Fig. 1). Overall, microbial biotechnology offers significant possible for reducing ecological impacts, minimizing waste management costs, and supporting sustainable mining practices (Gadd, 2010; Rawlings & Silver, 1995).

5 Value-Added Products from Microbial Upcycling

5.1 Biofuels and Energy

Biofuel production from mining waste through microbial upcycling is one of the promising ways to ensure sustainable energy and waste management. Mining activities produce large amounts of waste, which are rich in organic and inorganic compounds that can be used as feedstock for microbial processes. Certain microorganisms, such as bacteria and fungi, can metabolize these compounds convert them into biofuels, such as bioethanol, biodiesel, or biogas Das and Ghosh (2023d, 2023h). This process includes the digestion of complex organic matter by enzymatic reactions, which often leads to fermentation or another bioconversion technique (Abbadi & Mucsi, 2024; Yu, Zahidi, Fai, et al., 2024). However, microbial upcycling can also be applied towards environmental hazard reduction, that is, detoxification with a reduction in the metal toxicity of heavy metals included in mining waste while simultaneously producing valuables such as biofertilizers. These include enhanced genetic engineering and synthetic biology capabilities to achieve high-performance microorganisms that selectively produce biofuels under optimized conditions (Das, Mallavarapu, & Ghosh, 2023). Incorporation of these biotechnological approaches in the mining and waste management system will not only reduce the environmental footprint of mining but also diminish greenhouse gas emissions. However, the issues of scalability of microbial processes, variability of mining wastes, and economic feasibility need to be addressed. With research and development still ongoing, microbial upcycling could contribute significantly to renewable energy integration and sustainable resource utilization by facilitating a seamless transition process (Martin & Eklund, 2011).

5.2 *Biopolymers and Bioplastics*

It will turn out to be a very successful sustainable source for producing eco-friendly materials as well as for handling waste from the mining sectors. Organic chemicals and trace minerals found in mining wastes frequently act as nutrition for particular microorganisms. Fungi are the sort that metabolize waste materials to create biopolymers, while some of these are bacteria. Polyhydroxyalkanoates (PHAs), polylactic acid (PLA), and other biodegradable plastics are a few examples. Complex organic compounds are usually broken down into simpler substrates in this process, which is then followed by microbial fermentation and polymerisation (Abu-Thabit et al., 2022; Li et al., 2022; Pereyra-Camacho & Pardo, 2024). Microbial upcycling may purify mining wastes and lessen their environmental impact in addition to producing value-added biopolymers. Increasing the potential of microorganisms through developments in synthetic biology and metabolic engineering. These materials serve as a petroleum-free substitute for plastic in various industries, including packaging, agriculture, and medical (Das, Dey, Bala, et al., 2024; Das, Dutta, et al., 2023; Das & Ghosh, 2023a, 2023b; Ghosh, Sinha, et al., 2023; Priyadarsini & Das, 2024). By converting waste into valuable goods, microbial upcycling promotes a circular economy in mining operations. For widespread use, obstacles like the process's scalability, the economic feasibility of the mining waste, and its variability must be addressed. Microbial upcycling of mining waste could revolutionize waste management and sustainable material creation with continued research and development (Chwastowski et al., 2023).

5.3 Bio-Cement

Bio-cement production from mining waste is a sustainable method to address possible environmental issues, reduced carbon footprint and provide a greener alternate for eco-friendly construction materials. Biotechnology and material science aids in converting mining byproducts into valuable products (Ray et al., 2024; Sahu & Das, 2024; Swain et al., 2025). Biomineralization is a process simplified by bacteria species such as *Sporosarcina pasteurii* precipitate calcium carbonate, a key cement component of mining waste (Beatty et al., 2022; Konstantinou et al., 2023).

5.4 Fertilizers and Soil Amendments

Mining waste upcycling for the production of fertilizers and soil amendments provides a sustainable and innovative approach to agriculture and also results in reducing environmental footprint through diverting waste from landfills thus minimizing contamination as well as promotes resource utilization by methods such as formation of biochar, a nutrient-enriched, carbon-rich material made through pyrolyzing organic waste (Das & Ghosh, 2023a, 2023b; Ghosh, Tripathy, et al., 2023). It enhances the soil fertility, water retention capacity, and carbon sequestration (Table 1). Additionally, it reduces the demand for fertilizers through thereby upcy-

Table 1 Types of mining waste, their upcycling processes, value-added product recovery and their utilization

Waste type	Source	Conversion method	Valuable products	Application
Mine tailings	Metal ore processing	Geo-polymerization, smelting, bioremediation	Geopolymers, rare metals, bioleaching products	Construction, electronics, catalyst
Coal ash	Coal-fired power plants	Fly ash utilization, zeolite synthesis	Cement additive, zeolites	Cement industry, water purification
Slag	Steel and metal refining	Smelting, crushing, and sorting	Aggregates, metal recovery	Construction, metallurgy
Red mud	Bauxite refining	Acid/alkali treatment, polymer blending	Iron oxides, ceramics	Pigments, refractories
Acid mine drainage	Sulphide mineral oxidation	Neutralization, metal recovery	Sulfuric acid, recovered metals	Chemical industry, battery manufacturing
Waste rock	Open pit and underground mining	Crushing, sorting, carbon sequestration	Aggregates, carbonates	Road construction, carbon capture
Flotation waste	Mineral processing plants	Froth flotation, electrochemical processes	Recovered metals, industrial fillers	Metallurgy, construction
Spent ore	Heap leaching of metals	Secondary leaching, bioremediation	Extracted metals, fertilizers	Metal refining, agriculture

cling of mining tailings and thus enhancing food security and agricultural productivity by reducing toxicity (Abu-Thabit et al., 2022; Dey, Dash, & Das, 2024; Elmeknassi et al., 2024; Seow et al., 2022).

6 Technological Advancements in Microbial Upcycling

Around 20 billion tons of mining waste generated annually poses a significant environmental concern, with less than 15% recycled, mostly polluting landfills. Effective waste management strategies are crucial for recycling. Microbial biotechnology aids in biologically recycling and upcycling waste into valuable products, supporting a Circular Economy and Sustainable Development Goals. Microbes solubilize/precipitate metals through bioleaching, biomineralization, or biomining, generating or consuming energy for metal uptake and resistance.(Jawadand & Randive, 2021; Newsome & Falagán, 2021; Pereyra-Camacho & Pardo, 2024).

6.1 Genetic Engineering and Artificial Biology

Genetic engineering has advanced microorganisms like *Pseudomonas putida* and *Escherichia coli* for enhanced metal solubilization and pollutant degradation. These modifications increase resistance to mercury and arsenic, aid in gold and platinum recovery, and convert toxic substances into less harmful forms. Key bacteria in this process include *Pseudomonas*, *Rhodococcus*, *Ralstonia*, *Comamonas*, and *Dehalococcoides* (Pereira-García et al., 2024; Rafeeq et al., 2023; Saravanan et al., 2022; Shar et al., 2019; Xue et al., 2022).

Synthetic biology offers a promising approach to enhance bioremediation by increasing the expressions of genetically engineered microorganisms to achieve precise remediation targets. A key advancement in this field is the development of biosensors which are genetically modified microbes capable of detecting and indicating the presence of pollutants in mining waste through measurable indicators, such as fluorescence. These engineered microbes are designed to handle and degrade heavy metallic element and other pollutants in mining waste management (Jiménez-Díaz et al., 2022; Mishra et al., 2024; Yu et al., 2023).

6.2 Metagenomics and Metabolomics

Metagenomic and metabolomic are essential in bioremediation, addressing environmental challenges from mining waste as it provides comprehensive perceptions of metabolic pathways of microbial communities and helps in identifying novel microbial consortia for efficient bioleaching and bioremediation aiding in waste upcycling processes (Bharagava et al., 2019; Devarapalli et al., 2015; Gunjal et al., 2023).

6.3 *Nanotechnology*

Nanotechnology involves evolving techniques like biosynthetic nanoparticles to minimize environmental issues, which is also a cost-efficient solution. Nano-bioremediation offers insights to control mining pollutants present in air, soil, and water. Nanoparticles aid in increasing metal bioavailability, thereby helping in microbial leaching, which is a key factor for nano-bioremediation. Nanoparticles facilitate bioremediation by promoting microbial growth increasing microbial metabolism, reducing toxicity also stimulating biosurfactant production, improving contaminant solubility, reducing energy expenditure, and costs (Ningombam et al., 2024; Patel et al., 2020).

6.4 *Microbial Fuel Cells (MFCs)*

Microbial fuel cell for energy generation. Microbial Fuel Cells have surfaced as an innovative technology addressing ever-rising global energy demands and environmental concerns due to its unique ability of microorganisms to harness bioenergy. MFCs utilize bacteria to help convert organic and inorganic wastes into electrical energy. Through this method, mining waste can be used to not only create power but help treat the waste as well. Over the last recent decades, significant progress has been made in elucidating the extracellular electron transfer (EET) mechanisms in electroactive bacteria (EAB) such as *Geobacter* sp., have served as the impetus for the development and study of soil microbial fuel cells (MFCs), implicating them as an environmentally benign bio-electrochemical technology. Soil microbial fuel cells (MFCs) have been shown to facilitate heavy metal reduction and migration in soil by providing electron donors/acceptors that enhance bioremediation. This mechanism reduces oxygen or metal ions in the aerobic surface layer (cathode) by utilizing electrons generated from the oxidation of organic matter in the anaerobic subsoil (anode) by microorganisms via an external circuit. The electric field generated by the MFC induces heavy metals to migrate from the subsoil to the surface layer, which is in effect a redox reaction that generates current (Decesaro et al., 2017; Srivastava et al., 2022; Sun et al., 2022).

6.5 *Phytoremediation*

Soils contaminated with heavy metals can be treated by environmentally sustainable phytoremediation methods based on the use of plants. In phyto-mining, hyperaccumulating plants are grown on polluted substrates, with concentrations of valuable metals which accumulates in the plants that can then be harvested for recovery. Plant Species such as *Brassica napus*, *Brassica juncea*, *Festuca arundinacea*, *Pinus*

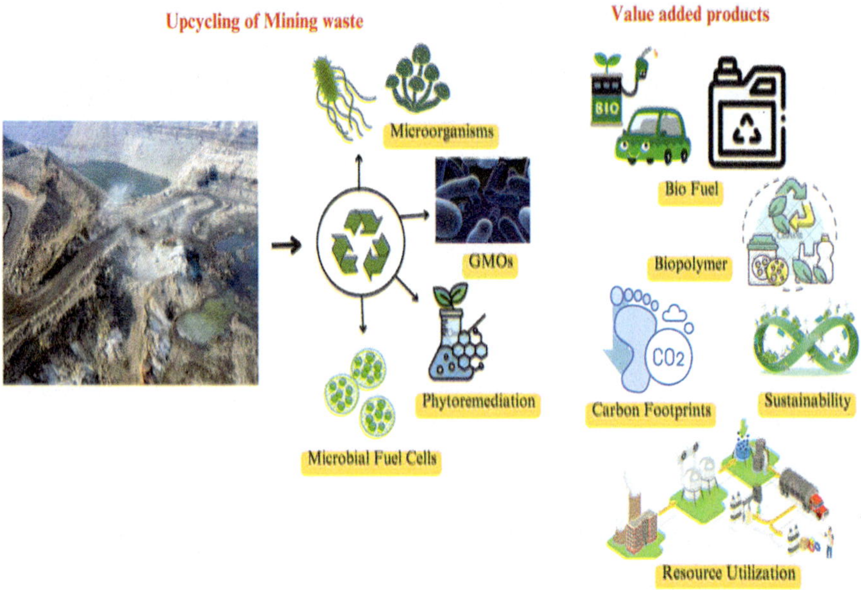

Fig. 2 Upcycling of mining waste and recovery of value-added products

massoniana, *Phragmites australis*, *Medicago sativa*, *Lolium perenne*, *Robinia pseudoacacia*, *Sedum alfredii*, *Chrysopogon zizanioides*, and *Helianthus annuus* are effective in absorb, accumulate, and transform pollutants, offering a green alternative to traditional remediation techniques. Research of heavy metal-contaminated soil with *Helianthus annuus*, *Zea mays*, and *Brassica juncea* was found to have improved phytoremediation potential (Fig. 2). This process produces extremely low levels of toxic effluents generated during metal extraction compared to traditional mining, which has low impact on the environment (Haghighizadeh et al., 2024).

7 Economic Viability and Scalability of Microbial Processes

Bioremediation is a sustainable strategy to minimize the environmental impacts of mining waste through the metabolic capability of autochthonous microorganisms to degrade, detoxify, or immobilize anthropogenic contaminants. Mining produces huge amounts of waste that is rich in heavy metals, sulphates and other toxic materials that possess a direct environmental and health risk. Bioremediation serves as a cost-effective and environmentally sound approach to the conventional chemical and physical treatment methods developed to solve these problems through the use of bacteria, fungi, and plants. Processes like bioleaching, biosorption or biomineralization allow microbial communities to transform toxic compounds into less harmful forms, while plants are used in phytoremediation to stabilize or extract contaminants (Cisternas et al., 2022; Elmeknassi et al., 2024; Kuppan et al., 2024). Such biological processes reduce the mobility and bioavailability of pollutants, decreasing their adverse effects in soil, water, and ecosystems. Recent developments in genetic engineering and biotechnology have improved ability of microbes to create optimize environmental conditions for tailored microbial strains with high efficiency and specificity to remove insoluble to complex waste matrices. Bioremediation efficacy is highly site-specific, resting on parameters (e.g. pH, temperature contaminants) needing to be optimized and carefully assessed. The translational and implementation potential of bioremediation as part of mining waste management strategies represents a significant step towards achieving ecological sustainability and minimizing the legacy of mining activity in terms of its impact on the environment (Elmeknassi et al., 2024; Velenturf & Purnell, 2021; Wu et al., 2021; Zhou et al., 2024).

The microbial bioeconomy which involves employing microorganisms to manufacture biodegradable and sustainable products offers a legitimate substitute for conventional production methodologies consistent with global economic tenets. Following up, new tendencies like circular economy, recyclability, recycling, and re-use are very important to the sustainability of the mining and metal extraction industries. This process provides an innovative and environmentally friendly method to convert low-grade resources to those that can be used in higher value industries through adapted microbial consortia (Cisternas et al., 2022; Velenturf & Purnell, 2021). Although microbial bioremediation has the potential to reduce the environmental impacts caused by mining waste, practical use is highly influenced by its economic viability and scale-up potential. Genetically Modified Organisms (GMOs) are essential building blocks to a sustainable bioeconomy, and are used to provide effective manufacturing of food and feed additives, pharmaceutical drugs, biomedicine, bioenergy, biofuels, bioplastics, environmental remediation, waste management, construction, transportation, forestry, agriculture, and other recyclable bio-based materials and products. The microbial bioeconomy expands economic sustainability through bio-based operations including biocatalysis and biotransformation and anaerobic digestion along with bioremediation since they utilize waste products and biomass which decreases environmental impact together with cost reduction from raw material expenses and value-added product generation (Akinsemolu et al., 2023; Dey, Dash, & Das, 2024; Nair et al., 2024).

8 Sustainable Mining Waste Management: Towards a Green Mining Future

The concern for the environmental management of mining waste containing heavy metals and toxic substances is deemed to be of paramount importance, is a huge challenge and has brought about massive pollution and the wasteful utilization of resources. In the field of green chemistry and sustainable methodologies, such as bio-based metal extraction in the mining industry, technological developments have illustrated the key role that they play in the preservation of the environment and efficient utilization of sustainable resources, and in achieving the high efficiency and economic feasibility of recycling processes, to enhance the efficiency and cost of recycling processes more effectively and at reduced costs (Zhou et al., 2023). Green mining, or sustainable mining, involves implementing mining performs that focus on environmental conservation, economic feasibility, and social accountability. The primary goal is to confirm minimal environmental impact from mining operations, safeguard miners and nearby communities, and avoid financial losses from irresponsible practices. Environmental conservation aims to limit habitat destruction and minimize pollution (Onifade et al., 2024). Future green mining will emphasize circular economy principles, promoting waste minimization, recycling, and material lifespan extension. These advanced recycling technologies extract valuable materials from mining waste, and provide insights of waste-to-energy solutions that will minimize environmental impact. These transformations not only help to conserve the environment, but also help the mining industry to deal with the challenges of resource constraints and face the document mining industry as socially responsible and sustainable one. Various measures have been implemented to ensure that sustainable approaches in the industry of mining help in minimalizing environmental impacts as well as optimal processes. There are some measures including biotechnologies for metal extraction or for pollutant removal, smart management systems for waste, renewable energy sources, energy saving in the mining processes, use of water in an effective way, recycling technology development, and safety and sustainability of mine waste. (Haghighizadeh et al., 2024).

9 Conclusion

Microbial upcycling is the use of microorganisms' metabolic capability to decompose, transfer, and immobile potentially dangerous mining debris, converting it into useful products. This unconventional approach goes beyond offsetting the detrimental environmental effects of mine waste and turns these wastes into valuable products, minimizing the use of raw materials, and reducing carbon footprint of the production. Bioprocesses by the microbial may convert mineral-rich substrates into biofuels, bioplastics and biofertilizers. The bioproducts are offered as sustainable alternatives to traditional materials, which mean that they minimize the use of non-renewable resources and do not have as great an environmental impact. However, the progress with current microbial upcycling is challenged by remaining issues that should be addressed with future developments in synthetic biology fermentation as well as waste processing technology. Sustainable mining waste management using microbial upcycling will soon be played by a growing role in the biotechnological innovations, supporting environmental sustainability and economic growth. Furthermore, this is a holistic approach that not only mitigates the impact of mining activities but it also increases the share of cleaner, more effective mining practices as well as global targets of sustainability.

References

Abbadi, A., & Mucsi, G. (2024). A review on complex utilization of mine tailings: Recovery of rare earth elements and residue valorization. *Journal of Environmental Chemical Engineering, 12*(3), 113118. https://doi.org/10.1016/J.JECE.2024.113118

Abu-Thabit, N. Y., Pérez-Rivero, C., Uwaezuoke, O. J., & Ngwuluka, N. C. (2022). From waste to wealth: Upcycling of plastic and lignocellulosic wastes to PHAs. *Journal of Chemical Technology & Biotechnology, 97*(12), 3217–3240. https://doi.org/10.1002/JCTB.6966

Agboola, O., Babatunde, D. E., Isaac Fayomi, O. S., Sadiku, E. R., Popoola, P., Moropeng, L., Yahaya, A., & Mamudu, O. A. (2020). A review on the impact of mining operation: Monitoring, assessment and management. *Results in Engineering, 8*, 100181. https://doi.org/10.1016/J.RINENG.2020.100181

Aggarwal, N., Pham, H. L., Ranjan, B., Saini, M., Liang, Y., Hossain, G. S., Ling, H., Foo, J. L., & Chang, M. W. (2023). Microbial engineering strategies to utilize waste feedstock for sustainable bioproduction. *Nature Reviews Bioengineering, 2*(2), 155–174. 10.1038/s44222-023-00129-2.

Akinsemolu, A., Onyeaka, H., Fagunwa, O., & Adenuga, A. H. (2023). Toward a resilient future: The promise of microbial bioeconomy. *Sustainability, 15*(9), 7251. https://doi.org/10.3390/SU15097251

Bal, B., Ghosh, S., & Das, A. P. (2018). Microbial recovery and recycling of manganese waste and their future application: A review. *Geomicrobiological Journal, 36*(1), 85–96. https://doi.org/10.1080/01490451.2018.1497731

Bala, K., Anshul, A., Verma, S., Dogra, A., Das, A. P., Kumar, R., & Nandini, A. (2024). Sustainable management for urban plastic waste generation. In *Sustainable management of urban plastic waste through circular economic approaches* (pp. 211–236). Routledge. https://doi.org/10.1201/9781003465447-13

Bala, K., Dogra, A., Lata, P., & Das, A. P. (2024). Discussing sources and biological-chemical remediation approaches for microplastic pollution. In *Sustainable microbial technology for synthetic and cellulosic microfiber bioremediation* (pp. 37–59). Springer. https://doi.org/10.1007/978-3-031-62600-5_3

Beatty, D. N., Williams, S. L., & Srubar, W. V. (2022). Biomineralized materials for sustainable and durable construction. *Annual Review of Materials Research, 52*, 411–439. https://doi.org/10.1146/ANNUREV-MATSCI-081720-105303/1

Belle, G., Fossey, A., Esterhuizen, L., & Moodley, R. (2021). Contamination of groundwater by potential harmful elements from gold mine tailings and the implications to human health: A case study in Welkom and Virginia, Free State Province, South Africa. *Groundwater for Sustainable Development, 12*, 100507. https://doi.org/10.1016/J.GSD.2020.100507

Bharagava, R. N., Purchase, D., Saxena, G., & Mulla, S. I. (2019). Applications of metagenomics in microbial bioremediation of pollutants: From genomics to environmental cleanup. In *Microbial diversity in the genomic era* (pp. 459–477). Elsevier. https://doi.org/10.1016/B978-0-12-814849-5.00026-5

Biswas, B., Qi, F., Biswas, J. K., Wijayawardena, A., Khan, M. A. I., & Naidu, R. (2018). The fate of chemical pollutants with soil properties and processes in the climate change paradigm—A review. *Soil systems, 2*(3), 51. https://doi.org/10.3390/SOILSYSTEMS2030051

Chwastowski, J., Guzik, M., Bednarz, S., & Staroń, P. (2023). Upcycling waste streams from a biorefinery process—A case study on cadmium and Lead biosorption by two types of biopolymer post-extraction biomass. *Molecules, 28*(17), 6345. https://doi.org/10.3390/MOLECULES28176345/S1

Cisternas, L. A., Ordóñez, J. I., Jeldres, R. I., & Serna-Guerrero, R. (2022). Toward the implementation of circular economy strategies: An overview of the current situation in mineral processing. *Mineral Processing and Extractive Metallurgy Review, 43*(6), 775–797. https://doi.org/10.1080/08827508.2021.1946690

Das, A., Dey, S., Bala, K., Gnanasekaran, R., & Das, A. P. (2024). Microbial bioremediation technology for sustainable treatment and management of synthetic microfiber waste. In *Sustainable microbial technology for synthetic and cellulosic microfiber bioremediation* (pp. 221–238). Springer. https://doi.org/10.1007/978-3-031-62600-5_12

Das, A., Dey, S., & Das, A. P. (2024). Ecotoxicological impacts of synthetic microfiber pollutants and development of sustainable mitigation strategies. *Environmental Chemistry and Ecotoxicology*. https://doi.org/10.1016/J.ENCECO.2024.12.003

Das, A., Dey, S., Gacem, A., Yadav, K. K., Nath, H., Ram, D. K., & Das, A. P. (2025). Revolutionizing solid hydrocarbon sourcing: Plastic's journey from waste to upcycled treasure. *Environmental Science and Engineering, 218*, 331–351. https://doi.org/10.1007/978-3-031-81615-4_14

Das, A., & Ghosh, S. (2022). Role of microorganisms in extenuation of mining and industrial wastes. *Geomicrobiology Journal, 39*, 173–175. https://doi.org/10.1080/01490451.2022.2038953

Das, A., & Mishra, S. (2010). Biodegradation of the metallic carcinogen hexavalent chromium Cr(VI) by an indigenously isolated bacterial strain. *Journal of Carcinogenesis, 9*, 6. https://doi.org/10.4103/1477-3163.63584

Das, A. P., Behera, I. D., & Das, N. P. (2024). *Renewable energy generation and value addition from environmental microfiber pollution through advanced greener solution* (p. 278). Springer.

Das, A., Ray, S. K., Mohanty, M., Mohanty, J., Dey, S., & Das, A. P. (2024). Ecotoxicity of microplastic wastes and their sustainable management: A review. *Environmental Chemistry and Ecotoxicology, 6*, 144–152. https://doi.org/10.1016/J.ENCECO.2024.05.003

Das, A. P., Dutta, K., Khatun, R., Behera, I. D., Singh, S., & Mishra, S. (2023). Microfiber pollution and its microbial mitigation: A review on current trends and future prospects. *Journal of the Taiwan Institute of Chemical Engineers, 166*, 105104. https://doi.org/10.1016/J.JTICE.2023.105104

Das, A. P., & Ghosh, S. (2018). Bioleaching of manganese from mining waste materials. *Materials Today: Proceedings, 5*(1), 2381–2390. https://doi.org/10.1016/J.MATPR.2017.11.459

Das, A. P., & Ghosh, S. (2023a). Cost-effectiveness of the biomining process. In *Manganese mining microorganisms* (pp. 143–151). Elsevier. https://doi.org/10.1016/B978-0-12-822147-1.00008-X

Das, A. P., & Ghosh, S. (2023b). Evolution of biomining technology. In *Manganese mining microorganisms* (pp. 5–19). Elsevier. https://doi.org/10.1016/B978-0-12-822147-1.00012-1

Das, A. P., & Ghosh, S. (2023c). Factors affecting bioleaching. In *Manganese mining microorganisms* (pp. 77–86). Elsevier. https://doi.org/10.1016/B978-0-12-822147-1.00004-2

Das, A. P., & Ghosh, S. (2023d). Introduction. In *Manganese mining microorganisms* (pp. 1–4). Elsevier. https://doi.org/10.1016/B978-0-12-822147-1.00005-4

Das, A. P., & Ghosh, S. (2023e). Manganese and its application. In *Manganese mining microorganisms* (pp. 29–40). Elsevier. https://doi.org/10.1016/B978-0-12-822147-1.00007-8

Das, A. P., & Ghosh, S. (2023f). Mechanism of manganese bioleaching. In *Manganese mining microorganisms* (pp. 87–98). Elsevier. https://doi.org/10.1016/B978-0-12-822147-1.00014-5

Das, A. P., & Ghosh, S. (2023g). Metagenomic insights into unculturable microbial diversity. In *Manganese mining microorganisms* (pp. 63–76). Elsevier. https://doi.org/10.1016/B978-0-12-822147-1.00002-9

Das, A. P., & Ghosh, S. (2023h). Microorganisms in manganese biomining processes. In *Manganese mining microorganisms* (pp. 55–61). Elsevier. https://doi.org/10.1016/B978-0-12-822147-1.00013-3

Das, A. P., & Ghosh, S. (2023i). Role of metabolites and metabolomics for understanding of manganese bioleaching. In *Manganese mining microorganisms* (pp. 117–127). Elsevier. https://doi.org/10.1016/B978-0-12-822147-1.00011-X

Das, A. P., Ghosh, S., Mohanty, S., & Sukla, L. B. (2015). Advances in manganese pollution and its bioremediation. In L. B. Sukla, N. Pradhan, S. Panda, & B. K. Mishra (Eds.), *Environmental microbe biotechnology* (1st ed., pp. 313–328). Springer. https://doi.org/10.1007/978-3-319-19018-1_16

Das, A. P., Mallavarapu, M., & Ghosh, S. (2023). Ecotoxicity of mining pollutants on the environment and their remediation. *Environmental Chemistry and Ecotoxicology, 5*, 165–167. https://doi.org/10.1016/J.ENCECO.2023.08.002

Das, A. P., Mohanty, A. K., van Hullebusch, E. D., Figueiredo, G., Das, A. P., Mohanty, A. K., van Hullebusch, E. D., & Figueiredo, G. (2023). Bioremediation as an emerging technology for the removal of synthetic microplastic pollutants from marine ecosystem. *MarPB, 194*, 115297. https://doi.org/10.1016/J.MARPOLBUL.2023.115297

Decesaro, A., Machado, T. S., Cappellaro, Â. C., Reinehr, C. O., Thomé, A., & Colla, L. M. (2017). Biosurfactants during in situ bioremediation: Factors that influence the production and challenges in evalution. *Environmental Science and Pollution Research, 24*(26), 20831–20843. https://doi.org/10.1007/S11356-017-9778-7/METRICS

Devarapalli, P., Kumavath, R. N., Devarapalli, P., & Kumavath, R. N. (2015). Metagenomics — A technological drift in bioremediation. In *Advances in bioremediation of wastewater and polluted soil*. IntechOpen.

Dey, S., Bissoyi, A., Koteswara Reddy, G., & Das, A. P. (2024). Synthetic microfibers: An urban emerging pollutant demanding for sustainable environmental solutions. In *Sustainable microbial technology for synthetic and cellulosic microfiber bioremediation* (pp. 107–121). Springer. https://doi.org/10.1007/978-3-031-62600-5_7

Dey, S., Das, A., Mallick, K., Sahu, A., & Das, A. P. (2023). *Environmental petroleum waste: Pollution, toxicity, sustainable remediation* (pp. 159–175). Springer. https://doi.org/10.1007/978-3-031-48220-5_7

Dey, S., Dash, A., & Das, A. P. (2024). Bioremediation as a sustainable and circular solution for mining waste management and bioenergy generation. In *Sustainable management of mining waste and tailings: A circular economy approach* (pp. 107–119). Routledge. https://doi.org/10.1201/9781003442455-6

Dey, S., Dash, N., Ray, S. K., Priyadarsini, S., Chhotaray, C., & Das, A. P. (2024). Occurrence and quantification of tire wear plastics pollutants from highway city dust of Bhubaneswar City of Odisha. In *Sustainable microbial technology for synthetic and cellulosic microfiber bioremediation* (pp. 77–89). Springer. https://doi.org/10.1007/978-3-031-62600-5_5

Dey, S., Tripathy, B., Kumar, M. S., & Das, A. P. (2023). Ecotoxicological consequences of manganese mining pollutants and their biological remediation. *Environmental Chemistry and Ecotoxicology, 5*, 55–61. https://doi.org/10.1016/J.ENCECO.2023.01.001

Elmeknassi, M., Elghali, A., de Carvalho, H. W. P., Laamrani, A., & Benzaazoua, M. (2024). A review of organic and inorganic amendments to treat saline-sodic soils: Emphasis on waste valorization for a circular economy approach. *Science of the Total Environment, 921*, 171087. https://doi.org/10.1016/J.SCITOTENV.2024.171087

Gadd, G. M. (2010). Metals, minerals and microbes: Geomicrobiology and bioremediation. *Microbiology, 156*(3), 609–643. https://doi.org/10.1099/MIC.0.037143-0/CITE/REFWORKS

Gao, X., Jiang, L., Mao, Y., Yao, B., & Jiang, P. (2021). Progress, challenges, and perspectives of bioleaching for recovering heavy metals from mine tailings. *Adsorption Science & Technology, 6*, 1–13. https://doi.org/10.1155/2021/9941979

Ghosh, S., & Das, A. P. (2020). Microbial metagenomics: Current advances in investigating microbial ecology and population dynamics. In S. K. Nayak & B. B. Mishra (Eds.), *Frontiers in soil and environmental microbiology* (1st ed., pp. 247–254). CRC Press.

Ghosh, S., Mohanty, S., Akcil, A., Sukla, L. B., & Das, A. P. (2016). A greener approach for resource recycling: Manganese bioleaching. *Chemosphere, 154*, 628–639. https://doi.org/10.1016/J.CHEMOSPHERE.2016.04.028

Ghosh, S., Sinha, J. K., Ghosh, S., Vashisth, K., Han, S., & Bhaskar, R. (2023). Microplastics as an emerging threat to the global environment and human health. *Sustainability, 15*(14), 10821. https://doi.org/10.3390/SU151410821

Ghosh, S., Tripathy, B., Dey, S., & Das, A. P. (2023). Comparative investigation of fungal and bacterial manganese biomining mechanisms. In *Manganese mining microorganisms* (pp. 99–115). Elsevier. https://doi.org/10.1016/b978-0-12-822147-1.00016-9

Godbole, P., Deshpande, K., Jawadand, S., Meshram, P., Dora, M. L., Meshram, R., Meshram, T., Nandi, A. K., & Randive, K. (2024). Innovative approach to transform mining waste into value added products. *Springer Proceedings in Earth and Environmental Sciences, 2412*, 217–239. https://doi.org/10.1007/978-3-031-50262-0_18

Gunjal, A., Gupta, S., Nweze, J. E., & Nweze, J. A. (2023). Metagenomics in bioremediation: Recent advances, challenges, and perspectives. In *Metagenomics to bioremediation: Applications, cutting edge tools, and future outlook* (pp. 81–102). Elsevier. https://doi.org/10.1016/B978-0-323-96113-4.00018-4

Haghighizadeh, A., Rajabi, O., Nezarat, A., Hajyani, Z., Haghmohammadi, M., Hedayatikhah, S., Asl, S. D., & Aghababai Beni, A. (2024). Comprehensive analysis of heavy metal soil contamination in mining environments: Impacts, monitoring techniques, and remediation strategies. *Arabian Journal of Chemistry, 17*(6), 105777. https://doi.org/10.1016/J.ARABJC.2024.105777

Jagatee, S., Priyadarshini, S., Rath, C. C., & Das, A. P. (2024). Synthetic microfiber: An enduring environmental problem linked to sustainable development. *Environmental Science and Engineering, 2333*, 93–112. https://doi.org/10.1007/978-3-031-51792-1_6

Jawadand, S., & Randive, K. (2021). A sustainable approach to transforming mining waste into value-added products. In *Innovations in sustainable mining*. Springer. https://doi.org/10.1007/978-3-030-73796-2_1

Jiménez-Díaz, V., Pedroza-Rodríguez, A. M., Ramos-Monroy, O., & Castillo-Carvajal, L. C. (2022). Synthetic biology: A new era in hydrocarbon bioremediation. *Processes, 10*(4), 712. https://doi.org/10.3390/PR10040712

John, A., & Das, A. P. (2024). Government strategies for achieving circular economy for plastic waste management in Southeast Asia. In *Sustainable management of urban plastic waste through circular economic approaches* (pp. 72–82). Routledge. https://doi.org/10.1201/9781003465447-6

Kalisz, S., Kibort, K., Mioduska, J., Lieder, M., & Małachowska, A. (2022). Waste management in the mining industry of metals ores, coal, oil and natural gas - A review. *Journal of Environmental Management, 304*, 114239. https://doi.org/10.1016/J.JENVMAN.2021.114239

Konstantinou, C., Wang, Y., Konstantinou, C., & Wang, Y. (2023). Unlocking the potential of microbially induced calcium carbonate precipitation (MICP) for hydrological applications: A review of opportunities, challenges, and environmental considerations. *Hydrology, 10*(9), 178. https://doi.org/10.3390/HYDROLOGY10090178

Kuppan, N., Padman, M., Mahadeva, M., Srinivasan, S., & Devarajan, R. (2024). A comprehensive review of sustainable bioremediation techniques: Eco friendly solutions for waste and pollution management. *Waste Management Bulletin, 2*(3), 154–171. https://doi.org/10.1016/J.WMB.2024.07.005

Lei, B., Li, X., Guo, Y., Qu, F., Zhao, C., Tam, V. W. Y., Wu, V., & Li, W. (2024). Recycling of copper tailing as filler material in asphalt paving mastic: A sustainable solution for mining waste recovery. *Case Studies in Construction Materials, 20*, e03237. https://doi.org/10.1016/J.CSCM.2024.E03237

Levett, A., Gleeson, S. A., & Kallmeyer, J. (2021). From exploration to remediation: A microbial perspective for innovation in mining. *Earth-Science Reviews, 216*, 103563. https://doi.org/10.1016/J.EARSCIREV.2021.103563

Li, Y., Shi, X., Ling, Q., Li, S., Wei, J., Xin, M., Xie, D., Chen, X., Liu, K., & Yu, F. (2022). Bacterial extracellular polymeric substances: Impact on soil microbial community composition and their potential role in heavy metal-contaminated soil. *Ecotoxicology and Environmental Safety, 240*, 113701. https://doi.org/10.1016/J.ECOENV.2022.113701

Mallick, K., Sahu, A., Dubey, N. K., & Das, A. P. (2023). Harvesting marine plastic pollutants-derived renewable energy: A comprehensive review on applied energy and sustainable approach. *Journal of Environmental Management, 348*, 119371. https://doi.org/10.1016/J.JENVMAN.2023.119371

Mallick, K., Sahu, S., Sahu, A., Dey, S., & Das, A. P. (2024). A critical review of marine microfiber pollution routes, toxicity, and its sustainable remediation. *Environmental Science and Engineering, 2333*, 189–211. https://doi.org/10.1007/978-3-031-51792-1_10

Martin, M., & Eklund, M. (2011). Improving the environmental performance of biofuels with industrial symbiosis. *Biomass and Bioenergy, 35*(5), 1747–1755. https://doi.org/10.1016/J.BIOMBIOE.2011.01.016

Matinde, E., Simate, G. S., & Ndlovu, S. (2018). Mining and metallurgical wastes: A review of recycling and re-use practices. *Journal of the Southern African Institute of Mining and Metallurgy, 118*(8), 825–844. 10.17159/2411-9717/2018/V118N8A5.

Mishra, S., & Das, A. P. (2022). Treatment of the wastewater polluted with synthetic microfiber released from washing machine. In B. B. Das, H. Hettiarachchi, P. K. Sahu, & S. Nanda (Eds.), *Lecture notes in civil engineering* (Vol. 207, pp. 109–117). Springer. https://doi.org/10.1007/978-981-16-7509-6_9

Mishra, S., Dash, D., Al-Tawaha, A. R. M. S., & Das, A. P. (2022). A review on heavy metal ion adsorption on synthetic microfiber surface in aquatic environments. *Applied Biochemistry and Biotechnology, 194*(10), 4639–4654. https://doi.org/10.1007/S12010-022-04029-W/METRICS

Mishra, S., Dash, D., & Das, A. P. (2023). Aquatic microbial diversity on plastisphere: Colonization and potential role in microplastic biodegradation. *Geomicrobiology Journal*. https://doi.org/10.1080/01490451.2023.2209750

Mishra, S., Ghosh, S., van Hullebusch, E. D., Singh, S., & Das, A. P. (2023). A critical review on the recovery of base and critical elements from electronic waste-contaminated streams using microbial biotechnology. *Applied Biochemistry and Biotechnology*. https://doi.org/10.1007/S12010-023-04440-X

Mishra, S., Patel, A., Bhatt, P., Chen, S., & Srivastava, P. K. (2024). Perspective evaluation of synthetic biology approaches for effective mitigation of heavy metal pollution. *Reviews of Environmental Contamination and Toxicology, 262*(1), 1–33. 10.1007/S44169-024-00072-2.

Moghimi Dehkordi, M., Pournuroz Nodeh, Z., Soleimani Dehkordi, K., Salmanvandi, H., Rasouli Khorjestan, R., & Ghaffarzadeh, M. (2024). Soil, air, and water pollution from mining and industrial activities: Sources of pollution, environmental impacts, and prevention and control methods. *Results in Engineering, 23*, 102729. https://doi.org/10.1016/J.RINENG.2024.102729

Mohanty, J., Dey, S., Bala, K., & Das, A. P. (2024). Recent advances in microplastic pollution for its sustainable management. In *Sustainable management of urban plastic waste through circular economic approaches* (pp. 27–48). Routledge. https://doi.org/10.1201/9781003465447-3

Mohanty, M., Mohanty, J., Dey, S., Dutta, K., Shah, M. P., & Das, A. P. (2024). The face mask: A tale from protection to pollution and demanding sustainable solution. *Emerging Contaminants, 10*(2), 100298. https://doi.org/10.1016/J.EMCON.2023.100298

Mohanty, M., Tripathy, B., Koteswara Reddy, G., & Das, A. P. (2024). Greener technologies for establishing restraint over microplastic pollution. In *Sustainable microbial technology for synthetic and cellulosic microfiber bioremediation* (pp. 239–258). Springer. https://doi.org/10.1007/978-3-031-62600-5_13

Mohanty, S., Ghosh, S., Bal, B., & Das, A. P. (2018). A review of biotechnology processes applied for manganese recovery from wastes. *Reviews in Environmental Science and Bio/Technology, 17*(4), 791–811. 10.1007/S11157-018-9482-1.

Mohanty, S., Ghosh, S., Nayak, S., & Das, A. P. (2017). Bioleaching of manganese by aspergillus sp. isolated from mining deposits. *Chemosphere, 172*, 302–309. https://doi.org/10.1016/J.CHEMOSPHERE.2016.12.136

Nair, V., Singh, A., & Gowda, N. K. (2024). Bioprocesses for sustainable management of mine waste in contaminated environmental matrices. In *Sustainable management of mining waste and tailings: A circular economy approach* (pp. 146–178). Springer. https://doi.org/10.1201/9781003442455-8

Nayak, N. P. (2023). Microorganisms and their application in mining and allied industries. *Materials Today: Proceedings, 72*, 2886–2891. https://doi.org/10.1016/J.MATPR.2022.07.392

Newsome, L., & Falagán, C. (2021). The microbiology of metal mine waste: Bioremediation applications and implications for planetary health. *GeoHealth, 5*(10), e2020GH000380. https://doi.org/10.1029/2020GH000380

Ningombam, L., Mana, T., Apum, G., Ningthoujam, R., & Disco Singh, Y. (2024). Nano-bioremediation: A prospective approach for environmental decontamination in focus to soil, water and heavy metals. *Environmental Nanotechnology, Monitoring & Management, 21*, 100931. https://doi.org/10.1016/J.ENMM.2024.100931

Onifade, M., Zvarivadza, T., Adebisi, J. A., Said, K. O., Dayo-Olupona, O., Lawal, A. I., & Khandelwal, M. (2024). Advancing toward sustainability: The emergence of green mining technologies and practices. *Green and Smart Mining Engineering, 1*(2), 157–174. https://doi.org/10.1016/J.GSME.2024.05.005

Patel, H. K., Kalaria, R. K., & Khimani, M. R. (2020). Nanotechnology: A promising tool for bioremediation. In *Removal of toxic pollutants through microbiological and tertiary treatment: New perspectives* (pp. 515–547). Elsevier. https://doi.org/10.1016/B978-0-12-821014-7.00020-4

Pattanaik, A., Samal, K., Sukla, L. B., & Pradhan, D. (2021). Advancements and use of OMIC technologies in the field of bioleaching: A review. *Biointerface Research in Applied Chemistry, 11*(3), 10185–10204. https://doi.org/10.33263/BRIAC113.1018510204

Paul, A., Dey, S., Ram, D. K., & Das, A. P. (2024). Hexavalent chromium pollution and its sustainable management through bioremediation. *Geomicrobiology Journal, 41*(4), 324–334. https://doi.org/10.1080/01490451.2023.2218377

Pereira-García, C., del Amo, E. H., Vigués, N., Rey-Velasco, X., Rincón-Tomás, B., Pérez-Cruz, C., Sanz-Sáez, I., Hu, H., Bertilsson, S., Pannier, A., Soltmann, U., Sánchez, P., Acinas, S. G., Bravo, A. G., Alonso-Sáez, L., & Sánchez, O. (2024). Unmasking the physiology of mercury detoxifying bacteria from polluted sediments. *Journal of Hazardous Materials, 467*, 133685. https://doi.org/10.1016/J.JHAZMAT.2024.133685

Pereyra-Camacho, M. A., & Pardo, I. (2024). Plastics and the sustainable development goals: From waste to wealth with microbial recycling and upcycling. *Microbial Biotechnology, 17*(4), e14459. https://doi.org/10.1111/1751-7915.14459

Perlatti, F., Martins, E. P., de Oliveira, D. P., Ruiz, F., Asensio, V., Rezende, C. F., Otero, X. L., & Ferreira, T. O. (2021). Copper release from waste rocks in an abandoned mine (NE, Brazil) and its impacts on ecosystem environmental quality. *Chemosphere, 262*, 127843. https://doi.org/10.1016/J.CHEMOSPHERE.2020.127843

Pradhan, G., Tripathy, B., Ram, D. K., Digal, A. K., & Das, A. P. (2024). Bauxite mining waste pollution and its sustainable management through bioremediation. *Geomicrobiology Journal, 41*(4), 335–344. https://doi.org/10.1080/01490451.2023.2235353

Priyadarshini, B., & Das, A. P. (2024). The adverse impact of synthetic microplastics released from disposable face masks pollutants on aquatic microbiota, aquatic species and humans: A review. *Environmental Science and Engineering, 2827*, 67–81. https://doi.org/10.1007/978-3-031-50840-0_5

Priyadarshini, B., Santhosh Kumar, M., & Das, A. P. (2024). Microbial degradation of synthetic plastics released from COVID-19 generated single use disposable face masks. *Environmental Science and Engineering, 2827*, 1–16. https://doi.org/10.1007/978-3-031-50840-0_1

Priyadarshini, S., Jagatee, S., & Das, A. P. (2024). Synthetic fabrics and microfiber pollution–an assessment of their global impact. *Environmental Science and Engineering, 2333*, 137–157. https://doi.org/10.1007/978-3-031-51792-1_8

Priyadarsini, S., & Das, A. P. (2024). Analytical and structural characterization of waste lithium-ion batteries for their effective recycling strategy. *Environmental Chemistry and Ecotoxicology*. https://doi.org/10.1016/j.enceco.2024.12.004

Priyadarsini, S., & Das, A. P. (2025). Lithium bioleaching: A review on microbial-assisted sustainable technology for lithium bio-circularity. *Journal of Water Process Engineering, 69*, 106744. https://doi.org/10.1016/j.jwpe.2024.106744

Rafeeq, H., Afsheen, N., Rafique, S., Arshad, A., Intisar, M., Hussain, A., Bilal, M., & Iqbal, H. M. N. (2023). Genetically engineered microorganisms for environmental remediation. *Chemosphere, 310*, 136751. https://doi.org/10.1016/J.CHEMOSPHERE.2022.136751

Rathna, R., & Nakkeeran, E. (2020). Biological treatment for the recovery of minerals from low-grade ores. In *Current developments in biotechnology and bioengineering: Resource recovery from wastes* (pp. 437–458). Elsevier. https://doi.org/10.1016/B978-0-444-64321-6.00022-7

Rawlings, D. E., & Silver, S. (1995). Mining with microbes. *Bio/Technology, 13*(8), 773–778. 10.1038/nbt0895-773.

Ray, S. K., Dey, S., Das, A. P., & John, A. (2024). Microplastic waste management through greener technologies: A review. In *Sustainable management of urban plastic waste through circular economic approaches* (pp. 237–261). Routledge. https://doi.org/10.1201/9781003465447-14

Reddy Koteswara, G., Bhargavi, S., Yamuna, D., Anjani Devi, M., Jeevana Lakshmi Kumar, G., Manas, T., Adarsh, S., Bhavya Sai, V., Nath, H., Alreshidi, M. A., Togun, H., Ram, D. K., & Das, A. P. (2025). Recent advancements in the remediation of marine hydrocarbon and plastic pollutants. *Environmental Science and Engineering, 218*, 127–142. https://doi.org/10.1007/978-3-031-81615-4_7

Ruiz-Fresneda, M. A., Martinez-Moreno, M. F., Povedano-Priego, C., Morales-Hidalgo, M., Jroundi, F., & Merroun, M. L. (2023). Impact of microbial processes on the safety of deep geological repositories for radioactive waste. *Frontiers in Microbiology, 14*, 1134078. https://doi.org/10.3389/FMICB.2023.1134078/BIBTEX

Sahu, A., Dey, S., Mallick, K., Priyadarsini, S., Behera, I. D., & Das, A. P. (2024). A review on advanced technology for sustainable management of synthetic microplastic waste. In *Sustainable management of urban plastic waste through circular economic approaches* (pp. 262–279). Routledge. https://doi.org/10.1201/9781003465447-15

Sahu, A., Mallick, K., & Das, A. P. (2024). *Environmental impact of single-use synthetic face mask and its recycling: A sustainable approach* (pp. 197–212). Springer. https://doi.org/10.1007/978-3-031-50840-0_11

Sahu, S., & Das, A. P. (2024). *Microplastics degradation mechanism by aquatic Plastiomes: A review* (pp. 213–226). Springer. https://doi.org/10.1007/978-3-031-50840-0_12

Samarasekere, P. W. (2024). Microbial remediation technologies for mining waste management. *Energy, Environment, and Sustainability, 2536*, 31–67. https://doi.org/10.1007/978-981-97-1152-9_3

Saravanan, A., Kumar, P. S., Ramesh, B., & Srinivasan, S. (2022). Removal of toxic heavy metals using genetically engineered microbes: Molecular tools, risk assessment and management strategies. *Chemosphere, 298*, 134341. https://doi.org/10.1016/J.CHEMOSPHERE.2022.134341

Satapathy, S., Kar, D., Priyadarsini, S., Das, A. P., & Kuanar, A. (2024). Application of microorganisms in emerging waste management strategies and remediation of synthetic and cellulosic microfibers. In *Sustainable microbial technology for synthetic and cellulosic microfiber bioremediation* (pp. 183–206). Springer. https://doi.org/10.1007/978-3-031-62600-5_10

Seow, Y. X., Tan, Y. H., Mubarak, N. M., Kansedo, J., Khalid, M., Ibrahim, M. L., & Ghasemi, M. (2022). A review on biochar production from different biomass wastes by recent carbonization technologies and its sustainable applications. *Journal of Environmental Chemical Engineering, 10*(1), 107017. https://doi.org/10.1016/J.JECE.2021.107017

Shar, S. S., Reith, F., Shahsavari, E., Adetutu, E. M., Nurulita, Y., Al-Hothaly, K., Haleyur, N., & Ball, A. S. (2019). Biomineralization of platinum by escherichia coli. *Meta, 9*(4), 407. https://doi.org/10.3390/MET9040407

Shome, S., Mhaske, S., Pathak, K., & Tiwari, S. (2021). Mine waste as resource: Indian mining scenario of coal and non coal mining sector. *International Journal of Recent Technology and Engineering, 9*, 2277–3878. https://doi.org/10.35940/ijrte.F5396.039621

Singh Jadaun, J., Bansal, S., Sonthalia, A., Rai, A. K., & Singh, S. P. (2022). Biodegradation of plastics for sustainable environment. *Bioresource Technology, 347*, 126697. https://doi.org/10.1016/J.BIORTECH.2022.126697

Smitha, M. S., Singh, S., & Singh, R. (2017). Microbial bio transformation: A process for chemical alterations. *Journal of Bacteriology & Mycology, 4*(2), 85. https://doi.org/10.15406/JBMOA.2017.04.00085

Song, Z., Song, G., Tang, W., Yan, D., Zhao, Y., Zhu, Y., Wang, J., & Ma, Y. (2021). Molybdenum contamination dispersion from mining site to a reservoir. *Ecotoxicology and Environmental Safety, 208*, 111631. https://doi.org/10.1016/J.ECOENV.2020.111631

Srivastava, R. K., Sarangi, P. K., Vivekanand, V., Pareek, N., Shaik, K. B., & Subudhi, S. (2022). Microbial fuel cells for waste nutrients minimization: Recent process technologies and inputs of electrochemical active microbial system. *Microbiological Research, 265*, 127216. https://doi.org/10.1016/J.MICRES.2022.127216

Sun, Y., Wang, H., Long, X., Xi, H., Biao, P., & Yang, W. (2022). Advance in remediated of heavy metals by soil microbial fuel cells: Mechanism and application. *Frontiers in Microbiology, 13*, 997732. https://doi.org/10.3389/FMICB.2022.997732/BIBTEX

Swain, S., Purnima Sudha, M., Dey, S., Badu, Y., Bhoi, S. R., Alreshidi, M. A., Qasim, M. T., Nath, H., Ram, D. K., & Das, A. P. (2025). Sources of micro and Nano plastic hydrocarbon pollutants in marine bodies and their treatment. *Environmental Science and Engineering, 218*, 45–67. https://doi.org/10.1007/978-3-031-81615-4_3

Thakare, M., Sarma, H., Datar, S., Roy, A., Pawar, P., Gupta, K., Pandit, S., & Prasad, R. (2021). Understanding the holistic approach to plant-microbe remediation technologies for removing heavy metals and radionuclides from soil. *Current Research in Biotechnology, 3*, 84–98. https://doi.org/10.1016/J.CRBIOT.2021.02.004

Tsapekos, P., Khoshnevisan, B., Alvarado-Morales, M., Zhu, X., Pan, J., Tian, H., & Angelidaki, I. (2021). Upcycling the anaerobic digestion streams in a bioeconomy approach: A review. *Renewable and Sustainable Energy Reviews, 151*, 111635. https://doi.org/10.1016/J.RSER.2021.111635

Velenturf, A. P. M., & Purnell, P. (2021). Principles for a sustainable circular economy. *Sustainable Production and Consumption, 27*, 1437–1457. https://doi.org/10.1016/J.SPC.2021.02.018

Vo, P. H. N., Danaee, S., Hai, H. T. N., Huy, L. N., Nguyen, T. A. H., Nguyen, H. T. M., Kuzhiumparambil, U., Kim, M., Nghiem, L. D., & Ralph, P. J. (2024). Biomining for sustainable recovery of rare earth elements from mining waste: A comprehensive review. *Science of the Total Environment, 908*, 168210. https://doi.org/10.1016/J.SCITOTENV.2023.168210

Wu, C., Li, F., Yi, S., & Ge, F. (2021). Genetically engineered microbial remediation of soils co-contaminated by heavy metals and polycyclic aromatic hydrocarbons: Advances and ecological risk assessment. *Journal of Environmental Management, 296*, 113185. https://doi.org/10.1016/J.JENVMAN.2021.113185

Xue, Y., Qiu, T., Sun, Z., Liu, F., & Yu, B. (2022). Mercury bioremediation by engineered pseudomonas putida KT2440 with adaptationally optimized biosecurity circuit. *Environmental Microbiology, 24*(7), 3022–3036. https://doi.org/10.1111/1462-2920.16038

Yu, H., Zahidi, I., Chow, M. F., Liang, D., & Madsen, D. Ø. (2024). Reimagining resources policy: Synergizing mining waste utilization for sustainable construction practices. *Journal of Cleaner Production, 464*, 142795. https://doi.org/10.1016/J.JCLEPRO.2024.142795

Yu, H., Zahidi, I., Fai, C. M., Liang, D., & Madsen, D. Ø. (2024). Mineral waste recycling, sustainable chemical engineering, and circular economy. *Results in Engineering, 21*, 101865. https://doi.org/10.1016/J.RINENG.2024.101865

Yu, W., Xu, X., Jin, K., Liu, Y., Li, J., Du, G., Lv, X., & Liu, L. (2023). Genetically encoded biosensors for microbial synthetic biology: From conceptual frameworks to practical applications. *Biotechnology Advances, 62*, 108077. https://doi.org/10.1016/J.BIOTECHADV.2022.108077

Zhou, C., Xin, Y., & Han, Y. (2023). Towards a green mining future: A dynamic evolutionary game model for collaborative waste recycling. *Heliyon, 9*(10), e20515. https://doi.org/10.1016/J.HELIYON.2023.E20515

Zhou, L., Contreras-Salgado, E. A., Georgina Sánchez-Morán, A., Yair Rodríguez-Preciado, S., Sifuentes-Franco, S., Rodríguez-Rodríguez, R., Macías-Barragán, J., & Díaz-Zaragoza, M. (2024). Multifaceted applications of synthetic microbial communities: Advances in biomedicine, bioremediation, and industry. *Microbiology Research, 15*(3), 1709–1727. 10.3390/MICROBIOLRES15030113.

Bioremediation Technology for Sustainable Mining Waste Management Sustainable Technology in the Development of Novel Processes and Products from Mining Wastes

Ayushi Singh

1 Introduction

The mining sector, as an industrial sector, is being made available most of the major resources upon which the contemporary world relies. The sector also presents a primary source of environmental problems, mostly because of the massive amounts of wastes generated during the extraction process. Management of mining wastes must be ensured to control the environmental and public health effects of the problem. Green technology, bioremediation, and emerging technology are providing new prospects for the recycling of mining waste in producing useful products and reducing harmful environmental effects (Das et al., 2025; Swain et al., 2025). The effort in this chapter has been made to define and categorize waste mining, to discuss the level of production of waste worldwide, to list the environmental and health hazards of mining waste, and to proclaim the significance of sustainable technologies. The book also discusses shifting toward developing a circular economy in mining and introduces the Sustainable Development Goals (SDGs) that can be used in managing mining waste.

Mining waste is residual waste that arises as a by-product of economic minerals extracted from ore deposits. Tailings, slag, overburden, and waste rock are the most common types of mining waste. Tailings are milled waste residual ground material produced as by-product of ore concentration and deposited usually in tailings ponds. The material is usually heavy with toxic chemicals and heavy metals and therefore constitutes major environmental risk. Slag is produced during smelting and comprises mainly impurities removed from the metallic content. Despite being reclaimed periodically as building material, it is most likely to pose environmental issues if

A. Singh (✉)
Department of Biological Science, Sri Venkateswara College, University of Delhi, New Delhi, India

© The Author(s), under exclusive license to Springer Nature Switzerland AG 2026

A. P. Das (ed.), *Bioremediation Technology for Sustainable Mining Waste Management*, Sustainable Environmental Waste Management Strategies,
https://doi.org/10.1007/978-3-032-03468-7_17

improperly disposed of (Das & Ghosh, 2023; Das, Ray, et al., 2024; Dey, Tripathy, et al., 2023; Mohanty, Dey, et al., 2024; Sahu et al., 2024; Singh et al., 2020). Overburden refers to earth and rock coverings which are transported to the surface in order to excavate mineral deposits; if improperly disposed of, it can lead to water quality degradation as well as soil erosion. Lastly, waste rock is rock mined but not utilized for economically valuable minerals that also cause pollution.

International consideration in the case of waste caused due to mining activity is extremely important. Mining generates more than 10 billion tons of waste each year and is among the largest waste producers worldwide (International Council on Mining and Metals, 2021). In Zambia's Copperbelt Province, for instance, mining generates approximately 791 million tons of tailings and 1899 million tons of overburden, the amount of waste generated (Kangwa, 2008). It is estimated that with a rise in demand for minerals worldwide, mining waste will rise proportionally as well (International Council on Mining and Metals, 2021).

Impacts of mining waste are long-lasting and extensive. Acid mine drainage (AMD) is merely one of the environmental impacts that is of significant importance. It is a state of chemical interaction between water and oxygen on sulfide minerals in mining wastes, producing sulfuric acid that percolates and leaches heavy metals in nearby aquatic systems. Such pollution does substantial damage to marine life, which translates into water of poor quality and low biodiversity (Das, Mohanty, et al., 2023; Ghosh & Das, 2020; John & Das, 2024). In addition, the heavy metals that would be leached out would preferentially be absorbed in wildlife and fish and thereby expose other communities using such resources as sources of food to harm (Doley et al., 2012).

Furthermore, the detrimental effect on human health by mining waste is far-reaching. Individuals who reside in communities surrounding mining operations are certain to be exposed to hazardous chemicals and heavy metals due to contamination of plants, water, and soil. Continued exposure to the chemicals is certain to cause a range of diseases from heavy metal poisoning diseases to neurological diseases, and respiratory diseases (Kuo et al., 2008). Short-term environmental and health implications of waste mining require the use of sustainable technologies for its handling. Conventional disposal by landfilling and use of tailings dams will likely create long-term environmental disruption and more risk of pollution. Sustainable technologies are aimed at reducing the volume of wastes from new ways of working, recovering valuables from waste streams, and reducing ecological footprints (Bala et al., 2024; Mohanty, Tripathy, et al., 2024; Tripathy et al., 2024). The recovery process, through the example of bioremediation, involves microbes and vegetables to restore degraded ecosystems with minimal environmental impact during the mining process (Gathuru, 2011). The sustainable development process ensures the recovery process provides positive effects both to the environment and local communities.

The move toward a circular economy in mining is gathering pace as businesses awaken to the need for environmental-friendly business models through wastage avoidance and reuse of materials. A circular economy model seeks to extend the lifespan of materials and minimize wastage through recycling, reuse, and upcycling processes. For the mining sector, it means creating new innovative ways of recycling mining wastes like tailings and slag into new products or raw materials for other sectors. For instance, metal extraction from tailings not only reduces environmental issues but also brings economic benefits to the mining corporations (Singh et al., 2020). United Nations Sustainable Development Goals (SDGs) provide a general guideline to address the issue of mining waste. All the SDGs, particularly Goal 12 (Responsible Consumption and Production), encourage adopting sustainable business practices. Adoption of responsible waste management practices is in line with the Sustainable Development Goals (SDGs) such that the exploitation of resources will never eliminate the sustainability of the environment or the health of people (United Nations, 2015). New technology introduction, for example, on bioremediation technology fostering more recovery of value content within wastes, should be introduced as part of efforts toward achieving such goals. Green technologies and bioremediation are the venues through which such promotion of mining by-product sustainability is fostered (Dey et al., 2024; Ray et al., 2024). The technologies apply natural processes to enable the remediation of polluted ecosystems with the general effect being degradation of harmful substances and recovery of the ecosystem (Doley et al., 2012). For instance, certain bacteria can be used in bioleaching of metal from tailings with benefits of resource recovery as well as detoxification of wastes (Doley et al., 2012). Furthermore, phytoremediation—plant usage to leach or stabilize pollutants—can be applied to revitalize derelict land and make the soil fertile (Haddaway et al., 2019). Application of these bioremediation technologies, among others, not only minimizes the environmental footprint of mining wastes but also sustains harmonious interaction between mining activity and the environment (Table 1, Fig. 1).

Table 1 Targets for mine tailings management for achieving sustainable development goals (Almeida et al., 2021)

SDGs	Targets for Mine Tailings Management and Valorization
SDG 6: Ensure availability and sustainable management of water and sanitation for all	– Minimize the impacts of tailings facilities and dams on water resources by minimizing acid mine drainage and the release of hazardous chemicals – Reduce the water consumption of continental water resources in tailings facilities by improving water recycling rate in tailings management, by using dewatering technologies
SDG 7: Ensure access to affordable, reliable, sustainable, and modern energy for all	Achieve energy efficiency in tailings management by improving water-energy nexus – Encourage the use of renewables energies in mine tailings management

(continued)

Table 1 (continued)

SDGs	Targets for Mine Tailings Management and Valorization
SDG 11: Make cities and human settlements inclusive, safe, resilient, and sustainable	– Reduce the impact of mine tailings deposits on communities nearby – Planning land use considering the end-of-life of the mine – Recovery of nearby areas of mine tailing's deposits – Support the development of processes to reuse, recycle, and/or re-processing tailings
SDG 12: Ensure sustainable consumption and production patterns	– Achieve sustainable management of tailings by reducing using of water, energy, and land – Minimize production of tailings and encourage sustainable practices and monitoring of chemicals and water management in tailings – Encourage tailing dewatering technologies, appropriate management of acid drainage, safe storage of tailings – Encourage and achieve valorization of tailings by re-processing, reusing, and recycling tailings
SDG 13: Take urgent action to combat climate change and its impacts	– Recognize climate change in the planning of managing and re-processing tailings – Improve energy and water efficiency in mine tailings management – Measure direct and indirect emissions of treatments of tailings – Encourage the use of renewable energies in mine sites
SDG 14: Conserve and sustainably use the oceans, seas, and marine resources for sustainable development	– Reduce marine pollution associated with mining tailings and acid mine drainage – Plan for responsible tailings management that avoid tailings disposal into the sea
SDG 15: Protect, restore, and promote sustainable use of terrestrial ecosystems, sustainably manage forests. Combat desertification, and halt and reverse land degradation and halt biodiversity loss	– Conduct environmental impact assessment of tailings facilities and dams – Support projects that connect communities and biodiversity and encourage the participation of governments, communities, researchers, and NGOs in mine planning – Minimize the impact of tailings on freshwater ecosystems by reducing the water consumption of tailings management – Ensure the conservation, restoration, and sustainable use of freshwater sources – Avoid impacts of tailings facilities and dams to critical habitats by applying the mitigation hierarchy

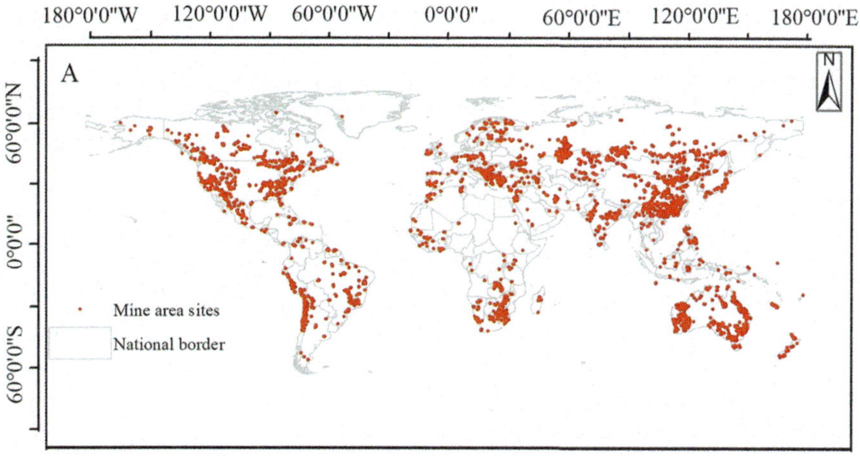

Fig. 1 Global distribution of mine area polygons assessed in this study (*N* = 24,605). All geographic data are georeferenced and downloadable as shapefiles in the supporting information (Xiang et al., 2021)

2 Technologies for Sustainable Processing of Mining Wastes

The mining sector is a dynamic player in the world economy as far as providing basic raw materials is concerned but, Ironically, produces a significant amount of waste that leads to severe environmental issues (Das et al., 2011; Das & Ghosh, 2023; Ghosh et al., 2016; Khatun et al., 2023; Mishra et al., 2019). The future direction to reclaim environmental degradation and enhance the recovery of resources is through the implementation of sustainable technology in waste management in the mining sector. This topic deals with the different bioremediation technologies such as bioleaching, microbial process for heavy metal recovery, phytoremediation, mycoremediation, and application of nanotechnology. These technologies not only treat pollution but also convert mine wastes into economically valuable products.

2.1 Bioremediation Techniques

Bioremediation is a technique of removal or breakdown of pollution using living organisms and utilizes different green technologies to treat mine wastes. Among the different bioremediation processes, bioleaching and biomineralization have gained much significance since they are efficient as well as environmentally friendly. Bioleaching is based on the application of microorganisms to leach metal from ores to be extracted without applying harmful chemicals, which are utilized under normal mining operations. For example, microorganisms like *Thiobacillus ferrooxidans* are employed to oxidize sulfide ores so that the leaching of metals is possible like copper and gold (Liu et al., 2012). Not only is it simple to extract the metal but also contamination by the low-grade ore returns. In addition, biomineralization or microbially influenced metal precipitation applies in immobilizing heavy metals within the environment so they would not move to groundwater (Jiang et al., 2016).

Microbial activity also has another significant role in the extraction and recovery of heavy metals from waste mining. Some microorganisms possess some characteristics in heavy metal tolerance and detoxification, and these can be utilized during bioremediation (Das et al., 2015; Das, Dey, et al., 2024; Nag et al., 2023; Paul et al., 2023; Pradhan et al., 2023). For example, biosorption or heavy metal adsorption on the microbial cell surface has proven to be an efficient technique of decontamination of water bodies that are contaminated. Experimental results showed that certain bacteria strains like *Bacillus subtilis* can be used in an attempt to biosorb lead and cadmium extremely efficiently as a method to lower the pollutant in contaminated mine areas (Khan et al., 2016). This ability of native microbial communities to be utilized in waste treatment is closer to more local, inexpensive, and environmentally friendly bioremediation processes.

There are several case studies which attest to the efficacy of microorganisms in treating mining wastes.

For example, *Acidithiobacillus* microbes are most often combined with the bioleaching process and have already been applied in a way that has significantly enhanced metal recovery from sulfide ore deposits by an enormously wide margin (Jiang et al., 2016). With the help of such a mixed culture of microbes, it has become easier for scientists to enhance the efficiency of extraction much more compared to the use of pure cultures alone. Besides these, lactic acid bacteria Lactobacillus plantarum have also been found to effectively mobilize lead and zinc among other heavy metals, which is an implication of their utilization as bioremediation agents within contaminated mining locations (Yuan et al., 2015). The outcomes call for enhanced research and emphasis on the distinctive interactions of the heavy metals and microorganisms involved that eventually render more efficient procedures of bioremediation.

2.2 Phytoremediation and Mycoremediation

Phytoremediation is a technology that utilizes plants for removal, uptake, and detoxification of toxic pollutants from water and soil (Das, Mallavarapu, & Ghosh, 2023; Das, Nayak, & Behera, 2024; Jagatee et al., 2024; Satapathy et al., 2024).

There are some plants referred to as hyperaccumulators with the uptake ability of heavy metals with high concentration, which is useful in the restoration of polluted mine lands. It is observed that certain plants like *Thlaspi* caerulescens and *Berkheya coddii* are effective lead, zinc, and cadmium hyperaccumulators from polluted soils. Certain uses are plant cultivation and harvesting such hyperaccumulator plants for polluted soil, where they retain contaminants and create biomass that can be harvested in order to further reclaim metals. Mycoremediation, or the use of fungi to clean up the environment, is a novel substitute for phytoremediation.

Fungi can degrade the majority of organic contaminants and heavy metal immobilization. For instance, fungus Pleurotus ostreatus mushroom form has been showing great promise when it comes to hydrocarbon pollutants degradation that is common on mine sites. Fungi mycelium has been found to have the ability to accumulate heavy metals and, therefore, enable polluted sites and soil to be remediated, as well as improve soil (Gadd, 2010). This dual function by fungi suggests that they are capable of pollutant degradation as well as immobilizing metals (Fig. 2).

Fig. 2 Hyperaccumulator plants: (**a**) Sedum plumbizincicola (Crassulaceae), Long-hua Wu, Institute of Soil Science, Chinese Academy of Sciences, China; (**b**) Alyssum murale (now Odontarrhena muralis) (Brassicaceae); (**c**) Neptunia amplexicaulis (Fabaceae); (**d**) Berkheya coddii (Asteraceae); (**e**) Pteris vittata (Pteridaceae), Mei Lei, Institute of Geographic Sciences and Natural Resources, Chinese Academy of Sciences, China; (**f**) Dicranopteris linearis (Gleicheniaceae), Wen-Shen Liu, School of Environmental Science and Engineering, Sun Yat-sen University, China; (**g**) Haumaniastrum katangense (Lamiaceae), Marta Sośnicka, University of the Witwatersrand, South Africa. (**h**) Noccaea caerulescens (Brassicaceae); (**i**) Phyllanthus rufuschaneyi (Phyllanthaceae). Adapted from Corzo Remigio et al. (2020)

2.2.1 Case Study: Application of Hyperaccumulators in Mining Rehabilitation

One such highly argumentative case study of hyperaccumulator use is Alyssum murale in an Australian zinc mine paddock.

Researchers had demonstrated that it was possible for this plant to take up huge amounts of zinc from the soil, thus cleansing the land. This is one feat to replicate in integrating phytoremediation and recovery practice of metals, a clean approach to waste legacy management from mining.

2.3 *Nanotechnology in Waste Processing*

Nanotechnology offers new techniques for the treatment of mining waste by taking advantage of the novel properties of nanomaterials to improve conventional treatment techniques.

The utilization of nanomaterials, whose elevated surface area and heightened reactivity have resulted in epochal advances in heavy metal adsorption. For instance, it has been established by research that nanostructured metal oxides and carbon nanotubes have enormous capacity to adsorb harmful pollutants like mercury, lead, and arsenic from water solutions (Ghafoor et al., 2018). Addition of in situ nanoscale zero-valent iron (nZVI) indicates a new direction toward remediation of heavy metals in soils contaminated with mining with high efficiency and low costs (López-Ramón et al., 2017). Furthermore, green nanomaterials producible from biowastes or plant extracts provide a potential solution to sustainable remediation technologies. Green nanotechnology is based on the philosophy to minimize environmental concerns arising due to the synthesis of nanomaterials, thus remediating safer alternatives. Biogenic nanoparticles that are produced through microbiological synthesis have been seen to remediate lower toxicity and stability heavy metals effectively as a possible means of addressing mining waste (Rafiee et al., 2017).

The use of green technology in addressing waste in mining is crucial in finding an environmental-friendly means of exploiting resources.

Microbial technologies in the shape of bacteria and fungi deployed through mycoremediation and phytoremediation contain unexploited possibilities of retrieving useful assets and rehabilitating contaminated land. New horizons also lie in the arrival of nanotechnology that supports highest remediation effectiveness at decreased environmental exposure via mining. Use of such technologies is able to revolutionize the mining business toward ecologically softer approaches while meeting economic requirements alongside conservation of nature (Figs. 3 and 4).

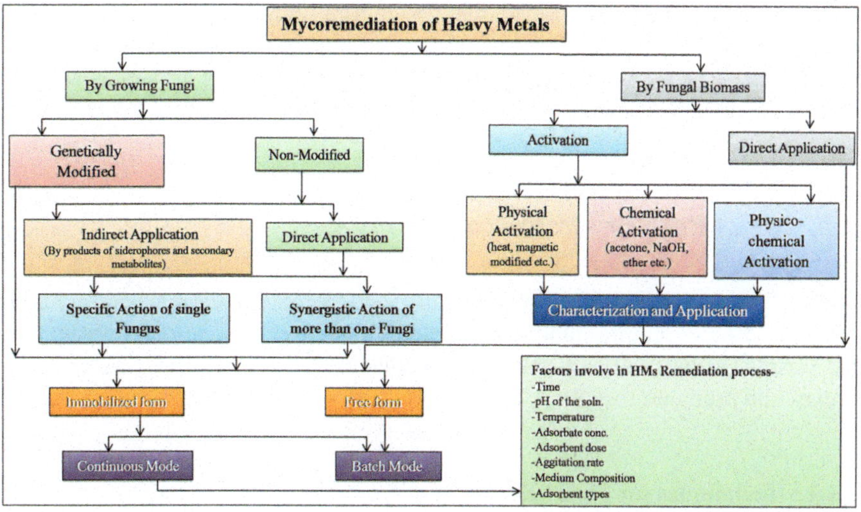

Fig. 3 Techniques in removal of heavy metals by growing fungi and fungal biomass from wastewater. (Adapted from Kumar and Dwivedi (2021). Environmental Science and Pollution Research, 28, 10375–10412)

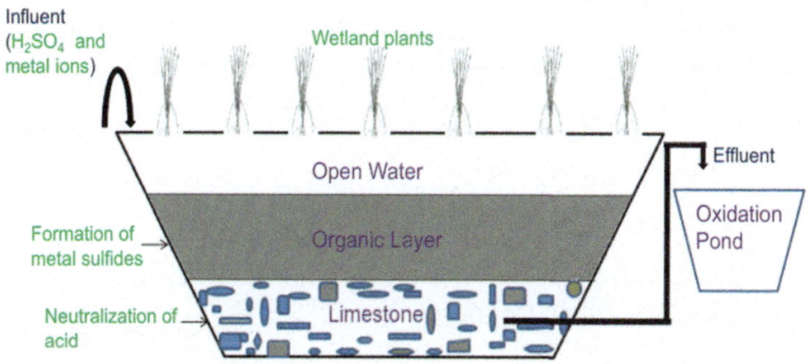

Fig. 4 Treatment of acid mine. (Reprinted from Mohapatra & Kirpalani (2017, p. 7))

3 Sustainable Technologies in the Development of Novel Processes and Products from Mining Wastes

Mining is pivotal for extracting minerals essential for modern technology and industry. However, mining activities generate a significant amount of waste, leading to environmental degradation and resource depletion. This research paper aims to explore innovative technologies and sustainable practices that convert mining waste from a liability into a valuable resource. By focusing on metal recovery, the creation of eco-friendly construction materials, energy generation from organic waste, and the production of bioplastics, the paper will illustrate pathways for a circular economy in the mining sector.

3.1 Metal Recovery from Mining Waste

The increasing demand for critical metals, such as lithium, cobalt, and rare earth elements (REEs), requires innovative methods to recover these resources from mining waste. The extraction of valuable minerals from waste reduces environmental impact and promotes recycling.

3.1.1 Techniques for Metal Recovery

Hydrometallurgy:

- Hydrometallurgy involves leaching metals from ores or waste products using aqueous solutions. The main sub-methods include:
- Acid Leaching: Sulfuric acid is commonly used to extract lithium from minerals like lepidolite. The process involves crushing the ore, treating it with acid, and allowing it to leach, which effectively dissolves the lithium ions (Li et al., 2019). The resultant lithium-rich solution can then undergo precipitation or crystallization to obtain lithium compounds.
- Alkaline Leaching: This method is effective for extracting metals such as nickel and cobalt. An example process includes treating lateritic ores with sodium hydroxide (García et al., 2020). Alkaline leaching can yield higher recovery rates compared to acid leaching for specific ores.

Bioleaching:

- Bioleaching utilizes microorganisms to solubilize metals from ores or mining waste. The advantages of this method include reduced chemical usage and lower operational costs. Microorganisms such as *Acidithiobacillus ferrooxidans* can oxidize sulfide minerals, leading to metal recovery (Deng et al., 2018).
- A case study by Ndlovu et al. (2020) illustrated that the bioleaching of copper from mining waste could achieve over 80% recovery in a period of several weeks, significantly outperforming traditional methods (Priyadarsini & Das, 2025).

Electrowinning:

- Following leaching, electrowinning can be employed to recover metals from solution. This process involves applying an electric current to a leach solution, causing metal ions to deposit on cathodes. The method efficiently recovers metals like copper and zinc.

Sustainability of Metal Recovery:

- The recovery processes are aligned with circular economy principles, which prioritize minimizing waste and optimizing resource use. By extracting metals from mining waste, not only is the environmental burden reduced, but the lifecycle of materials is extended.
- Recycling efforts can help stabilize raw material supply chains, crucial for industries reliant on critical metals.

Case Study: Gold Recovery from Mine Tailings:

- Gold mining operations commonly generate tailings with substantial unrecovered gold content, representing both an economic opportunity and an environmental challenge.

3.1.2 Historical Context and Conventional Methods

Historically, cyanide leaching has been the preferred method for gold recovery. While effective, it poses risks of toxicity and has fallen out of favor. The challenges associated with cyanide include potential leaching into groundwater and harm to local ecosystems.

3.1.3 Alternative Recovery Methods

Thiosulfate Leaching:

- Thiosulfate is emerging as a viable alternative to cyanide for gold recovery in acidic environments. A key study by Akbulut et al. (2018) demonstrated that thiosulfate could recover 90% of gold from sulfide concentrates within a short duration, showcasing its efficiency and reduced environmental impact.

Gravity and Flotation Separation:

- Gravity methods, such as sluicing and jigs, exploit density differences between gold and gangue minerals.
- Flotation techniques can separate gold from other minerals using chemical reagents that alter the surface properties of the particles, enabling their separation based on buoyancy.

3.2 Eco-Friendly Construction Materials from Mining Wastes

Mining operations produce immense amounts of waste, but innovative approaches allow for the recycling of mining waste into sustainable construction materials.

3.2.1 Manufacturing Geopolymer Cement

The Role of Geopolymer Cement

Geopolymer cement, derived from aluminosilicate materials like fly ash and slag, provides a sustainable alternative to conventional Portland cement, which is responsible for significant CO2 emissions (Palacios et al., 2021). Geopolymerization occurs through a process involving the chemical reaction of a precursor material with alkaline activators, resulting in a binder that exhibits comparable or superior mechanical properties.

Characteristics of Geopolymer Cement

Lower Carbon Footprint: Geopolymer cements can reduce greenhouse gas emissions by up to 80% when compared to traditional cements.

Durability: Geopolymer cement exhibits excellent resistance to chemical attack and thermal stability, making it suitable for various construction applications.

Applications

Geopolymer cement can be used for constructing structural elements, non-fired bricks, and other building materials, substantially improving sustainability in the construction industry. Researchers have explored its use in road construction and precast elements (Wang et al., 2019).

Sustainable Bricks and Aggregates

Properties and Performance
Mining waste can be processed into bricks and aggregates, mitigating the environmental impact associated with traditional clay bricks. The compressive strength and thermal resistance of bricks made from industrial byproducts can often exceed that of conventional bricks, making them suitable for usage in various applications (Almeida et al., 2021).

3.2.2 The Production Process

The production process generally entails:

- Collection and Preparation: Mining waste is processed to remove impurities.
- Mixing: The waste material is blended with appropriate binders to enhance stability.

- Shaping and Curing: The mixture is molded into the desired shapes and subjected to curing methods to achieve the target strength.

3.3 Energy Recovery and Biofuels from Mining Waste

Mining processes often generate organic waste that can be harnessed for energy production.

3.3.1 Biochar Production from Mining Residues

Biochar is produced through pyrolysis, a thermal decomposition process that converts organic materials into a charcoal-like substance under limited oxygen conditions. The benefits of biochar include carbon sequestration, soil improvement, and waste reduction.

Pyrolysis Process: The process involves:

- Heating organic mining waste, which releases volatile compounds while the carbon-rich solid residue remains.
- The resulting biochar can enhance soil properties, including nutrient retention and microbial activity, making it beneficial in land rehabilitation efforts.

Biochar can be used in various applications:

- Soil Amendment: Its addition improves soil fertility, water retention, and carbon storage in agricultural settings.
- Phytoremediation: Biochar helps mitigate pollution and improve soil health in areas affected by mining activities.

Biogas generation through anaerobic digestion of organic residues provides an effective waste management and energy recovery solution.

3.3.2 Anaerobic Digestion Process

Anaerobic digestion involves converting organic matter into biogas in the absence of oxygen.

Organic mining waste, such as food scraps from mining camps, can serve as a substrate for this process. The main byproducts of anaerobic digestion are methane (a renewable energy source) and digestate (a nutrient-rich agricultural input). Research conducted by Mata et al. (2021) indicates that using organic mining waste can significantly enhance biogas yields, supporting the viability of anaerobic digestion in optimizing waste management practices.

The produced biogas can be used for electricity generation, heating, or as a vehicle fuel. Leveraging this renewable energy source reduces reliance on fossil fuels, enhances energy security, and can provide financial benefits to mining operations.

3.4 Bioplastics and Sustainable Composites from Mining Byproducts

The shift toward bioplastics made from mining waste has gained traction as companies look for sustainable alternatives to traditional petroleum-based plastics (Dey, Das, et al., 2023; Lahiri et al., 2021).

3.4.1 Use of Mining Waste in Polymer Composites

Integrating mining byproducts into polymer composites aims to reduce environmental impacts while enhancing material properties.

Mining waste, when utilized as fillers in polymer matrices, can improve the mechanical properties of the final material. The bio-based composites developed can possess substantial strength, durability, and thermal resistance compared to conventional plastics.

3.4.2 Applications in Automotive and Packaging

In transportation, weight reduction is crucial for improving fuel efficiency. Composites containing mining waste can help achieve tensile strength while decreasing overall vehicle weight. In the packaging industry, the application of bioplastics made from mining residues serves as an eco-friendly alternative to plastic packaging, addressing the pressing issue of plastic waste.

3.4.3 Challenges and Future Directions

While significant progress has been made in utilizing mining waste sustainably, various challenges exist that require attention:

- Technical Feasibility: Although the potential for mining waste conversion is substantial, many processes are still emerging and need optimization for large-scale application. Continued research is essential to refine methods and improve recovery efficiency.
- Economic Viability: The economic aspects of recycling mining waste must ensure that end products are competitive in the market. Investment in technology

and infrastructure, as well as innovative business models, will be critical in overcoming economic barriers.
- Environmental Regulations: Navigating complex legal frameworks around environmental protection can sometimes inhibit innovative approaches. Establishing clear guidelines and standards is necessary to foster the safe and productive use of mining waste.
- Interdisciplinary Collaboration: Engaging researchers from varied fields, including material science, engineering, and environmental science, can enhance the development of novel processes and technologies tailored around mining waste utilization.

The sustainable management of mining waste presents numerous opportunities to transition from a linear to a circular economy. By focusing on metal recovery, eco-friendly construction materials, energy recovery from organic waste, and the development of bioplastics, we can reduce waste and environmental impacts while creating valuable products. Continued innovation and collaboration in research will be fundamental in advancing these technologies, paving the way for a sustainable future in mining and related sectors.

4 Case Studies of Sustainable Mining Waste Management

This section explores sustainable technologies for the development of novel processes and products from mining wastes, focusing on successful industrial implementations in Australia, Chile, and India.

4.1 Successful Industrial Implementations

4.1.1 Sustainable Tailings Repurposing in Australia

In Australia, the initiative undertaken by Newmont Boddington Gold (NBG) is a notable example of sustainable tailings repurposing. The NBG mine is one of the largest gold mines in the country, generating significant quantities of tailings. Traditional disposal methods for tailings pose considerable environmental risks, including contamination of groundwater and soil. Recognizing this challenge, NBG has adopted a sustainable approach to tailings management.

Instead of conventional storage methods, NBG has implemented a process to convert tailings into bricks for use in construction. The process involves drying the tailings and mixing them with binders and additives to create a composite material that meets structural standards. This innovative solution not only reduces the volume of waste but also diminishes the need for virgin materials in construction, effectively providing a circular solution to the mining industry's waste cycle. The endeavor has yielded bricks that are not only environmentally friendly but also competitive in terms of strength and durability compared to traditional bricks.

Industry collaboration has played a crucial role in the success of this project. NBG partnered with local construction companies to trial the use of tailings-based bricks in real construction projects. This collaboration provided invaluable feedback for improving the product and fostered stakeholder trust in using mining waste for construction purposes. By innovating tailings utilization, NBG exemplifies how mining companies can leverage technology to transition toward a circular economy while fulfilling environmental and community expectations.

4.1.2 Eco-Friendly Mining Waste Treatment in Chile

Chile, known for its copper mining, faces considerable challenges associated with mine waste, particularly acid mine drainage (AMD). AMD is a significant environmental problem that arises when sulfide minerals are exposed to air and water, resulting in acid formation that can leach heavy metals into the environment. A leading initiative focusing on eco-friendly mining waste treatment is the use of phytoremediation techniques at the El Teniente mine, one of the world's largest underground copper mines.

The El Teniente mine has adopted a strategy that employs native plant species capable of thriving in heavy metal-laden soils. These plants can accumulate metals while mitigating the acidic conditions of AMD. For instance, indigenous species such as "Eucalyptus spp." have been utilized to reduce soil acidity and stabilize heavy metals in the substrates. The remediation process not only cleans up contaminated environments but also promotes biodiversity by reintroducing native plant species into the area.

Moreover, the Chilean government has facilitated the implementation of this eco-friendly treatment method through financial support and policies encouraging responsible mining practices. The collaboration between academia, industry, and government has resulted in the successful scaling up of this technique across various mining sites in Chile. This multifaceted approach demonstrates how integrating bioremediation with sustainable agricultural practices can significantly enhance the management of mining waste in a manner that benefits both the environment and local communities.

4.1.3 Bioremediation-Based Metal Recovery in India

India has a rich reserve of minerals and metals and has utilized the land extensively for mining. However, the country grapples with the dual challenge of managing mining waste while also recovering valuable metals. Bioremediation-based metal recovery is gaining traction as a sustainable technology to tackle these issues. For example, the use of bacteria for the recovery of copper from mining tailings at the Indian Bureau of Mines showcases such an application.

Research has shown that specific strains of bacteria, such as *Acidithiobacillus ferrooxidans*, thrive in acidic environments and can solubilize metal ions effectively. A pilot project implemented by the Indian Bureau of Mines demonstrated the feasibility of bioleaching technologies to extract copper from tailings. In this process, bacteria are introduced into the tailings, where they facilitate the transformation of copper ions from insoluble to soluble forms, making them recoverable through leaching solutions.

The results from this biotechnological approach were exceptional; the recovery rate of copper reached around 70%, significantly higher than traditional methods, which are often less efficient and hazardous. The by-products generated from this recovery process also have potential applications in various sectors, thereby adding value to what was previously considered waste.

Furthermore, the successful implementation of bioremediation technologies in India has spurred interest in developing similar projects across other regions, supported by the government's initiatives to promote sustainable mining practices. This experience highlights the potential of integrating microbial technologies into mining waste management strategies to foster environmental sustainability while enhancing resource recovery.

4.2 Challenges and Lessons Learned

4.2.1 Economic Feasibility and Policy Roadblocks

The journey toward sustainable technologies in mining waste management is fraught with challenges. A significant hurdle lies in the economic feasibility of implementing innovative processes. Many sustainable technologies require substantial upfront investments in research, development, and infrastructure. For instance, the initial costs associated with the transformation of waste into construction materials, such as in the case of Newmont Boddington Gold, can be high, leading to reluctance from industry stakeholders who prioritize short-term profitability over long-term environmental benefits.

Moreover, mining companies often operate under tight profit margins, particularly in volatile global commodities markets. As such, they may prioritize projects with immediate financial returns rather than those that promise long-term sustainability. This economic hesitance can hinder investment in environmentally sound technologies and practices, essentially locking companies into traditional but environmentally harmful waste management practices.

Further compounding economic issues are the policy roadblocks that can arise at both regulatory and operational levels. Regulatory frameworks in many regions may still adhere to outdated models of waste disposal that do not incentivize sustainable practices or may lack provisions for new technologies. In Australia, for example, while state governments are beginning to create policies that support sustainable mining practices, there are inconsistencies and gaps that may lead to confusion or hesitation among companies looking to transition to greener technologies. The lack of standardized regulations can create a compliance burden for mining companies, discouraging them from adopting innovative solutions.

Community engagement is another critical dimension that can impact the economic feasibility of adopting sustainable technologies. If the benefits of repurposing or bioremediating mining waste are not clearly communicated to local communities, there may be resistance due to fears of negative environmental impacts or lack of perceived benefits. Bridging these gaps in understanding between mining companies, policymakers, and communities is essential for facilitating the broader adoption of sustainable technologies.

4.2.2 Scaling Up Sustainable Technologies

Scaling up sustainable technologies faces multifaceted challenges that go beyond mere implementation. A key issue is the variability in technology maturity; while some methods—such as phytoremediation—are relatively well-developed and understood, others, such as advanced bioremediation techniques, are still in the research and development phase. This disparity often results in an uneven playing field for technology deployment. For example, scaling up bioremediation techniques in India, such as the bioleaching of metals, requires not only technological advancements but also the establishment of standardized processes and assessments for effectiveness and safety.

Technical challenges also pose significant barriers to scaling. For instance, the efficiency of bioremediation methods can vary widely based on site-specific characteristics, such as soil type, the presence of contaminants, and climate conditions. This variability necessitates tailored solutions that may require bespoke research and development for different locations. Consequently, companies may find it challenging to justify the costs involved in scaling technologies that lack generalized applicability.

4.2.3 Future Directions in Research and Industry Adoption

Looking ahead, there are significant opportunities for advancing sustainable technologies in mining waste management. Collaborative research initiatives can be pivotal in developing more effective bioremediation techniques, particularly those that utilize native species or innovative microbial processes. For example, governments and universities can work with mining companies to conduct field trials of new technologies, facilitating a knowledge exchange that can lead to improved methodologies tailored to specific contaminants and geographies. Moreover, the role of technology in enhancing the circular economy in mining cannot be overstated. Future research should focus on life cycle analyses of mining waste management technologies to provide a comprehensive view of the environmental and economic benefits of these innovations over traditional practices. This analysis would be crucial for stakeholders seeking to justify investments in sustainable technologies, particularly in regions where regulatory frameworks are still evolving.

Industry adoption will also require a shift in corporate culture toward sustainability and innovation. Mining companies should be encouraged to adopt a more holistic view that prioritizes long-term environmental stewardship alongside profitability. Creating a reporting culture that tracks sustainability metrics—not just production and profit—can engender greater accountability and encourage the adoption of sustainable practices.

In addition, leveraging advancements in technology, such as digital monitoring and data analytic tools, can support more effective and transparent tracking of waste management processes. Real-time data collection can enhance decision-making, enabling companies to swiftly adapt to changing environmental conditions or regulatory demands.

Furthermore, engaging with local communities to educate and involve them in sustainable practices can cultivate a sense of ownership and partnership, which is invaluable for the successful implementation of innovative technologies. Initiatives that stress community benefits, whether through job creation or environmental restoration, can help ease public apprehension and gather support for new sustainable technologies.

Another promising avenue for future research and technology implementation lies in public-private partnerships. By collaborating with government bodies, NGOs, and academic institutions, mining companies can pool resources and knowledge, facilitating access to funding, expertise, and networks necessary for the successful deployment of sustainable technologies.

The transition toward more sustainable mining waste management practices represents both a challenge and an opportunity. The combined efforts of researchers, industry stakeholders, and policymakers will be crucial in navigating this landscape, fostering innovations that can significantly mitigate the environmental impacts of mining activities while enhancing economic viability.

5 Policy, Regulations, and Future Perspectives on Global Mining Waste Management

5.1 Global Mining Waste Regulations and Sustainability Standards

Mining is increasingly recognized as a significant source of waste and environmental impact, necessitating comprehensive regulation. Several global frameworks govern mining waste, emphasizing sustainable practices to minimize adverse effects on the environment and human health.

In the USA, the management of mining waste is primarily overseen by the Environmental Protection Agency (EPA) under the Resource Conservation and

Recovery Act (RCRA). This legislation distinguishes between hazardous and non-hazardous wastes, leading to potentially less stringent regulations for mining waste. The EPA has also delineated exemptions for certain mining wastes, which has fueled controversy and prompted calls for reform (EPA). For a more detailed understanding of the legislative and regulatory landscape surrounding mining waste, the EPA's timeline provides insights into evolving policies, emphasizing key developments and regulatory frameworks relevant to mining (EPA).

Globally, the United Nations Environment Programme (UNEP) serves as a pivotal agency promoting sustainable mining practices through guidelines and frameworks that address environmental conservation and the socio-economic dimensions of mining activities. In Europe, the Mining Waste Directive (MWD) mandates the implementation of comprehensive waste management plans and emphasizes the disposal and treatment of mining waste to mitigate its environmental impact (European Commission). This directive asserts the principle of "waste hierarchy," encouraging the reduction, recovery, and recycling of waste materials to minimize environmental degradation.

5.1.1 Role of Environmental Agencies

The role of environmental agencies like UNEP and the EPA is critical in ensuring compliance with regulations that govern mining waste. The EPA's approach involves regulating the management of hazardous waste, including tailings and leftover materials from mining operations. The agency's guidelines aim to provide clear standards to ensure safe disposal and treatment of mining waste, thereby protecting human health and the environment (EPA).

In the EU, the MWD articulates similar goals but with a more prescriptive approach regarding waste management. Member states are required to adopt measures that reduce the volume and impact of mining waste, reinforcing principles of sustainability and community engagement in planning processes. This regulation is crucial for promoting transparency and accountability in mining waste management across the EU (European Commission).

5.1.2 Sustainable Mining Policies in the EU, US, and China

In the European Union, sustainable mining policies are increasingly being integrated into broader environmental strategies, encompassing the European Green Deal, which aims to reduce greenhouse gas emissions and promote sustainable resource use. The EU emphasizes circular economy principles, aiming to minimize resource extraction and waste, ensuring that mining operations align with climate and sustainability goals (European Commission). Policymakers advocate for the development of sustainable mining technologies that meet environmental standards while supporting economic growth.

In the USA, state-level regulations often vary, with some states maintaining more stringent requirements for mining waste management compared to others. For example, states like Colorado and Arizona have implemented rigorous guidelines concerning mine reclamation and waste disposal, reflecting growing public awareness and regulatory pressure regarding the environmental impacts of mining (Ogungbangbe).

In China, the government has made significant strides toward improving regulatory frameworks by establishing stricter environmental standards and promoting technological innovation in the mining sector. Policies now reflect a commitment to minimizing environmental impacts through enhanced waste management practices, showcasing a comprehensive approach to addressing the socio-environmental challenges posed by mining (Ogungbangbe).

5.2 *Future of Sustainable Mining Waste Technologies*

Technological advancements hold great promise for revolutionizing mining waste management practices. The integration of AI-driven technologies enables real-time monitoring of waste generation and environmental impact, allowing for rapid response to potential hazards. Machine learning algorithms can analyze vast datasets from mining operations, facilitating the early detection of issues related to waste management and environmental compliance (Global Mining Review).

Blockchain technology also emerges as a transformative tool for tracking and managing mining waste. By ensuring transparent and verifiable waste disposal paths, blockchain enhances accountability across the supply chain. This technology provides stakeholders, including regulatory agencies and local communities, with the ability to audit waste management practices, thus fostering trust and collaboration (Global Mining Review).

Moreover, nanotechnology offers innovative solutions for resource recovery from mining waste, effectively transforming by-products into valuable materials. This technological application not only reduces waste but also contributes to increased resource efficiency in mining operations. Studies indicate that the utilization of nanomaterials in waste processing could lead to significant improvements in recovery rates and reductions in environmental contaminants (Ogungbangbe).

Startups are increasingly playing a vital role in the development of green mining solutions. Many emerging companies focus on engineering and developing technologies that enhance waste management in the mining sector. These startups often leverage funding and innovative thinking to provide sustainable alternatives to traditional waste disposal methods, aligning with global sustainability goals (Global Mining Review).

Collaboration between established mining companies and startups fosters an environment conducive to innovation. Strategic partnerships can lead to the integration of cutting-edge technologies that not only reduce waste but also enhance productivity and sustainability. Notably, companies like Anglo American and Rio Tinto are incorporating Environmental, Social, and Governance (ESG) criteria into their operations, which often includes investing in sustainable waste management solutions (Ogungbangbe).

5.3 Concluding Remarks

Addressing the complexities of mining waste management involves a multidisciplinary approach that integrates technological innovation, regulatory compliance, and community engagement. As the mining industry evolves, policymakers must embrace new technologies, including AI and blockchain, while prioritizing sustainable practices that mitigate environmental impact and support local communities.

The integration of nanotechnology and biotechnology presents exciting opportunities to transform mining waste from a liability into a resource, highlighting the role of innovation in achieving sustainability goals. Ultimately, a collaborative framework involving various stakeholders, including governments, industries, and communities, will be crucial to advancing responsible waste management practices in the mining sector.

References

Akbulut, A., Altun, N., & Yalçın, A. (2018). Thiosulfate leaching of gold and silver from a refractory sulfide concentrate. *Hydrometallurgy, 177*, 182–188. https://doi.org/10.1016/j.hydromet.2017.10.006

Almeida, J., Rosa, J., & Fortunato, E. (2021). Sustainable bricks produced from mining waste: A comparative study of mechanical properties. *Journal of Cleaner Production, 280*, 124396. https://doi.org/10.1016/j.jclepro.2020.124396

Bala, K., Anshul, A., Verma, S., Dogra, A., Das, A. P., Kumar, R., & Nandini, A. (2024). *Sustainable management for urban plastic waste generation. Sustainable management of urban plastic waste through circular economic approaches* (pp. 211–236). Routledge. https://doi.org/10.1201/9781003465447-13

Corzo Remigio, A., Chaney, R. L., Baker, A. J. M., Edraki, M., Erskine, P. D., Echevarria, G., & van der Ent, A. (2020). Phytoextraction of high-value elements and contaminants from mining and mineral wastes: Opportunities and limitations. *Plant and Soil, 449*, 11–37. https://doi.org/10.1007/s11104-020-04486-w

Das, A., Dey, S., Bala, K., Gnanasekaran, R., & Das, A. P. (2024). Microbial bioremediation technology for sustainable treatment and management of synthetic microfiber waste. In *Sustainable microbial technology for synthetic and cellulosic microfiber bioremediation* (pp. 221–238). Springer. https://doi.org/10.1007/978-3-031-62600-5_12

Das, A., Dey, S., Gacem, A., Yadav, K. K., Nath, H., Ram, D. K., & Das, A. P. (2025). Revolutionizing solid hydrocarbon sourcing: Plastic's journey from waste to upcycled treasure. *Environmental Science and Engineering, 218*, 331–351. https://doi.org/10.1007/978-3-031-81615-4_14

Das, A., Ray, S. K., Mohanty, M., Mohanty, J., Dey, S., & Das, A. P. (2024). Ecotoxicity of microplastic wastes and their sustainable management: A review. *Environmental Chemistry and Ecotoxicology, 6*, 144–152. https://doi.org/10.1016/J.ENCECO.2024.05.003

Das, A. P., & Ghosh, S. (2023). Microorganisms in manganese biomining processes. In *Manganese mining microorganisms* (pp. 55–61). Elsevier. https://doi.org/10.1016/B978-0-12-822147-1.00013-3

Das, A. P., Ghosh, S., Mohanty, S., & Sukla, L. B. (2015). Advances in manganese pollution and its bioremediation. In L. B. Sukla, N. Pradhan, S. Panda, & B. K. Mishra (Eds.), *Environmental microbe biotechnology* (1st ed., pp. 313–328). Springer. https://doi.org/10.1007/978-3-319-19018-1_16

Das, A. P., Mallavarapu, M., & Ghosh, S. (2023). Ecotoxicity of mining pollutants on the environment and their remediation. *Environmental Chemistry and Ecotoxicology, 5*, 165–167. https://doi.org/10.1016/J.ENCECO.2023.08.002

Das, A. P., Mohanty, A. K., van Hullebusch, E. D., Figueiredo, G., Das, A. P., Mohanty, A. K., van Hullebusch, E. D., & Figueiredo, G. (2023). Bioremediation as an emerging technology for the removal of synthetic microplastic pollutants from marine ecosystem. *MarPB, 194*, 115297. https://doi.org/10.1016/J.MARPOLBUL.2023.115297

Das, A. P., Nayak, B., & Behera, I. D. (2024). Applications of microorganisms in advanced Geomicrobiological and biotechnological exploration. *Geomicrobiology Journal, 41*(4), 309–311. https://doi.org/10.1080/01490451.2024.2327059

Das, A. P., Sukla, L. B., Pradhan, N., & Nayak, S. (2011). Manganese biomining: A review. *Bioresource Technology, 102*(16), 7381–7387. https://doi.org/10.1016/J.BIORTECH.2011.05.018

Deng, P., Zhang, K., & Qiu, G. (2018). Application of bioleaching technologies for metal recovery from mining waste: A review. *Minerals Engineering, 128*, 198–211. https://doi.org/10.1016/j.mineng.2018.07.020

Dey, S., Das, A., Mallick, K., Sahu, A., & Das, A. P. (2023). *Environmental petroleum waste: Pollution, toxicity, sustainable remediation* (pp. 159–175). Springer. https://doi.org/10.1007/978-3-031-48220-5_7

Dey, S., Dash, A., & Das, A. P. (2024). Bioremediation as a sustainable and circular solution for mining waste management and Bioenergy generation. In *Sustainable management of mining waste and tailings: A circular economy approach* (pp. 107–119). CRC Press. https://doi.org/10.1201/9781003442455-6

Dey, S., Tripathy, B., Kumar, M. S., & Das, A. P. (2023). Ecotoxicological consequences of manganese mining pollutants and their biological remediation. *Environmental Chemistry and Ecotoxicology*. https://doi.org/10.1016/J.ENCECO.2023.01.001

Doley, D., Audet, P., & Mulligan, D. R. (2012). Examining the Australian context for post-mined land rehabilitation: Reconciling a paradigm for the development of natural and novel ecosystems among the post-disturbance landscape. *Agriculture, Ecosystems & Environment, 163*, 85–93. https://doi.org/10.1016/j.agee.2012.04.022

Gadd, G. M. (2010). Fungi in biogeochemical cycles. *Nutrient Cycling in Agroecosystems, 87*(2), 255–263. https://doi.org/10.1007/s10705-010-9359-4

García, A., Martínez, M., & Pérez, M. (2020). Durability of concrete containing mine tailings: An experimental approach. *Construction and Building Materials, 260*, 120104. https://doi.org/10.1016/j.conbuildmat.2020.120104

Gathuru, G. (2011). *The performance of selected tree species in the rehabilitation of a limestone quarry at east African Portland cement company land Athi River, Kenya*. Kenyatta University. http://ir-library.ku.ac.ke/handle/123456789/3849

Ghafoor, A., Khedher, N. B., & Khemiri, H. (2018). Green nanotechnology for environmental remediation. *Critical Reviews in Environmental Science and Technology, 48*(2), 172–218. https://doi.org/10.1080/10643389.2018.1502956

Ghosh, S., & Das, A. P. (2020). Microbial metagenomics: Current advances in investigating microbial ecology and population dynamics. In S. K. Nayak & B. B. Mishra (Eds.), *Frontiers in soil and environmental microbiology* (1st ed., pp. 247–254). CRC Press.

Ghosh, S., Mohanty, S., Akcil, A., Sukla, L. B., & Das, A. P. (2016). A greener approach for resource recycling: Manganese bioleaching. *Chemosphere, 154*, 628–639. https://doi.org/10.1016/J.CHEMOSPHERE.2016.04.028

Haddaway, N. R., Macura, B., & Collins, A. (2019). The role of bioremediation in sustainable mining: A systematic review. *Environmental Research, 172*, 468–481.

International Council on Mining and Metals. (2021). *Global mining waste*. https://www.icmm.com/en-gb

Jagatee, S., Priyadarshini, S., Rath, C. C., & Das, A. P. (2024). Synthetic microfiber: An enduring environmental problem linked to sustainable development. *Environmental Science and Engineering, 2333*, 93–112. https://doi.org/10.1007/978-3-031-51792-1_6

Jiang, J., He, Y., Wang, W., & Zhou, L. (2016). A novel bioleaching approach for heavy metals from mining tailings using *Bacillus cereus*. *Biotechnology Reports, 9*, 1–6. https://doi.org/10.1016/j.btre.2016.06.002

John, A., & Das, A. P. (2024). Government strategies for achieving circular economy for plastic waste Management in Southeast Asia. In *Sustainable management of urban plastic waste through circular economic approaches* (pp. 72–82). Routledge. https://doi.org/10.1201/9781003465447-6

Kangwa, K. P. (2008). *An assessment of the economic, social and environmental impacts of the mining industry: A case study of copper mining in Zambia*. University of Lund.

Khan, M. I., Bhosale, R. R., & Patil, S. (2016). Use of *Bacillus subtilis* for biosorption of heavy metals from wastewater. *Journal of Environmental Chemical Engineering, 4*(1), 305–313. https://doi.org/10.1016/j.jece.2015.12.012

Khatun, R., Singh, S., Dubey, N. K., & Prasad Das, A. (2023). A review on marine-based phytochemicals and their application in biomedical research. In *Recent frontiers of phytochemicals: Applications in food, pharmacy, cosmetics, and biotechnology* (pp. 383–395). Elsevier. https://doi.org/10.1016/B978-0-443-19143-5.00013-X

Kumar, V., Dwivedi, S. K. (2021). Mycoremediation of heavy metals: processes, mechanisms, and affecting factors. *Environ Sci Pollut Res, 28*, 10375–10412. https://doi.org/10.1007/s11356-020-11491-8

Kuo, Y., Yang, T., & Huang, G. W. (2008). The use of grey relational analysis in solving multiple attribute decision-making problems. *Computers & Industrial Engineering, 55*(1), 80–93. https://doi.org/10.1016/j.cie.2007.12.002

Lahiri, D., Nag, M., Dey, A., Sarkar, T., Joshi, S., Pandit, S., Das, A. P., Pati, S., Pattanaik, S., Tilak, V. K., & Ray, R. R. (2021). Biofilm mediated degradation of petroleum products. *Geomicrobiology Journal*. https://doi.org/10.1080/01490451.2021.1968979

Li, S., Ma, Y., Wang, H., & Liu, R. (2019). Extraction of lithium from lepidolite using sulfuric acid. *Hydrometallurgy, 185*, 93–101. https://doi.org/10.1016/j.hydromet.2019.02.012

Liu, G., et al. (2012). Nanoenhanced materials for reclamation of mine lands and other degraded soils. *Journal of Nanotechnology, 2012*, 957432. https://doi.org/10.1155/2012/957432

López-Ramón, M. V., et al. (2017). Nanoscale zero-valent iron for the remediation of contaminated soils: A green approach. *Science of the Total Environment, 605*, 440–451. https://doi.org/10.1016/j.scitotenv.2017.06.152

Mata, A., Santos, E., & Ferreira, L. F. (2021). Biogas production from organic mining waste: A review. *Waste Management, 120*, 92–109. https://doi.org/10.1016/j.wasman.2020.09.019

Mishra, S., Rath, C., & Das, A. P. (2019). Marine microfiber pollution: A review on present status and future challenges. *Marine Pollution Bulletin, 140*, 188–197. https://doi.org/10.1016/J.MARPOLBUL.2019.01.039

Mohanty, J., Dey, S., Bala, K., & Das, A. P. (2024). Recent advances in microplastic pollution for its sustainable management. In *Sustainable management of urban plastic waste through circular economic approaches* (pp. 27–48). Routledge. https://doi.org/10.1201/9781003465447-3

Mohanty, M., Tripathy, B., Koteswara Reddy, G., & Das, A. P. (2024). Greener technologies for establishing restraint over microplastic pollution. In *Sustainable microbial technology for synthetic and cellulosic microfiber bioremediation* (pp. 239–258). Springer. https://doi.org/10.1007/978-3-031-62600-5_13

Mohapatra, D. P., Kirpalani, D. M. (2017). Process effluents and mine tailings: sources, effects and management and role of nanotechnology. Nanotechnol. *Environ. Eng, 2*, 1. https://doi.org/10.1007/s41204-016-0011-6

Nag, M., Lahiri, D., Ghosh, S., Sarkar, T., Pati, S., Das, A. P., Ram, D. K., Bhattacharya, D., & Ray, R. R. (2023). Application of microorganisms in biotransformation and bioremediation of environmental contaminant: A review. *Geomicrobiology Journal*. https://doi.org/10.1080/01490451.2023.2261443

Ndlovu, T., et al. (2020). Biotechnological advances in bioleaching of metals: A review. *Minerals, 10*(11), 993. https://doi.org/10.3390/min10110993

Palacios, M., et al. (2021). Geopolymer binder from recycled materials: Properties and future applications. *Sustainable Materials and Technologies, 28*, e00261. https://doi.org/10.1016/j.susmat.2021.e00261

Paul, A., Dey, S., Ram, D. K., & Das, A. P. (2023). Hexavalent chromium pollution and its sustainable management through bioremediation. *Geomicrobiology Journal*. https://doi.org/10.1080/01490451.2023.2218377

Pradhan, G., Tripathy, B., Ram, D. K., Digal, A. K., & Das, A. P. (2023). Bauxite mining waste pollution and its sustainable management through bioremediation. *Geomicrobiology Journal*. https://doi.org/10.1080/01490451.2023.2235353

Priyadarsini, S., & Das, A. P. (2025). Lithium bioleaching: A review on microbial-assisted sustainable technology for lithium bio-circularity. *Journal of Water Process Engineering, 69*, 106744. https://doi.org/10.1016/j.jwpe.2024.106744

Rafiee, M., et al. (2017). Biogenic synthesis of silver nano-particles using Penicillium sp.: With focus on evaluation of their antibacterial properties. *Journal of Chemical Health Risks, 7*(1), 81–88. https://doi.org/10.22034/jchr.2017.237936.1058

Ray, S. K., Dey, S., Das, A. P., & John, A. (2024). Microplastic waste management through greener technologies: A review. In *Sustainable management of urban plastic waste through circular economic approaches* (pp. 237–261). Routledge. https://doi.org/10.1201/9781003465447-14

Sahu, A., Dey, S., Mallick, K., Priyadarsini, S., Behera, I. D., & Das, A. P. (2024). A review on advanced technology for sustainable management of synthetic microplastic waste. In *Sustainable management of urban plastic waste through circular economic approaches* (pp. 262–279). Routledge. https://doi.org/10.1201/9781003465447-15

Satapathy, S., Kar, D., Priyadarsini, S., Das, A. P., & Kuanar, A. (2024). Application of microorganisms in emerging waste management strategies and remediation of synthetic and cellulosic microfibers. In *Sustainable microbial technology for synthetic and cellulosic microfiber bioremediation* (pp. 183–206). Springer. https://doi.org/10.1007/978-3-031-62600-5_10

Singh, S., Sukla, L. B., & Goyal, S. K. (2020). Mine waste & circular economy. *Journal of Environmental Management, 270*, 111623. https://doi.org/10.1016/j.matpr.2020.01.616

Swain, S., Purnima Sudha, M., Dey, S., Badu, Y., Bhoi, S. R., Alreshidi, M. A., Qasim, M. T., Nath, H., Ram, D. K., & Das, A. P. (2025). Sources of micro and Nano plastic hydrocarbon pollutants in marine bodies and their treatment. *Environmental Science and Engineering, 218*, 45–67. https://doi.org/10.1007/978-3-031-81615-4_3

Tripathy, B., Sahoo, P. P., Sundaray, H., & Das, A. P. (2024). In-vitro biodegradation of discarded marine microplastics across the eastern coast of the bay of Bengal, India using Exiguobacterium sp. *Environmental Chemistry and Ecotoxicology, 6*, 236–247. https://doi.org/10.1016/J.ENCECO.2024.07.003

United Nations. (2015). *Transforming our world: The 2030 agenda for sustainable development*. https://sdgs.un.org/2030agenda

Wang, H., Zhang, J., & Xu, Z. (2019). Sustainable geopolymer cement: A review on the environmental impact and the future perspectives. *Renewable and Sustainable Energy Reviews, 103*, 23–32. https://doi.org/10.1016/j.rser.2018.11.038

Xiang, J., Liu, Q., Su, Y., Yang, J., Liu, H., & Wang, H. (2021). A global-scale spatial assessment and geodatabase of mine areas. *Global and Planetary Change, 204*, 103578. https://doi.org/10.1016/j.gloplacha.2021.103578

Yuan, S., et al. (2015). Phytoremediation of heavy metals with lactic acid bacteria: A new approach for sustainable mining waste management. *Environmental Technology, 36*(14), 1931–1941. https://doi.org/10.1080/09593330.2015.1024441

Current Biotechnologies for Bio-Upcycling of Mining Waste to Critical Minerals

Subhashree Parida, Sudeshna Dey, Amel Gacem, Maha Awjan Alreshidi, Krishna Kumar Yadav, and Alok Prasad Das

1 Introduction

India has emerged as the second largest country that produces wastes through coal mining, and around 170 million tons of mining-related solid waste has been on generation that is expected to increase to 300 million tonnes in size. Implementation of sustainable practices in salvage and upgrading processes should help to minimize the adverse impacts associated with the production of materials and their disposal (Das & Ghosh, 2023a, 2023b; Priyadarshini, Jagatee, & Das, 2024). Diversity of native microorganisms, bacteria, fungi, and archaea is being implemented to remediate different environmental contaminants (Bala, Dogra, et al., 2024; Dey, Bissoyi, Koteswara Reddy, & Das, 2024; Dey, Dash, Ray, et al., 2024; Ray et al.,

S. Parida · A. P. Das (✉)
Department of Life Sciences, Rama Devi Women's University, Bhubaneswar, Odisha, India

S. Dey
Department of Life Sciences, Rama Devi Women's University, Bhubaneswar, Odisha, India

Centre for Biotechnology, Siksha 'O' Anusandhan University, Bhubaneswar (Deemed to be University), Bhubaneswar, Odisha, India

A. Gacem
Department of Physics, Faculty of Sciences, University 20 Août 1955, Skikda, Algeria

M. A. Alreshidi
Department of Chemistry, College of Science, University of Ha'il, Ha'il, Saudi Arabia

K. K. Yadav
Department of Environmental Science, Parul Institute of Applied Sciences, Parul University, Vadodara, Gujarat, India

Environmental and Atmospheric Sciences Research Group, Scientific Research Center, Al-Ayen University, Nasiriyah, Iraq

© The Author(s), under exclusive license to Springer Nature Switzerland AG 2026

A. P. Das (ed.), *Bioremediation Technology for Sustainable Mining Waste Management*, Sustainable Environmental Waste Management Strategies,
https://doi.org/10.1007/978-3-032-03468-7_18

2024). Different microorganisms may have different inherent mechanisms of degradation for the specific pollutant and may also be tolerant of higher levels of metals. A promising approach for this growing pollution through mining could be bioremediation of mining sludge during bioenergy formation as a product of biofuels and organic acids (Das, Dey, & Das, 2024). Upcycling is the act of recycling waste streams into higher-value products. The by-products from landfilling such materials as plastics, rubber, and electronics are alarming, especially as the global population increases, in addition to production costs being cheaper across the board. Furthermore, batteries, fly ash, solar cells, automobiles, and appliances, as well as perfluoroalkyl substances, all produce hazardous byproducts (Ghosh, Sinha, et al., 2023; Mohanty, Dey, et al., 2024; Sahu & Das, 2024; Sahu, Dey, et al., 2024). However, these waste streams exhibit high resource potential, containing significant concentrations of valuable metals and commodity chemicals that are extractable. For example, to put things into perspective, 140 million tons of solid waste were put into the United States' landfills, and only 67 million tons of plastic were recovered in 2018 (Wyss et al., 2023). There is a great need for sectors to be quick in reducing their emissions of greenhouse gases, a move to prevent catastrophic climate change. The sector with a central role in its contribution as a greenhouse gas contributor is the mining sector. It will try as much to cut down its greenhouse gas emissions, but it will increase its emissions as much as it can. This is due to an increasing requirement for a few precious metals around the globe, which will triple. The mining industry is a critical source of providing key metals for any shift towards a zero-carbon economy. Some of these are vanadium (V), tungsten (W), tin (S), tellurium (Te), rhenium (Re), cobalt (Co), gallium (Ga), germanium (Ge), indium (In), lithium (Li), nickel (Ni) which, on their side, play a central role in alternative energy technologies including wind turbines, Li-ion batteries or solar panels, whereas their high-tech applications such as for smartphone devices, smart televisions or electric vehicles. Such elements are crucial to other countries due to the input that they make into society and the vulnerability of disruption of supply (Priyadarsini & Das, 2024, 2025). For instance, in the automobile sector, the converter is utilized in the form of rhodium-based catalysts for decreasing the level of GHG emissions. This accounts for 69% of the production capacity worldwide, and the prominent suppliers of this metal include Russia and South Africa (Villa Gomez et al., 2024). Such an approach that can ensure critical metals come into supply while reducing emissions of GHG is metal recovery from mining waste. It might also help minimize the buildup of waste stockpiles, an issue that traditionally always seems to exist in this economy of lines. The number of reductions in GHG release arises from the erosion in the amount of energy needed to recover metals from waste is significantly lower compared to the large amounts of energy the mining industry consumes to transform raw materials into marketable metals. For instance, nearly all the power expended (practically all is fossil fuel-based) is in extracting, crushing, and grinding ore to recover only 1% or even 1 ppm of the mass of mined material. There is a valid reason to extract metals from waste. Critical metals make up tremendous quantities in processing waste materials and mining for several reasons. Some of the causes are that the infrastructure either lacks efficiency or is inapt for recovery, the economy, and many mineralogical effects (Das, Dey, & Das, 2024; Das, Mallavarapu, & Ghosh, 2023; Das, Ray, Mohanty, et al., 2024; Dey, Tripathy, et al., 2023). Still, it has been seen that value chains for mine-waste valorisation processes, such as tailings, are still incomplete. Among others, expressly obscured by technological and research constraints and poor transfer of technology interventions. New technologies today can recover metals from waste at a fraction of the cost (Villa Gomez et al., 2024). These are the metals, which are also used for various purposes that also exist in our daily life, like arsenic, nickel, cobalt, copper, zinc, silver, gold, platinum, iron, manganese, aluminium, cadmium, lead, and so on. A similar case on copper production that (Elshkaki et al.) suggests in the period 2010–2050 anticipated that requirement for "Cu" will rise to (275–350)% by 2050, and most of the Cu-producing countries by 2050 won't be in a position to support their production, taking into account their real share of worldwide production of copper (Paul et al., 2023, 2024). Today, Bloomberg NEF Company reports that seven million electric automobiles are presently on the production line all over the world; expected to rise by 30% by 2040. Consequently, if each of these electrified vehicles includes 85 kg of Cu, an estimate indicates that Cu miners will have to triple their production of Cu, which now stands at 20 million tons globally, to serve the increased requirement as take-up of electric cars shoots by 30%. A few of those are strategic primary materials, highly essential to the economy of a country (Bala, Anshul, et al., 2024; Das, Dey, Bala, et al., 2024; Das & Ghosh, 2023c; Mohanty, Tripathy, et al., 2024). The REEs with antimony, indium, lithium, cobalt, and copper are considered rare earth metals due to its risk, economic, and source importance. As an example, many applications of copper are seen, especially containing circuit boards, batteries, conductors of heat and infrastructure, plating, electricity, construction, and electrical and electronic equipment. In addition to this, an organism requires copper as a micronutrient (Cozma et al., 2024). The other one is waste groups, such as mobiles, metal waste: printed circuit boards, batteries, computers, and other E-devices; coal fly ash, mine tailings/waste, contaminated dust, soil slag, or municipal solid waste incineration fly ash, that are everything together known as solid wastes which may have a potential for being a valuable resource for future metal recycling (Bal et al., 2018; Ghosh et al., 2016; Sahu, Mallick, & Das, 2024). For example, apart from these toxic materials (mercury, cadmium, arsenic, PAHs, chromium and dioxins, lead), electronic waste can also hold some other valuable minerals like silver, gold, palladium, platinum, silver, and gold or some of precious metals including iron, copper, zinc, nickel, and aluminium that may be collected and thus from its decay. The statistics revealed that in 2019, the worldwide electronic waste production was evaluated at 53.6 million tons and material recovery at about USD 57 billion. Umicore, a pyrometallurgical smelter based in Belgium, is the biggest e-waste processing facility, which treats about 350,000 tons of e-waste to recover over 100 tons of gold and over 2400 tons of silver each year. This has paramount importance in providing metal and material sources while reducing the requirement for finite mineral and natural resources, in addition to treating waste with minimum extents of environmental damage caused by disposal (Mishra & Das, 2021; Mishra, Dash, Al-Tawaha, et al., 2022; Pradhan et al., 2024; Priyadarshini & Das, 2024). Bioremediation of mining waste pollutants is the least expensive, most environmentally friendly, and the safest approach among all (Dey, Dash, & Das, 2024). On the contrary, another most commonly used alternative for metal bio-extraction/bio-recovery from solid waste matrices are biomining or bioleaching, and biosorption (Das & Ghosh, 2023e, 2023h; Ghosh & Das, 2018). Bioleaching or bioleaching technology, as it is now defined, is the extraction of metals from their ores, concentrates, and solid compounds with various microbial processes (Das & Ghosh, 2023a, 2023j; Mallick et al., 2023; Priyadarshini, Santhosh Kumar, & Das, 2024; Singh Jadaun et al., 2022). Recent discoveries have made bioleaching an increasingly important technique in the mining industry for extracting copper. From hydrometallurgy standpoint, biomining process is the ability of only a few microorganisms to naturally dissolve some of the mineral constituents through microbial attack that results in metal solubilization in industrially relevant settings starting from low-grade ores of gold, manganese, and uranium (Das & Ghosh, 2018). Bioleaching is partly used for the leaching of copper sulphides and also has operation so; Nickel, zinc, and cobalt ought to be recorded as having also been recovered by this method during oxidative pretreatment of refractory gold ores (Mishra et al., 2024; Priyadarshini & Das, 2024; Sahu, Dey, et al., 2024). Recent advances include use of extremely acidophilus organisms in the treatment of electronic waste, metal extraction procedures from oxidized ores and selective metal recovery from aqueous streams (Ghosh, Tripathy, et al., 2023). Some of those bacteria include the fully developed acidophilic chemo lithotrophic bacteria and archaea found to accelerate the oxidative dissolution of sulphide minerals for use in the development and subsequent application in extracting metals using recently developed biotechnology for sulphide ores and concentrates (Dey, Das, et al., 2023; Khatun et al., 2023; Mallick et al., 2024; Mishra et al., 2019). For example, microorganisms in the process of biomining, dissolve gold, other noble metals, minerals enclothing target metals are removed through the first process and dissolved by mining using thiosulfate or cyanide for gold. For such examples, the stage of biological pretreatment is called bio-oxidation (Das et al., 2012, 2015a, 2015b; Das & Ghosh, 2023i; Mohanty et al., 2016). However, the fundamental concept and consortium of microorganisms used in both bioleaching and bio-oxidation, to treat concentrates works and mineral ores, is about similar. The estimates suggest that around 5% of gold, 15% of copper and even lesser percentages of remaining minerals such as nickel, zinc etc. are being formed in the world through technologies of biomining (Das & Ghosh, 2018, 2023b, 2023f; Dey, Dash, & Das, 2024; Ghosh, Tripathy, et al., 2023; Mohanty et al., 2017). The application of targeted biomineralization extends to both active and closed mine sites, enabling the recovery of metals from process waters and waste streams (Johnson, 2014). These include the nature-based biological processes that are biomining, biosorption, bioprecipitation, and phytomining processes; all these are more advanced as compared to the usual approaches of pyrometallurgy and hydrometallurgy: use lesser energy, less natures-harming approaches, lesser temperatures, no metals' contamination, and even carbon footprint (Cozma et al., 2024). This book chapter defines how advance technologies help through various processes (like biomining, bioleaching, phytomining) for bio-upcycling and recycling the critical minerals from ores and waste materials.

2 Extraction of Metals from Mining and Metallurgical Residues

Waste streams processed through metal recovery can generally be classified as leachate-concentrated streams, in other words, concentration greater than 1 g/L, from extraction processes applied to solid mining and metallurgical wastes; or diluted streams from liquid mining and metallurgical wastes grouped together as mine-impacted waters, with concentrations between 0.001 and 1 g/L. These waters may be appropriately referred to as: effluent from mineral processing and metal refining operations, post-mine acid waters, and acid mine drainage (Mishra & Das, 2022, 2023; Mishra, Dash, & Das, 2022). The latter terms captivate one's interest for the sheer reasons of their extreme acidity and enormous metal and sulphate content that poses threats to ecosystems. Mine-impacted waters until very recently have been regarded as non-remunerative sources for recovery of critical elements due to their low metal concentrations. However, with extremely high volumes often experienced and changing demands and prices for the metals contained, these waters can be identified for recovery as economically viable and environmentally favourable by available technologies (Ghosh et al., 2017, 2020; Ghosh & Das, 2017; Mohanty et al., 2018). Another point at which the critical metals shine is that in the streams of solid waste-the methods of their recovery demand less energy and chemicals than other streams of metals, according to (Dey, Dash, Ray, et al., 2024). AMD generally contains dissolved metals and metalloids such as cadmium, lead, copper, iron, zinc, aluminium, arsenic, and finally, rare earth elements. The specific metals found and their concentrations are dependent on specific variables, including the reagent that reacted with water, geological and environmental conditions, and hydrological regimes. For instance, while the chemical properties and quantity of effluent produced during metal refining processes may vary with ore deposit types, the concentration and extraction technology will also play a very important role. These include, among others, non-ferrous metallurgical waste waters from extraction and purification with strong specificity to metal speciation, industrial waste waters from the metalworking industries that must be treated to eliminate fluoride, chloride, and sulfonate pollutants; acid/alkaline waste waters from rare earth hydrometallurgy; and concentrated effluents containing ammonia nitrogen (Villa Gomez et al., 2024). The recent reports are indicating increasing global production of waste rock and tailings due to the growing demand for minerals. According to the report World Mining Data 2024, mining activities all over the world produce tens of billions of tons annually. A significant quantity is produced for waste rock and tailings from the mines of copper, gold, and nickel, which are expected to increase in the future decades significantly (Siddiqui et al. (2021), Torell et al. (1993), Zhuang et al.

(2015)). Global coal consumption in 2022 reached eight billion tons, a truly gigantic volume of production that is impressive, but also troubling, at the same time. Although this level of output is undeniably a huge source of revenue, waste rock and tailings from mining activities that are replete with toxic substances do not help the environment (Dey, Bissoyi, Koteswara Reddy, & Das, 2024). Mineral extraction waste can be solid, slurry with tailings, waste rock, ash, and tail ends being the most frequent, but vegetation and sediment may also be termed trash in some cases. To minimize environmental impact, waste is stored in water bodies(ponds), reservoirs, or tips in line with local waste control treatment regulations relevant to each mining region, as well as recycling when technically feasible (Ghosh, Tripathy, et al., 2023).

Impacts on Human Health Groundwater quality and mining operations interact in multiple ways. "The potential impact of water on the long-term evolution. Mining activities may cause them to move into adjacent surface and underground waters). Heavy metals and other dissolved heavy metals are serious health risks that can result in, among others, devastating medical conditions like cancer (Das, Mohanty, et al., 2023; John & Das, 2024). Certain rivers have the potential to fail to reach a globally excellent chemical and ecological state due to past mining activities. The most important sources of metal discharge here are cadmium, copper, iron, and zinc. In some locales, chloride and sulphate plumes flow into or risk saturating threatened drinking water resources.

Impact on Waters and Aquatic System The impact of these on the aquatic community can cause various serious environmental effects, such as decreased numbers and species diversity in benthic invertebrates (thus loss of ecosystem services), increased fish mortalities, and reproductive failure leading to reduced abundance and biodiversity throughout rivers. Iron-rich mine waters can kill fish and invertebrates as they pass along the riverbed, coating it with iron hydroxides, also known as ochre. Aquatic life is particularly vulnerable to this type of pollution as fish need open, well-aerated gravels in which they go upstream and lay their eggs. Subsequently, water may be very toxic due to its low pH level, which has adverse effects on biodiversity (Bal et al., 2016; Ghosh et al., 2020). Metal solubilities of aluminium, copper, lead (Pb), zinc, and cadmium may be increased under acidic conditions leading to these metals becoming highly available for toxic effects (Das, Mallavarapu, & Ghosh, 2023). In the light of this scenario, recovery of metals through bioleaching will play a significant part in changing the present scenario, as the rapidly depleting natural resources and substitute for metal recovery techniques are the current need (Das, Dey, Bala, et al., 2024).

3 Recovering Rare and Critical Metals

The complexity of the composition and varying physicochemical properties of waste streams make metal recovery difficult to carry out efficiently. That is why such innovative technologies are required for flexibility as well as high selectivity to differentiate metals in heterogeneous wastes, both high and low concentrations (Spooren et al., 2020). It offers an attractive pretreatment opportunity where physical separation techniques can concentrate valuable metals before chemical processing and allow easier recovery with minimum environmental impacts. However, the optimum mineral liberation normally requires more comminution steps to adequately expose the valuable metals. In that case, advanced and sustainable methods need to be developed and implemented to improve the outcomes of recovery against the complexities of modern waste streams (Villa Gomez et al., 2024). Hydrometallurgy, which is the extraction of metals from liquid streams, can be divided into the recovery of metals in solutions by reduction, chemical precipitation, cementation, adsorption, solvent extraction, electrochemical techniques, and ion exchange. Some methods have gone into the canonical hydrometallurgical techniques, those suited for the dilute streams must also include biological-based processes already applied at full-scale in the recovery of major metals from waste streams. Examples include sulphate-reducing bioreactors designed and operated to recover Zn from Zn-contaminated groundwater; the successful recovery of Cu from hydrometallurgical streams produced by Au/Cu mineral processing sites(Das, Dey, & Das, 2024; Das, Dey, Bala, et al., 2024, Paul et al., 2023) Then next shall follow a presentation of recent advancements in critical metal recovery technologies. These methods are based on existing technologies that have been proven on an industrial scale for the recovery of other minerals or for the removal of metals without subsequent resource recovery. Solvent extraction, electrochemical methods, membrane technologies, and bio-based systems are the most significant breakthroughs in the field (Villa Gomez et al., 2024).

4 Advancement in Critical Metal Recovery Technologies

4.1 Electrochemical Process

For decades, these processes have been employed to recover metals from concentrated aqueous streams (Kim, 2021). While some metallurgies are recovered industrially through electrowinning, where these metals are prepared from solution at a reasonably economical cost of aqueous ore-derived solutions. For example, electrowinning along with solvent extraction is a hybrid process and the highly pure Cu obtained from this process is used in various plants across the globe and was able to figure out a method of extracting Co from ores via electrowinning by Dehaine et al. (2021). This basically means, besides copper electrowinning from ores, that aqueous metal electrowinning is performed exclusively in parallel plate geometries using inert anodes. High-rate electrowinning cells can make for that sort of fabrication of more or less dilute solutions into cylindrical metal products (Das & Ghosh, 2023c, 2023d, 2023g). There exist a few review papers on copper electrowinning in the set of their works, especially those handling the recycling of e-wastes . Co-forms lump formations rather than plates and, thus, these flake off and fall off the deposition site on the cathode due to deposits usually being under tensile stress. The most in demand to extract important metals from the present time for diluted and leachate streams is electrochemical methods, as well as electronic wastes, metallurgical and mining wastes, states (Dehaine et al., 2021; Kim, 2021). It is mostly used because no toxic chemicals are involved; there are no secondary wastes, and there only tiny alterations to the pH or temperatures (Das, Dey, Bala, et al., 2024). Electrochemical methods include the addition of an external source of energy in the form of an electrochemical cell, to drive the chemical reactions. They can be divided into four classes: electrowinning, electrosorption, electrodialysis, and electrocoagulation.

4.1.1 Electrowinning

Electrowinning is an electrochemical process for metal recovery through deposition from primary metal sources or ores in dissolved form with passing a constant current in between an anode and a cathode. It is also termed electrodeposition (Hu & Cheng (2016), Keyhani et al. (2010)). Electrorefining is the process in which impure metal is removed out of an anode. The technique, in general, takes up less energy since it has more negative potentials when compared to the chlorine or oxygen gas emission. Enhanced inert catalytic anodes illustrated major focus of interest in electrowinning.

4.1.2 Electrosorption

It refers to increases the rate of adsorption for charged species that comprises ions by forcing them towards an oppositely charged electrode stripping them from the liquid due to a field (Kim, 2021). In electrodialysis, the combination of electrical potential with the use of an ion-exchange membrane facilitates the separation of ions because the ions possess polarity in the charge. In electrocoagulation, Fe or Al sacrificial anode produces ionic coagulants: $Fe^{2+/3+}$ or Al^{3+}; anode decomposes water to oxygen and H^+; cathode decomposes water to hydrogen and OH^-; reduction of other metal ions may take place at cathode. Formation of different metal hydroxides and complexes depends on pH of solution which will adhere to the particles and provoke coagulation. This further makes the hydrogen and oxygen gases easy to separate pollutants using flotation where the coagulants separate from the liquid (Kim, 2021; Villa Gomez et al., 2024).

4.1.3 Critical Metal Recovery

Among such techniques are: electrodeposition, electrodialysis, electro filtration, and electrocoagulation in the extraction of metals. Among them include the following Te, In, Eu, W, Co, Li, V, and Ga metals extracted from wastes in the waters. It is highly applied in recovering non-ferrous metals, and recovery of noble metals also includes: Ag, Pt, Co, Ni, Sn, Au,. This can be seen in critical metals like Ga, In, and Te whose redox potential is near that of the standard hydrogen electrode; these are thus recoverable selectively. However, unwanted metals can be stripped through either potentiostatic or galvanostatic operations, but mostly alloy co-deposit as such when their redox potential is within the time range, in which metals similar to Co (E0 = −0.28 V) and Ni (E0 = −0.26 V) are often close within redox (Golodnitsky et al., 2002; Kim, 2021). There is recent growing interest in the extraction of electrochemically Li from Salt-Lake brines as well as from seawater (Golodnitsky et al., 2002; Liu et al., 2017). In this route of electrochemical extractions, it has utilized Li-selective material such as $FePO_4$ to intercalate within the Li-ions then recovered through passing positive currents (Palagonia et al., 2019). The use of selective electrodes and counter electrodes must increase efficiency. However, Mn results in water pollution particularly if Li originates from lithium manganese oxide minerals (Xue et al., 2019). This process is a similar function of a lithium-ion battery in which the Li intercalates in the structure of Ni-Co-Mn oxide or Co oxide.

4.2 *Supported Liquid Membranes*

They are damped with a hydrous or electrolytic solvent, being the interface between two liquid phases: the feed phase and the stripping phase. They are known as supported ionic liquid membranes (SILMs) if the use involves an ionic solvent (Lozano et al., 2011). The extraction process of SLM consists of three stages: first, the removal of molecules from the feed phase into the membrane; second, diffusion through the membrane; and, third, re-extraction into the stripping phase. There are two kinds of SLMs available: flat sheet SLM and hollow fibre SLM. The simple lab setup of FSSLM is not industrially used. The outer shell and hollow fibre module with an inner compartment were used in HFSLM. The shell passes the stripping phase and uses thin fibres in the feed phase (Hidalgo-Ulloa et al. (2022), Hlihor et al. (2022)). These are the advantages of SLMs: they operate at low cost, very few require energy use, it is friendly to the environment, avoid the generation of sludge, separation efficiency highly high, and possibly total recovery may be affected (Villa Gomez et al., 2024).

4.2.1 Metals Recovery

Supported liquid membranes have been used for the recovery of metal, primarily Ni, Te, W, V, Ga, Ge, In, and Co, mainly in flat sheet SLM experiments. Most of these works used artificially prepared aqueous solutions, but others were concerned with industrial waste, and fortified environmental matrices samples such as river water and soil (Ghanim et al., 2020). Ionized liquids-doped polymeric barrier film has also been prepared in the recovery of lithium, with one disadvantage in their short-term application being poor durability (Chen et al., 2021; Zante et al., 2019).

Thin-Film Composite Anisotropic Membrane

The composition of multilayer thin-film composite membranes includes the thin layer acting as an overlayer selectivity placed at the top of the thicker porous support. Usually, every overlayer and/or support consists of polymers of different types (Baker et al., 2004;). Thin-film composite anisotropic membranes can be seen with their applications in several process applications like pervaporation, reverse osmosis, gas separation, and nanofiltration (Baker et al., 2004). Their design allows independent layer optimization. The thin surface layer has selectivity enhancement properties while the porous support provides strength, resistance to compression, and conductivity (Lau et al., 2012;).

Critical Metals Recovery

Meschke et al. (2020) studied that Ge and Re were finally recovered from the bioleached Theisen sludge through TFC membrane separation, following other steps in the process chain. Before the treatment, leachate from the slag was pre-treated with microfiltration membranes. A ceramic membrane disk for greater barium removal, produced in China, had a nominal size of 0.2 μm. This one should theoretically have the potential for superior recyclability because it combines the removal of sludge particles and bacteria. After sterilization, the solution was passed through a commercial TFC nanofiltration membrane, NF99HF. This NF99HF retained ≥99% of the cationic metals Zn Cu, Co, and Fe and about ≥97% of Mo and ≤14% of Ge (Mo8O264-) and Re (ReO4-). Finally, a TFC reverse osmosis commercial membrane, RO90, was used, allowing to retain ≥97% Ge and >4% Re. TFC membranes together achieved 73% Ge and 76% Re recovery. It was therefore proved that membrane filtration is applicable for concentrating and separating critical metals from waste streams (Villa Gomez et al., 2024).

4.3 Solvent Extraction and Synergistic Solvent Extraction

It is a key process in hydrometallurgy. Hydrometallurgy mainly applies solvent extraction, and metals that can be separated are often those containing Cu, Co, Ni, and U. It involves the exchange of chemical species through mixed organic and aqueous liquid solutions with selective transfer between two phases. Flexibility in the design of molecules that form extractants could help enhance selectivity. The most practical method to employ in liquid–liquid extraction is when the aqueous solution has a high enough concentration of metal. In solutions that are very dilute, other methods may be more useful since the engineering factors begin to outweigh the high loadings of metal. Finally, the replacement of organic liquid and constituents including the extractant and modifier can be very costly. Another established and efficient technique for extraction of metal from both primary and secondary sources of metals is the technique of solvent extraction is synergistic solvent extraction. It is a process which involves mixing of leaching solution containing a particular desired metal extraction into an organic phase is achieved by mixing with a suitable organic solvent. Solvent extraction success depends on properties regarding the extractant molecules which have to be prepared on specific separation requirements (Love et al., 2019). Recently, a technique devised is called synergistic solvent extraction (SSX), which includes using two or more reagents to produce metal recovery, which results in efficiency higher than any of its individual counterparts. SSX systems are mixtures, including acid and neutral extractants or chelating and neutral extractants (Bond et al., 2000). The most simply understood and the most frequently used system consists of an acid extractant and a neutral extractant (Atanassova et al., 2016). Separation of the metal from the extractants is carried out in the stripping process after extraction. For example, SSX in its application was utilized within the Boleo operation in Mexico where Lix63 and Versatic 10 were applied as a blend of reagents for the separation of Co and Zn from a complex feed stream of other metals except that there were issues that involved Mn-induced degradation of the extractant and problematic pH control (Marte et al., 2017).

Critical Metal Recovery Synergistic solvent extraction, or SSX, has successfully been demonstrated to separate and recycle the important metals: Co, Ga, V, Ge, In, Ni, Te, W, and Re (Cheng et al., 2015). Co and Ni are targeted primarily because no commercial extractant has proven highly specific towards these metals (Cheng et al., 2015). In the last 30 years, several SSX systems have been evolved, like Versatic 10/4PC and Versatic 10/LIX63, which are used for recovery of Ni and Co from chloride-leach solutions and sulphuric acid containing metals like Mg, Mn, and Ca (Chen et al., 2021; Cheng et al., 2015). The extraction of Li from brines and seawater is difficult because Li has a high hydration tendency (Nguyen et al., 2018). D2EHPA and MEHPA extractors can selectively extract Li from geothermal water, but commercial acidic extractors have very low extraction efficiencies (Hano et al., 1992). SSX has been tested on real mining waste solutions and synthetic mining waste solutions to be a potential tool for separating and recovering metals from mining waste (Liu et al., 2017; Zeng et al., 2020). Most experiments, however, are still being carried out at the lab scale, and so large-scale trials are needed to determine whether it can be commercially viable. More work remains in this area, but on the operational challenges-wherein organic liquid chemistry will be monitored in situ as it forms during the reaction process (Cheng et al., 2015; Liu et al., 2017).

4.4 Biometallurgy

The term biometallurgy is employed by integrating all bio-based processes involving the action of microorganisms and biological substrates in recovering metals through metal solubilization and immobilization (Gavrilescu, 2019), these are processes highly appealing to operators handling pouring assets in a small, high-volume waste stream that is energy-demanding and having low energy input to run (Nancharaiah et al., 2016a). In these ways, the recovery of metals occurs by immobilization of the metal and the interactions between the metals and the microorganism's allowing precipitation, reduction, and adsorption that result in concentration and separation of metals from liquid streams. These interactions are mediated either by the microorganisms themselves or by the application of electric current (Nancharaiah et al., 2016b; Ostermeyer et al., 2022). One of the options for metal recovery is an exopolysaccharide (EPS), which is produced as a protective agent for survival in metals by microorganisms in harsh periods and exhibits different sequestering potentials. Thus, it may be one of the alternative options to separate or recover critical metals, as reported in Gavrilescu (2019).

4.5 Bioprecipitation

Bioprecipitation—the process by which some bacteria, through enabling the metals to precipitate out from the waters, form the latter as carbonates or hydroxide and sulphide. In fundamental terms, this involves the reaction carried out using sulphate-reducing microorganisms with organic compounds being the reducers or using hydrogen as the electron carrier. When the conditions do not favour the presence of oxygen, sulphur or sulphate is reduced directly to be a sulphide. This "biogenic sulphide" supports the formation of insoluble metal sulphides even under acidic or low-metal conditions, and this can be complemented well with sulphate in mine-impacted waters. Since then, it has been commercially applied for metal extraction from mining and industrial wastes, including Zn and Cu (Ghosh et al., 2023; Pradhan et al., 2023).

Critical Metal Recovery Some of these include critical metals, for which applied for the removal of mixed metal from simulated and actual AMD. There are a few papers researcher might lead to selective extraction of critical metals by bioprecipitation. Some examples have succeeded in the performance of restoration based on the adjustment of the concentration and pH of sulphide, that gives the solubility of sulphide precipitates of present metals. Therefore, the method has found its utility in the separation of a great difference between the solubilities of metals, for example, Fe and Ni (Bijmans et al., 2009). Possible metals recoverable by this method include any of the critical metal elements with an affinity for precipitating as metal sulphides, such as Co or Ni, but not critical metals like Ga and Ge (Fan et al. (2012), Fireman et al. (2016)). During biosorption and biological sulphate reduction through other methods including biosorption besides selective by above, scientists endeavoured to bridge some of these challenges during the critical metal recovery process. According to Liu et al. (2021), during the biosorption of EPS, this enabled the selective complexation of Ni with efficiencies that remained up to levels >99% in the recovery process, up to precipitate Co from the solution of >84%. This sulphur (S),

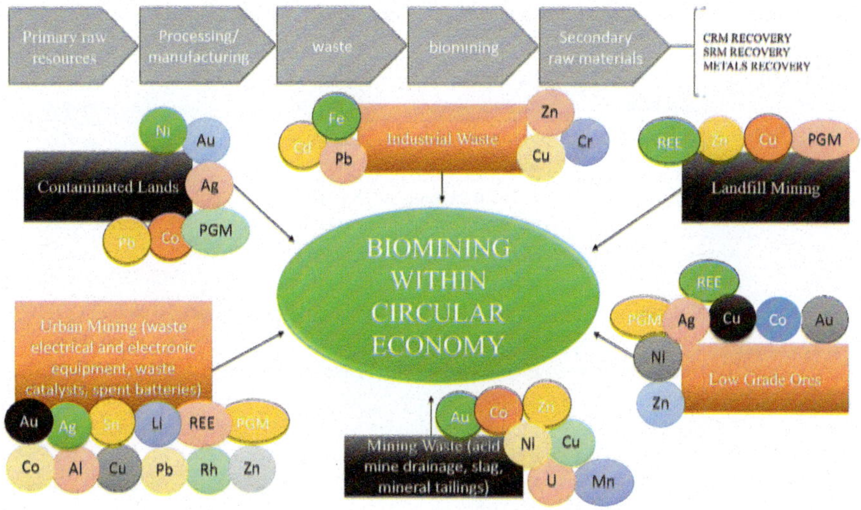

Fig. 1 Bio-upcycling of mining waste

selenium (Se), and tellurium (Te) produced by microbes, which was accompanied by bioprecipitation with biogenic sulphide, had biosorption of gallium (Ga), germanium (Ge), and zinc (Zn) as reported by. Ga and Zn had efficient removal at 69% using a biosorbent for tellurium and 60% using biogenic sulphide. Recovery of germanium was not reported. These studies, therefore, indicate that there may be a combination of biosorption and bioprecipitation in such an effective recovery of metals.(Villa Gomez et al., 2024). The bio-upcycling of mining waste is represented in Fig. 1.

Table 1 presents information about sources, methods, and technologies of metals and critical minerals:

4.6 *Biosorption and Bioaccumulation*

There are two techniques for the removal of substances from solutions by biomass: biosorption and bioaccumulation. In biosorption, the process takes place with non-living biomass, whereas in bioaccumulation, the substance is first adsorbed on the surface, and then the living biomass takes it into the cell. Chojnacka (2010) states that. both are physicochemical operations that contain mechanisms of chelation, absorption, adsorption, ion exchange, and diffusion across the cell walls. These processes are more rapid and reversible due to aspects such as the types of temperature, biomass, pH, and the existence of interfering ions (Gadd, 2010; Gavrilescu, 2004). Bioaccumulation happens slowly; its rate is tied to the growth of the biomass and has energy requirements and is limited due to the toxicity of the substance

Table 1 Heavy metals, source types, technology, and process for upcycling

Critical metals and metals	Source waste	Various technologies	Process type	Advantages
Co	Tungsten carbide scrap	Electrochemical method	Electrowinning process	High purity recovery
Zn	Mine tailings	Solvent extraction method	Synergistic solvent extraction process	Efficient at large scales
Ni	Aqueous solutions	Biometallurgy method	Biosorption process	Cost effective
Li	Salt like brine	Electrochemical method	Electrodialysis process	Selective separation
Cu	Hydrometallurgical streams	Electrochemical method	Electrowinning process	Commercially scalable
Ga	Ga-in solution	Membrane based technique	Supported liquid membranes process	High selectivity
V	Acid leaching solution	Electrochemical method	Electrosorption process	No secondary waste
Re	Bioleached sludge	Membrane based technique	Thin-film composite membranes process	Selective recovery
Ge	Industrial waste leachate	Solvent extraction method	Supported ionic liquid phases process	High selective
Te	Leachates	Biometallurgy method	Bio-electrochemical systems process	Energy harvesting possible
In	Electronic waste	Electrochemical method	Electrocoagulation process	No chemical reagents

accumulated (Aksu, 2005; Chojnacka, 2010). It is used in treating large quantities of wastewater with comparatively lower levels of contaminants, (Gavrilescu, 2004.)

Critical Metal Recovery In various biosorption studies, seagrass, algae, bacteria, and crab shells have been employed for the recovery of Sn, Co, Ni, Re, V, and W from an aqueous medium. Only few used bacteria and yeast; this was based on the ability for examining further the potentiality of Ni and Co recovery through bioaccumulation from aqueous solutions. Even though the numbers of studies are not the same, these studies show that metals, critical as they are, can be recovered from multiple metal-containing liquid solutions by biosorption and bioaccumulation, and the biomasses used show different levels of attraction to the metals dissolved in the solution. This technology has a drawback: so far, most studies have been small-scale and have focused on getting metals from man-made solutions. Examples include work by Akhtar et al. (2004), and Ghanim et al. (2020). To make the most of these technologies, we need studies that use large amounts of solution from actual mining waste.

4.7 Bio-Electrochemical Systems

Bio-electrochemical systems use microorganisms, metals, and electrodes to obtain redox reactions in recovering metals. The recovery of metals is enhanced through the oxidation and reduction process, where the breakdown of organic matter could be through the transfer of electrons by microbes. They can operate using one electrode: bioanode or both, the bioanode & the biocathode (Dominguez-Benetton et al., 2018; Pat-Espadas & Cervantes, 2018; Albero et al. (2019), Curigliano et al. (2017)). One of the biggest advantages of BES is that it allows selective metal recovery due to the changes in the potential of the electrodes. Power is generated when positive reduction potential metals are used, although external power would be needed to generate power using metals with a negative reduction potential. Systems of BES are grouped into four types, mainly based on their mechanisms of metal recovery:

1. Abiotic cathodes, either with or without external power.
2. Bio-cathodes, with or without external power.

These mechanisms are constituted by direct metal recovery or metal conversion mechanisms. These have been studied and elaborately discussed in various literatures (Dominguez-Benetton et al., 2018; Nancharaiah et al., 2016a). The advantages include the consumption of less energy compared to electrochemical methods, cheaper microbial catalysts instead of costly metals, and selectivity to target metals with the ability to drive thermodynamically unfavourable reactions using specific micro-organisms (Dominguez-Benetton et al., 2018). On the other hand, the disadvantages are that the metal concentrations in the microbe are fluctuating and showing instability within electron transfer channels, and BES systems do not scale well into commercial applications (Dominguez-Benetton et al., 2018).

Critical Metal Recovery BES has inspired interest in critical metal recovery, and there are many configurations of bioanodes and bio-cathodes and microbial communities that have been conceptualized and fabricated (Dominguez-Benetton et al., 2018; Nancharaiah et al., 2016b). BES can be used to recover molybdenum (Mo), vanadium (V), (Co), nickel (Ni), tin (Sn), indium (In), rhenium (Re), and tungsten (W). However, most of the works are limited to laboratory-scale experiments and are not specific to mine waters, with further research and development needing to be applied practically. The earlier BES research primarily focused on the bioanode where the organic compounds get oxidized to generate power or drive target reactions. Earlier steps of metal extraction relied on electrochemical reduction or precipitation of dissolved metals in an abiotic catholyte, in which metals Co and Ni are recovered (Ostermeyer et al., 2022; Sethurajan & Gaydardzhiev, 2021; Yu et al., 2020). The other approach to integration is through the integration of bioanodes with metal extraction in cathodes through precipitation of metals as hydroxides or carbonates with a potential that permits selective recovery also. Another promising method that also seems to gain momentum is the retrieval of critical metals as Co,

V, and Te, even though a bioreduction mechanism in biologically active cathodes (Ostermeyer et al., 2022) (Ghosh et al., 2016).

5 Conclusion

Bio-upcycling can present a sustainable and environmental-friendly solution to the recovery of critical minerals from mining waste, including REEs, lithium, and cobalt, in addressing environmental and resource challenges. By using biological processes bioleaching, biosorption, and bioprecipitation, this method is done through synthetic biology and genetic engineering, which enhance the efficiency of metal recovery with low emissions of greenhouse gases and waste. One means to achieve metal recovery with superior efficiency and scale-up capability can be accomplished with complementary technologies of bio-electrochemical systems, electrochemical methods, and advanced membranes. Waste streams like e-waste or mine tailings may thus become sources of valuables by utilizing approaches contributing towards circular economy development, avoiding more significant depletion of natural, limited reserves. Although promising in nature, areas of scalability in process optimization with cost efficiency and further research/innovation need to be combined with policy supports to industrialize these methods into a sustainable future. Bio-upcycling is found to be coherent with global goals for sustainability support in the new zero-carbon economy.

References

Akhtar, N., Iqbal, J., & Iqbal, M. (2004). Removal and recovery of nickel (II) from aqueous solution by loofa sponge-immobilized biomass of Chlorella sorokiniana: Characterization studies. *Journal of Hazardous Materials, 108*(1–2), 85–94. https://doi.org/10.1016/j.jhazmat.2004.01.002

Aksu, Z. (2005). Application of biosorption for the removal of organic pollutants: A review. *Process Biochemistry, 40*(3–4), 997–1026. https://doi.org/10.1016/j.procbio.2004.04.008

Albero, B., Tadeo, J. L., & Pérez, R. A. (2019). Ultrasound-assisted extraction of organic contaminants. *TrAC Trends in Analytical Chemistry, 118*, 739–750. https://doi.org/10.1016/j.trac.2019.07.007

Atanassova, M., Kurteva, V., & Dukov, I. (2016). The interaction of extractants during synergistic solvent extraction of metals. Is it an important reaction? *RSC Advances, 6*(84), 81250–81265. https://doi.org/10.1039/C6RA18478B

Baker, C., Shah, S. I., & Hasanain, S. K. (2004). Magnetic behavior of iron and iron-oxide nanoparticle/polymer composites. *Journal of Magnetism and Magnetic Materials, 280*(2-3), 412–418. https://doi.org/10.1016/j.jmmm.2004.03.037

Bal, B., Armstrong, P. B., & Das, A. P. (2016). Development of indigenous bio-sensing methodology for rapid and low cost endotoxin detection system. *Sensor Network Data Communication, 1*, 5.

Bal, B., Ghosh, S., & Das, A. P. (2018). Microbial recovery and recycling of manganese waste and their future application: A review. *Geomicrobiological Journal, 36*(1), 85–96. https://doi.org/10.1080/01490451.2018.1497731

Bala, K., Anshul, A., Verma, S., Dogra, A., Das, A. P., Kumar, R., & Nandini, A. (2024). Sustainable management for urban plastic waste generation. In *Sustainable management of urban plastic waste through circular economic approaches* (pp. 211–236). Springer. https://doi.org/10.1201/9781003465447-13

Bala, K., Dogra, A., Lata, P., & Das, A. P. (2024). Discussing sources and biological-chemical remediation approaches for microplastic pollution. In *Sustainable microbial technology for synthetic and cellulosic microfiber bioremediation* (pp. 37–59). Springer. https://doi.org/10.1007/978-3-031-62600-5_3

Bijmans, M. F., van Helvoort, P. J., Dar, S. A., Dopson, M., Lens, P. N., & Buisman, C. J. (2009). Selective recovery of nickel over iron from a nickel–iron solution using microbial sulfate reduction in a gas-lift bioreactor. *Water Research, 43*(3), 853–861. https://doi.org/10.1016/j.watres.2008.11.023

Bond, A. H., Dietz, M. L., & Chiarizia, R. (2000). Incorporating size selectivity into synergistic solvent extraction: A review of crown ether-containing systems. *Industrial & Engineering Chemistry Research, 39*(10), 3442–3464. https://doi.org/10.1021/ie000356j

Chen, J., Lin, S., & Yu, J. (2021). High-selective cyclic adsorption and magnetic recovery performance of magnetic lithium-aluminum layered double hydroxides (MLDHs) in extracting Li+ from ultrahigh Mg/Li ratio brines. *Separation and Purification Technology, 255*, 117710. https://doi.org/10.1016/j.seppur.2020.117710

Cheng, C. Y., Urbani, M. D., Davies, M. G., Pranolo, Y., & Zhu, Z. (2015). Recovery of nickel and cobalt from leach solutions of nickel laterites using a synergistic system consisting of Versatic 10 and Acorga CLX 50. *Minerals Engineering, 77*, 17–24. https://doi.org/10.1016/j.mineng.2015.01.015

Chojnacka, K. (2010). Biosorption and bioaccumulation–the prospects for practical applications. *Environment International, 36*(3), 299–307. https://doi.org/10.1016/j.envint.2009.12.001

Cozma, P., Bețianu, C., Hlihor, R.-M., Simion, I. M., & Gavrilescu, M. (2024). Bio-recovery of metals through biomining within circularity-based solutions. *Processes, 12*(9), 1793. https://doi.org/10.3390/PR12091793

Curigliano, G., Burstein, H. J., Winer, E. P., Gnant, M., Dubsky, P., Loibl, S., Colleoni, M., Regan, M. M., Piccart-Gebhart, M., Senn, H.-J., Thürlimann, B., André, F., Baselga, J., Bergh, J., Bonnefoi, H., Brucker, S. Y., Cardoso, F., Carey, L., Ciruelos, E., Cuzick, J., Denkert, C., Di Leo, A., Ejlertsen, B., Francis, P., Galimberti, V., Garber, J., Gulluoglu, B., Goodwin, P., Harbeck, N., Hayes, D. F., Huang, C.-S., Huober, J., Khaled, H., Jassem, J., Jiang, Z., Karlsson, P., Morrow, M., Orecchia, R., Osborne, K. C., Pagani, O., Partridge, A. H., Pritchard, K., Ro, J., Rutgers, E. J. T., Sedlmayer, F., Semiglazov, V., Shao, Z., Smith, I., Toi, M., Tutt, A., Viale, G., Watanabe, T., Whelan, T. J., & Xu, B. (2017). De-escalating and escalating treatments for early-stage breast cancer: The St. Gallen International Expert Consensus conference on the primary therapy of early breast cancer. *Annals of Oncology, 28*(8), 308. https://doi.org/10.1093/annonc/mdx308

Das, A., Dey, S., Bala, K., Gnanasekaran, R., & Das, A. P. (2024). Microbial bioremediation technology for sustainable treatment and management of synthetic microfiber waste. In *Sustainable microbial technology for synthetic and cellulosic microfiber bioremediation* (pp. 221–238). Springer. https://doi.org/10.1007/978-3-031-62600-5_12

Das, A., Dey, S., & Das, A. P. (2024). Ecotoxicological impacts of synthetic microfiber pollutants and development of sustainable mitigation strategies. *Environmental Chemistry and Ecotoxicology*. https://doi.org/10.1016/J.ENCECO.2024.12.003

Das, A., Ray, S. K., Mohanty, M., Mohanty, J., Dey, S., & Das, A. P. (2024). Ecotoxicity of microplastic wastes and their sustainable management: A review. *Environmental Chemistry and Ecotoxicology, 6*, 144–152. https://doi.org/10.1016/J.ENCECO.2024.05.003

Das, A. P., & Ghosh, S. (2018). Bioleaching of manganese from mining waste materials. *Materials Today: Proceedings, 5*(1), 2381–2390. https://doi.org/10.1016/J.MATPR.2017.11.459

Das, A. P., & Ghosh, S. (2023a). Cost-effectiveness of the biomining process. In *Manganese mining microorganisms* (pp. 143–151). Elsevier. https://doi.org/10.1016/B978-0-12-822147-1.00008-X

Das, A. P., & Ghosh, S. (2023b). Evolution of biomining technology. In *Manganese mining microorganisms* (pp. 5–19). Elsevier. https://doi.org/10.1016/B978-0-12-822147-1.00012-1

Das, A. P., & Ghosh, S. (2023c). Factors affecting bioleaching. In *Manganese mining microorganisms* (pp. 77–86). Elsevier. https://doi.org/10.1016/B978-0-12-822147-1.00004-2

Das, A. P., & Ghosh, S. (2023d). Introduction. In *Manganese mining microorganisms* (pp. 1–4). Elsevier. https://doi.org/10.1016/B978-0-12-822147-1.00005-4

Das, A. P., & Ghosh, S. (2023e). Manganese and its application. In *Manganese mining microorganisms* (pp. 29–40). Elsevier. https://doi.org/10.1016/B978-0-12-822147-1.00007-8

Das, A. P., & Ghosh, S. (2023f). Mechanism of manganese bioleaching. In *Manganese mining microorganisms* (pp. 87–98). Elsevier. https://doi.org/10.1016/B978-0-12-822147-1.00014-5

Das, A. P., & Ghosh, S. (2023g). Metagenomic insights into unculturable microbial diversity. In *Manganese mining microorganisms* (pp. 63–76). Elsevier. https://doi.org/10.1016/B978-0-12-822147-1.00002-9

Das, A. P., & Ghosh, S. (2023h). Microorganisms in manganese biomining processes. In *Manganese mining microorganisms* (pp. 55–61). Elsevier. https://doi.org/10.1016/B978-0-12-822147-1.00013-3

Das, A. P., & Ghosh, S. (2023i). Mixed culture bioleaching: An insight into manganese biomining process and efficacy. In *Manganese mining microorganisms* (pp. 129–141). Elsevier. https://doi.org/10.1016/B978-0-12-822147-1.00001-7

Das, A. P., & Ghosh, S. (2023j). Role of metabolites and metabolomics for understanding of manganese bioleaching. In *Manganese mining microorganisms* (pp. 117–127). Elsevier. https://doi.org/10.1016/B978-0-12-822147-1.00011-X

Das, A. P., Ghosh, S., Mohanty, S., & Sukla, L. B. (2015a). Advances in manganese pollution and its bioremediation. In L. B. Sukla, N. Pradhan, S. Panda, & B. K. Mishra (Eds.), *Environmental microbe biotechnology* (1st ed., pp. 313–328). Springer. https://doi.org/10.1007/978-3-319-19018-1_16

Das, A. P., Ghosh, S., Mohanty, S., & Sukla, L. B. (2015b). Consequences of manganese compounds: A review. *Toxicological & Environmental Chemistry, 96*(7), 981–997. https://doi.org/10.1080/02772248.2015.1005428

Das, A. P., Mallavarapu, M., & Ghosh, S. (2023). Ecotoxicity of mining pollutants on the environment and their remediation. *Environmental Chemistry and Ecotoxicology, 5*, 165–167. https://doi.org/10.1016/J.ENCECO.2023.08.002

Das, A. P., Mohanty, A. K., van Hullebusch, E. D., Figueiredo, G., Das, A. P., Mohanty, A. K., van Hullebusch, E. D., & Figueiredo, G. (2023). Bioremediation as an emerging technology for the removal of synthetic microplastic pollutants from marine ecosystem. *MarPB, 194*, 115297. https://doi.org/10.1016/J.MARPOLBUL.2023.115297

Das, A. P., Swain, S., Panda, S., Pradhan, N., & Sukla, L. B. (2012). Reductive acid leaching of low grade manganese ores. *Geomaterials, 2012*(04), 70–72. https://doi.org/10.4236/GM.2012.24011

Dehaine, Q., Tijsseling, L. T., Glass, H. J., Törmänen, T., & Butcher, A. R. (2021). Geometallurgy of cobalt ores: A review. *Minerals Engineering, 160*, 106656. https://doi.org/10.1016/J.MINENG.2020.106656

Dey, S., Bissoyi, A., Koteswara Reddy, G., & Das, A. P. (2024). Synthetic microfibers: An urban emerging pollutant demanding for sustainable environmental solutions. In *Sustainable microbial technology for synthetic and cellulosic microfiber bioremediation* (pp. 107–121). Springer. https://doi.org/10.1007/978-3-031-62600-5_7

Dey, S., Das, A., Mallick, K., Sahu, A., & Das, A. P. (2023). *Environmental petroleum waste: Pollution, toxicity, sustainable remediation* (pp. 159–175). Springer. https://doi.org/10.1007/978-3-031-48220-5_7

Dey, S., Dash, A., & Das, A. P. (2024). Bioremediation as a sustainable and circular solution for mining waste management and bioenergy generation. In *Sustainable management of mining waste and tailings: A circular economy approach* (pp. 107–119). Routledge. https://doi.org/10.1201/9781003442455-6

Dey, S., Dash, N., Ray, S. K., Priyadarsini, S., Chhotaray, C., & Das, A. P. (2024). Occurrence and quantification of tire wear plastics pollutants from highway city dust of Bhubaneswar City of Odisha. In *Sustainable microbial technology for synthetic and cellulosic microfiber bioremediation* (pp. 77–89). Springer. https://doi.org/10.1007/978-3-031-62600-5_5

Dey, S., Tripathy, B., Kumar, M. S., & Das, A. P. (2023). Ecotoxicological consequences of manganese mining pollutants and their biological remediation. *Environmental Chemistry and Ecotoxicology*. https://doi.org/10.1016/J.ENCECO.2023.01.001

Dominguez-Benetton, X., Varia, J. C., Pozo, G., Modin, O., Ter Heijne, A., Fransaer, J., & Rabaey, K. (2018). Metal recovery by microbial electro-metallurgy. *Progress in Materials Science, 94*, 435–461. https://doi.org/10.1016/j.pmatsci.2018.01.007

Fan, F. L., Qin, Z., Bai, J., Rong, W. D., Fan, F. Y., Tian, W., & Zhao, L. (2012). Rapid removal of uranium from aqueous solutions using magnetic Fe3O4@ SiO2 composite particles. *Journal of Environmental Radioactivity, 106*, 40–46. https://doi.org/10.1016/j.jenvrad.2011.11.003

Fireman, E., Shai, A. B., Alcalay, Y., et al. (2016). Identification of metal sensitization in sarcoid-like metal-exposed patients by the MELISA® lymphocyte proliferation test — A pilot study. *Journal of Occupational Medicine and Toxicology, 11*, 18. https://doi.org/10.1186/s12995-016-0101-1

Gadd, G. M. (2010). Metals, minerals and microbes: Geomicrobiology and bioremediation. *Microbiology, 156*(3), 609–643. https://doi.org/10.1099/mic.0.037143-0

Gavrilescu, C. (2019). An analysis of the trade balance for the main agri-food products. In *Papers series agrarian economy and rural development–Realities and perspectives for Romania* (pp. 26–34).

Gavrilescu, M. (2004). Removal of heavy metals from the environment by biosorption. *Engineering in Life Sciences, 4*(3), 219–232. https://doi.org/10.1002/elsc.200420026

Ghanim, B., Murnane, J. G., O'Donoghue, L., Courtney, R., Pembroke, J. T., & O'Dwyer, T. F. (2020). Removal of vanadium from aqueous solution using a red mud modified saw dust biochar. *Journal of Water Process Engineering, 33*, 101076. https://doi.org/10.1016/j.jwpe.2019.101076

Ghosh, S., Bal, B., & Das, A. P. (2017). Enhancing manganese recovery from low-grade ores by using mixed culture of indigenously isolated bacterial strains. *Geomicrobiology, 35*(3), 242–246. https://doi.org/10.1080/01490451.2017.1362080

Ghosh, S., & Das, A. P. (2017). Bioleaching of manganese from mining waste residues using Acinetobacter sp. *Geology, Ecology, and Landscapes, 1*(2), 77–83. https://doi.org/10.1080/24749508.2017.1332847

Ghosh, S., & Das, A. P. (2018). Metagenomic insights into the microbial diversity in manganese-contaminated mine tailings and their role in biogeochemical cycling of manganese. *Scientific Reports, 8*(1), 1–12. https://doi.org/10.1038/s41598-018-26311-w

Ghosh, S., Gandhi, M., van Hullebusch, E. D., & Das, A. P. (2020). Proteomic insights into Lysinibacillus sp.-mediated biosolubilization of manganese. *Environmental Science and Pollution Research, 28*(30), 40249–40263. https://doi.org/10.1007/S11356-020-10863-4

Ghosh, S., Mohanty, S., Akcil, A., Sukla, L. B., & Das, A. P. (2016). A greener approach for resource recycling: Manganese bioleaching. *Chemosphere, 154*, 628–639. https://doi.org/10.1016/J.CHEMOSPHERE.2016.04.028

Ghosh, S., Sinha, J. K., Ghosh, S., Vashisth, K., Han, S., & Bhaskar, R. (2023). Microplastics as an emerging threat to the global environment and human health. *Sustainability, 15*(14), 10821. https://doi.org/10.3390/SU151410821

Ghosh, S., Tripathy, B., Dey, S., & Das, A. P. (2023). Comparative investigation of fungal and bacterial manganese biomining mechanisms. In *Manganese mining microorganisms* (pp. 99–115). Elsevier. https://doi.org/10.1016/B978-0-12-822147-1.00016-9

Golodnitsky, D., Rosenberg, Y., & Ulus, A. (2002). The role of anion additives in the electrodeposition of nickel–cobalt alloys from sulfamate electrolyte. *Electrochimica Acta, 47*, 17. https://doi.org/10.1016/S0013-4686(02)00135-4

Hano, T., Matsumoto, M., Ohtake, T., Egashir, N., & Hori, F. (1992). Recovery of lithium from geothermal water by solvent extraction technique. *Solvent Extraction and Ion Exchange, 10*(2), 195–206. https://doi.org/10.1080/07366299208918100

Hidalgo-Ulloa, A., Buisman, C., & Weijma, J. (2022). Metal sulfide precipitation mediated by an elemental sulfur-reducing thermoacidophilic microbial culture from a full-scale anaerobic reactor. *Hydrometallurgy*. https://doi.org/10.1016/j.hydromet.2022.105950

Hlihor, R. M., Petronela, C., & Gavrilescu, M. (2022). Removal of heavy metals from the environment by phytoremediation and microbial remediation. In *Sustainable solutions for environmental pollution: Air, water and soil reclamation* (pp. 95–146). Wiley. https://doi.org/10.1002/9781119827665.ch2

Hu, Y., & Cheng, H. (2016). A method for apportionment of natural and anthropogenic contributions to heavy metal loadings in the surface soils across large-scale regions. *Environmental Pollution, 214*, 400–409. https://doi.org/10.1016/j.envpol.2016.04.028

John, A., & Das, A. P. (2024). Government strategies for achieving circular economy for plastic waste management in Southeast Asia. In *Sustainable management of urban plastic waste through circular economic approaches* (pp. 72–82). Springer. https://doi.org/10.1201/9781003465447-6

Johnson, D. B. (2014). Biomining—Biotechnologies for extracting and recovering metals from ores and waste materials. *Current Opinion in Biotechnology, 30*, 24–31. https://doi.org/10.1016/J.COPBIO.2014.04.008

Keyhani, A., Ghasemi-Varnamkhasti, M., Khanali, M., & Abbaszadeh, R. (2010). An assessment of wind energy potential as a power generation source in the capital of Iran, Tehran. *Energy, 35*(1), 188–201. https://doi.org/10.1016/j.energy.2009.09.009

Khatun, R., Singh, S., Dubey, N. K., & Prasad Das, A. (2023). A review on marine-based phytochemicals and their application in biomedical research. In *Recent frontiers of phytochemicals: Applications in food, pharmacy, cosmetics, and biotechnology* (pp. 383–395). Elsevier. https://doi.org/10.1016/B978-0-443-19143-5.00013-X

Kim, K. (2021). Impacts of COVID-19 on transportation: Summary and synthesis of interdisciplinary research. *Transportation Research Interdisciplinary Perspectives., 9*, 100305. https://doi.org/10.1016/j.trip.2021.100305

Lau, W. J., Ismail, A. F., Misdan, N., & Kassim, M. A. (2012). A recent progress in thin film composite membrane: A review. *Desalination, 287*, 190–199. https://doi.org/10.1016/j.desal.2011.04.004

Lestari, L., Harmesa, H., Taufiqurrahman, E., Budiyanto, F., & Wahyudi, A. A. J. (2021). Assessment of potential variability of cadmium and copper trace metals using hindcast estimates. *Environmental Monitoring and Assessment, 193*, 1–22. https://doi.org/10.1007/s10661-021-09501-4

Liu, Q., Zhang, A., Wang, R., Zhang, Q., & Cui, D. (2021). A review on metal-and metal oxide-based nanozymes: Properties, mechanisms, and applications. *Nano-Micro Letters, 13*, 1–53. https://doi.org/10.1007/s40820-021-00674-8

Liu, Y., Deng, Y., Dong, H., Liu, K., & He, N. (2017). Progress on sensors based on nanomaterials for rapid detection of heavy metal ions. *Science China Chemistry, 60*, 329–337. https://doi.org/10.1007/s11426-016-0253-2

Love, J. B., Miguirditchian, M., & Chagnes, A. (2019). New insights into the recovery of strategic and critical metals by solvent extraction: The effects of chemistry and the process on performance. In *Ion exchange and solvent extraction* (pp. 1–44). CRC Press.

Lozano, L. J., Godínez, C., De Los Rios, A. P., Hernández-Fernández, F. J., Sánchez-Segado, S., & Alguacil, F. J. (2011). Recent advances in supported ionic liquid membrane technology. *Journal of Membrane Science, 376*(1–2), 1–14. https://doi.org/10.1016/j.memsci.2011.03.036

Mallick, K., Sahu, A., Dubey, N. K., & Das, A. P. (2023). Harvesting marine plastic pollutants-derived renewable energy: A comprehensive review on applied energy and sustainable approach. *Journal of Environmental Management, 348*, 119371. https://doi.org/10.1016/J.JENVMAN.2023.119371

Mallick, K., Sahu, S., Sahu, A., Dey, S., & Das, A. P. (2024). A critical review of marine microfiber pollution routes, toxicity, and its sustainable remediation. *Environmental Science and Engineering, 2333*, 189–211. https://doi.org/10.1007/978-3-031-51792-1_10

Marte, K. D., Gamarano, T. V., & Barnard, K. R. (2017). Operation of a direct solvent extraction circuit at Boleo. *Proceedings of COM, 2017*, 2017.

Meschke, K., Hofmann, R., Haseneder, R., & Repke, J. U. (2020). Membrane treatment of leached mining waste–A potential process chain for the separation of the strategic elements germanium and rhenium. *Chemical Engineering Journal, 380*, 122476. https://doi.org/10.1016/j.cej.2019.122476

Mishra, S., & Das, A. P. (2021). Current treatment technologies for removal of microplastic and microfiber pollutants from wastewater. In *Wastewater treatment: Cutting-edge molecular tools, techniques and applied aspects* (pp. 237–251). Elsevier. https://doi.org/10.1016/B978-0-12-821881-5.00011-8

Mishra, S., & Das, A. P. (2022). Treatment of the wastewater polluted with synthetic microfiber released from washing machine. In B. B. Das, H. Hettiarachchi, P. K. Sahu, & S. Nanda (Eds.), *Lecture notes in civil engineering* (Vol. 207, pp. 109–117). Springer. https://doi.org/10.1007/978-981-16-7509-6_9

Mishra, S., & Das, A. P. (2023). Microbial remediation of synthetic microfiber contaminated wastewater. In M. P. Shah (Ed.), *Microbial technologies in industrial wastewater treatment* (pp. 337–350). Springer. https://doi.org/10.1007/978-981-99-2435-6_17

Mishra, S., Dash, D., Al-Tawaha, A. R. M. S., et al. (2022). A review on heavy metal ion adsorption on synthetic microfiber surface in aquatic environments. *Applied Biochemistry and Biotechnology, 194*, 4639–4654. https://doi.org/10.1007/s12010-022-04029-w

Mishra, S., Dash, D., & Das, A. P. (2022). Detection, characterization and possible biofragmentation of synthetic microfibers released from domestic laundering wastewater as an emerging source of marine pollution. *Marine Pollution Bulletin, 185*, 114254. https://doi.org/10.1016/J.MARPOLBUL.2022.114254

Mishra, S., Dash, D., & Das, A. P. (2024). Aquatic microbial diversity on plastisphere: Colonization and potential role in microplastic biodegradation. *Geomicrobiology Journal, 41*(4), 312–323. https://doi.org/10.1080/01490451.2023.2209750

Mishra, S., Rath, C., & Das, A. P. (2019). Marine microfiber pollution: A review on present status and future challenges. *Marine Pollution Bulletin, 140*, 188–197. https://doi.org/10.1016/J.MARPOLBUL.2019.01.039

Mohanty, J., Dey, S., Bala, K., & Das, A. P. (2024). Recent advances in microplastic pollution for its sustainable management. In *Sustainable management of urban plastic waste through circular economic approaches* (pp. 27–48). Routledge. https://doi.org/10.1201/9781003465447-3

Mohanty, M., Tripathy, B., Koteswara Reddy, G., & Das, A. P. (2024). Greener technologies for establishing restraint over microplastic pollution. In *Sustainable microbial technology for synthetic and cellulosic microfiber bioremediation* (pp. 239–258). Springer. https://doi.org/10.1007/978-3-031-62600-5_13

Mohanty, S., Ghosh, S., Bal, B., & Das, A. P. (2018). A review of biotechnology processes applied for manganese recovery from wastes. *Reviews in Environmental Science and Bio/Technology, 17*(4), 791–811. https://doi.org/10.1007/S11157-018-9482-1

Mohanty, S., Ghosh, S., Nayak, S., & Das, A. P. (2016). Isolation, identification and screening of manganese solubilizing fungi from low-grade manganese ore deposits. *Geomicrobiological Journal, 34*(4), 309–316. https://doi.org/10.1080/01490451.2016.1189016

Mohanty, S., Ghosh, S., Nayak, S., & Das, A. P. (2017). Bioleaching of manganese by aspergillus sp. isolated from mining deposits. *Chemosphere, 172*, 302–309. https://doi.org/10.1016/J.CHEMOSPHERE.2016.12.136

Nancharaiah, Y. V., Mohan, S. V., & Lens, P. N. L. (2016a). Biological and bioelectrochemical recovery of critical and scarce metals. *Trends in Biotechnology, 34*(2), 137–155. https://doi.org/10.1016/j.tibtech.2015.11.003

Nancharaiah, Y. V., Mohan, S. V., & Lens, P. N. L. (2016b). Recent advances in nutrient removal and recovery in biological and bioelectrochemical systems. *Bioresource Technology, 215*, 173–185. https://doi.org/10.1016/j.biortech.2016.03.129

Nguyen, H. P., Matsuoka, M., Kim, T. H., & Lee, S. W. (2018). Iron (III)-based metal-organic frameworks as potential visible light-driven catalysts for the removal of NOx: A solution for urban air purification. *Journal of Photochemistry and Photobiology A: Chemistry, 367*, 429–437. https://doi.org/10.1016/j.jphotochem.2018.09.008

Ostermeyer, P., Van Landuyt, J., Bonin, L., Folens, K., Williamson, A., Hennebel, T., & Rabaey, K. (2022). High rate production of concentrated sulfides from metal bearing wastewater in an expanded bed hydrogenotrophic sulfate reducing bioreactor. *Environmental Science and Ecotechnology, 11*, 100173. https://doi.org/10.1016/j.ese.2022.100173

Palagonia, M. S., Erinmwingbovo, C., Brogioli, D., & La Mantia, F. (2019). Comparison between cyclic voltammetry and differential charge plots from galvanostatic cycling. *Journal of Electroanalytical Chemistry, 847*, 201. https://doi.org/10.1016/j.jelechem.2019.05.052

Pat-Espadas, A. M., & Cervantes, F. J. (2018). Microbial recovery of metallic nanoparticles from industrial wastes and their environmental applications. *Journal of Chemical Technology & Biotechnology, 93*(11), 3091–3112. https://doi.org/10.1002/jctb.5681

Paul, A., Dey, S., Ram, D. K., & Das, A. P. (2023). Hexavalent chromium pollution and its sustainable management through bioremediation. *Geomicrobiology Journal*. https://doi.org/10.1080/01490451.2023.2218377

Paul, A., Priyadarsini, S., Karakuş, S., Behera, J. K., Gnanasekaran, R., & Das, A. P. (2024). Fate and accumulation of synthetic microfibers in the environment. In *Sustainable microbial technology for synthetic and cellulosic microfiber bioremediation* (pp. 61–76). Springer. https://doi.org/10.1007/978-3-031-62600-5_4

Pradhan, G., Tripathy, B., Ram, D. K., Digal, A. K., & Das, A. P. (2023). Bauxite Mining Waste Pollution and Its Sustainable Management through Bioremediation. Geomicrobiology Journal, 1–10. https://doi.org/10.1080/01490451.2023.2235353

Pradhan, G., Dey, S., Priyadarsini, S., Kumar, M. S., & Das, A. P. (2024). Characterization and quantification of microplastics pollutants in sediment samples from Daya River of Odisha state in India for their appropriate management. *Environmental Science and Engineering, 2333*, 75–92. https://doi.org/10.1007/978-3-031-51792-1_5

Priyadarshini, B., & Das, A. P. (2024). The adverse impact of synthetic microplastics released from disposable face masks pollutants on aquatic microbiota, aquatic species and humans: A review. *Environmental Science and Engineering, 2827*, 67–81. https://doi.org/10.1007/978-3-031-50840-0_5

Priyadarshini, B., Santhosh Kumar, M., & Das, A. P. (2024). Microbial degradation of synthetic plastics released from COVID-19 generated single use disposable face masks. *Environmental Science and Engineering, 2827*, 1–16. https://doi.org/10.1007/978-3-031-50840-0_1

Priyadarshini, S., Jagatee, S., & Das, A. P. (2024). Synthetic fabrics and microfiber pollution–an assessment of their global impact. *Environmental Science and Engineering, 2333*, 137–157. https://doi.org/10.1007/978-3-031-51792-1_8

Priyadarsini, S., & Das, A. P. (2024). Analytical and structural characterization of waste lithium-ion batteries for their effective recycling strategy. *Environmental Chemistry and Ecotoxicology*. https://doi.org/10.1016/j.enceco.2024.12.004

Priyadarsini, S., & Das, A. P. (2025). Lithium bioleaching: A review on microbial-assisted sustainable technology for lithium bio-circularity. *Journal of Water Process Engineering, 69*, 106744. https://doi.org/10.1016/j.jwpe.2024.106744

Ray, S. K., Dey, S., Das, A. P., & John, A. (2024). Microplastic waste management through greener technologies: A review. In *Sustainable management of urban plastic waste through circular economic approaches* (pp. 237–261). Routledge. https://doi.org/10.1201/9781003465447-14

Sahu, A., Dey, S., Mallick, K., Priyadarsini, S., Behera, I. D., & Das, A. P. (2024). A review on advanced technology for sustainable management of synthetic microplastic waste. In

Sustainable management of urban plastic waste through circular economic approaches (pp. 262–279). Springer. https://doi.org/10.1201/9781003465447-15

Sahu, A., Mallick, K., & Das, A. P. (2024). *Environmental impact of single-use synthetic face mask and its recycling: A sustainable approach* (pp. 197–212). Springer. https://doi.org/10.1007/978-3-031-50840-0_11

Sahu, S., & Das, A. P. (2024). *Microplastics degradation mechanism by aquatic plastiomes: A review* (pp. 213–226). Springer. https://doi.org/10.1007/978-3-031-50840-0_12

Sethurajan, M., & Gaydardzhiev, S. (2021). Bioprocessing of spent lithium ion batteries for critical metals recovery–A review. *Resources, Conservation and Recycling, 165*, 105225. https://doi.org/10.1016/j.resconrec.2020.105225

Siddiqui, M. N., Chanbasha, B., Al-Arfaj, A. A., Konkova, T., & Ali, I. (2021). Super-fast removal of cobalt metal ions in water using inexpensive mesoporous carbon obtained from industrial waste material. *Environmental Technology & Innovation, 21*, 101257. https://doi.org/10.1016/j.eti.2020.101257

Singh Jadaun, J., Bansal, S., Sonthalia, A., Rai, A. K., & Singh, S. P. (2022). Biodegradation of plastics for sustainable environment. *Bioresource Technology, 347*, 126697. https://doi.org/10.1016/J.BIORTECH.2022.126697

Spooren, J., Binnemans, K., Björkmalm, J., Breemersch, K., Dams, Y., Folens, K., González-Moya, M., Horckmans, L., Komnitsas, K., Kurylak, W., Lopez, M., Mäkinen, J., Onisei, S., Oorts, K., Peys, A., Pietek, G., Pontikes, Y., Snellings, R., Tripiana, M., Varia, J., Willquist, K., Yurramendi, L., & Kinnunen, P. (2020). Near-zero-waste processing of low-grade, complex primary ores and secondary raw materials in Europe: Technology development trends. *Resources, Conservation and Recycling, 160*, 104919. https://doi.org/10.1016/j.resconrec.2020.104919

Torell, L. M., Jacobsson, P., & Petersen, G. (1993). A Raman study of ion solvation and association in polymer electrolytes. *Polymers for Advanced Technologies, 4*(2–3), 152–163. https://doi.org/10.1002/pat.1993.220040215

Villa Gomez, D., Whitworth, A. J., Vaughan, J., Sultana, U., Ledezma, P., & Parbhakar-Fox, A. (2024). Review on developments in technologies for critical metal recovery from mining and processing wastes. *Mineral Processing and Extractive Metallurgy Review*. https://doi.org/10.1080/08827508.2024.2408015

Wyss, K. M., Deng, B., & Tour, J. M. (2023). Upcycling and urban mining for nanomaterial synthesis. *Nano Today, 49*, 101781. https://doi.org/10.1016/J.NANTOD.2023.101781

Xue, J., Wu, T., Dai, Y., & Xia, Y. (2019). Electrospinning and electrospun nanofibers: Methods, materials, and applications. *Chemical Reviews, 119*(8), 5298–5415. https://doi.org/10.1021/acs.chemrev.8b00593

Yu, F., Li, Y., Huang, G., Yang, C., Chen, C., Zhou, T., & Ma, J. (2020). Adsorption behavior of the antibiotic levofloxacin on microplastics in the presence of different heavy metals in an aqueous solution. *Chemosphere, 260*, 127650. https://doi.org/10.1016/j.chemosphere.2020.127650

Zante, G., Boltoeva, M., Masmoudi, A., Barillon, R., & Trebouet, D. (2019). Lithium extraction from complex aqueous solutions using supported ionic liquid membranes. *Journal of Membrane Science, 580*, 62–76. https://doi.org/10.1016/j.memsci.2019.03.013

Zhuang, W. Q., Fitts, J. P., Ajo-Franklin, C. M., Maes, S., Alvarez-Cohen, L., & Hennebel, T. (2015). Recovery of critical metals using biometallurgy. Current Opinion in Biotechnology, 33, 327–335. https://doi.org/10.1016/j.copbio.2015.03.019

Zeng, Y., Bi, C., Jia, J., Deng, L., & Chen, Z. (2020). Impact of intensive land use on heavy metal concentrations and ecological risks in an urbanized river network of Shanghai. *Ecological Indicators, 116*, 106501. https://doi.org/10.1016/j.ecolind.2020.106501

Adverse Effect of Emerging Mining Pollutants on the Environment and Novel Methods for Its Environmental-Friendly Remediation

Ankita Mohanty, Aditya Kishore Dash, and Himansu Sekhar Patra

1 Introduction

Minerals are considered as one of the important parameters for the economic growth of a country. Mining has remained quintessentially integral quite deeply within human societal fabric from remarkably ancient eras onward very profoundly. It furnishes raw materials necessary for technological advancements and burgeoning industries rapidly nowadays with significant impact on infrastructure development. Mining extracts valuable minerals and metals from earth essential for manufacturing myriad products rapidly across diverse industries worldwide quite frequently. Iron ore serves as a fundamental constituent of steel used pretty extensively in various construction projects and manufacturing processes nationwide. Copper plays a crucial role in electrical wiring owing largely to exceptionally high conductivity. Rare earth elements play a crucial role in the production of high-tech gadgets like smart phones and electric vehicles remarkably well nowadays. Mining industry also serves as major economic driver pretty much nationwide. It contributes significantly toward GDP of many nations pretty substantially every year. Economic development through mining comes at the cost of environmental pollution (Li et al., 2023). Infrastructure development entails upgrading transportation networks and energy systems thereby increasing efficiency in production and swift delivery of various goods and services. High productivity ensues and job creation gets a boost alongside foreign investment thereby fostering economic growth pretty rapidly.

A. Mohanty · A. K. Dash (✉)
Department of Chemistry, ITER, Siksha O Anusandhan (Deemed to be University), Bhubaneswar, Odisha, India
e-mail: adityadash@soa.ac.in

H. S. Patra
Environment and Sustainability, MSP Steel & Power Ltd, Jamgaon, Chhattisgarh, India

© The Author(s), under exclusive license to Springer Nature Switzerland AG 2026

A. P. Das (ed.), *Bioremediation Technology for Sustainable Mining Waste Management*, Sustainable Environmental Waste Management Strategies,
https://doi.org/10.1007/978-3-032-03468-7_19

Mineral deposits are distinct to specific locations, necessitating ecofriendly operational practices during extraction. The selection of mining methods requires careful consideration to maintain environmental balance and promote sustainable development. Mining operations have diverse impacts on local environments, including the removal of vegetation and topsoil (Patra & Dash, 2024; Pattajoshi & Dash, 2024), which increases vulnerability to wind and water erosion and can result in sedimentation in nearby water bodies. Additionally, dust and particulate matter released from mining contribute to air pollution, affecting both human health and local wildlife. Mining generates significant waste, including overburden, tailings, and slag, which if not managed properly can leach harmful chemicals into soil and groundwater, leading to lasting environmental repercussions (Das et al., 2011; Das & Ghosh, 2023b, 2023c, 2023d, 2023f). The presence of heavy metals such as As, Pb, and Hg in mine waste is particularly troubling due to their toxicity and persistence. To address these challenges, mine reclamation aims to restore mined land to a natural or usable state. Techniques like phytoremediation, which employs plants to absorb contaminants, are also being explored to tackle soil and water pollution issues. The success of these reclamation efforts relies on meticulous planning and ongoing monitoring, ensuring that restored ecosystems are resilient and self-sustaining. However, mining and allied activities significantly releases various environmental pollutants causing air, water, and soil pollution and also causes human health risks and ecosystem degradation. Mineral tailing accumulation releases heavy metals into the environment (Mohanty et al., 2023). Mining near dry areas emits more heavy metals because of lack of humidity and vegetation (Pradhan et al., 2020). Sulfide contain in tailings can produce acid drainage and releases heavy metals (Orellana Mendoza et al., 2021). Soil contaminated with heavy metals is found to be common near mining and smelting areas as the metals are extracted and refined from ores (Mathieu et al., 2021). Hence, there is a need to address the various emerging environmental pollutants from mining and allied activities and to find our environmentally friendly remedial measures.

2 Impacts of Mining

Mining of ores and minerals causes environmental degradation. The environmental impact can be either transitory, short term, or long term. The significant environmental impact caused due to mining operation includes; water availability and standard, quality of air, land interruption, production of waste, loss of biological diversity, inconvenience, and disruption. Furthermore, mining also have the social impacts like displacement from housing, resettlement issues, employment, health and safety concerns, ecosystem services, socio political conflicts, etc.

Impact on Air Quality Mining operations severely degrade air quality quite badly around affected areas. Blasting and crushing operations release copious amounts of dust into air alongside sundry particulate matter during mining activities. Activities like drilling, blasting over burden transportation, crushing, and screening generates gaseous pollutants like SO_2, NO_x, CO, dusts, and suspended particles (Khobragade, 2020). PM_{10} and $PM_{2.5}$ particles traverse considerable distances quite rapidly and burrow deep into respiratory systems causing maladies such as asthma and cardiovascular disease (Das et al., 2015a, 2015b; Das & Ghosh, 2018; Dey, Tripathy, et al., 2023; Ghosh & Das, 2017). Ambient air monitoring and future prediction plays a vital role for establishing suitable control measures for air pollution (Dash & Dash, 2017, 2018a, 2018b; Kumar et al., 2018). Heavy metals and other toxic pollutants are contained within these particles quite frequently alongside various other nasty stuff. The particles formed from mining operations are the major sources of airborne emissions and pollution (Mishra & Das, 2008; Paul et al., 2023). During the mining cycle, the ore extraction, generation of solid waste, and ore refining and processing create air pollution which has a cumulative effect on the environment and public health. According to Tripathi et al., 2016, some waste disposal methods such as conical dumping by draglines result in conditions that are susceptible to gully erosion and dump slope failure, which leads to air and water pollution, reduced aesthetic values, blockage of pit access, and filling of sump. The mine waste dumped over the ground contains particles that are easily dispersible by the wind. Air pollution causes various health hazards and also degrades natural ecosystem (Dash et al., 2019, 2023; Parida et al., 2015; Sahoo et al., 2017; Sahu et al., 2015). The particulate matters are the major sources of air pollution in mining areas. This is the major pollutant in and around open pit mine areas.

Impact on Water Quality Mining usually has the greatest impact on water quality and the availability of water resources. It disrupts hydrological regime, reduces the ground water table and affects the ground water regime (Khobragade, 2020). Environmental repercussions arising from acid mine drainage and tailings are substantial issues of concern. A major contributor to and peril for water quality is acid mine drainage, chiefly via the leaching of dissolved metals and pollutants (Das & Ghosh, 2023a, 2023e; Priyadarsini & Das, 2025). Acid mine drainage result in the release of toxic heavy metals due to the presence of the pyrite bands in the coal seams (Khobragade, 2020). Nevertheless, in some cases, the leaching of toxic substances such as As, Se, and other heavy metals can occur. Processes like heap leaching and blasting lead to heightened levels of cyanide and nitrogen compounds like ammonia, nitrate, and nitrite in water resources (https://www.elaw.org/files/mining-eia-guidebook/Chapter1.pdf). Due to the development of cracks near the surface, rain water enters underground which may carry various toxic pollutants. Water polluted due to mining activities is of poor quality, that it is unfit for human consumption. Overburden dumping needs to be avoided in valleys or tracts near the mined area that constitute the water supply source either from ground water or surface water bodies (Nriagu & Pacyna, 1988). In addition, it has devastating effects on organisms and ecosystems in freshwater bodies. Surface water quality negatively impacts terrestrial wildlife too. Table 1 provides the health impact of various heavy metals with their symptoms and controlling measures at different countries. The common heavy metals reviewed include Pb, Hg, As (Dehkordi et al., 2024).

Table 1 Health effect of heavy metals in different Nations with their symptoms and preventive measures

Sl. No.	Country (city)	Metal type	Symptoms	Controlling measures	References
1	India (Delhi)	Pb	Irritability, abdominal pain, headache	Water purifier, air filter, catalytic converter	Jain et al. (2022)
2	Mexico (Mexico city)	Pb	Memory impairment, fatigue, headache	Water purifier, air filter, wearing mask	Sanchez-Landero et al. (2024)
3	Japan (Minamata)	Hg	Muscle weakness, hearing problem	Water purifier, not to eat contaminate fish	Matsuyama et al. (2021)
4	China (Urumqi)	Pb	Vomiting, deficiency in iron and copper, fever	Water purifier, air filter, proper washing of vegetables and fruits	Zhou et al. (2020)
5	China (Lanzhou)	As	Diarrhea, nausea, pain in abdomen	Water purifier, air filter, drinking clean water	Liu et al. (2022)
6	Thailand (Thailand)	As	Vomiting, hand and feet with numbness and tingling	Clean fruits and vegetables, clean water consumption	Kaya et al. (2024)
7	Turkey (Antalya)	Pb	Diarrhea, tremors, nausea	Consumption of uncontaminated vegetables and fruits, air filter	Coban et al. (2023)

Impact on Soil Quality Mining activities have a serious effect on soil fertility due to the removal of topsoil and contamination of the soil. Change in top soil characteristics reduces soil fertility in the nearby areas due to the factors like siltation and surface run off from overburden dumps. Exposed soils, extracted mineral ores, tailings, and fine material in waste rock piles can undergo erosion, leading to significant sediment loading in surface water and alterations in drainage pattern. Moreover, the release of hazardous substances and the accumulation of contaminated dust and water can result in soil contamination. (https://www.elaw.org/files/mining-eia-guidebook/Chapter1.pdf). One of the most prominent effects of mining and associated activities is soil degradation near mining sites due to contamination from multi metal pollutants like Cd, Cu, Zn, and Pb, which greatly limits land use (El Hamiani et al., 2010; El Khalil et al., 2008). Due to subsidence, there is a change in topography as well as drainage pattern. Mining can affect soil over large areas, resulting in detrimental effects on soil properties and components, such as the degradation of soil structure and layers, disruption of soil microbial communities, and nutrient cycles (Benidire et al., 2020), which results in the destruction of natural vegetation (Sheoran et al., 2010). Absence of vegetation leaves contaminated soil and sediments is vulnerable to soil erosion by wind and rain naturally over time quite rapidly (Sheoran et al., 2010). During the mining process and even after operations have stopped, the tailings ponds remaining onsite, serve as a significant source of contamination and disruption to the surrounding ecosystem.

Impact on Wildlife Mining heavily ravages wild fauna by stripping away vegetation and topsoil and destroying ecosystems with pollutants and ear shattering noise generation. Mining ravages wildlife habitats pretty badly through destruction and pollutes both water and air and makes a lot of noise. Noise and vibration caused due to mining activities drive away various faunal species from the region (Khobragade, 2020). Open pit and strip mining operations ravage vegetation utterly and disrupt land precipitating habitat loss fairly extensively and fragmentation occurs subsequently. It gives short term and long term impact on habitat causing habitat loss. During the excavation or mine spoil piling, the wildlife species along with the sedentary species such as invertebrates, reptiles, burrowing rodents and small mammals are destructed. Similarly, habitat fragmentation is one of the most ecological impacts (Das et al., 2023; Das, Dey, & Das, 2024; Das, Ray, Mohanty, et al., 2024; Priyadarshini et al., 2024; Reddy Koteswara et al., 2025). It occurs due to breaking of a large land area into smaller patches. This process divides large continuous habitats into smaller isolated patches, which can negatively impact biodiversity and wildlife population. Further development of infrastructure, complexes, township requires clearing of vast areas that drives away the animals and birds from the area (Khobragade, 2020).

Impact Due to Noise and Vibration Movement of heavy machinery and drilling operations breed noise pollution significantly in mining industries mainly from blasting and transportation activities. Installed machinery, compressor house and DG set create noise that becomes ambient noise (Khobragade, 2020). Loud noise can precipitously cause hearing loss in workers and stress and also decreasing overall productivity in workplace environments. Activities such as drilling and blasting produce extremely loud noises pretty disruptive especially for workers operating nearby and local residents. Blast energy transmission occurs through ground creating vibration waves that propagate wildly through various soil strata and rock beneath nearby building foundations (ERM, 2013). Vibration affects the infrastructure such as buildings and resistance of people living in and around the mining areas. It also provides bad impact on domestic animals and wild fauna.

3 Innovative Methods for Remediation of Mine Pollution

Emerging mining sites, mainly those implicated in heavy metal contamination, can be improved using a combination of approaches such as physical isolation, chemical treatments, and biological methods. Physical methods, such as discarding and concealing, aim to confine contaminants, chemical methods like electro kinetic and soil washing remediation focus on removing pollutants. Phytoremediation and other biological methods use plants to take in or immobilize heavy metals.

4 Water Pollution Management in Mines

Mining operations not only consume significant amounts of water but also alter the hydrological balance of an area, by lowering the natural groundwater table in deep underground and opencast mines, leading to reduced groundwater availability. It often causes water pollution through the release of heavy metals, acids, and suspended solids. Mining wastewater discharge into the environment has been a major public concern as it can easily pollute water resources and negatively affect living organisms, including humans, aquatic animals, and plants (Thomas et al., 2022). Treatment of water and wastewater through low cost biological method is found to be acceptable in comparison to the costlier physiochemical methods (Pradhan et al., 2013, 2015, 2017). Some of the key concerns with respect to water pollution in mining areas include groundwater inflow into mines, contamination from mining pollutants, land subsidence, and the loss of surface water features like lakes and streams. Many of the processed mineral resources have sulfidic minerals in their related waste rock. Acid and/or metalliferous drainage ("AMD") occurs when sulfides are oxidized due to exposure to alternating cycles of infiltration and oxygen (Taylor & Pape, 2007). Extreme environmental consequences can occur when AMD escapes into the surrounding environment, often for tens of kilometers downstream from a mine site. Implementation of effective water management and pollution control in mining operations, with their relevance depending on the location, type of mining, and mineral characteristics is required for control of water pollution. The overall water management of mines can be categorized into surface runoff control and treatment, treatment of mines drainage and effluent followed by monitoring of water quality. A study conducted by Erasmus et al., 2022 explored human health risks associated with the consumption of fish in the Hex River, Rustenburg, South Africa, which is contaminated with trace elements from the surrounding mines. Mining extraction activities conducted underwater, open-pit, or underground produce large quantities of waste containing heavy metals, organic compounds, and unacceptable levels of salinity in wastewater, which can pollute local streams, wetlands, and groundwater, can affect aquatic life and humans through the food chain (Kurniawan et al., 2022).

Surface Runoff Treatment and Control The drainage planning in mining aims at preserving the pre mining runoff patterns to the greatest extent possible. The drainage at the core mining zone should be carefully planned to minimize impacts on existing surface water bodies and maintain the area's natural drainage pattern. Garland drains are to be constructed around all quarries and dumping site thus channeling surface runoff through properly designed catch pits or settling pits for removing suspended solids. The catch pits/settling pits should be developed based on expected silt load, slope, and required detention time. Retaining walls with weep holes should be constructed outside the dumps thus allowing storm water to flow into the garland drains. The treated surplus mine water are discharged into nearby ponds or agricultural fields to aid in groundwater recharge. Rainwater harvesting through check dams on natural streams and the development of water bodies should be implemented in and around the mines to further support groundwater recharging. Before use of mine drainage water, the toxic elements of the water should be treated to meet the standard after which it can be used in agriculture or for recharging purpose. In order to minimize erosion risks, overburden, and topsoil should be stored on flat areas, and settling tanks should be used to treat mine water. In order to regulate water flow and control erosion, stone barriers across drains and stone pitching at strategic locations should be made. The settling pits and drainage systems must be de-silted regularly.

Treatment of Mines Drainage and Effluent Mine water, depending on its chemical composition, must be properly treated before being discharged. To handle site specific contaminants, treatment plants should be designed accordingly. With progress in mines operation, a flexible and well-designed network should be established to pump mine water to the treatment facility. Wastewater from vehicle workshops should be properly treated to remove suspended solids, oil, and grease through effluent treatment plants comprising of oil and grease trapper. Similarly, effluents from canteens, rest areas, and residential townships should undergo adequate treatment in sewage treatment plant before being discharged outside the site. Neutralization, ion exchange, membrane technology, bioremediation are some of the important techniques for treatment of mine wastewater.

Neutralization is an effective method for treating acid mine drainage (AMD) by using alkaline substances to counteract acidity, raise pH levels, and precipitate heavy metals from contaminated water. This process is widely used due to its high removal efficiency and rapid reaction rates, with alkaline agents like lime, limestone, or other alkali reagents added to react with sulfuric acid in AMD. As pH increases, many heavy metals become insoluble and precipitate as hydroxides or carbonates, facilitating their removal. While neutralization is simple and cost effective for large scale AMD treatment, it generates significant sludge that can be challenging to dispose of, and its effectiveness varies based on the specific composition of AMD and the type of alkaline substance used. Highly reactive lime is effective but produces more sludge, while slower reacting limestone is cheaper and easily available. Use of industrial alkaline by-products like red mud and coal ash can act as alternatives to traditional chemicals.

Ion exchange remediation technique in mining operations cans efficiently removing heavy metals and other contaminants from wastewater. This process swaps unwanted ions with those attached to resin beads or zeolites, effectively cleaning pollutants from the solution. Contaminants like As and NO^{3-} are often present in mining influenced water. For example, when mining influenced water passes through a column containing ion exchangers, pollutants bind readily to the resin or zeolite, while sodium and hydrogen ions are released. Regeneration occurs after saturation, where a concentrated solution of replacement ions is used to free the pollutants for later disposal. Ion exchange resins are particularly effective in removing heavy metals such as Cu, Pb, Zn, and Cd from mining wastewater, while zeolites can effectively eliminate uranium. These techniques also remove nitrates from water in both underground settings and treatment facilities. In the process of mining wastewater remediation, zeolites facilitate the exchange of toxic or surplus metal ions (such as Pb^{2+}, Cd^{2+}, Zn^{2+}) into less harmful ones, like sodium (Na^+) or hydrogen (H^+). Because of this flexibility, zeolites can be customized to eliminate specific ions according to environmental requirements.

$$\text{Na-zeolite} + \text{H}^+ \rightarrow \text{H-zeolite} + \text{Na}$$

or in reverse, depending on the application:

$$\text{H-zeolite} + \text{Na}^+ \rightarrow \text{Na-zeolite} + \text{H}^+$$

Membrane technology is an effective remediation approach for mining wastewater, utilizes various membrane based processes to separate contaminants underground. Pressure driven and thermal driven systems efficiently remove heavy metals, dissolved ions, and other pollutants from mine effluents. Reverse osmosis forces water through a semipermeable membrane under substantial pressure, effectively rejecting smaller molecular contaminants, including heavy metals. Nanofiltration operates similarly but features slightly larger pore sizes, allowing it to remove specific pollutants in mining effluent treatment. Ultrafiltration, with its larger pore sizes, effectively separates larger molecules and suspended solids, often used after pretreatment. Microfiltration targets larger particles and microorganisms, while membrane distillation takes advantage of temperature differences to vaporize water, concentrating contaminants efficiently. This diverse array of membrane techniques makes treating highly concentrated mine effluents remarkably effective and manageable.

Bioremediation in mining effectively utilizes microbes to address contamination, particularly in AMD, by degrading pollutants in controlled environments. This eco-friendly approach harnesses microbial power to break down stubborn contaminants, such as sulfur bound materials and heavy metals, often resulting from the extraction of hard rock or sulfur bound ores. Diverse microorganisms including bacteria, algae, fungi, and yeasts transform pollutants efficiently under varying environmental conditions. Bioremediation techniques can be classified as in situ, treating contamination onsite, or ex situ, which involves transporting contaminated materials for treatment elsewhere. Strategies such as biostimulation, which enhances native microbial activity through nutrient addition, and bioaugmentation, involving the introduction of specific microbial strains, are commonly employed (Das, Dey, Bala, et al., 2024; Mohanty, Dey, et al., 2024; Pradhan et al., 2023). Additional methods like bioventing, bioleaching, and biosorption are also utilized in mining operations, showcasing the notable efficacy and sustainability of bioremediation for cleaning contaminated sites and significantly improving water quality.

Water Quality Monitoring Program Regular monitoring of water quality both of mine discharge into nearby rivers or streams and of domestic water sources should be carried out to ascertain the effectiveness of mitigation efforts. Treatment of wastewater employing biological methods are found to be cost effective (Jena et al.,

2022; Mahapatra et al., 2019; Parida et al., 2023; Shreya et al., 2022). A detailed water management plan should be developed which will be visually represented in site layout drawings of the mining area. Piezometers, both shallow and deep, should be installed near and around the mining areas in consultation with the Central Ground Water Board to monitor aquifer levels. Additionally, a comprehensive water audit, including a water balance diagram, should be conducted to monitor water usage across all plant operations.

5 Air Pollution Management in Mines

Air quality in mining areas is predominantly influenced by the types and concentrations of emissions as well as meteorological conditions. The primary air pollutants associated with mining operations include particulate matter of varying sizes, gases pollutants such as SO_2, NO_x, and CO from vehicular exhaust. Dust is the most prevalent air pollutant identified in iron ore mining activities. In the open cast iron ore mining areas, dust is the single largest air pollutant (Chaudhari & Gajghate, 2000; Saha et al., 2011). Emissions of NO_x, SO_2, and CO generally occur at low levels as a result of diesel powered equipment, diesel operated drilling machines, blasting activities, and the movement of heavy earthmoving machinery and vehicles. Though the concentration of gases may not be excessively high, dust can pose a significant nuisance to nearby land users and could present potential health risks under certain circumstances. They also cause damage to flora and fauna by reducing photosynthesis leading to reduction in reproduction capabilities and productivity (Wheeler et al., 2000). The air pollution in mines can be controlled and managed through following ways;

Dust controlling at paved surfaces of mines: Paved surfaces have less dust generation potential. Vehicle movement causes generation of dust in mining areas. Dust emissions from paved surfaces can be minimized through adopting various activities such as control of spillages of materials on the paved surfaces during materials transportation, minimize mud and dust track out from unpaved areas through use of wheel wash facilities, regular cleaning of paved surfaces, using mobile vacuum sweeper or a water flushing system and speed control on vehicle movements.

Dust controlling at unpaved surfaces of mines: Dust emissions from unpaved surfaces are significantly higher than those from paved areas. Dust controlling measures such as water sprinkling during dry and windy conditions using tankers or fixed sprinklers, use of chemical stabilization in combination with wet suppression are adopted for controlling the emissions. These chemicals form a surface crust, bind dust particles and reduce airborne dispersion. Revegetation of exposed areas, surface improvements using concrete, asphalt, gravel, or chemical dust suppressants, speed control of vehicles, and roadside plantations also contribute to minimizing dust generation. Wet suppression alone can reduce dust emissions by approximately 70%, which can increase up to 95% with chemical additives. Use of surface miners, fog cannons, mist sprayers, wheel washing systems and mechanized road sweepers can control the dust emission to larger extent. Drilling equipment is fitted with dust suppression systems, and greater use of surface miners and bucket wheel excavators thus reduces the need for drilling and blasting. Regular maintenance of vehicles, as per manufacturer guidelines, helps minimize emissions (Dey, Bissoyi, et al., 2024; Dey, Das, et al., 2023; Mohanty, Tripathy, et al., 2024; Sahu et al., 2024). Additionally, dust suppression systems are installed at loading, unloading, and transfer points, as well as water spraying setups across washeries, coal handling plants, feeder breakers, crushers, belt conveyors, haul roads, and coal stockyards can control fugitive dust effectively.

Dust control from conveyors: Dust emissions from open conveyor systems primarily occur due to wind exposure, as well as during material loading, discharge, and transfer between belts. Several effective control measures such as enclosed conveyors significantly reduce dust escape. Water spray or sprinklers at transfer points suppress airborne particles. Drop heights at these points should be minimized to prevent excessive dust generation. Regular cleaning of spillage around transfer areas is essential to prevent loose material from being lifted by the wind. Additionally, enclosing transfer points and installing dust extraction systems further enhance dust control. An advance air pollution sensor based technology is the latest advancement in monitoring air quality and generates real time data inside mines.

Pollution control from vehicles: Vehicle movement in mines contributes significantly to dust emissions as they crush surface particles and debris, lifting dust into the air and the turbulent airflow they generate. Several measures such as preventing spillage from loaded trucks, optimize travel distances through efficient site layout, and install wheel and truck wash facilities at site exits can be implemented to minimize dust emissions from vehicles. Enforcing speed limits is particularly effective (Dey, Dash, et al., 2024; Moharana et al., 2024; Pradhan et al., 2024; Sahu et al., 2024). Through proper vehicle maintenance and operator training, vehicular emissions of particulates, SO_2, NO_x, and hydrocarbons can be significantly reduced. At present, MoEF and CC are giving conditions to mines owner to conduct the pollution under control test for all running vehicles inside the mines. Large mines group like Tata, JSW are promoting the use of electric vehicles inside the mines to cut carbon emission and to control the pollution.

Pollution control at material stockpiles: Topsoil and overburden stockpiles are vulnerable to wind erosion, when wind speeds exceeding 5 m/s. Dust emissions can also occur during material handling activities, dropping material from conveyors, loading/unloading with trucks, shovels, or frontend loaders. Wet suppression using water sprinklers is one of the most common methods. Designing stockpiles with lower height and slopes can help to reduce wind turbulence and dust lift. Factors such as prevailing wind direction and surrounding site features should be taken into account during the designing of stockpiles. Additional dust control measures such as minimizing material drop heights from conveyors, installing windbreaks around stockpiles to reduce wind speed, and using sprinklers or fogging systems with polyelectrolyte chemicals for enhanced dust binding should be adopted at mines. Installing hoods at discharge points also helps to contain dust during material transfer.

6 Noise Pollution Control in Mines

Mining activities such as excavation, drilling, blasting, ore and overburden handling, and operation of processing equipment lead to increased noise levels, which may affect nearby habitats. The impacts of the noise can be measured and simulated using computerized noise modeling tools. Mining companies are now adopting both engineering and administrative controls for mitigation of noise and vibrations such as using low noise equipment, modifying older machinery, implementing effective preventive maintenance which can reduce noise by over 50%, installing noise barriers or enclosures around the sources like DG sets and compressors, and properly mounting vibrating structures. When source and path reduction are insufficient, administrative measures such as altering work schedules, relocating personnel, shortening sight lengths, modifying operations, and using warning signs can be implemented. Personal hearing protection devices can minimize the impact of noise. Noise and ground vibrations from blasting can be minimized by designing optimal blast patterns considering burden, spacing, hole depth, and stemming height, using shock tube initiation systems to control delays and reduce impacts, and ensuring public safety.

7 Biological Stability and Greenbelt Development in Mines

Biological stability plays a crucial role in the long term environmental rehabilitation of mine sites. While physical and chemical stability depends on site conditions, biological stability helps in controlling erosion, leaching, and runoff. Establishing a vegetation cover is a key objective in mine rehabilitation, offering a long term solution for site stabilization. Successful revegetation requires managing soil nutrients, especially when topsoil has lower fertility. High nutrient plants are intended, or rapid growth of native flora is needed during non moisture limiting periods.

Greenbelt Development It is vital for controlling dust, noise, and improving the microclimate, ecological balance, and aesthetics. It not only stabilizes slopes of overburden dumps but, attract wildlife back to the area during post-mining. Using native, fast growing species with large crowns and small, smooth leaves, the green belt should be planned from the inception period of the project. A mix of species is recommended to meet these criteria covering mine lease boundaries, office and colony areas, road sides, external overburden dumps, backfilled areas, and upper mine benches. Indigenous species that are better adapted to the local climate and ecosystem should be planted with proper spacing (Chaney et al., 2007). Table 2 gives the list of plant species suitable for plantation near mining areas with their common name and major utilization potential.

To provide effective dust control and windbreaks in open cast mines, a layered greenbelt should be established along the boundary, with smaller plant species planted on the inside and taller species on the outside. Dense plantation around buildings and along roads not only improves air quality but also helps in reducing noise levels. For external dumps, once dumping activity ceases, topsoil is spread and leveled, trenches are dug on the tops of the dumps, filled with a mixture of topsoil, manure, and fertilizer to facilitate the planting of saplings. In backfilled areas and mine benches, after leveling and spreading topsoil, contour trenches and pits are created for planting drought resistant, fast growing plant species. Initial

Table 2 Details of plant species suitable for mine overburden stabilization

Botanical name	Family	Common name	Utilization
Acacia catechu	Fabaceae	Khaira	Medicinal value, as a dye and tanning agent
Aegle marmelas	Rutaceae	Bael	Food preparation, traditional medicine, therapeutic application
Anogeissus latifolia		Dhou	Antimicrobial, antioxidant, anti-diabetic, wound healing
Anthocephalus chinensis	Rubiaceae	Kadamba	Match factory, particleboard industry
Artocarpus heterophyllus	Moraceae	Panasa	Food, fodder, timber, medicine
Azadirachta indica	Meliaceae	Neem	Traditional medicine, insecticidal and pesticidial application
Bambusa tulda	Poaceae	Bamboo	Construction, pulp paper, furniture
Bauhinia variegata	Fabaceae	Kanchana	Ornamental, edible part, medicinal value
Bridelia retusa	Phyllanthaceae	Kasi	Medicinal, wound hilling, fodder
Buchanania lanzan	Anacardiaceae	Char	Digestive, wound treatment
Butea monosperma	Fabaceae	Palas	Medicinal use, pigment
Cassia fistula	Fabaceae	Sumari	Medicinal, timber value
Careya arborea	Lecythidaceae	Kumbhi	Medicinal value, decorative
Dalbergia sissoo	Fabaceae	Sissoo	Traditional medicine, timber, fuel
Diospyros melanoxylon	Ebenaceae	Kendu	Medicinal importance, ecological value
Ficus religiosa	Moraceae	Papal	Medicinal value
Gmelina arborea	Lamiaceae	Gambari	Construction, furniture, pulp
Haldinia cordifolia	Rubiaceae	Kumm	Timber production
Lannea coromandelica	Anacardiaceae	Mai	Medicinal, gum, sprouts
Madhuca indica	Sapotaceae	Mahula	Medicinal, food, fuel, industry
Phyllanthus emblica	Phyllanthaceae	Anola	Food, medicine

cover of grasses should allow for the development of different, self-sustaining plant diversity (Shu et al., 2002). Plants can further improve soil conditions through various processes, such as increasing soil organic matter, biological N_2 fixation, nutrient uptake, and enhancing infiltration and storage of water. Indigenous species like Shorea *robusta, Pongamia pinnata, Syzygium cumini, Dalbergia sissoo, Croton roxburghii, Phyllanthus emblica,* and *Azadirachta indica* are more tolerant to adverse soil moisture, nutrients content, pH, and stress of soil in comparison to grasses and legumes. Indigenious species possess root adaptations that enhance drought tolerance (Lowry et al., 1994). The accumulation of plant litter and root exudates supports nutrient cycling, contributing to the formation of new, self-sustaining topsoil (Padmavathiamma & Li, 2007).

8 Land Use Pattern Study

The major impact on the land use during the pre-mining phase is removal of vegetation, alternation in land use pattern and resettlement of displaced population. During mining and post mining phases, drastic changes in landscape with landform take place. Mining activities have a significant impact on land, especially when the core area includes forests, agricultural land, or human settlements, leading to biodiversity loss or displacement of communities. Improper storage and management of these waste materials have severe environmental consequences, including deforestation and landscape alteration (Patra & Sethy, 2014). The extent of land impact depends on the area acquired and its existing land use, which can be analyzed using satellite imagery and baseline data. A cost benefit analysis should be conducted to weigh the economic benefits of the mining project against the ecological and economic value of the current land use, including degradation impacts on both core and surrounding areas. Mitigation measures can greatly reduce land degradation and include compensation and rehabilitation for displaced populations, topsoil conservation and protection plans, segregation and safe disposal of high risk waste, and strategic planning of waste and topsoil storage areas based on site specific slope and runoff conditions.

9 Solid Waste Management in Mines

Mine solid waste management is a critical aspect of mining operations, as the extraction of ore generates huge quantities of waste, including overburden, soil, debris, and low grade ore. Open cast mining tends to produce about 8–10 times more waste compared to underground mining, though the volume can vary depending on the type of mineral and local geological conditions. A comprehensive waste audit is essential to assess the sources, types, characteristics, and generation patterns of waste, as well as the associated costs of storage, treatment, and disposal.

The primary environmental impacts of mine waste include runoff from overburden, topsoil, and ore stockpiles, which can lead to siltation and contamination of water bodies. This runoff can also cause the loss of vegetation and wildlife habitat, harm surrounding agricultural land, and degrade groundwater quality due to leachate from stockpiles and tailings. The overburden should be assessed for heavy metal leachability, to ensure minimal environmental risk. Topsoil should be preserved before blasting, stored in embanked areas to prevent erosion, and both overburden and topsoil dumps should be stabilized with vegetation and coir netting. Effective mine waste management requires a phased approach, with surface plans for waste disposal outlined for each 5-year period. Hazards wastes from mine drainage treatment should be disposed of according to regulations in secure landfills. All waste management practices must be aligned with the approved mine plan to ensure compliance with environmental standards and regulations. For controlling soil erosion and stabilizing dump slopes, revegetation is a well-established method (Akers & Muter, 1974), thus helping to restore the ecological balance (Jorgensen, 1994). Vegetation growth on dump slopes brings positive impact on both hydro geological as well as mechanical processes (Cherubini & Geiasi, 1997). Plant roots play an important role in managing rainwater interception, evaporation, reducing the pore pressure thus enhancing the dump stability (Hussain, 1995). Vegetation growth also reinforces the loose dump material through intense root structure thus increasing the strength of the material. The Miyawaki method of plantation is more effective than the conventional method of plantation for dump stabilization which involves the planting of a number of different indigenous plants in a small pit or dump slope. Grasses are commonly used as an early vegetation establishment on mine lands as they are tolerant of drought, less soil nutrient, and environmental stresses as their fibrous root systems help in slow erosion, and over time, they contribute to soil formation, stabilize the soil, and conserve moisture. Initial cover of grasses should allow for the development of different, self-sustaining plant diversity (Shu et al., 2002). The design of the waste dumps should allow for progressive rehabilitation thus minimizing the disturbed area at any given time and facilitating the completion of final rehabilitation as soon as possible. Figure 1 gives the sketch of ultimate dump design near mining area for soil stabilization.

9.1 Physical Remediation Methods

The physical remediation method employs physical and mechanical barriers to isolate, recover, and separate soil pollutants. Techniques encompass capping soil washing and thermal desorption pretty extensively (Ossai et al., 2020) and also soil replacement occasionally. Soil replacement involves excavating tainted dirt and swapping it out with pristine soil, whereas soil washing is an ex situ pretreatment method that mechanically teases contaminants out of soil. Thermal desorption process entails heating contaminated soil via methods like microwaves infrared

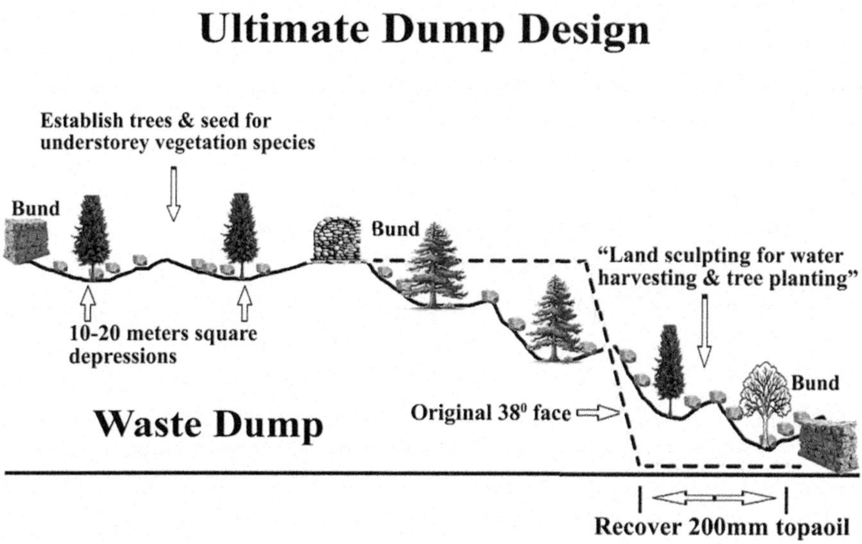

Fig. 1 The ultimate dump design near mining area for soil stabilization

radiation and steam making contaminants volatile rapidly underground. Soil treated via physical remediation can be reused afterwards and this benefit manifests rather obviously after quite lengthy treatment processes. Soil replacement, dumping and covering, solidification/stabilization, soil washing, thermal treatment are some of the techniques for physical remediation in mining areas.

9.2 Chemical Remediation Methods

Chemical remediation provides rapid and powerful options for soil remediation, independent of contaminant concentration, and can serve as alternatives to bioremediation. These methods can be performed in situ or ex situ and involve using various chemicals to enhance soil treatment, including augmentation, leaching, and oxidation. Chemical augmentation adds substances to promote the activity of degrading microbes, while commonly used chemicals include phosphoric acid, hydrogen chloride, sulfuric acid, potassium phosphate, and nitric acid (Hong et al., 2002; Yao et al., 2012). Treatment mechanisms often include ion exchange, precipitation, and adsorption. Chemical oxidation utilizes oxidants like permanganate and sodium persulfate, and advanced oxidation processes employ agents such as hydrogen peroxide in conjunction with ultraviolet light. However, the chemical method faces challenges, including high toxicity and costs associated with the use of chemicals. Ammonium oxalate extraction, hydroxylamine hydrochloride extraction and mixed acid extraction are some of the common extraction techniques.

Ammonium oxalate extraction occurs where ammonium oxalate, when acidified, can dissolve amorphous and poorly crystalline iron oxides that may adsorb arsenic. The reaction leads to the creation of soluble iron (III)-oxalate complexes as follows;

$$Fe_2O_3 + 6C_2O_4^{2-} + 6H^+ \rightarrow 2Fe\left(C_2O_4\right)_3^{3-} + 3H_2O$$

Hydroxylamine hydrochloride extraction occurs where hydroxylamine hydrochloride acts as a reducing agent, transforming iron (III) to iron (II) and thus dissolving iron oxides as follows:

$$Fe_2O_3 + 6NH_2OH \rightarrow 2Fe^{2+} + 3H_2O + 3NH_3$$

Short range ordered iron oxides can be extracted effectively using this method.

Mixed acid extraction usually employs a mixture of hydrochloric acid and acetic acid, occasionally supplemented with other substances such as ammonium acetate. These acids eliminate iron oxides by protonating surface sites and disintegrating the oxide lattice as follows;

$$Fe_2O_3 + 6H^+ \rightarrow 2Fe^{2+} + 3H_2O$$

This approach works well for the dissolution of iron oxides in both crystalline and amorphous forms.

Solidification and stabilization is a widely used method for remediating soils contaminated with heavy metals or mining tailings, involving the blending of contaminated materials with binding agents like cement or geopolymers. This process effectively encapsulates contaminants within a robust matrix, significantly reducing their leachability and toxicity. Benefits of these techniques include enhanced stability of treated materials, which lowers their susceptibility to erosion and displacement under environmental conditions. Commonly utilized binding agents such as geopolymers and flyash create a durable encapsulation, ensuring that contaminants are effectively trapped from rainwater and groundwater seepage. It can be applied both in situ and ex situ at processing plants.

Electro kinetic mining is an innovative method for extracting rare earth elements and other critical metals from ore deposits using an electric field to accelerate ion movement. This sustainable extraction technique significantly reduces environmental degradation compared to traditional mining practices and minimizes resource depletion. It utilizes electromigration and electroosmosis to drive rare earth elements and metal ions toward a collection well, coupled with rapid water movement. The method offers numerous advantages, including increased efficiency, reduced leaching agent requirements, and shorter mining durations, often achieving recovery rates of over 95%. This technique promotes sustainable mining by minimizing waste and the use of chemical agents, proving effective for various deposit types, including ion-adsorption rare earth deposits.

9.3 Biological Remediation Methods

Bioleaching effectively extracts valuable metals from low grade ores and mine tailings using biological remediation methods and microorganisms, primarily bacteria. These bacteria transform metal compounds into soluble forms, enabling efficient extraction from the resultant mixtures. This ecofriendly process surpasses traditional methods in cost effectiveness and aids in revitalizing contaminated areas. Key bacteria, such as *Thiobacillus ferroxidans* and *Thiobacillus thioxidans*, produce acids that dissolve metal bearing minerals, converting insoluble compounds into soluble forms like sulfates or chlorides. Various bioleaching techniques exist, including direct bioleaching, where microbes attack metal bearing minerals, and heap or dump leaching, where bacterial solutions are applied to ore piles or larger areas. Bioleaching is notably less polluting and requires significantly less energy compared to traditional smelting methods, making it a sustainable option for metal recovery. Biological methods utilize living organisms, such as plants and microbes, to effectively remediate soil contaminants by breaking them down rapidly. It has also been studied that, microorganisms can grow in polluted water and reduces the pollution load (Dash, 2012; Dash & Mishra, 1996a, 1996b, 1997, 1999a, 1999b). According to Enyiukwu et al. (2021) these methods are characterized by biological processes that detoxify contaminated areas. The goal in soil remediation is to optimize parameters like temperature, pH, porosity, nutrient content, moisture levels, C:N ratio, and microbial diversity (O'Brien et al., 2017). Techniques such as bioaugmentation, biostimulation, verm remediation, and phytoremediation are commonly employed in various contaminated environments (Samanta et al., 2002).

Phytoremediation in mining employs plants to remediate heavily contaminated soil and water, providing an ecofriendly and cost effective solution. This method leverages the ability of plants to absorb and degrade pollutants, making it particularly effective in areas heavily impacted by mining activities. Soil physico-chemical study (Dash et al., 2016) and assessment of floristic composition and biological spectrum (Dash & Patra, 2016) near the mining cluster plays a key role in phytoremediation. Hyperaccumulator plants, which can accumulate high concentrations of heavy metals, play a crucial role in soil pollutant removal. Phytostabilization, a key aspect of phytoremediation, reduces pollutant mobility and bioavailability by using plants with robust root systems that stabilize the soil and effectively filter contaminants from groundwater. Additionally, certain plants can degrade pollutants through their root systems, enhancing soil fertility and promoting the growth of other vegetation. Plant species is also potential of extracting the pollutants as nutrients from the fly ash (Dash et al., 2015; Pradhan et al., 2015). Phytoremediation is advantageous as it significantly reduces soil erosion, requires minimal equipment, and capitalizes on natural processes, thus supporting sustainability and minimizing ecological footprints. Table 3 gives the comparative study of phytoremediation potential of different species for heavy metal removal with their merits and limitations (Dehkordi et al., 2024).

Table 3 Comparative study of phytoremediation techniques for heavy metal removal

Technique	Plant type	Metal type (% removal)	Merits	Method adopted	Limitations	References
Phytoremediation	Sunflower	Pb (75%)	Larger applicability, less cost, ecofriendly	Stabilization, absorption	Metal specific, require more time, skilled knowledge	Zehra et al. (2020)
	Lettuce	Cd (65%)	Ecofriendly, less cost, broad applicability	Absorption, stabilization	Require more time, type of soil, and metal matters skilled knowledge	Guo et al. (2024)
	Salicornia	Cu (70%)	Ecofriendly, less cost, broad applicability, soil improvement	Absorption, extraction	Require more time, type of soil and metal matters, skilled knowledge	Sordes et al. (2023)
Root stabilization	Alfalfa	Cd (70%)	Soil metal stabilization	Deposition around roots	Ineffective for higher mobile metals	Kareem et al. (2023)
Phytoextraction	*Brassica juncea*	Pb (80%)	Soil metal separation and recycling	Storage of metal in plant aerial parts	Requires more time	Di Gregorio et al. (2006)
Root surface absorption	Maize	Ni (65%)	Widely applicable	Root surface absorption	Absorption capacity is limited	Vaculík et al. (2021)

Constructed wetland serves as a biological remediation method treating mine drainage effectively by harnessing natural processes like filtration and microbial activity removing heavy metals. Such systems replicate natural wetland ecosystems allowing breakdown and absorption of pollutants by flora and microbes within a wetland matrix ordinarily. Mine drainage wastewater trickles through constructed wetlands were coarse gravel and sand filter out larger particles and suspended solids effectively.

10 Mining Waste and Circular Economy

The National Resource Efficiency Policy (NREP), spearheaded by the MoEF and CC, has taken a crucial step toward building a circular economy which ensures environmental sustainability, equitable economic growth, resource security, and biodiversity conservation. The NREP aims to reduce primary resource consumption to sustainable levels in aligned with the sustainable development goals, create more value with fewer materials by adopting resource efficient and circular approaches, minimize waste generation while ensuring material security and promoting employment opportunities and innovative business models which support environmental protection. As the initial link in many product value chains, the mining industry plays a critical role in both resource consumption and waste generation. Its environmental footprint can be significantly mitigated through circular economy principles. The construction sector can drive greater efficiency in the mining and refining processes through focusing on smarter exploration, sustainable design, and reduced waste generation throughout the mining life cycle. Mining waste can be repurposed into valuable resources through circular practices, keeping materials in use for as long as possible and retaining their value throughout their life cycle. For example, waste from metallurgical processes can be transformed into useful products using advanced techniques and technologies. A notable example is the extraction and conversion of slurries from mine disposal sites into pigment-grade iron oxides, which are in high demand in the pigment industry. Hedin (2003) demonstrated the feasibility of recovering iron from acid mine drainage in abandoned coal mines. His work illustrated the potential to repurpose mining by products into commercially viable materials. Iron oxides derived from such processes have multiple applications, including cosmetics, pottery glazes, soil remediation, filtration media, phosphate retention, coatings, and even animal feed additives. By promoting the profitable reuse of industrial waste, the policy encourages collaborative industrial ecosystems where manufacturers benefit from material recovery and industrial symbiosis. Such synergies are not only environmentally beneficial but also economically rewarding. Complementing NREP, the Extended Producer Responsibility framework mandates that producers take accountability for the entire lifecycle of their products, particularly their contribution to landfill waste. Together, these policies form a robust framework aimed at accelerating India's transition to a resource efficient, low carbon, and circular economy.

11 Conclusion

Mineral plays an important role for a country to progress. It also causes environmental issues like land degradation, soil contamination, air and water pollution, acid mine drainage, loss of biodiversity, noise pollution including impact on human health. To mitigate these issues, an integrated approach is needed that includes improved technology, community participation, strict regulation and suitable strategies for post closure management. Present mining practices need to be changed and should be focused for community development aiming to more equity, well protection of natural resources which should be environment friendly and should meet the goal of sustainable development. Now a day, the mining companies are looking after to reduce both environmental and social impacts and minimize the footprint of their activities. Novel technologies like IoT and AI, proper air and water treatment, green energy sources, reprocessing the tailings with advanced equipment, digitalizing mines, green energy sources and mine electrification are proved to be cost effective toward reducing environmental impacts. Mineral resource management of a country should be closely associated with the overall environmental protection and economic development.

References

Akers, J. D., & Muter, B. R. (1974). Gob pile stabilization and reclamation. In *Proceeding of the forth mineral waste utilization symposium* (pp. 229–239). Springer.

Benidire, L., Pereira, S. I., Naylo, A., Castro, P. M., & Boularbah, A. (2020). Do metal contamination and plant species affect microbial abundance and bacterial diversity in the rhizosphere of metallophytes growing in mining areas in a semiarid climate? *Journal of Soils and Sediments, 20*(2), 1003–1017. https://doi.org/10.1007/s11368-019-02475-4

Chaney, R. L., Angle, J. S., Broadhurst, C. L., & Peters, C. A. (2007). Improved understanding of hyperaccumulation yields commercial phytoextraction and phytomining technologies. *Journal of Environmental Quality, 36*(5), 1429–1443. https://doi.org/10.2134/jeq2006.0514

Chaudhari, P. R., & Gajghate, D. G. (2000). Assessment of air pollution effect on plants-a review. *Indian Journal of Environmental Protection, 20*(12), 925–933.

Cherubini, C., & Geiasi, C. (1997). The influence of vegetation on slope stability. In P. G. Marinos, G. C. Koukis, G. C. Tsiambaos, & G. C. Stournaras (Eds.), *Engineering geology and environment* (Vol. 1, pp. 67–71). Balkema.

Coban, M., Karakas, F., & Akbulut Coban, N. (2023). Quantitative analysis of healthcare waste generation and composition in Antalya, Turkey. *Waste Management, 160*, 80–89. https://doi.org/10.1016/j.wasman.2023.02.008

Das, A., Dey, S., Bala, K., Gnanasekaran, R., & Das, A. P. (2024). Microbial bioremediation technology for sustainable treatment and management of synthetic microfiber waste. In *Sustainable microbial technology for synthetic and cellulosic microfiber bioremediation* (pp. 221–238). Springer. https://doi.org/10.1007/978-3-031-62600-5_12

Das, A., Dey, S., & Das, A. P. (2024). Ecotoxicological impacts of synthetic microfiber pollutants and development of sustainable mitigation strategies. *Environmental Chemistry and Ecotoxicology*. https://doi.org/10.1016/J.ENCECO.2024.12.003

Das, A., Ray, S. K., Mohanty, M., Mohanty, J., Dey, S., & Das, A. P. (2024). Ecotoxicity of microplastic wastes and their sustainable management: A review. *Environmental Chemistry and Ecotoxicology, 6*, 144–152. https://doi.org/10.1016/J.ENCECO.2024.05.003

Das, A. P., & Ghosh, S. (2018). Bioleaching of manganese from mining waste materials. *Materials Today: Proceedings, 5*(1), 2381–2390. https://doi.org/10.1016/J.MATPR.2017.11.459

Das, A. P., & Ghosh, S. (2023a). Factors affecting bioleaching. In *Manganese mining microorganisms* (pp. 77–86). Elsevier. https://doi.org/10.1016/B978-0-12-822147-1.00004-2

Das, A. P., & Ghosh, S. (2023b). Manganese and its application. In *Manganese mining microorganisms* (pp. 29–40). Elsevier. https://doi.org/10.1016/B978-0-12-822147-1.00007-8

Das, A. P., & Ghosh, S. (2023c). Mechanism of manganese bioleaching. In *Manganese mining microorganisms* (pp. 87–98). Elsevier. https://doi.org/10.1016/B978-0-12-822147-1.00014-5

Das, A. P., & Ghosh, S. (2023d). Microorganisms in manganese biomining processes. In *Manganese mining microorganisms* (pp. 55–61). Elsevier. https://doi.org/10.1016/B978-0-12-822147-1.00013-3

Das, A. P., & Ghosh, S. (2023e). Mixed culture bioleaching: An insight into manganese biomining process and efficacy. In *Manganese mining microorganisms* (pp. 129–141). Elsevier. https://doi.org/10.1016/B978-0-12-822147-1.00001-7

Das, A. P., Ghosh, S., Mohanty, S., & Sukla, L. B. (2015a). Advances in manganese pollution and its bioremediation. In L. B. Sukla, N. Pradhan, S. Panda, & B. K. Mishra (Eds.), *Environmental microbe biotechnology* (1st ed., pp. 313–328). Springer. https://doi.org/10.1007/978-3-319-19018-1_16

Das, A. P., Ghosh, S., Mohanty, S., & Sukla, L. B. (2015b). Consequences of manganese compounds: A review. *Toxicological & Environmental Chemistry, 96*(7), 981–997. https://doi.org/10.1080/02772248.2015.1005428

Das, A. P., & Ghosh, S. (2023f). *Manganese mining microorganisms*. Elsevier.

Das, A. P., Mallavarapu, M., & Ghosh, S. (2023). Ecotoxicity of mining pollutants on the environment and their remediation. *Environmental Chemistry and Ecotoxicology, 5*, 165–167. https://doi.org/10.1016/J.ENCECO.2023.08.002

Das, A. P., Sukla, L. B., Pradhan, N., & Nayak, S. (2011). Manganese biomining: A review. *Bioresource Technology, 102*(16), 7381–7387. https://doi.org/10.1016/J.BIORTECH.2011.05.018

Dash, A. K. (2012). Impact of domestic wastewater on seed germination and physiological parameters of wheat and rice. *International Journal of Research and Review on Applied Science, 12*(2), 280–286.

Dash, A. K., & Dash, S. K. (2017). Atmospheric pollution load assessment through air quality index: A case study. *Indian Journal of Environmental Protection, 37*(9), 736–741.

Dash, A. K., & Mishra, P. C. (1996a). Changes in pigment and protein content of Westiellopsis prolifica, a blue-green alga, grown in paper mill waste water. *Microbios, 85*, 257–266.

Dash, A. K., & Mishra, P. C. (1996b). Changes in biomass, pigment and protein content of Westiellopsis prolifica, a blue-green alga, grown in nutrient manipulated paper mill wastewater. *Cytobios, 88*, 11–16.

Dash, A. K., & Mishra, P. C. (1997). Blue–green alga in sewage—amended paper mills waste water. *International Journal of Environmental Studies, 53*, 9–10.

Dash, A. K., & Mishra, P. C. (1999a). Growth response of the blue-green alga, Westiellopsis prolifica in sewage enriched paper mill waste water. *Revista Internacional de Contaminación Ambiental, 15*(2), 79–83.

Dash, A. K., & Mishra, P. C. (1999b). Role of the blue—Green alga, Westiellopsis prolifica in reducing pollution load from paper mill wastewater. *Indian Journal of Environmental Protection, 19*(1), 1–15.

Dash, A. K., & Patra, H. S. (2016). Floristic composition and biological Spectrum study near the mining cluster at Keonjhar District, Odisha, India. *International Journal of Engineering Sciences & Research Technology, 5*(2), 441–447. https://www.researchgate.net/publication/313526474

Dash, A. K., Pradhan, A., Das, S., & Mohanty, S. S. (2015). Fly ash as a potential source of soil amendment in agriculture and a component of integrated plant nutrient supply system. *Journal of Industrial Pollution Control, 31*(2), 251–259.

Dash, A. K., Pradhan, A., & Patra, H. S. (2016). Soil physico-chemical characteristics near the mining areas of Keonjhar District, Odisha, India. *International Journal of Current Research, 8*(5), 30306–30309.

Dash, S. K., & Dash, A. K. (2018a). Air pollution tolerance index to assess the pollution tolerance level of plant species in industrial areas. *Asian Journal of Chemistry, 30*(1), 219–222. https://doi.org/10.14233/ajchem.2018.20991

Dash, S. K., & Dash, A. K. (2018b). Seasonal variation of air quality index and assessment. *Global Journal of Eviznmental Science and Management, 4*(4), 483–492. https://doi.org/10.22034/gjesm.2018.04.008

Dash, S. K., Pradhan, A., Panda, B. P., & Dash, A. K. (2023). Ambient air quality monitoring and assessment of an industrial area as a tool for sustainable environmental management. *Materials Today Proceeding*. https://doi.org/10.1016/j.matpr.2023.10.024

Dash, S. K., Sahu, H. B., & Dash, A. K. (2019). Atmospheric dispersion modeling using AERMOD to predict the impact of PM_{10} near Bileipada, Odisha. *Indian Journal of Environmental Protection, 39*(4), 299–306.

Dehkordi, M. M., Nodeh, Z. P., Dehkordi, K. S., Salmanvandi, H., Khorjestan, R. R., & Ghaffarzadeh, M. (2024). Soil, air, and water pollution from mining and industrial activities: Sources of pollution, environmental impacts, and prevention and control methods. *Results in Engineering, 23*, 102729. https://doi.org/10.1016/j.rineng.2024.102729

Dey, S., Bissoyi, A., Koteswara Reddy, G., & Das, A. P. (2024). Synthetic microfibers: An urban emerging pollutant demanding for sustainable environmental solutions. In *Sustainable microbial technology for synthetic and cellulosic microfiber bioremediation* (pp. 107–121). Springer. https://doi.org/10.1007/978-3-031-62600-5_7

Dey, S., Das, A., Mallick, K., Sahu, A., & Das, A. P. (2023). Environmental petroleum waste: Pollution, toxicity, sustainable remediation. In *Impact of petroleum waste on environmental pollution and its sustainable management through circular economy* (pp. 159–175). Springer. https://doi.org/10.1007/978-3-031-48220-5_7

Dey, S., Dash, N., Ray, S. K., Priyadarsini, S., Chhotaray, C., & Das, A. P. (2024). Occurrence and quantification of tire wear plastics pollutants from highway city dust of Bhubaneswar City of Odisha. In *Sustainable microbial technology for synthetic and cellulosic microfiber bioremediation* (pp. 77–89). Springer. https://doi.org/10.1007/978-3-031-62600-5_5

Dey, S., Tripathy, B., Kumar, M. S., & Das, A. P. (2023). Ecotoxicological consequences of manganese mining pollutants and their biological remediation. *Environmental Chemistry and Ecotoxicology*. https://doi.org/10.1016/J.ENCECO.2023.01.001

Di Gregorio, S., Barbafieri, M., Lampis, S., Sanangelantoni, A. M., Tassi, E., & Vallini, G. (2006). Combined application of triton X-100 and Sinorhizobium sp. Pb002 inoculum for the improvement of lead phytoextraction by Brassica juncea in EDTA amended soil. *Chemosphere, 63*, 293–299. https://doi.org/10.1016/j.chemosphere.2005.07.020

El Hamiani, O., El Khalil, H., Lounate, K., Sirguey, C., Hafidi, M., Bitton, G., Schwartz, C., & Boularbah, A. (2010). Toxicity assessment of garden soils in the vicinity of mining areas in southern Morocco. *Journal of Hazardous Materials, 177*(1-3), 755–761. https://doi.org/10.1016/j.jhazmat.2009.12.096

El Khalil, H., El Hamiani, O., Bitton, G., Ouazzani, N., & Boularbah, A. (2008). Heavy metal contamination from mining sites in South Morocco: Monitoring metal content and toxicity of soil runoff and groundwater. *Environmental Monitoring and Assessment, 136*(1-3), 147–160. https://doi.org/10.1007/s10661-007-9671-9

Enyiukwu, D. N., Bassey, I. N., Nwaogu, G. A., Asamudo, N. U., Chukwu, L. A., Monday, E. A., & Maranzu, J. O. (2021). Pig (sus scrofa domestic us) dung and their associated fungi: Potential candidates for bioremediation and nutrients improvement of crude oil contaminated soils biodiverse. *Journal of Biological Diversity, 22*(8), 822. https://doi.org/10.13057/biodiv/d220822

Erasmus, J. H., Zimmermann, S., Smit, N. J., Malherbe, W., Nachev, M., Sures, B., & Wepener, V. (2022). Human health risks associated with consumption of fish contaminated with trace elements from intensive mining activities in a peri-urban region. *Science of Total Environment., 825*, 154011. https://doi.org/10.1016/J.SCITOTENV.2022.154011

ERM. (2013). Environmental and social impact assessment (ESIA) for the proposed gamsberg zinc mine and associated infrastructure in Northern Cape noise impact assessment. Report No GAM-NVI-R02 March 2013.

Ghosh, S., & Das, A. P. (2017). Bioleaching of manganese from mining waste residues using Acinetobacter sp. *Geology, Ecology, and Landscapes, 1*(2), 77–83. https://doi.org/10.1080/24749508.2017.1332847

Guo, T., He, D., Liu, Y., Li, J., & Wang, F. (2024). Lanthanum promotes Solanum nigrum L. growth and phytoremediation of cadmium and lead through endocytosis: Physiological and biochemical response, heavy metal uptake and visualization. *Science of Total Environment, 912*, 168915. https://doi.org/10.1016/j.scitotenv.2023.168915

Hedin, S. R. (2003). Recoverable of marketable iron oxide from mine drainage in USA. *Land Contamination and Reclamation, 11*(2), 2003.

Hong, K. J., Tokunaga, S., & Kajiuchi, T. (2002). Evaluation of remediation process with plant-derived biosurfactant for recovery of heavy metals from contaminated soils. *Chemosphere, 49*(4), 379–387. https://doi.org/10.1016/S0045-6535(02)00321-1

Hussain, A. (1995). Fill compaction-erosion study in reclaimed areas. *Indian Mining Engineering Jornal, 34*(6), 19–21.

Jain, V., Tripathi, S. N., Tripathi, N., Sahu, L. K., Gaddamidi, S., Shukla, A. K., Bhattu, D., & Ganguly, D. (2022). Seasonal variability and source apportionment of nonmethane VOCs using PTR-TOF-MS measurements in Delhi, India. *Atmospheric Environment, 283*, 119163. https://doi.org/10.1016/j.atmosenv.2022.119163

Jena, P. S., Pradhan, A., Nanda, S. P., Dash, A. K., & Naik, B. (2022). Biosorption of heavy metals from wastewater using Saccharomyces cerevisiae as a biosorbent: A mini review. *Materials Today Proceeding, 67*(8), 1140–1146. https://doi.org/10.1016/j.matpr.2022.07.306

Jorgensen, S. E. (1994). Models as instruments for combination of ecological theory and environmental practice. *Ecological Modelling, 75-76*, 5–20. https://doi.org/10.1016/0304-3800(94)90003-5

Kareem, H. A., Adeel, M., Azeem, M., Ahmad, M. A., Shakoor, N., Hassan, M. U., Saleem, S., Irshad, A., Niu, J., Guo, Z., & Branko, ´C., Hołubowicz, R., Wang, Q. (2023). Antagonistic impact on cadmium stress in alfalfa supplemented with nano-zinc oxide and biochar via upregulating metal detoxification. *Journal of Hazardous Material, 443*, 130309. https://doi.org/10.1016/j.jhazmat.2022.130309

Kaya, C., Ugurlar, F., Ashraf, M., Hou, D., Kirkham, M. B., & Bolan, N. (2024). Microbial consortia-mediated arsenic bioremediation in agricultural soils: Current status, challenges, and solutions. *Science of Total Environment, 917*, 170297. https://doi.org/10.1016/j.scitotenv.2024.170297

Khobragade, K. (2020). Impact of mining activity on environment: An overview. *International Journal of Scientific and Research Publications, 10*(5), 784–791. https://doi.org/10.29322/IJSRP.10.05.2020.p10191

Kumar, D. S., Bhushan, S. H., & Kishore, D. A. (2018). Atmospheric dispersion model to predict the impact of gaseous pollutant in an industrial and mining cluster. *Global Journal of Environmental Science and Management, 4*(3), 351–358. https://doi.org/10.22034/gjesm.2018.03.008

Kurniawan, S. B., Ramli, N. N., Said, N. S. M., Alias, J., Imron, M. F., Abdullah, S. R. S., Othman, A. R., Purwanti, I. F., & Hasan, H. A. (2022). Practical limitations of bioaugmentation in treating heavy metal contaminated soil and role of plant growth promoting approach. *Heliyon, 8*, e08995. https://doi.org/10.1016/J.HELIYON.2022.E08995

Li, H., Yao, J., Sunahara, G., Min, N., Li, C., & Duran, R. (2023). Quantifying ecological and human health risks of metal(loid)s pollution from non-ferrous metal mining and smelting activities in Southwest China. *Science of Total Environment, 873*, 162364. https://doi.org/10.1016/j.scitotenv.2023.162364

Liu, S., Zhan, C., Zhang, J., Guo, J., Liu, H., Liu, T., Zheng, J., Yao, R., Liu, X., Xing, X., Cao, J., & Qi, S. (2022). Source-specific health risk assessment of PM2.5 bound heavy metal in re-suspended fugitive dust: A case study in Wuhan metropolitan area, Central China. *Journal of Cleaner Production, 379*, 134480. https://doi.org/10.1016/j.jclepro.2022.134480

Lowry, J. B., Prinsen, H., & Burrows, D. M. (1994). Albizia lebbeck: A promising forage tree for semiarid regions. In R. C. Gutteridge & H. M. Shelton (Eds.), *Forage tree legumes in tropical agriculture* (pp. 75–83). CAB International.

Mahapatra, B., Dhal, N. K., Dash, A. K., Panda, B. P., Panigrahi, K. C. S., & Pradhan, A. (2019). Perspective of mitigating atmospheric heavy metal pollution: Using mosses as biomonitoring and indicator organism. *Environmental Science and Pollution Research.*, *26*(29), 29620–29638. https://doi.org/10.1007/s11356-019-06270-z

Mathieu, L., Riller, U., Gibson, L., & Lightfoot, P. (2021). Data on the geology and structure of the copper cliff embayment and offset dyke, Sudbury igneous complex, Canada. *Data in Brief, 35*, 106957. https://doi.org/10.1016/j.dib.2021.106957

Matsuyama, A., Yano, S., Taniguchi, Y., Kindaichi, M., Tada, A., & Wada, M. (2021). Trends in mercury concentrations and methylation in Minamata Bay, Japan, between 2014 and 2018. *Marine Pollution Bulletin, 173*, 112886. https://doi.org/10.1016/j.marpolbul.2021.112886

Mishra, S., & Das, A. P. (2008). Hexavalent chromium reduction and 16S rDNA identification of bacteria isolated from a Cr (VI) contaminated site. *The Internet Journal of Microbiology, 7*, 1.

Mohanty, J., Dey, S., Bala, K., & Das, A. P. (2024). Recent advances in microplastic pollution for its sustainable management. In *Sustainable management of urban plastic waste through circular economic approaches* (pp. 27–48). Routledge. https://doi.org/10.1201/9781003465447-3

Mohanty, M., Tripathy, B., Koteswara Reddy, G., & Das, A. P. (2024). Greener Technologies for Establishing Restraint over microplastic pollution. In *Sustainable management of urban plastic waste through circular economic approaches* (pp. 239–258). Routledge. https://doi.org/10.1007/978-3-031-62600-5_13

Mohanty, S., Benya, A., Hota, S., Kumar, M. S., & Singh, S. (2023). Eco-toxicity of hexavalent chromium and its adverse impact on environment and human health in Sukinda Valley of India: A review on pollution and prevention strategies. *Environmental Chemistry and Ecotoxicology, 5*, 46–54. https://doi.org/10.1016/j.enceco.2023.01.002

Moharana, S., Dey, S., Priyadarsini, S., Kumar, M. S., & Das, A. P. (2024). Identification and characterization of microplastic pollutants from the marine sediments of Paradeep coast of bay of Bengal, India for their sustainable management. *Environmental Science and Engineering, 2333*, 57–73. https://doi.org/10.1007/978-3-031-51792-1_4

Nriagu, J. O., & Pacyna, J. M. (1988). Quantitative assessment of worldwide contamination of air, water and soils by trace metals. *Nature, 333*, 134–139.

O'Brien, P. L., DeSutter, T. M., Casey, F. X. M., Wick, A. F., & Khan, E. (2017). Evaluation of soil function following remediation of petroleum hydrocarbons—A review of current remediation techniques current pollution. *Report, 3*(3), 192–205. https://doi.org/10.1007/s40726-017-0063-7

Orellana Mendoza, E., Cuadrado, W., Yallico, L., Zarate, R., Quispe-Melgar, H. R., Limaymanta, C. H., Sarapura, V., & Bao-Condor, D. (2021). Heavy metals in soils and edible tissues of Lepidium meyenii (maca) and health risk assessment in areas influenced by mining activity in the central region of Peru. *Toxicology Report, 8*, 1461–1470. https://doi.org/10.1016/j.toxrep.2021.07.016

Ossai, C., Ahmed, A., Hassan, A., & Hamid, F. S. (2020). Remediation of soil and water contaminated with petroleum hydrocarbon: A review. *Environmental Technology and Innovation., 17*, 10052. https://doi.org/10.1016/j.eti.2019.100526

Padmavathiamma, P. K., & Li, L. Y. (2007). Phytoremediation technology: Hyper accumulation metals in plants. *Water Air and Soil Pollution, 184*(1-4), 105–126.

Parida, B., Dash, A. K., & Verma, A. (2023). Technologies for treatment of landfill leachate: A brief review. *Recent Innovations in Chemical Engineering., 16*(4), 260–284. https://doi.org/10.2174/0124055204276847231013040437

Parida, S., Das, M., & Dash, A. K. (2015). Ambient air study near the industrial complex 'nisha'at Angul, Odisha, India. In *Advances in environmental sciences and engineering* (pp. 514–537). Daya Publishing House.

Patra, H. S., & Dash, A. K. (2024). Management and recycling of mining wastes with zero waste adaptation technology as a tool for sustainable environmental management. In A. P. Das, E. D. Van Hullebusch, & A. Akçil (Eds.), *Sustainable management of mining waste and tailings* (1st ed.). CRC Press. https://doi.org/10.1201/9781003442455

Patra, H. S., & Sethy, K. M. (2014). Assessment of impact of opencast mine on surrounding forest : A case study from Keonjhar district of Odisha, India. *Journal of Environmental Research and Development, 9*(1), 249–254.

Pattajoshi, P. K., & Dash, A. K. (2024). Sustainable management of bauxite residue: Opportunities and challenges. In A. P. Das, E. D. Van Hullebusch, & A. Akçil (Eds.), *Sustainable management of mining waste and tailings* (1st ed.). CRC Press. https://doi.org/10.1201/9781003442455

Paul, A., Dey, S., Ram, D. K., & Das, A. P. (2023). Hexavalent chromium pollution and its sustainable management through bioremediation. *Geomicrobiology Journal*. https://doi.org/10.1080/01490451.2023.2218377

Pradhan, A., Dash, A. K., Mohanty, S. S., & Das, S. (2015). Potential use of fly ash in floriculture: A case study on the photosynthetic pigments content and vegetative growth of Tagetes erecta (MARIGOLD). *Ecology Environment and Conservation, 21*, 369–376.

Pradhan, A., Kumari, S., Dash, S., Biswal, D. P., Dash, A. K., & Panigrahi, K. C. S. (2017). Heavy metal absorption efficiency of two species of mosses (Physcomitrella patens and Funaria hygrometrica) studied in mercury treated culture under laboratory condition. IOP publishing, IOP Conf. Series. *Materials Science and Engineering, 225*, 012225. https://doi.org/10.1088/1757-899X/225/1/012225

Pradhan, A., Sahu, S. K., & Dash, A. K. (2013). Changes in pigment content (chlorophyll and carotenoid), enzyme activities (catalase and peroxidase), biomass and yield of rice plant (Oriza sativa. L) Following irrigation of rice mill wastewater under pot culture conditions. *International Journal of Scientific and Engineering Research, 4*(6), 2706–2717.

Pradhan, G., Dey, S., Priyadarsini, S., Kumar, M. S., & Das, A. P. (2024). Characterization and quantification of microplastics pollutants in sediment samples from Daya River of Odisha state in India for their appropriate management. *Environmental Science and Engineering, 2333*, 75–92. https://doi.org/10.1007/978-3-031-51792-1_5

Pradhan, G., Tripathy, B., Ram, D. K., Digal, A. K., & Das, A. P. (2023). Bauxite mining waste pollution and its sustainable management through bioremediation. *Geomicrobiology Journal*. https://doi.org/10.1080/01490451.2023.2235353

Pradhan, S. K., Singh, N. R., Kumar, U., Mishra, S. R., Perumal, R. C., Benny, J., & Thatoi, H. (2020). Illumina MiSeq based assessment of bacterial community structure and diversity along the heavy metal concentration gradient in Sukinda chromite mine area soils, India. *Ecological Genetics and Genomics, 15*, 100054. https://doi.org/10.1016/j.egg.2020.100054

Priyadarshini, S., Jagatee, S., & Das, A. P. (2024). Synthetic fabrics and microfiber pollution–an assessment of their global impact. *Environmental Science and Engineering, 2333*, 137–157. https://doi.org/10.1007/978-3-031-51792-1_8

Priyadarsini, S., & Das, A. P. (2025). Lithium bioleaching: A review on microbial-assisted sustainable technology for lithium bio-circularity. *Journal of Water Process Engineering, 69*, 106744. https://doi.org/10.1016/j.jwpe.2024.106744

Reddy Koteswara, G., Bhargavi, S., Yamuna, D., Anjani Devi, M., Jeevana Lakshmi Kumar, G., Manas, T., Adarsh, S., Bhavya Sai, V., Nath, H., Alreshidi, M. A., Togun, H., Ram, D. K., & Das, A. P. (2025). Recent advancements in the remediation of marine hydrocarbon and plastic pollutants. *Environmental Science and Engineering, 218*, 127–142. https://doi.org/10.1007/978-3-031-81615-4_7

Saha, S., Pattanayak, S. K., Sills, E. O., & Singha, A. K. (2011). Under-mining health: Environmental justice and mining in India. *Health & Place, 17*(1), 140–148. https://doi.org/10.1016/j.healthplace.2010.09.007

Sahoo, D., Dash, A. K., & Sahu, S. K. (2017). Ambient air quality monitoring and health impact study of air pollution near Joda of Keonjhar, Odisha, India. *International Journal of Engineering Sciences & Research Technology, 6*(1), 429–434. https://doi.org/10.5281/zenodo.260069

Sahu, A., Dey, S., Mallick, K., Priyadarsini, S., Behera, I. D., & Das, A. P. (2024). A review on advanced technology for sustainable management of synthetic microplastic waste. In *Sustainable management of urban plastic waste through circular economic approaches* (pp. 262–279). Routledge. https://doi.org/10.1201/9781003465447-15

Sahu, S. K., Mishra, A., & Dash, A. K. (2015). Ambient air quality around OCL India limited at Rajgangpur, Odisha, India. In *Advances in environmental sciences and engineering* (pp. 459–479). Daya Publishing House.

Samanta, S. K., Singh, O. V., & Jain, R. K. (2002). Polycyclic aromatic hydrocarbons: Environmental pollution and bioremediation. *Trends in Biotechnology, 20*(6), 243–248. https://doi.org/10.1016/s0167-7799(02)01943-1

Sanchez-Landero, L. A., Benítez-Badillo, G., Sangabriel-Conde, W., Alvarado-Castillo, G., & Lagunes-Diaz, E. G. (2024). Surface deposits and intracellular concentration of atmospheric particles in foliage of urban forests in the valley of Mexico. *Trees, for People, 15*, 100492. https://doi.org/10.1016/j.tfp.2023.100492

Sheoran, V., Sheoran, A. S., & Poonia, P. (2010). Soil reclamation of abandoned mine land by revegetation: A review. *International Journal of Soil, Sediment and Water, 3*(2), 13.

Shreya, Verma, A. K., Gilron, J., Oren, Y., Ronen, Z., Dash, A. K., Bhunia, P., & Dash, R. R. (2022). Effect of biological treatment on Perfluoroalkyl and Polyfluoroalkyl (PFASs) substances degradation. In *Current developments in biotechnology and bioengineering: Sustainable treatment technologies for pre and poly-flourakyl substances*. Elsevier. https://doi.org/10.1016/B978-0-323-99906-9.00004-8

Shu, W. S., Xia, H. P., Zhang, Z. Q., & Wong, M. H. (2002). Use of vetiver and other three grasses for revegetation of Pb/Zn mine tailings: Field experiment. *International Journal of Phytoremediation, 4*(1), 47–57.

Sordes, F., Pellequer, E., Sahli, S., Sarzynski, T., Denes, M., & Techer, I. (2023). Phytoremediation of chloride from marine dredged sediments: A new model based on a natural vegetation recolonization. *Journal of Environmental Management, 344*, 118508. https://doi.org/10.1016/j.jenvman.2023.118508

Taylor, J., & Pape, S. (2007). *Managing acid and metalliferous drainage. Leading practice sustainable development program for the mining industry, commonwealth* (p. 107). Department of Industry, Tourism and Resources.

Thomas, G., Sheridan, C., & Holm, P. E. (2022). A critical review of phytoremediation for acid mine drainage-impacted environments. *Science of Total Environment, 811*, 152230. https://doi.org/10.1016/J.SCITOTENV.2021.152230

Tripathi, N., Choppala, G., & Singh, R. S. (2016). Evaluation of modified chitosan for remediation of zinc contaminated soils. *Journal of Geochemical Exploration*. https://doi.org/10.1016/j.gexplo.2016.08.011

Vaculík, M., Kovac, J., Fialova, I., Fiala, R., Jaskova, K., & Luxova, M. (2021). Multiple effects of silicon on alleviation of nickel toxicity in young maize roots. *Journal of Hazardous Material., 415*, 125570. https://doi.org/10.1016/j.jhazmat.2021.125570

Wheeler, A. J., Williams, I., Beaumont, R. A., & Manilton, R. S. (2000). Characterisation of particulate matter sampled during a study of children's personal exposure to air borne particulate matter in a UK urban environment. *Environmental Monitoring and Assessment, 65*(1-2), 69–77.

Yao, Z., Li, J., Xie, H., & Yu, C. (2012). Review on remediation technologies of soil contaminated by heavy metals. *Procedia Environmental Sciences, 16*, 722–729. https://doi.org/10.1016/j.proenv.2012.10.099

Zehra, A., Sahito, Z. A., Tong, W., Tang, L., Hamid, Y., Khan, M. B., Ali, Z., Naqvi, B., & Yang, X. (2020). Assessment of sunflower germplasm for phytoremediation of lead polluted soil and production of seed oil and seed meal for human and animal consumption. *Journal of Environmental Science, 87*, 24–38. https://doi.org/10.1016/j.jes.2019.05.031

Zhou, M., Cao, Y., & Mielke, H. W. (2020). Comprehensive Urumqi screening for potentially toxic metals in soil-dust-plant total environment and evaluation of children's (0–6 years) risk-based blood lead levels prediction. *Chemosphere, 258*, 127342. https://doi.org/10.1016/j.chemosphere.2020.127342

Index

© The Editor(s) (if applicable) and The Author(s), under exclusive license to Springer Nature Switzerland AG 2026
A. P. Das (ed.), *Bioremediation Technology for Sustainable Mining Waste Management*, Sustainable Environmental Waste Management Strategies,
https://doi.org/10.1007/978-3-032-03468-7

The manufacturer's authorised representative in the EU is Springer Nature Customer Service Centre GmbH, Europaplatz 3, 69115 Heidelberg, Germany. If you have any concerns regarding our products, please contact ProductSafety@springernature.com

Printed and bound by CPI Group (UK) Ltd, Croydon, CR0 4YY

21/08/2026

02202034-0010